Numerical Methods and Methods of Approximation

in Science and Engineering

Numerical Methods and Methods of Approximation

in Science and Engineering

Karan S. Surana

Department of Mechanical Engineering
The University of Kansas
Lawrence, Kansas

CRC Press
Taylor & Francis Group
Boca Raton London New York

CRC Press is an imprint of the
Taylor & Francis Group, an **informa** business

CRC Press
Taylor & Francis Group
6000 Broken Sound Parkway NW, Suite 300
Boca Raton, FL 33487-2742

Printed on acid-free paper

International Standard Book Number-13: 978-0-367-13672-7 (Hardback)

Visit the Taylor & Francis Web site at
http://www.taylorandfrancis.com

and the CRC Press Web site at
http://www.crcpress.com

To my granddaughter Riya,

who has filled my life with joy.

Contents

Preface

Numerical methods and numerical analysis are an integral part of applied mathematics. With the shift in engineering education over the last fifty years from formulae, design, and synthesis oriented curriculum to one in which basic sciences, mechanics, and applied mathematics constitute the core of the engineering education, numerical methods and methods of approximation have become an integral part of the undergraduate engineering curriculum. At present most engineering curricula incorporate study of numerical methods and methods of approximation in some form, generally during the third (junior) year of the four year undergraduate study leading to baccalaureate degree in engineering. Adopting the text books and writings on this subject that are mathematically rigorous with theorems, lemmas, corollaries, and their proofs with very little illustrative examples in engineering curriculum was not very beneficial in terms of good understanding of the methods and their applications. This spurred a host of new text books on numerical methods that are specifically designed for engineering students. The progression and evolution of such writings at present has reached a stage that specifically caters the study of numerical methods to software packages and their use. Such writings lack theoretical foundation, deeper understanding of methods, discussion of pros, cons, and limitations of the methods.

The author has taught the numerical methods subject at the University of Kansas for over twenty years using his own class notes, which have evolved into the manuscript of this text book. The author's own research in computational mathematics and computational mechanics and his own graduate level text books on these subjects have been instrumental in designing the unique presentation of the material on the numerical methods and methods of approximation in this text book. The material in this book focuses on sound theoretical foundation, yet is presented with enough clarity, simplicity, and worked out illustrative examples to facilitate thorough understanding of the subject and its applications. This manuscript and its earlier versions have successfully been used at the University of Kansas mechanical engineering department since 1984 by the author and his colleagues.

The study of numerical methods and the methods of approximation using this text book requires that the students have knowledge of a computer programming language and also know how to structure a sequence of operations into a program using a programming language of their choice. For this reason, this book contains no material regarding any of the programming languages or instructions on how to structure a sequence of operations into a computer program.

In this book, all numerical methods are clearly grouped in two categories:

(i) The numerical methods that do not involve any approximations. In such methods the calculated numerical solutions are exact solutions of the mathematical models within the accuracy of computations on the computer. We refer to such methods as *numerical methods* or numerical methods without approximation.

(ii) Those methods in which the numerically calculated solution is always approximate. We refer to such methods as *methods of approximation* or numerical methods with approximations. In such methods often we can progressively approach (converge to) the true solution, but can never obtain precise theoretical solution.

In the numerical calculations of the solutions of the mathematical models, it is important to know whether the computed solutions are exact or true solutions of the mathematical models or if they are approximations of the exact solution. In approximate solutions, some assessment of error, computed or estimated, is highly meritorious as it helps in establishing the accuracy of the solution. Throughout the book in all chapters we keep this aspect of the computed solution in mind.

The book consists of eleven chapters. Chapters 2 and 3 consider methods of solutions of linear and nonlinear simultaneous algebraic equations. Standard and general eigenvalue problems, properties of eigenvalue problems, and methods of calculating eigenpairs are presented in Chapter 4. Chapter 5 contains interpolation theory and mapping in $\mathbb{R}^1$, $\mathbb{R}^2$, and $\mathbb{R}^3$ in the physical domain as well as the natural coordinate space $\xi\eta\zeta$. Numerical integration or quadrature methods: trapezoid rule, Simpson's 1/3 and 3/8 rules are presented in Chapter 6. Gauss quadrature in $\mathbb{R}^1$, $\mathbb{R}^2$, and $\mathbb{R}^3$ is also presented in Chapter 6 using physical and natural coordinate spaces. Curve fitting methods and numerical differentiation techniques are considered in Chapters 7 and 8. Methods of obtaining numerical solutions of boundary value problems (BVPs) and initial value problems (IVPs) are presented in Chapters 9 and 10. Time integration techniques are described in Chapter 10. Chapter 11 is devoted to the Fourier series and its applications in approximate analytical representation of functions that may or may not be analytic.

I am grateful to my former M.S. student, Mr. Tommy Hirst, for his interest in initiating the typing of the earlier preliminary version of the current manuscript. My very special thanks to Dr. Aaron D. Joy, my former Ph.D. student, for typesetting the current manuscript, preparing tables and graphs, performing some numerical studies, and for bringing the original preliminary version of the manuscript of the book to significant level of completion. Aaron's interest in the subject, hard work, and commitment to this book project are instrumental in the completion of the major portion of this book. Also my very sincere and special thanks to Mr. Dhaval Mysore, my current Ph.D. student for completing the typing and type setting of much of the newer material in Chapters 7 through 11. His interest in the subject, hard work and commitment have helped in the completion of final manuscript of this book. My sincere thanks to many of my colleagues of the mechanical engineering department at the University of Kansas, and in particular to my colleague and good friend Professor Peter TenPas, for valuable suggestions and many discussions that have helped me in improving the manuscript of the book.

This book contains many equations, derivations, mathematical details, and tables of solutions that it is hardly possible to avoid some typographical and other errors. The author would be grateful to those readers who are willing to draw attention to the errors using the email kssurana@ku.edu.

Karan S. Surana, *Lawrence, KS*

About the Author

Karan S. Surana, born in India, went to undergraduate school at Birla Institute of Technology and Science (BITS), Pilani, India, and received a B.E. degree in Mechanical Engineering in 1965. He then attended the University of Wisconsin, Madison, where he obtained M.S. and Ph.D. degrees in Mechanical Engineering in 1967 and 1970, respectively. He worked in industry, in research and development in various areas of computational mechanics and software development, for fifteen years: SDRC, Cincinnati (1970–1973), EMRC, Detroit (1973–1978); and McDonnell-Douglas, St. Louis (1978–1984). In 1984, he joined the Department of Mechanical Engineering faculty at University of Kansas, where he is currently the Deane E. Ackers University Distinguished Professor of Mechanical Engineering.

His areas of interest and expertise are computational mathematics, computational mechanics, and continuum mechanics. He is author of over 350 research reports, conference papers, and journal articles. He has served as advisor and chairman of 50 M.S. students and 22 Ph.D. students in various areas of Computational Mathematics and Continuum Mechanics. He has delivered many plenary and keynote lectures in various national and international conferences and congresses on computational mathematics, computational mechanics, and continuum mechanics. He has served on international advisory committees of many conferences and has co-organized mini-symposia on k-version of the finite element method, computational methods, and constitutive theories at U.S. National Congresses of Computational Mechanics organized by the U.S. Association of Computational Mechanics (USACM). He is a member of International Association of Computational Mechanics (IACM) and USACM, and a fellow and life member of ASME.

Dr. Surana's most notable contributions include: large deformation finite element formulations of shells, the k-version of the finite element method, operator classification and variationally consistent integral forms in methods of approximations for BVPs and IVPs, and ordered rate constitutive theories for solid and fluent continua. His most recent and present research work is in non-classical internal polar continuum theories and non-classical Cosserat continuum theories for solid and fluent continua and associated ordered rate constitutive theories. He is the author of recently published textbooks: *Advanced Mechanics of Continua*, CRC/Taylor & Francis, *The Finite Element Method for Boundary Value Problems: Mathematics and Computations*, CRC/Taylor & Francis, and *The Finite Element Method for Initial Value Problems: Mathematics and Computations*, CRC/Taylor & Francis.

1 Introduction

Numerical methods and methods of approximation play a significant role in engineering, mathematical and applied physics, and engineering science. The mathematical descriptions of physical systems leads to mathematical models that may be in differential, integral, or algebraic form. The specific form depends upon the basic principles and formulation strategy utilized in deriving them. Regardless of the specific forms of the mathematical models we can possibly choose either of two approaches in obtaining their solutions. In the first approach we seek analytic or theoretical solutions of the equations constituting the mathematical model. Unfortunately, this approach can only be used for simple and often trivial mathematical models. In practical applications the complexity of the mathematical models prohibit the use of this approach. However, in cases where this approach can be used, we obtain analytical expressions for the solution that are highly meritorious.

1.1 Numerical Solutions

In the second approach we resort to numerical methods or methods of approximation for obtaining the solutions of the mathematical models. In general, when using such methods we obtain numerical values of the solution. In some cases, the union of piecewise analytical expressions and numerical solutions constitute the entire solution, as in the finite element method. On the other hand, in finite difference methods we only have numerical values of the solution at a priori chosen locations in the spatial domain. Broadly speaking, the method of obtaining numerical solutions can be classified in the following two categories.

1.1.1 Numerical Methods without any Approximation

These are a class of numerical methods that yield a numerical solution, but the numerical solution is not an approximation of the true solution of the mathematical models. In these methods we obtain the exact solution of the mathematical model but in numerical form. The only errors in this solution are those due to truncations in the computations due to limited word size of the computers. We simply refer to these methods as *numerical methods*.

1.1.2 Numerical Methods with Approximations

These are a class of numerical methods in which we only obtain an approximate solution of the mathematical models. Such numerical methods are called *methods of approximation.* Obviously the solutions obtained using this class of methods contain error compared to the exact or analytical solution.

Remarks.

(a) For a given class of mathematical models, some methods of obtaining numerical solutions may be numerical methods (no approximation), while others may be methods of approximation. For example, if the mathematical model consists of a system of linear simultaneous algebraic equations (Chapter 2), then methods like Gauss elimination, Gauss-Jordan method, and Cramer's rule for obtaining their solution are numerical methods without any approximation, while Gauss-Seidel and Jacobi methods are methods of approximation.

(b) Some methods of obtaining numerical solutions are always methods of approximation. Numerical integration techniques (such as Simpson's rules or Gauss quadrature) for integrands that are not algebraic polynomials are always approximate. Solutions of nonlinear equations (algebraic or otherwise) are always iterative, hence fall into the category of methods of approximation.

(c) Methods of calculating eigenvalues (characteristic polynomial) are numerical methods when the degree of the characteristic polynomial is three or less, but methods of approximation are typically employed when the degree is higher than three.

(d) Methods of obtaining numerical solutions of boundary value problems and initial value problems such as finite element method, finite difference method, etc. are methods of approximation.

1.2 Accuracy of Numerical Solution, Error

Obviously in numerical methods without approximation, the errors are only due to truncation because of the word size during computations. Such errors when performing computations with word size of 64 bits or greater are very small and generally not worthy of quantification. On the other hand, in methods of approximation the calculated numerical solution is an approximation of the true solution. Thus, in such methods:

(i) If the true solution is known, the error can be measured as the difference between the true solution and the calculated solution in the pointwise sense, or if possible in the sense of L_2-norm.

(ii) When the theoretical solution is not known, as is the case with most practical applications, we can possibly consider some of the following.

   (a) We can attempt to estimate the error bounds. This provides the least upper bound of the error in the solution, i.e., the true error is less than or equal to the estimated error bound. In many cases (but not always), this estimation of the error bound is possible.

   (b) There are methods of approximation in which errors can be computed based on the current numerical solution without knowledge of the theoretical solution. The residual functional or L_2-norms of residuals in the finite element methods with minimally conforming approximation spaces are examples of this approach. This approach is highly meritorious as it provides a quantitative measure of error in the computed solution without knowldge of the theoretical solution, hence can be used to compute errors in practical applications.

   (c) There are methods in which the solution error can neither be estimated nor computed but there is some vague indication of improvement. Order of truncation errors in finite difference processes fall under this category. With increasing order of truncation, the solution errors are expected to reduce.

We remark that a comprehensive treatment of these topics is beyond the scope of this book. However, brief discussions are included wherever felt necessary.

1.3 Concept of Convergence

In general, the concept of convergence means approaching the desired goal. Thus, precisely what we are accomplishing through the process of convergence depends upon what our objective or goal is. In the case of nonlinear mathematical models, the numerical solutions are obtained iteratively. That is, we assume a solution (initial starting solution for the iterative process) and iterate using a recursive scheme established using the mathematical model to obtain progressively improved solutions. When two successive solutions are within some pre-specified tolerance, we consider the interative process to be converged, i.e., we have an approximate numerical solution of the mathematical model that is no longer changing as we continue to iterate.

In many applications, the mathematical models used in the iterative procedure are themselves an approximation of the true physics. Nonlinear algebraic equations obtained by finite element or finite difference methods are approximations of the true physics due to choice of a characteristic length used in obtaining them. In such cases, for a choice of discretization we obtain a converged solution from the iterative solution procedure. This is repeated for progressively refined discretizations leading to a sequence of progressively improved solutions (hence convergence) of the actual mathematical model. Figures 1.1 and 1.2 show schematic block diagrams of the convergence concepts for linear and nonlinear physical processes. We observe that in the case of linear processes (Figure 1.1), the convergence concept only implies convergence to the correct solution. If Figure 1.2 for nonlinear processes, there is a concept of convergence of the iterative solution procedure as well as the concept of progressively refined discretization solutions converging to the true solution of the problem.

1.4 Mathematical Models

The mathematical models describing the physical systems are derived using various methodologies depending upon the requirements of the physics at hand. In this book we do not dwell on the derivations of the mathematical models, but rather use representative mathematical models with desired features to present the numerical solution techniques suitable for them. However, whenever and wherever appropriate, enough description and insight is provided regarding the origins and applications of the mathematical models so that the significance and usefulness of the methods presented in this book are realized.

1.5 A Brief Description of Topics and Methods

Chapter 2 contains a review of linear algebra followed by solution methods for linear simultaneous algebraic equations. These consist of numerical methods such as Gauss elimination, Gauss-Jordan method, Cholesky decomposition, and Cramer's rule, as well as methods of approximation such as Gauss-Seidel method, Jacobi method, and relaxation method. Details of each method are followed by model problem solutions.

Chapter 3 contains methods of solution for nonlinear single or simultaneous equations. Using $f(x) = 0$, a single nonlinear function in independent variable x, various methods of finding the solution x are introduced with numerical examples. These consist of graphical method, incremental search method, bisection method, method of false position, Newton-Raphson method, secant method, fixed point method, and basic iteration method.

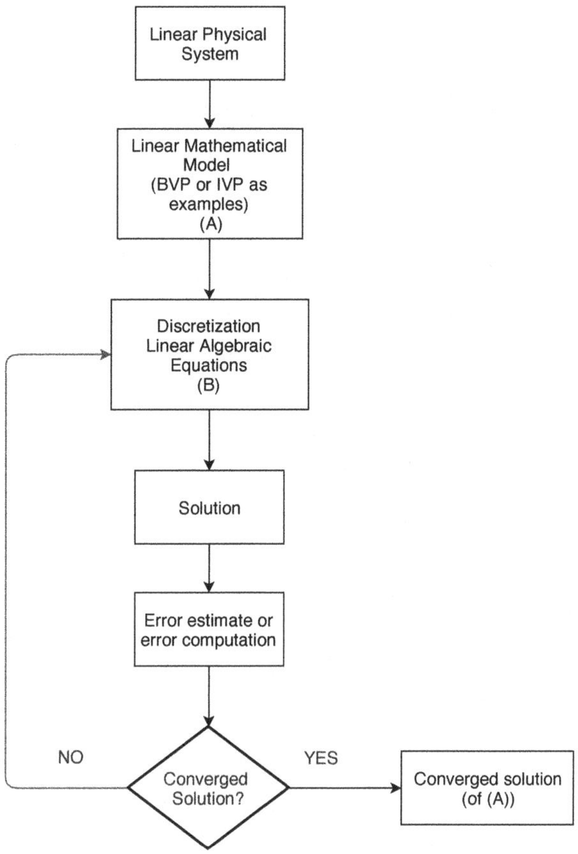

Figure 1.1: Concepts of convergence in linear systems

Newton-Raphson method is extended to a system of simultaneous nonlinear equations.

Chapter 4 presents treatment of algebraic eigenvalue problems. Basic properties of eigenvalue problems, the characteristic polynomial and efficient methods of constructing it, standard eigenvalue problems (SEVP) as well as general eigenvalue problems (GEVP) are considered. Inverse and forward iteration methods with Gram-Schmidt orthogonalization are presented for determining eigenpairs of the SEVP. Jacobi, Generalized Jacobi, QR Householder method, subspace iteration method and inverse iteration methods of determining eigenpairs are presented.

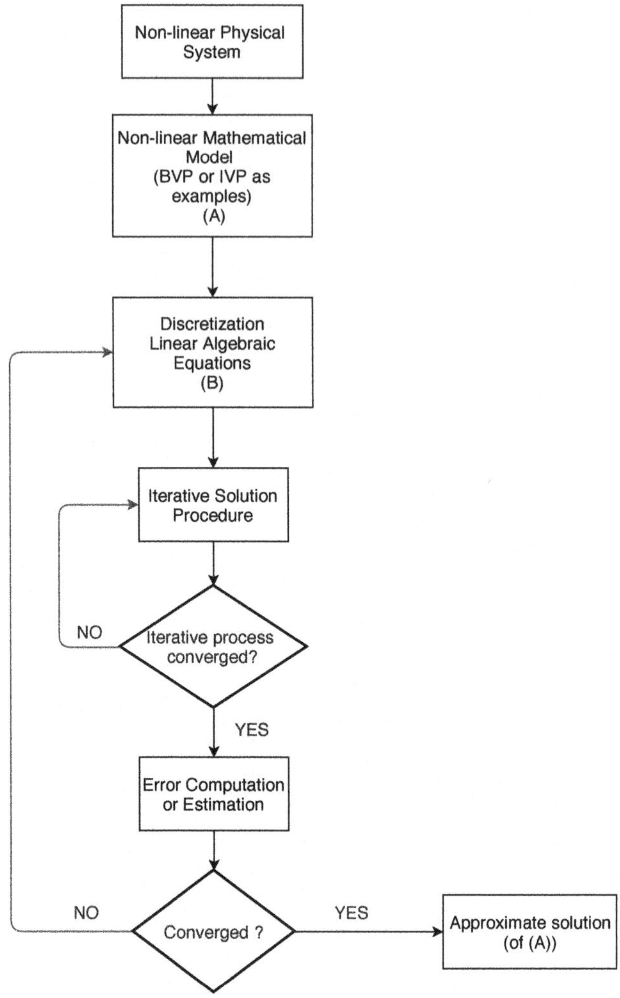

Figure 1.2: Concepts of convergence in non-linear systems

Interpolation theory and mapping in $\mathbb{R}^1$, $\mathbb{R}^2$, and $\mathbb{R}^3$ are presented in Chapter 5.

Various techniques of numerical integration such as trapezoid rule and Simpson's $1/3$ and $3/8$ rules are presented in Chapter 6 for numerical integration in $\mathbb{R}^1$. Gauss quadrature in $\mathbb{R}^1$, $\mathbb{R}^2$, and $\mathbb{R}^3$ is presented using physical coordinates (x, y, z) and natural coordinates (ξ, η, ζ).

Curve fitting using least squares fit, weighted least squares fit, and least squares fit for nonlinear case are given in Chapter 7.

Numerical differentiation and model problem solutions are contained in Chapter 8.

Numerical solutions of boundary value problems (BVPs) and Initial Value Problems (IVPs) using finite element and finite difference methods are considered in Chapters 9 and 10.

Chapter 11 contains Fourier series representation of analytic as well as non-analytic functions with model problem.

2

Linear Simultaneous Algebraic Equations and Methods of Obtaining Their Solutions

2.1 Introduction, Matrices, and Vectors

Linear simultaneous algebraic equations arise in all branches of engineering, physics, applied mathematics, and in many other disciplines. In some cases the mathematical representation of the physics may naturally result in these while in other applications these may arise, for example, when considering solutions of differential and partial differential equations using methods of approximation such as finite difference, finite element methods, etc. In obtaining the solutions of linear simultaneous algebraic equations, one could employ methods that are not methods of approximation. In such methods the sources of errors are not due to the method used, but rather due to computational inaccuracies. The solutions resulting from these methods are exact within the computational precision. On the other hand, if methods of approximation are employed in obtaining the solutions of linear simultaneous algebraic equations, then obviously the calculated solutions are approximate and are only accurate within some tolerance. In this chapter we consider both methods of obtaining solutions of linear simultaneous algebraic equations.

First we introduce the concept of simultaneous equations in a more general form. Consider

$$f_i(x_1, x_2, \dots, x_n) = b_i \qquad ; \qquad i = 1, 2, \dots, n \tag{2.1}$$

in which x_j; $j = 1, 2, \dots, n$ are unknown and b_i are known (numbers). Each $f_i(\cdot)$ defines a functional relationship between x_j; $j = 1, 2, \dots n$ that satisfies (2.1). It is rather obvious that in doing so we cannot consider each $f_i(\cdot)$ individually as each $f_i(\cdot)$ is a function of x_j; $j = 1, 2, \dots, n$. Instead we must consider them all simultaneously.

2.1.1 Basic Definitions

Definition 2.1 (Nonlinear System). The system of equations (2.1) is called a *system of nonlinear simultaneous algebraic equations* if some or all $f_i(\cdot)$ are nonlinear functions of some or all x_j.

Definition 2.2 (Linear System). The system of equations (2.1) is called a *system of linear simultaneous algebraic equations* if each $f_i(\cdot)$ is a linear combination of x_j; $j = 1, 2, \dots n$ in which the coefficients in the linear combination are known (numbers). For this case we can express (2.1) as

$$\begin{aligned} f_1(x_j; j=1,2,\dots,n) - b_1 &= a_{11}x_1 + a_{12}x_2 + \dots + a_{1n}x_n - b_1 = 0 \\ f_2(x_j; j=1,2,\dots,n) - b_2 &= a_{21}x_1 + a_{22}x_2 + \dots + a_{2n}x_n - b_2 = 0 \\ &\vdots \\ f_n(x_j; j=1,2,\dots,n) - b_n &= a_{n1}x_1 + a_{n2}x_2 + \dots + a_{nn}x_n - b_n = 0 \end{aligned} \tag{2.2}$$

We note that each $f_i(\cdot)$ is a linear combination of x_j using a_{ij}; $i, j = 1, 2, \dots, n$. a_{ij} and b_{ij} are known coefficients.

Remarks.

(1) When (2.1) represents a system of nonlinear simultaneous algebraic equation, a form like (2.2) is also possible, but in this case the coefficients (some or all) may be functions of unknowns x_j; $j = 1, 2, \dots, n$. Thus, in general we can write (2.2) with the following definitions of the coefficients a_{ij}:

$$a_{ij} = a_{ij}(x_j; j = 1, 2, \dots, n) \qquad i, j = 1, 2, \dots, n \tag{2.3}$$

(2) In this chapter we consider methods of determining solutions of linear simultaneous algebraic equations that are in the form (2.2).

(3) If the number of equations are large (large value of n in equation (2.1)), then the representation (2.2) is cumbersome, i.e., not very compact. We use *matrix and vector* notations to represent (2.2).

Definition 2.3 (Matrix). A matrix is an ordered rectangular (in general) arrangement of elements and is generally denoted by a symbol. Thus, $n \times m$ elements a_{ij}; $i = 1, 2, \dots, n$; $j = 1, 2, \dots, m$ can be represented by a symbol $[A]$ called the *matrix A* as follows:

$$[A] = \begin{bmatrix} a_{11} & a_{12} & \dots & a_{1m} \\ a_{21} & a_{22} & \dots & a_{2m} \\ \vdots & & & \\ a_{n1} & a_{n2} & \dots & a_{nm} \end{bmatrix} \tag{2.4}$$

The elements along each horizontal line are called *rows* whereas the elements along each vertical line are called *columns*. Thus, the matrix $[A]$ has n rows and m columns. We refer to $[A]$ as an $n \times m$ matrix. We identify each element of $[A]$ by row and column location. Thus, the element a_{ij} of $[A]$ is located at row i and column j. The first subscript in a_{ij} is the row location and the second subscript is the column location. This is a standard notation and is used throughout the book.

Definition 2.4 (Rectangular Matrix). In the matrix $[A]$ when $n \neq m$, i.e., the number of rows and columns are not the same, then $[A]$ is called a rectangular matrix.

Definition 2.5 (Square Matrix). In a square matrix, the number of rows is the same as the number of columns, i.e., $n = m$. The square matrices are of special significance in representing coefficients a_{ij}; $i, j = 1, 2, \ldots, n$ appearing in the linear simultaneous equations (2.2).

Definition 2.6 (Row Matrix). In (2.1), if $n = 1$, then the matrix $[A]$ will contain only one row, hence we can represent its elements by a single subscript only. Thus, a row matrix containing m columns can be represented by

$$[A] = [a_1 \; a_2 \; \ldots \; a_m] \tag{2.5}$$

Definition 2.7 (Column Matrix or Vector). In (2.4), if $m = 1$ then the matrix $[A]$ will contain only one column, hence we can also represent its elements by a single subscript. A matrix containing only one column is called a *vector*. Thus a column matrix or a vector containing n elements can be represented by

$$\{A\} = \begin{Bmatrix} a_1 \\ a_2 \\ \vdots \\ a_n \end{Bmatrix} \tag{2.6}$$

Definition 2.8 (Symmetric Matrix). A square matrix $[A]$ is symmetric if each row of the matrix is identical to the corresponding column.

$$a_{ij} = a_{ji} \qquad i, j = 1, 2, \ldots, n \tag{2.7}$$

The elements a_{ii}; $i = 1, 2, \ldots, n$ are called *diagonal elements* of matrix $[A]$. Thus, in a symmetric matrix the elements of the matrix below the diagonal are a mirror reflection of the elements above the diagonal and vice versa.

$$[A] = \begin{bmatrix} a_{11} & a_{12} & a_{13} \\ a_{12} & a_{22} & a_{23} \\ a_{13} & a_{23} & a_{33} \end{bmatrix} = \begin{bmatrix} & a_{11} & a_{12} & a_{13} \\ & & a_{22} & a_{23} \\ symm. & & & a_{33} \end{bmatrix} \tag{2.8}$$

$[A]$ is a (3×3) symmetric square matrix.

Definition 2.9 (Skew-Symmetric or Antisymmetric Matrix). A square matrix $[A]$ is called *skew-symmetric* or *antisymmetric* if its elements above the diagonal are negative of the elements below the diagonal or vice versa and if its diagonal elements are zero, i.e., $a_{ji} = -a_{ij}$ or $a_{ij} = -a_{ji}$; $j \neq i$ and $a_{ii} = 0$.

$$[A] = \begin{bmatrix} 0 & a_{12} & a_{13} \\ -a_{12} & 0 & a_{23} \\ -a_{13} & -a_{23} & 0 \end{bmatrix} \tag{2.9}$$

The matrix $[A]$ is a (3×3) skew-symmetric square matrix.

Definition 2.10 (Diagonal Matrix). The elements a_{ij}; $i \neq j$ of a square matrix $[A]$ are called *off-diagonal elements* and the elements a_{ij}; $j = i$, i.e., a_{ii}, are called *diagonal elements* of $[A]$. If all off-diagonal elements of a matrix $[A]$ are zero ($a_{ij} = 0$; $j \neq i$), then the matrix $[A]$ is called a *diagonal matrix.*

$$[A] = \begin{bmatrix} a_{11} & 0 & 0 \\ 0 & a_{22} & 0 \\ 0 & 0 & a_{33} \end{bmatrix} \tag{2.10}$$

The matrix $[A]$ is a (3×3) diagonal matrix.

Definition 2.11 (Identity Matrix). An *identity matrix* is a diagonal matrix whose diagonal elements are unity (one). We denote an identity matrix by $[I]$. Thus

$$[I] = \begin{bmatrix} 1 & 0 & 0 \\ 0 & 1 & 0 \\ 0 & 0 & 1 \end{bmatrix} \tag{2.11}$$

is a (3×3) identity matrix.

Definition 2.12 (Kronecker Delta (δ_{ij})). The elements of an identity matrix $[I]$ can be identified as

$$\delta_{ij} = \begin{cases} 1 & \text{if } j = i \\ 0 & \text{if } j \neq i \end{cases} \qquad i, j = 1, 2, \dots, n \tag{2.12}$$

The notation (2.12) is helpful when expressing $[I]$ in terms of its components (Einstein notation). Thus δ_{ij} is in fact the identity matrix expressed in Einstein notation. If we consider the product of $[A]$ and $[I]$, then we can write:

$$[A][I] = a_{ij}\delta_{jk} = a_{ik} = [A]\,; \quad i, j, k = 1, 2, \dots, n \tag{2.13}$$

Likewise:

$$[I][I] = \delta_{ij}\delta_{jk} = \delta_{ik} = [I] \tag{2.14}$$

Definition 2.13 (Upper Triangular Matrix). If all elements below the diagonal of a square matrix $[A]$ are zero, then $[A]$ is called an *upper triangular matrix.* For such matrices $a_{ij} = 0$ for $i > j$ holds. Thus

$$[A] = \begin{bmatrix} a_{11} & a_{12} & a_{13} \\ 0 & a_{22} & a_{23} \\ 0 & 0 & a_{33} \end{bmatrix} \tag{2.15}$$

is a (3×3) upper triangular matrix.

Definition 2.14 (Lower Triangular Matrix). If all elements above the diagonal of a square matrix $[A]$ are zero, then $[A]$ is called a *lower triangular matrix.* For such matrices $a_{ij} = 0$ for $i < j$ holds. Thus

$$[A] = \begin{bmatrix} a_{11} & 0 & 0 \\ a_{21} & a_{22} & 0 \\ a_{31} & a_{32} & a_{33} \end{bmatrix} \tag{2.16}$$

is a (3×3) lower triangular matrix.

Definition 2.15 (Banded Matrix). All elements of a *banded matrix* are zero, with the exception of a band about the diagonal. Thus

$$[A] = \begin{bmatrix} a_{11} & a_{11} & 0 & 0 \\ a_{21} & a_{22} & a_{23} & 0 \\ 0 & a_{32} & a_{33} & a_{34} \\ 0 & 0 & a_{43} & a_{44} \end{bmatrix} \tag{2.17}$$

has a *bandwidth* of three. All non-zero elements are within a band whose width is three elements. Such matrices with a bandwidth of three centered on the diagonal are called *tridiagonal matrices.*

2.1.2 Matrix Algebra

2.1.2.1 Addition and Subtraction of Two Matrices

The addition and subtraction of two matrices $[A]$ and $[B]$ results in a matrix $[C]$.

$$[A] \pm [B] = [C] \tag{2.18}$$

The matrix $[C]$ is defined by:

$$c_{ij} = a_{ij} \pm b_{ij} \; ; \quad i = 1, 2, \ldots, n \; ; \quad j = 1, 2, \ldots, m \tag{2.19}$$

Obviously for the addition or subtraction of $[A]$ and $[B]$ to be valid, both $[A]$ and $[B]$ must have the same number of rows and columns. The resulting matrix $[C]$ has the same number of rows and columns as well. We note that $[A] \pm [B] = \pm[B] + [A]$ holds for addition and subtraction of matrices, that is, matrix addition is commutative.

2.1.2.2 Multiplication by a Scalar

Multiplication of a matrix $[A]$ by a scalar s results in a matrix $[D]$.

$$s[A] = [D] \tag{2.20}$$

$[D]$ is defined by

$$d_{ij} = sa_{ij} \; ; \quad i = 1, 2, \ldots, n \; ; \quad j = 1, 2, \ldots, m \tag{2.21}$$

That is, every element of $[A]$ gets multiplied by the scalar s.

2.1.2.3 Product of Matrices

A matrix $[A]_{(n \times m)}$ can be multiplied with a matrix $[B]_{(m \times l)}$. The resulting matrix is $[C]_{(n \times l)}$.

$$[A]_{(n \times m)}[B]_{(m \times l)} = [C]_{(n \times l)} \tag{2.22}$$

$[C]_{(n \times l)}$ is defined by

$$c_{ij} = a_{ik}b_{kj}; \; i = 1, 2, \ldots, n; \; j = 1, 2, \ldots, l; \; k = 1, 2, \ldots, m \tag{2.23}$$

We note that the number of columns in $[A]$ must be the same as the number of rows in $[B]$, otherwise the product of $[A]$ and $[B]$ is not valid. Consider

$$[A] = \begin{bmatrix} a_{11} & a_{12} \\ a_{21} & a_{22} \\ a_{31} & a_{32} \end{bmatrix} \qquad [B] = \begin{bmatrix} b_{11} & b_{12} \\ b_{21} & b_{22} \end{bmatrix} \tag{2.24}$$

Then

$$[C] = [A][B] = \begin{bmatrix} a_{11} & a_{12} \\ a_{21} & a_{22} \\ a_{31} & a_{32} \end{bmatrix} \begin{bmatrix} b_{11} & b_{12} \\ b_{21} & b_{22} \end{bmatrix} = \begin{bmatrix} (a_{11}b_{11} + a_{12}b_{21}) & (a_{11}b_{12} + a_{12}b_{22}) \\ (a_{21}b_{11} + a_{22}b_{21}) & (a_{21}b_{12} + a_{22}b_{22}) \\ (a_{31}b_{11} + a_{32}b_{21}) & (a_{31}b_{12} + a_{32}b_{22}) \end{bmatrix} \tag{2.25}$$

We note that $[A]_{(n \times n)}[I]_{(n \times n)} = [I]_{(n \times n)}[A]_{(n \times n)} = [A]_{(n \times n)}$.

2.1.2.4 Algebraic Properties of Matrix Multiplication

Associative Property:

A product of matrices is invariant of the order of multiplication.

$$[A][B][C] = [A]([B][C]) = ([A][B])[C] = [D] \tag{2.26}$$

Distributive Property:

The sum of $[A]$ and $[B]$ multiplied with $[C]$ is the same as $[A]$ and $[B]$ multiplied with $[C]$, then summed.

$$([A] + [B])[C] = [A][C] + [B][C] \tag{2.27}$$

Commutative Property:

The product of $[A]$ and $[B]$ is not the same as product of $[B]$ and $[A]$. Thus, in taking the product of $[A]$ and $[B]$, their positions cannot be changed.

$$[A][B] \neq [B][A] \tag{2.28}$$

Definition 2.16 (Trace of a Matrix). The *trace* of a square matrix $[A]$ is the sum of its diagonal elements.

$$\text{tr}[A] = \sum_{i=1}^{n} a_{ii} \tag{2.29}$$

The trace is only defined for a square matrix.

Definition 2.17 (Inverse of a Matrix). For every *non-singular* (defined later) square matrix $[A]$ there exists another matrix $[A]^{-1}$ (inverse of $[A]$) such that the following holds:

$$[A]^{-1}[A] = [A][A]^{-1} = [I] \tag{2.30}$$

A *singular* matrix is one for which its inverse does not exist. The inverse is only defined for a square matrix.

Definition 2.18 (Transpose of a Matrix). The *transpose* of a matrix $[A]$ is denoted by $[A]^T$ and is obtained by interchanging rows with the corresponding columns. If a matrix $[A]$ has elements a_{ij}; $i = 1, 2, \ldots, n$; $j = 1, 2, \ldots, m$, then the elements of $[A]^T$ are a_{ji}; $i = 1, 2, \ldots, n$; $j = 1, 2, \ldots, m$. We note that the matrix $[A]$ is $(n \times m)$ where the matrix $[A]^T$ is $(m \times n)$. If

$$[A] = \begin{bmatrix} a_{11} & a_{12} & a_{13} \\ a_{21} & a_{22} & a_{23} \end{bmatrix}_{(2\times3)} \tag{2.31}$$

then

$$[A]^T = \begin{bmatrix} a_{11} & a_{21} \\ a_{12} & a_{22} \\ a_{13} & a_{23} \end{bmatrix}_{(3\times2)} \tag{2.32}$$

Row one of $[A]$ is the same as column one of $[A]^T$. Likewise row one of $[A]^T$ is the same as column one of $[A]$ and so on. That is, rows of $[A]$ are same as columns of $[A]^T$ and vice versa.

Transpose of a Row Matrix:

If the row matrix $[A]$ is defined by

$$[A]_{(1\times m)} = \big[a_1\ a_2\ \dots\ a_m\big]_{(1\times m)} \tag{2.33}$$

then

$$[A]^T_{(m\times 1)} = \begin{Bmatrix} a_1 \\ a_2 \\ \vdots \\ a_m \end{Bmatrix}_{(m\times 1)} \tag{2.34}$$

That is, the transpose of a row matrix is a column matrix or vector.

Transpose of a Vector:

If $\{A\}$ is a vector defined by

$$\{A\}_{(n\times 1)} = \begin{Bmatrix} a_1 \\ a_2 \\ \vdots \\ a_n \end{Bmatrix}_{(n\times 1)} \tag{2.35}$$

then

$$\{A\}^T_{(1\times n)} = \big[a_1\ a_2\ \dots\ a_n\big]_{(1\times n)} \tag{2.36}$$

That is, the transpose of a column vector is a row matrix.

Transpose of a Product of Matrices:

Let $[A]_{m\times n}$ and $[B]_{n\times p}$ be rectangular matrices, then:

$$\Big([[A][B]]_{m\times p}\Big)^T = [[B]^T[A]^T]_{p\times m} \tag{2.37}$$

Likewise:

$$([A][B][C])^T = [C]^T[B]^T[A]^T \tag{2.38}$$

and

$$([A]_{m\times n}\{c\}_{n\times 1})^T = (\{c\}^T[A]^T)_{1\times m} \tag{2.39}$$

Thus, the transpose of the product of matrices is the product of their transposes in reverse order.

Transpose of a Symmetric Matrix: If $[A]$ is a $(n \times n)$ symmetric matrix, then:

$$a_{ij} = a_{ji}\ ; \quad i,j = 1,2,\dots,n \tag{2.40}$$

or

$$[A]^T = [A] \tag{2.41}$$

That is, the transpose of a symmetric matrix is the matrix itself.

Transpose of a Skew-Symmetric Matrix: If $[A]$ is a $(n \times n)$ skew-symmetric matrix, then:

$$a_{ij} = -a_{ji},\ i \neq j \quad \text{and} \quad a_{ii} = 0\ ; \quad i,j = 1,2,\dots,n \tag{2.42}$$

or

$$[A]^T = -[A] \tag{2.43}$$

Transpose of the Products of Symmetric and Skew-Symmetric Matrices: If $[A]$ is a $(n\times n)$ symmetric matrix and $[B]$ is a $(n\times n)$ skew-symmetric matrix, then:

$$\begin{aligned} a_{ij} = a_{ji}\ ; \quad i,j = 1,2,\dots,n \\ b_{ij} = -b_{ji},\ i \neq j \quad \text{and} \quad b_{ii} = 0\ ; \quad i,j = 1,2,\dots,n \end{aligned} \tag{2.44}$$

Therefore, we have:

$$\left([A][B]\right)^T = [B]^T[A]^T = -[B][A] \tag{2.45}$$

Likewise:

$$\left([B][A]\right)^T = [A]^T[B]^T = [A](-[B]) = -[A][B] \tag{2.46}$$

One can conclude from this that the product of a symmetric matrix and a skew-symmetric matrix is a skew-symmetric matrix.

Definition 2.19 (Orthogonal Matrix). A matrix $[R]$ is *orthogonal* if its transpose is the same as its inverse.

$$[R]^{-1} = [R]^T \tag{2.47}$$

$$\therefore \quad [R]^{-1}[R] = [R][R]^{-1} = [R]^T[R] = [R][R]^T = [I] \tag{2.48}$$

Rotation matrices defining rotation of a frame of reference into another frame are examples of such matrices. Orthogonality in this sense is only defined for a square matrix.

Definition 2.20 (Positive-Definite Matrix). A square matrix $[A]$ is positive-definite if and only if

$$\{x\}^T[A]\{x\} > 0 \quad \forall \{x\} \neq \{0\} \tag{2.49}$$

If $\{x\}^T[A]\{x\} \leq 0$ then $[A]$ is not positive-definite. All positive-definite matrices are symmetric. Eigenvalues of a positive-definite matrix are real and strictly greater than zero, and the associated eigenvectors are real (see Chapter 4).

Definition 2.21 (Positive-Semidefinite Matrix). A square matrix $[A]$ is positive-semidefinite if and only if

$$[A] = [B]^*[B] \tag{2.50}$$

for some square matrix $[B]$. Neither $[A]$ nor $[B]$ are necessarily symmetric. When $[A]$ is not symmetric, $[B]$ and $[B]^*$ are complex. If $[B]$ is not complex, then $[B]^* = [B]^T$. This is only ensured if $[A]$ is symmetric. Thus, if $[A]$ is symmetric then $[B]$ is also symmetric and in this case (see Chapter 4 for proof):

$$\{x\}^T[A]\{x\} = \{x\}^T[B]^T[B]\{x\} \geq 0 \quad \forall \{x\} \neq \{0\} \tag{2.51}$$

and

$$\{x\}^T[A]\{x\} = 0 \quad \text{for some } \{x\} \neq \{0\} \tag{2.52}$$

Definition 2.22 (Orthogonality of Vectors). If $\{x\}_i$ and $\{x\}_j$ are two vectors of unit norm or length in an n-dimensional space, then $\{x\}_i$ and $\{x\}_j$ are orthogonal if and only if

$$\{x\}_i^T\{x\}_j = \delta_{ij} \tag{2.53}$$

where δ_{ij} is the Kronecker delta.

Definition 2.23 (Orthogonality of Vectors with Respect to a Matrix). If $\{x\}_i$ and $\{x\}_j$ are two vectors that are normalized with respect to a matrix $[M]$, i.e.,

$$\begin{aligned} \{x\}_i^T[M]\{x\}_i &= 1 \\ \{x\}_j^T[M]\{x\}_j &= 1 \end{aligned} \tag{2.54}$$

then $\{x\}_i$ and $\{x\}_j$ are $[M]$-orthogonal if and only if

$$\{x\}_i^T[M]\{x\}_j = \delta_{ij} \tag{2.55}$$

Definition 2.24 (Orthogonality of Vectors with Respect to Identity $[I]$). Definition 2.23 implies:

$$\{x\}_i^T[I]\{x\}_j = \delta_{ij} \tag{2.56}$$

when $\{x\}_i$ and $\{x\}_j$ are orthogonal with respect to $[I]$. Thus, when (2.56) holds, so does (2.53). We note that (2.56) is a special case of (2.55) with $[M] = [I]$.

2.1.2.5 Decomposition of a Square Matrix into Symmetric and Skew-Symmetric Matrices

Consider a square matrix $[A]$.

$$[A] = \frac{1}{2}[A] + \frac{1}{2}[A] \tag{2.57}$$

Add and subtract $\frac{1}{2}[A]^T$ to right side of (2.57).

$$[A] = \frac{1}{2}[A] + \frac{1}{2}[A] + \frac{1}{2}[A]^T - \frac{1}{2}[A]^T \tag{2.57}$$

or

$$[A] = \frac{1}{2}([A] + [A]^T) + \frac{1}{2}([A] - [A]^T) \tag{2.58}$$

We define

$$[D] = \frac{1}{2}([A] + [A]^T) \qquad [W] = \frac{1}{2}([A] - [A]^T) \tag{2.59}$$

$$\therefore \quad [A] = [D] + [W] \tag{2.60}$$

We note that

$$\begin{aligned} [D]^T &= \frac{1}{2}([A]^T + [A]) = [D] \\ [W]^T &= \frac{1}{2}([A]^T - [A]) = -[W] \end{aligned} \tag{2.61}$$

Thus the matrix $[D]$ is symmetric and $[W]$ is skew-symmetric (or antisymmetric) with zeros on the diagonal. Equation (2.60) is the decomposition of the square matrix $[A]$ into a symmetric matrix $[D]$ and the skew-symmetric matrix $[W]$.

2.1.2.6 Augmenting a Matrix

If a new matrix is formed from the original matrix $[A]$ by adding an additional column or columns to it, then the resulting matrix is an *augmented matrix* $[A_{ag}]$. Consider

$$[A] = \begin{bmatrix} a_{11} & a_{12} & a_{13} \\ a_{12} & a_{22} & a_{23} \\ a_{13} & a_{23} & a_{33} \end{bmatrix} \tag{2.62}$$

Then

$$[A_{ag}] = \left[\begin{array}{ccc|ccc} a_{11} & a_{12} & a_{13} & 1 & 0 & 0 \\ a_{12} & a_{22} & a_{23} & 0 & 1 & 0 \\ a_{13} & a_{23} & a_{33} & 0 & 0 & 1 \end{array}\right] \tag{2.63}$$

is the (3×6) matrix obtained by augmenting $[A]$ with the (3×3) identity matrix. We separate the original matrix $[A]$ from $[I]$ (in this case) by a vertical line in defining the augmented matrix $[A_{ag}]$.

Consider

$$[A_{ag}] = \left[\begin{array}{ccc|c} a_{11} & a_{12} & a_{13} & b_1 \\ a_{12} & a_{22} & a_{23} & b_2 \\ a_{13} & a_{23} & a_{33} & b_3 \end{array}\right] \tag{2.64}$$

$[A_{ag}]$ in this case is the (3×4) matrix defined by augmenting $[A]$ by a vector whose components are b_1, b_2, and b_3.

Definition 2.25 (Linear Dependence and Independence of Rows). If a row of a matrix can be generated by a linear combination of the other rows of the matrix, then this row is called *linearly dependent.* Otherwise, the row is called *linearly independent.*

Definition 2.26 (Linear Dependence and Independence of Columns). If a column of a matrix can be generated by a linear combination of the other columns of the matrix, then this column is called *linearly dependent.* Otherwise, the column is called *linearly independent.*

Definition 2.27 (Rank of a Matrix). The *rank* of a square matrix is the number of linearly independent rows or columns. In a $(n \times n)$ square matrix, if all rows and all columns are linearly independent, then n is the rank of the matrix.

Definition 2.28 (Rank Deficient Matrix). In a *rank deficient* $(n \times n)$ square matrix, there is at least one row and one column that can be expressed as a linear combination of the other rows and columns. Thus, in a $(n \times n)$ matrix of rank $(n-m)$ there are m rows and columns that can be expressed as linear combinations of the others. In such matrices, a reduced $(n-m \times n-m)$ matrix can be formed by removing the linearly dependent rows and columns that would have a rank of $(n - m)$.

2.1.2.7 Determinant of a Matrix

The *determinant* of a square matrix $[A]$ is a scalar, i.e., a real number if the elements of $[A]$ are real numbers, and is denoted by $det[A]$ or $|A|$. If

$$[A] = \begin{bmatrix} a_{11} & a_{12} & \dots & a_{1n} \\ a_{12} & a_{22} & \dots & a_{2n} \\ \vdots & & & \vdots \\ a_{n1} & a_{n2} & \dots & a_{nn} \end{bmatrix} \tag{2.65}$$

then $det[A] = |A|$ can be obtained by using the following:

(i) *Minor of a_{ij}:*
The minor of a_{ij} is defined as the determinant of $[A]$ obtained after deleting row i and column j from $[A]$ and is denoted by m_{ij}.

col. j

$m_{ij} =$ 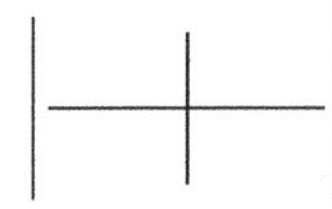 row i (2.66)

(ii) *Cofactor of a_{ij}:*
The cofactor of a_{ij} is a scalar denoted by $\bar{a}_{ij}$. It is the *signed minor* of a_{ij}, i.e., the cofactor of a_{ij} is obtained by assigning a sign to the minor of a_{ij} and is defined by

$$\bar{a}_{ij} = (-1)^{i+j} m_{ij} \tag{2.67}$$

(iii) *Computation of Determinant:*
The determinant of $[A]$ is obtained by multiplying each element of any one row or any one column of $[A]$ with its associated cofactor and summing the products. This is called Laplace expansion. Thus, if we use the first row of $[A]$ then

$$|A| = a_{11}\bar{a}_{11} + a_{12}\bar{a}_{12} + \cdots + a_{1n}\bar{a}_{1n} \tag{2.68}$$

Using the second column of $[A]$ we obtain

$$|A| = a_{12}\bar{a}_{12} + a_{22}\bar{a}_{22} + \cdots + a_{n2}\bar{a}_{n2} \tag{2.69}$$

The determinant computed using (2.68) is identical to that found using (2.69). Typically, the row or column with the most 0 elements is chosen for ease of calculation.

The determinant is only defined for a square matrix. Obviously, the calculation of $\det[A]$ is facilitated by choosing a row or a column containing zeros.

Definition 2.29 (Singular Matrix). A matrix $[A]$ is *singular* if it is non-invertible (i.e., if $[A]^{-1}$ does not exist). This is equivalent to $|A| = 0$, linear dependence of any rows or columns, and rank deficiency. If any one of these conditions hold, then they all do. A matrix $[A]$ is *non-singular* if and only if none of the previously mentioned conditions hold.

Example 2.1 (Determinant of a 2×2 Matrix). Consider a (2×2) matrix $[A]$.

$$[A] = \begin{bmatrix} a_{11} & a_{12} \\ a_{21} & a_{22} \end{bmatrix}$$

Find $|A|$.

Solution:
Determine $|A|$ using the first row of $[A]$.

(i) The minors m_{11} and m_{12} of a_{11} and a_{12} are given by

$$m_{11} = |a_{22}| = a_{22} \; ; \qquad m_{12} = |a_{21}| = a_{21}$$

(ii) The cofactors of a_{11} and a_{12} are given by the signed minors of a_{11} and a_{12}.

$$\bar{a}_{11} = (-1)^{1+1} m_{11} = a_{22} \; ; \qquad \bar{a}_{12} = (-1)^{1+2} m_{12} = -a_{21}$$

(iii) The determinant of $[A]$ is given by

$$|A| = a_{11}\bar{a}_{11} + a_{12}\bar{a}_{12}$$

Substituting for the cofactors, we have

$$|A| = a_{11}a_{22} - a_{12}a_{21}$$

Example 2.2. Consider a (3×3) matrix $[A]$.

$$[A] = \begin{bmatrix} a_{11} & a_{12} & a_{13} \\ a_{21} & a_{22} & a_{23} \\ a_{31} & a_{32} & a_{33} \end{bmatrix}$$

Find $|A|$.

Solution:
Determine $|A|$ using the first row of $[A]$.

(i) Minors m_{11}, m_{12}, and m_{13} of a_{11}, a_{12}, and a_{13} are given by

$$m_{11} = \begin{vmatrix} a_{22} & a_{23} \\ a_{32} & a_{33} \end{vmatrix} ; \quad m_{12} = \begin{vmatrix} a_{21} & a_{23} \\ a_{31} & a_{33} \end{vmatrix} ; \quad m_{13} = \begin{vmatrix} a_{21} & a_{22} \\ a_{31} & a_{32} \end{vmatrix}$$

(ii) Cofactors $\bar{a}_{11}$, $\bar{a}_{12}$, and $\bar{a}_{13}$ are given by

$$\bar{a}_{11} = (-1)^{1+1} m_{11} \; ; \quad \bar{a}_{12} = (-1)^{1+2} m_{12} \; ; \quad \bar{a}_{13} = (-1)^{1+3} m_{13}$$

(iii)

$$|A| = a_{11}\bar{a}_{11} + a_{12}\bar{a}_{12} + a_{13}\bar{a}_{13}$$

Substituting for $\bar{a}_{11}$, $\bar{a}_{12}$, and $\bar{a}_{13}$:

$$|A| = a_{11}(1)m_{11} + a_{12}(-1)m_{12} + a_{13}(1)m_{13}$$

Further substituting for m_{11}, m_{12}, and m_{13}:

$$|A| = a_{11}(1)\begin{vmatrix} a_{22} & a_{23} \\ a_{32} & a_{33} \end{vmatrix} + a_{12}(-1)\begin{vmatrix} a_{21} & a_{23} \\ a_{31} & a_{33} \end{vmatrix} + a_{13}(1)\begin{vmatrix} a_{21} & a_{22} \\ a_{31} & a_{32} \end{vmatrix}$$

Expanding determinants in the above expression using the first row in each case:

$$\begin{aligned} \begin{vmatrix} a_{22} & a_{23} \\ a_{32} & a_{33} \end{vmatrix} &= a_{22}\bar{a}_{22} + a_{23}\bar{a}_{23} \\ &= a_{22}(-1)^{2+2}m_{22} + a_{23}(-1)^{2+3}m_{23} \\ &= a_{22}(-1)^{2+2}a_{33} + a_{23}(-1)^{2+3}a_{32} \\ &= a_{22}a_{33} - a_{23}a_{32} \end{aligned}$$

Similarly:

$$\begin{vmatrix} a_{21} & a_{23} \\ a_{31} & a_{33} \end{vmatrix} = a_{21}a_{33} - a_{23}a_{31}$$

$$\begin{vmatrix} a_{21} & a_{22} \\ a_{31} & a_{32} \end{vmatrix} = a_{21}a_{32} - a_{22}a_{31}$$

Substituting these in the expression for $|A|$, the determinant is given by

$$|A| = a_{11}(a_{22}a_{33} - a_{23}a_{32}) - a_{12}(a_{21}a_{33} - a_{23}a_{31}) + a_{13}(a_{21}a_{32} - a_{22}a_{31})$$

Example 2.3. Consider a (2×2) matrix $[A]$.

$$[A] = \begin{bmatrix} -2 & 3 \\ -2 & 3 \end{bmatrix}$$

$$\begin{aligned} |A| &= (-2)(-1)^{1+1}(3) + (3)(-1)^{1+2}(-2) \\ &= (-2)(3) + (3)(-1)(-2) \\ &= -6 + 6 = 0 \end{aligned}$$

In matrix $[A]$, row two is identical to row one. It can be shown that if $[A]$ is a square matrix $(n \times n)$ and if any two rows are the same, the $|A| = 0$, regardless of n.

Example 2.4. Consider a (2×2) matrix $[A]$.

$$[A] = \begin{bmatrix} -2 & -2 \\ 3 & 3 \end{bmatrix}$$

$$\begin{aligned} |A| &= (-2)(-1)^{1+1}(3) + (-2)(-1)^{1+2}(3) \\ &= -6 + 6 = 0 \end{aligned}$$

In this case column one is identical to column two. It can be shown that if $[A]$ is any square matrix $(n \times n)$ and if any two columns are the same, the $|A| = 0$, regardless of n.

Example 2.5. Consider a (2×2) matrix $[A]$.

$$[A] = \begin{bmatrix} 4 & 4 \\ 4a & 4a \end{bmatrix}$$

$$\begin{aligned} |A| &= (4)(-1)^{1+1}(4a) + (4)(-1)^{1+2}(4a) \\ &= 16a - 16a = 0 \end{aligned}$$

In matrix $[A]$, row two is a multiple of row one (by a) or row one is a multiple of row two (by $1/a$).

Remarks.

(1) We note that the matrix $[A]$ in Example 2.4 is the transpose of the matrix $[A]$ in Example 2.3, hence we can conclude that if $|A| = 0$, then $|A^T| = 0$.

(2) In general, for an $(n \times n)$ matrix $[A]$, if any two rows are multiples of each other, then $|A| = 0$. We note that in Example 2.5, the two columns are the same, but this is not the case in general.

(3) It also holds that for any $(n \times n)$ matrix $[A]$, if any two columns are multiples of each other, then $|A| = 0$.

(4) As an illustration, $|A| = 0$ in Example 2.3 and column two can be obtained by multiplying column one by $-3/2$.

2.2 Matrix and Vector Representation of Linear Simultaneous Algebraic Equations

Consider equation (2.1):

$$f_i(x_1, x_2, \dots, x_n) = b_i \qquad i = 1, 2, \dots, n \tag{2.70}$$

When each $f_i(\cdot)$ is a linear combination of x_j; $j = 1, 2, \dots, n$, then we can write (2.70) as

$$\begin{aligned} a_{11}x_1 + a_{12}x_2 + \dots + a_{1n}x_n &= b_1 \\ a_{21}x_1 + a_{22}x_2 + \dots + a_{2n}x_n &= b_2 \\ &\vdots \\ a_{n1}x_1 + a_{n2}x_2 + \dots + a_{nn}x_n &= b_n \end{aligned} \tag{2.71}$$

Equations (2.71) represent a system of n linear simultaneous algebraic equations. These equations are linear in x_j and each equation simultaneously depends on all x_j; $j = 1, 2, \dots, n$. The coefficients a_{ij},b_i; $i, j = 1, 2, \dots, n$ are known. Our objective is to find x_j; $j = 1, 2, \dots, n$ that satisfy (2.71). Equations (2.71) can be represented more compactly using matrix and vector notation. If we define the coefficients a_{ij} by a matrix $[A]_{(n\times n)}$, b_i by a vector $\{b\}_{(n\times 1)}$, and x_j by a vector $\{x\}_{(n\times 1)}$, then (2.71) can be written as

$$[A]\{x\} = \{b\} \tag{2.72}$$

in which

$$[A] = \begin{bmatrix} a_{11} & a_{12} & \dots & a_{1n} \\ a_{12} & a_{22} & \dots & a_{2n} \\ \vdots & & & \vdots \\ a_{n1} & a_{n2} & \dots & a_{nn} \end{bmatrix} \; ; \qquad \{b\} = \begin{Bmatrix} b_1 \\ b_2 \\ \vdots \\ b_n \end{Bmatrix} \; ; \qquad \{x\} = \begin{Bmatrix} x_1 \\ x_2 \\ \vdots \\ x_n \end{Bmatrix} \tag{2.73}$$

The matrix $[A]$ is called the *coefficient matrix*, $\{b\}$ is called the *right-hand side* or *non-homogeneous* part, and $\{x\}$ is a vector of unknowns to be determined such that (2.72) holds. Sometimes we augment $[A]$ by $\{b\}$ by including it as $(n+1)^{th}$ column in $[A]$. Thus augmented matrix $[A_{ag}]$ would be:

$$[A_{ag}] = \left[\begin{array}{cccc|c} a_{11} & a_{12} & \dots & a_{1n} & b_1 \\ a_{12} & a_{22} & \dots & a_{2n} & b_2 \\ \vdots & \vdots & \vdots & \vdots & \vdots \\ a_{n1} & a_{n2} & \dots & a_{nn} & b_n \end{array}\right] \tag{2.74}$$

2.2.1 Elementary Row Operations

The augmented matrix $[A_{ag}]$ is a compact representation of the coefficients of $[A]$ and $\{b\}$ in the linear simultaneous equations (2.72). We note that in (2.72) if an equation is multiplied by a constant c, the solution of the new equations is the same as those of (2.72). Likewise, if an equation of (2.72) is multiplied by a constant and then added to another equation of (2.72), the solution of the new system of equations is the same as that of (2.72). These operations are called *elementary row operations*. This is more effectively used with $[A_{ag}]$. In $[A_{ag}]$, a row R_i can be multiplied by a constant c and added to another row R_m to form a new row $R_m^1 = R_m + cR_i$. The equations defined by the new $[A_{ag}]$ have the same solutions as (2.72).

It is important to note that in elementary row operations, when the coefficients of a row of $[A]$ are multiplied by a constant, the corresponding element of $\{b\}$ must also be multiplied by the same constant. The same holds true when adding or subtracting two rows. Thus, the elementary row operations should be performed on $[A_{ag}]$ and not $[A]$, as only $[A_{ag}]$ includes the right-hand side vector $\{b\}$.

2.3 Methods of Obtaining Solutions of Linear Simultaneous Algebraic Equations

Consider a system of n linear simultaneous algebraic equation in n unknowns $\{x\}_{(n\times 1)}$.

$$[A]_{(n\times n)}\{x\}_{(n\times 1)} = \{b\}_{(n\times 1)} \tag{2.75}$$

The coefficient matrix $[A]$ and the right-hand side vector $\{b\}$ (equation (2.73)) are known. Broadly speaking, the methods of obtaining the solution $\{x\}$ of (2.75) can be classified into two groups. In the first group of methods, one only obtains approximations of the true value of $\{x\}$. Graphical methods and iterative methods such as Gauss-Seidel or Jacobi methods fall into this category. With the second group of methods we seek the solution $\{x\}$ that satisfies (2.75) within the precision of the computations (i.e., the exact solution), for example the word size of a computer. Even though the second group of methods are superior to the first group in terms of accuracy of the solution $\{x\}$, the first group of methods are sometimes preferred due to ease of their use (specifically iterative methods). In the following we list various methods of obtaining the solution $\{x\}$ of (2.75) based on the fundamental concept involved in the design of the method.

(A) Direct methods

- (a) Graphical methods
- (b) Cramer's rule

(B) Elimination methods

(a) Gauss elimination

i. Naive Gauss elimination
ii. Gauss elimination with partial pivoting
iii. Gauss elimination with full pivoting

(b) Gauss-Jordan method

(c) $[L][U]$ Decomposition

i. Classical or Cholesky $[L][U]$ decomposition
ii. Crout $[L][U]$ decomposition
iii. $[L][U]$ decomposition using Gauss elimination

(C) Using the Inverse of $[A]$, i.e., $[A]^{-1}$

(a) Direct method of obtaining $[A]^{-1}$

(b) Inverse of $[A]$ by elementary row operations

(c) Inverse of $[A]$ using $[L][U]$ decomposition

(D) Iterative methods (methods of approximation)

(a) Gauss-Seidel method

(b) Jacobi method

(c) Relaxation techniques

We remark that Cramer's rule, Gauss elimination, Gauss-Jordan elimination, $[L][U]$ decomposition, and use of the inverse of $[A]$ to solve linear systems are numerical methods (when $[A]^{-1}$ is not approximate). Graphical methods, Gauss-Seidel method, Jacobi method, and relaxation methods are methods of approximation. In the former methods, the computed solutions are the theoretical solutions, whereas in the latter the calculated solutions are always approximate. We present details of each of these methods in the following sections and provide numerical examples illustrating how to use them.

2.4 Direct Methods

The direct methods are only helpful in obtaining solutions of a very small system of linear simultaneous algebraic equations, generally $n = 2$ and $n = 3$. For n greater than three, these methods are either not usable or become impractical due to complexity in their use.

2.4.1 Graphical Method

In this method we use graphical representations of the equations, hence this method is difficult to use for $n > 3$. We plot a graph corresponding to each equation. These are naturally straight lines (or flat planes) as the equations are linear. The common point of intersection of these straight lines or planes is the solution of the system of equations. The solution (x_1, x_2) (or (x, y)) is the only ordered pair that satisfies both equations; graphically, this is the only coordinate point that lies on both lines. Consider:

$$\begin{aligned} a_{11}x + a_{12}y &= b_1 \\ a_{21}x + a_{22}y &= b_2 \end{aligned} \tag{2.76}$$

We rewrite (2.76) by dividing the first equation by a_{11} and the second equation by a_{22} (provided $a_{11} \neq 0$ and $a_{22} \neq 0$).

$$\begin{aligned} y &= \left(-{}^{a_{11}}\!/_{a_{12}}\right) x + \left({}^{b_1}\!/_{a_{12}}\right) \\ y &= \left(-{}^{a_{21}}\!/_{a_{22}}\right) x + \left({}^{b_2}\!/_{a_{22}}\right) \end{aligned} \tag{2.77}$$

If we define

$$\begin{aligned} m_1 &= \left(-{}^{a_{11}}\!/_{a_{12}}\right) \quad & m_2 &= \left(-{}^{a_{21}}\!/_{a_{22}}\right) \\ c_1 &= \left({}^{b_1}\!/_{a_{12}}\right) \quad & c_2 &= \left({}^{b_2}\!/_{a_{22}}\right) \end{aligned} \tag{2.78}$$

Then, (2.77) can be written as:

$$\begin{aligned} y &= m_1 x + c_1 \\ y &= m_2 x + c_2 \end{aligned} \tag{2.79}$$

If we consider two-dimensional xy-space (x being the abscissa and y being the ordinate) then (2.79) are equations of straight lines in which m_1, m_2 are their slopes and c_1,c_2 are the corresponding intercepts with the y-axis. Thus, (2.79) can be plotted in the xy-plane. Their intersection is the solution of (2.79) as it would naturally satisfy both equations in (2.79). Figure 2.1 shows the details.

Remarks.

(1) When the determinant of the coefficient matrix in (2.76), i.e., $(a_{11}a_{22} - a_{12}a_{22})$, is not equal to zero, the intersection of the straight lines is distinct and we clearly have a unique solution (as shown in Figure 2.1)

(2) If $a_{21} = a_{11}$ and $a_{22} = a_{12}$ in (2.76), we have

$$\begin{aligned} a_{11}x + a_{12}y &= b_1 \\ a_{11}x + a_{12}y &= b_2 \end{aligned} \tag{2.80}$$

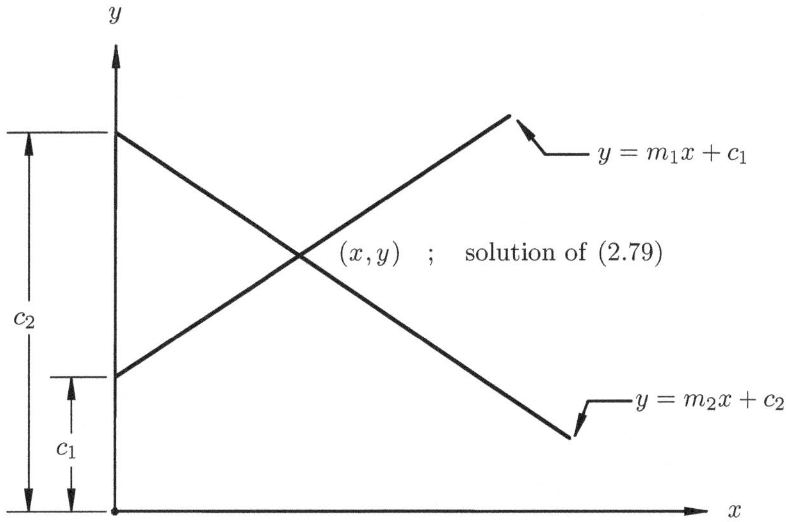

Figure 2.1: Graphical method of obtaining solution of two linear simultaneous equations

or

$$\begin{aligned} y &= \left(-{}^{a_{11}}/{}_{a_{12}}\right) x + \left({}^{b_1}/{}_{a_{12}}\right) \\ y &= \left(-{}^{a_{11}}/{}_{a_{12}}\right) x + \left({}^{b_2}/{}_{a_{13}}\right) \end{aligned} \tag{2.81}$$

Let

$$m_1 = -{}^{a_{11}}/{}_{a_{12}} \qquad c_1 = {}^{b_1}/{}_{a_{12}} \qquad c_2 = {}^{b_2}/{}_{a_{12}} \tag{2.82}$$

Hence, (2.81) can be written as

$$\begin{aligned} y &= m_1x + c_1 \\ y &= m_1x + c_2 \end{aligned} \tag{2.83}$$

Equations (2.83) are the equations of straight lines that are parallel. Parallel lines have the same slopes but different intercepts, thus these will never intersect. In this case we obviously cannot find a solution (x, y) of (2.83). We also note that the determinant of the coefficient matrix of (2.80) is zero. In (2.80) row one of the coefficient matrix is the same as row two and the columns are multiples of each other. This system of equations (2.80) is rank deficient. Figure 2.2 shows plots of (2.83).

(3) Consider a case in which column two of the coefficient matrix in (2.76) is a multiple of column one, i.e., for a scalar s we have

$$\begin{aligned} a_{12} &= sa_{11} \\ a_{22} &= sa_{21} \end{aligned} \tag{2.84}$$

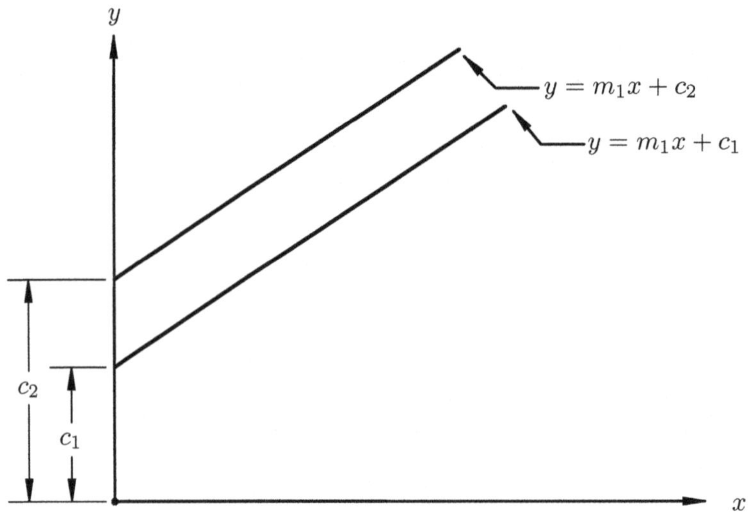

Figure 2.2: An ill-conditioned system ($det[A] = 0$); no solution

Thus, for this case (2.76) reduces to

$$\begin{aligned} a_{11}x + sa_{11}y &= b_1 \\ a_{21}x + sa_{21}y &= b_2 \end{aligned} \tag{2.85}$$

Divide the first equation by sa_{11} and the second equation by sa_{21}.

$$\begin{aligned} y &= -\left(1/s\right)x + \left(b_1/sa_{11}\right) \\ y &= -\left(1/s\right)x + \left(b_2/sa_{21}\right) \end{aligned} \tag{2.86}$$

or

$$\begin{aligned} y &= m_1x + c_1 \\ y &= m_1x + c_2 \end{aligned} \tag{2.87}$$

which is the same as (2.83). We can also arrive at (2.87) if row two of the coefficient matrix in (2.76) is a multiple of row one or vice versa. When the coefficients are such that $c_2 \neq c_1$, graphs of (2.87) are the same as those in Figure 2.2. But for some choice of coefficients if we have $c_2 = c_1$, then both equations in (2.87) are identical. Their xy plots naturally coincide (Figure 2.3), hence this case the two straight lines intersect at infinitely many locations, implying infinitely many solutions. The solution of (2.85) is not unique in such a case.

(4) From the above remarks, we conclude that whenever the determinant of the coefficient matrix is a system of algebraic equations is zero, their solution either does not exist or is not unique.

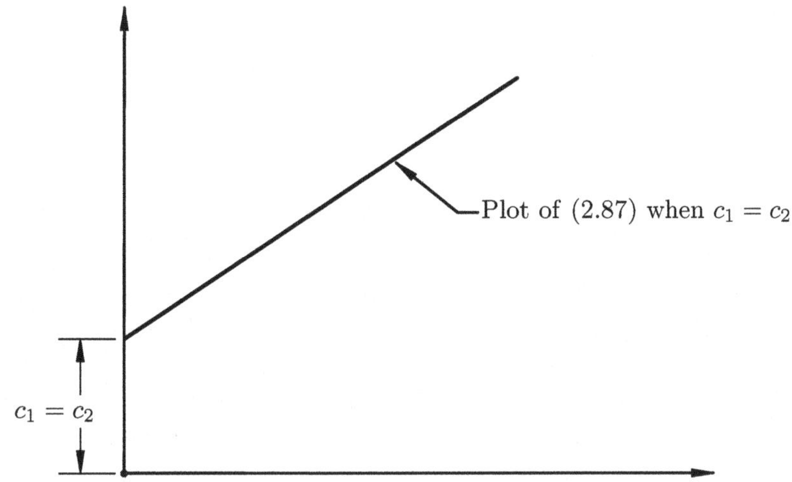

Figure 2.3: Infinitely many solutions when two equations are identical

(5) Consider system of equations (2.76). It could happen that the coefficients a_{ij}; $i, j = 1, 2$ are such that the determinant of the coefficient matrix $(a_{11}a_{22} - a_{12}a_{21})$ may not be zero but may be close to zero. In this case the straight lines defined by the two equations in (2.76) do have an intersection but their intersection may not be distinct (Figure 2.4).

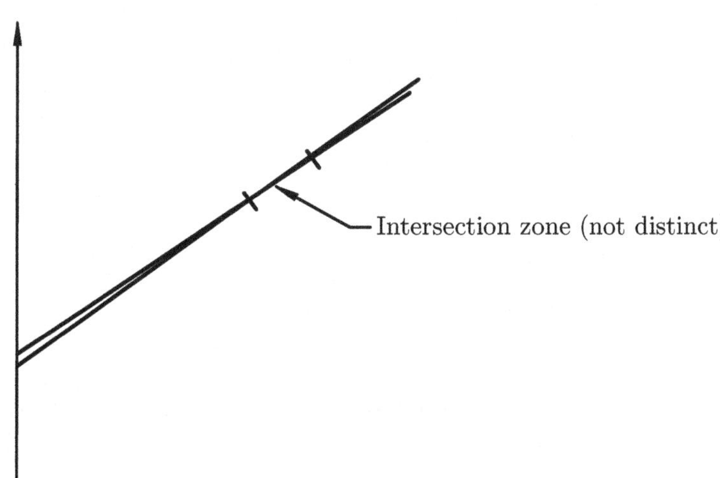

Figure 2.4: Non-distinct intersection: when $det[A] \approx 0$

(6) Consider $n = 3$ in (2.75). In this case we have three linear simultaneous algebraic equations.

$$\begin{aligned} a_{11}x_1 + a_{12}x_2 + a_{13}x_3 &= b_1 \\ a_{21}x_1 + a_{22}x_2 + a_{23}x_3 &= b_2 \\ a_{31}x_1 + a_{32}x_2 + a_{33}x_3 &= b_3 \end{aligned} \tag{2.88}$$

In $x_1x_2x_3$ orthogonal coordinate space (or xyz-space), each equation in (2.88) represents a plane. Graphically we can visualize this as follows (assuming $det[A] \neq 0$). Consider the first two equations in (2.88). The intersection of the planes defined by these two is a straight line. Intersection of this straight line with the plane defined by the third equation in (2.88) is a point (x_1^*, x_2^*, x_3^*) in $x_1x_2x_3$-space, which is the solution of (2.88). The remarks given for a system of two equations apply here as well and are not repeated.

(7) We clearly see that for $n > 3$, the graphical approach is difficult and impractical. However, the graphical approach gives deeper insight into the meaning of the solutions of linear simultaneous equations.

2.4.2 Cramer's Rule

Let

$$[A]_{(n\times n)}\{x\}_{(n\times 1)} = \{b\}_{(n\times 1)} \tag{2.89}$$

be a system of n simultaneous algebraic equations. To illustrate the details of this method, let us consider $n = 3$. For this case

$$[A] = \begin{bmatrix} a_{11} & a_{12} & a_{13} \\ a_{21} & a_{22} & a_{23} \\ a_{31} & a_{32} & a_{33} \end{bmatrix} \qquad \{x\} = \begin{Bmatrix} x_1 \\ x_2 \\ x_3 \end{Bmatrix} \qquad \{b\} = \begin{Bmatrix} b_1 \\ b_2 \\ b_3 \end{Bmatrix} \tag{2.90}$$

Then

$$x_1 = \frac{\begin{vmatrix} b_1 & a_{12} & a_{13} \\ b_2 & a_{22} & a_{23} \\ b_3 & a_{32} & a_{33} \end{vmatrix}}{|A|} \qquad x_2 = \frac{\begin{vmatrix} a_{11} & b_2 & a_{13} \\ a_{21} & b_2 & a_{23} \\ a_{31} & b_3 & a_{33} \end{vmatrix}}{|A|} \qquad x_3 = \frac{\begin{vmatrix} a_{11} & a_{12} & b_1 \\ a_{21} & a_{22} & b_2 \\ a_{31} & a_{32} & b_3 \end{vmatrix}}{|A|} \tag{2.91}$$

Thus, to calculate x_1, we replace the first column of $[A]$ by $\{b\}$, then divide its determinant by the determinant of $[A]$. For x_2 and x_3 we use the second and third columns of $[A]$ with $\{b\}$ respectively, with the rest of the procedure remaining the same as for x_1.

Remarks.

(1) If $det[A]$ is zero then x_j; $j = 1, 2, 3$ are infinity, i.e., they are not defined.

(2) When n is large, calculations of determinants is tedious and time consuming. Hence, this method is not preferred for large systems of linear simultaneous algebraic equations. However, unlike the graphical method, this method can be used for $n \geq 3$.

Example 2.6 (Cramer's Rule). Consider the following system of three linear simultaneous algebraic equations:

$$
\begin{aligned}
x_1 + x_2 + x_3 &= 6 \\
0.1x_1 + x_2 + 0.2x_3 &= 2.7 \\
x_1 + 0.2x_2 + x_3 &= 4.4
\end{aligned}
$$

In this case

$$
[A] = \begin{bmatrix} 1 & 1 & 1 \\ 0.1 & 1 & 0.2 \\ 1 & 0.2 & 0.2 \end{bmatrix} \qquad \{x\} = \begin{Bmatrix} x_1 \\ x_2 \\ x_3 \end{Bmatrix} \qquad \{b\} = \begin{Bmatrix} 6 \\ 2.7 \\ 4.4 \end{Bmatrix}
$$

in

$$
[A]\{x\} = \{b\}
$$

We use Cramer's Rule to obtain the solution of $\{x\}$. Following (2.91):

$$
det[A] = \begin{vmatrix} 1 & 1 & 1 \\ 0.1 & 1 & 0.2 \\ 1 & 0.2 & 0.2 \end{vmatrix} = 0.08
$$

$$
x_1 = \frac{\overset{\{b\}}{\begin{vmatrix} 6 & 1 & 1 \\ 2.7 & 1 & 0.2 \\ 4.4 & 0.2 & 1 \end{vmatrix}}}{|A|} \qquad x_2 = \frac{\overset{\{b\}}{\begin{vmatrix} 1 & 6 & 1 \\ 0.1 & 2.7 & 0.2 \\ 1 & 4.4 & 1 \end{vmatrix}}}{|A|} \qquad x_3 = \frac{\overset{\{b\}}{\begin{vmatrix} 1 & 1 & 6 \\ 0.1 & 1 & 2.7 \\ 1 & 0.2 & 4.4 \end{vmatrix}}}{|A|}
$$

or

$$
x_1 = \frac{0.08}{0.08} = 1 \qquad x_2 = \frac{0.16}{0.08} = 2 \qquad x_3 = \frac{0.24}{0.08} = 3
$$

Hence, the solution $\{x\}$ is

$$
\{x\} = \begin{Bmatrix} x_1 \\ x_2 \\ x_3 \end{Bmatrix} = \begin{Bmatrix} 1 \\ 2 \\ 3 \end{Bmatrix}
$$

2.5 Elimination Methods

Consider

$$[A]_{(n\times n)}\{x\}_{(n\times 1)} = \{b\}_{(n\times 1)} \tag{2.92}$$

In the elimination methods we reduce the number of variables by $(i-1)$, where i is the equation number for each equation in (2.92), beginning with the first equation. Thus for equation one ($i=1$), we maintain all n variables (x_j; $j=1,2,\dots,n$). In equation two ($i=2$) we reduce the number of variables by one ($i-1=1$), leaving $n-1$ variables. Thus, in the last equation ($i=n$), we reduce the number of variables by $(n-1)$, hence it will only contain one variable x_n. This reduction in the number of variables in the equations is accomplished by elementary row operations, which obviously changes $[A]$. Hence, to ensure that the solution $\{x\}$ from the reduced system is the same as that of (2.92), we must augment $[A]$ by $\{b\}$ before performing the reduction. This allows $\{b\}$ to be modified accordingly during the reduction process. When the reduction process is finished, the last equation has only one variable, x_n, and the reduced $[A]$ is in upper triangular form, hence we can solve for x_n using the last equation.

Knowing x_n, we use $(n-1)^{th}$ equation that contains x_{n-1} and x_n variables, hence we can solve for x_{n-1} using this equation. This process is continued until we have obtained solutions for all of variables $\{x\}$. Due to the fact that in this method we eliminate variables from equations, the method is called a *elimination method.* This approach is the basis for *Gauss elimination.* We note this is a two-step process: in the first step the variables are eliminated from the augmented equations to make $[A]$ upper triangular, and in the second step the numerical values are calculated for the variables beginning with the last and proceeding in backward fashion. This process is called triangulation (upper) and back substitution.

2.5.1 Gauss Elimination

As mentioned earlier, in the process of eliminating variables from the equations (2.92), we operate on the coefficients of the matrix $[A]$ as well as the vector $\{b\}$. This process can be made systematic by augmenting $[A]$ with $\{b\}$ and then performing elementary row operations on this augmented $[A_{ag}]$ matrix. We discuss various elimination methods and their details in the following section.

2.5.1.1 Naive Gauss Elimination

In this method we augment $[A]$ by $\{b\}$ to construct $[A_{ag}]$. We perform elementary row operations on $[A_{ag}]$ to make the portion corresponding to $[A]$

upper triangular, without switching rows or columns. The row and column locations in $[A_{ag}]$ are preserved during the elimination process.

Consider (2.92) with $n = 3$, i.e., three linear simultaneous algebraic equations in three unknowns: x_1, x_2, and x_3. $[A]\{x\} = \{b\}$ is given by:

$$\begin{bmatrix} a_{11} & a_{12} & a_{13} \\ a_{21} & a_{22} & a_{23} \\ a_{31} & a_{32} & a_{33} \end{bmatrix} \begin{Bmatrix} x_1 \\ x_2 \\ x_3 \end{Bmatrix} = \begin{Bmatrix} b_1 \\ b_2 \\ b_3 \end{Bmatrix} \tag{2.93}$$

We augment the coefficient matrix $[A]$ by $\{b\}$.

$$[A_{ag}] = \begin{matrix} R1- \\ R2- \\ R3- \end{matrix} \left[\begin{array}{ccc|c} a_{11} & a_{12} & a_{13} & b_1 \\ a_{21} & a_{22} & a_{23} & b_2 \\ a_{31} & a_{32} & a_{33} & b_3 \end{array}\right] \tag{2.94}$$

Our objective is to make $[A]$ upper triangular, hence row one remains unaltered. From row two (the second equation) we eliminate x_1 and from row three (the third equation) we eliminate x_1 and x_2. This can be done if we can make a_{21}, a_{31}, and a_{32} zero. We do this by elementary row operations, in which we multiply a row of (2.94) by a scalar and then add or subtract to any desired row. These operations are valid due to the fact that the transformed system has the same solution as the original system (2.93) because we are operating on thje augmented matrix $[A_{ag}]$. Let us denote the rows of (2.94) by $R1$, $R2$, and $R3$.

Step 1: Making $[A]$ Upper Triangular

To make a_{21} and a_{31} zero, we perform the following two elementary row operations:

$$\begin{matrix} R1 \\ R2 - \left(\frac{a_{21}}{a_{11}}\right)R1 \\ R3 - \left(\frac{a_{31}}{a_{11}}\right)R1 \end{matrix} \left[\begin{array}{ccc|c} a_{11} & a_{12} & a_{13} & b_1 \\ 0 & a'_{22} & a'_{23} & b'_2 \\ 0 & a'_{32} & a'_{33} & b'_3 \end{array}\right] \tag{2.95}$$

The coefficients in (2.95) with primes are the new values due to elementary row operations shown in (2.95). The diagonal element a_{11} in (2.95) is called the *pivot* element. After these elementary row operations in (2.95), column one is in upper triangular form.

Next, in column two of (2.95), we make a'_{32} zero by the following elementary row operation using rows of (2.95):

$$\begin{matrix} \\ \\ R3 - \left(\frac{a'_{32}}{a'_{22}}\right)R2 \end{matrix} \left[\begin{array}{ccc|c} a_{11} & a_{12} & a_{13} & b_1 \\ 0 & a'_{22} & a'_{23} & b'_2 \\ 0 & 0 & a''_{33} & b''_3 \end{array}\right] \tag{2.96}$$

In (2.96), we note that all elements below the diagonal are zero in $[A]$, i.e., $[A]$ in (2.96) is in upper triangular form. This the main objective of elimination of variables: to make matrix $[A]$ upper triangular in the augmented form. In the elementary row operations in (2.96), a'_{22} is the pivot element. We note that pivot elements are used to divide the other elements, hence these cannot be zero.

Step 2: Back Substitution

In this step, known as back substitution, we calculate the solution in the reverse order, i.e., x_3, then x_2, then x_1 using the upper triangular form of $[A]$ in $[A_{ag}]$, hence the name back substitution. We note that (2.96) represents the following system of equations.

$$\begin{aligned} a_{11}x_1 + a_{12}x_2 + a_{13}x_3 &= b_1 \\ a'_{22}x_2 + a'_{23}x_3 &= b'_2 \\ a''_{33}x_3 &= b''_3 \end{aligned} \tag{2.97}$$

In (2.97), we can solve for x_3 using the last equation.

$$x_3 = \frac{b''_3}{a''_{33}} \tag{2.98}$$

In this case a''_{33} is the pivot element. Next we can use the second equation in (2.97) to solve for x_2, as x_3 is already known.

$$x_2 = \frac{(b'_2 - a'_{23}x_3)}{a'_{22}} \tag{2.99}$$

Now using the first equation in (2.97), we can solve for x_1 as x_2 and x_3 are already known.

$$x_1 = \frac{(b_1 - a_{21}x_2 - a_{13}x_3)}{a_{11}} \tag{2.100}$$

Thus, the complete solution $[x_1 \;\; x_2 \;\; x_3]^T$ is known.

Remarks.

(1) The elements a_{11}, a'_{22}, and a''_{33} are pivots that are used to divide other coefficients. These cannot be zero otherwise this method will fail.

(2) In this method we maintain the positions of rows and columns in the augmented matrix, i.e., we do not perform row and column interchanges even if zero pivots are encountered, hence the name *naive Gauss elimination.*

(3) It is a two step process: in the first step we make the matrix $[A]$ in the augmented form upper triangular using elementary row operations. In

the second step, called back substitution, we calculate x_3, x_2, and x_1 in this order.

(4) When a solution is required for more than one $\{b\}$ (i.e., more than one right side), then the matrix $[A]$ can be augmented by all of the right side vectors before performing elementary row operations to make $[A]$ upper triangular. As an example consider (2.92), a (3×3) system in which we desire solutions $\{x\}$ for $\{b\} = \{p\}$ and $\{b\} = \{q\}$.

$$\{b\} = \{p\} = \begin{Bmatrix} p_1 \\ p_2 \\ p_3 \end{Bmatrix} \qquad \{b\} = \{q\} = \begin{Bmatrix} q_1 \\ q_2 \\ q_3 \end{Bmatrix} \tag{2.101}$$

We augment $[A]$ by both $\{p\}$ and $\{q\}$.

$$\left[\begin{array}{ccc|c|c} a_{11} & a_{12} & a_{13} & p_1 & q_1 \\ a_{12} & a_{22} & a_{23} & p_2 & q_2 \\ a_{13} & a_{32} & a_{33} & p_3 & q_3 \end{array}\right] \tag{2.102}$$

Using the details in Step 1, we make $[A]$ upper triangular in (2.102).

$$\left[\begin{array}{ccc|c|c} a_{11} & a_{12} & a_{13} & p_1 & q_1 \\ 0 & a'_{22} & a'_{23} & p'_2 & q'_2 \\ 0 & 0 & a''_{33} & p''_3 & q''_3 \end{array}\right] \tag{2.103}$$

Equation (2.103) clearly implies that we have the following:

$$\begin{aligned} a_{11}x_1 + a_{12}x_2 + a_{13}x_3 &= p_1 \\ a'_{22}x_2 + a'_{23}x_3 &= p'_2 \\ a'_{33}x_3 &= p''_3 \end{aligned} \tag{2.104}$$

and

$$\begin{aligned} a_{11}x_1 + a_{12}x_2 + a_{13}x_3 &= q_1 \\ a'_{22}x_2 + a'_{23}x_3 &= q'_2 \\ a'_{33}x_3 &= q''_3 \end{aligned} \tag{2.105}$$

Now we can use back substitution for (2.104) and (2.105) to find solutions for $\{x\}$ for $\{b\} = \{p\}$ and $\{b\} = \{q\}$.

Example 2.7 (Naive Gauss Elimination). Consider the same system of equations as used in Example 2.6 for Cramer's Rule:

$$\begin{aligned} x_1 + x_2 + x_3 &= 6 \\ 0.1x_1 + x_2 + 0.2x_3 &= 2.7 \\ x_1 + 0.2x_2 + x_3 &= 4.4 \end{aligned}$$

which can be written as

$$[A]\{x\} = \{b\}$$

where

$$[A] = \begin{bmatrix} 1 & 1 & 1 \\ 0.1 & 1 & 0.2 \\ 1 & 0.2 & 1 \end{bmatrix} \qquad \{x\} = \begin{Bmatrix} x_1 \\ x_2 \\ x_3 \end{Bmatrix} \qquad \{b\} = \begin{Bmatrix} 6 \\ 2.7 \\ 4.4 \end{Bmatrix}$$

We augment $[A]$ by adding $\{b\}$ as a fourth column to $[A]$.

$$[A_{ag}] = \begin{matrix} R1- \\ R2- \\ R3- \end{matrix} \left[\begin{array}{ccc|c} 1 & 1 & 1 & 6 \\ 0.1 & 1 & 0.2 & 2.7 \\ 1 & 0.2 & 1 & 4.4 \end{array}\right]$$

Upper Triangular Form of $[A]$ in $[A_{ag}]$

Make column one in $[A_{ag}]$ upper triangular by using the elementary row operations shown below.

$$\begin{matrix} \\ R2 - (\frac{0.1}{1})R1 \\ R3 - (\frac{1}{1})R1 \end{matrix} \left[\begin{array}{ccc|c} 1 & 1 & 1 & 6 \\ 0 & 0.9 & 0.1 & 2.1 \\ 0 & -0.8 & 0 & -1.6 \end{array}\right]$$

Next we make column two in the modified $[A_{ag}]$ upper triangular using the elementary row operations shown below.

$$\begin{matrix} \\ \\ R3 - (\frac{-0.8}{0.9})R2 \end{matrix} \left[\begin{array}{ccc|c} 1 & 1 & 1 & 6 \\ 0 & 0.9 & 0.1 & 2.1 \\ 0 & 0 & (\frac{-0.8}{0.9})(0.1) & -1.6 - (\frac{0.8}{0.9})2.1 \end{array}\right]$$

$$\left[\begin{array}{ccc|c} 1 & 1 & 1 & 6 \\ 0 & 0.9 & 0.1 & 2.1 \\ 0 & 0 & \frac{0.8}{9} & \frac{0.8}{3} \end{array}\right]$$

Back Substitution

From the third row in the upper triangular form:

$$\frac{0.8}{9}x_3 = \frac{0.8}{3}$$

$$\therefore \quad x_3 = 3$$

Using the second row of the upper triangular form:

$$0.9x_2 = 2.1 - 0.1x_3 = 2.1 - 0.1(3) = 1.8$$

$$\therefore \quad x_2 = 2$$

Using the first row of the upper triangular form:

$$x_1 = 6 - x_2 - x_3 = 6 - 2 - 3 = 1$$

Hence,

$$\{x\} = \begin{Bmatrix} x_1 \\ x_2 \\ x_3 \end{Bmatrix} = \begin{Bmatrix} 1 \\ 2 \\ 3 \end{Bmatrix}$$

The solution $\{x\}$ is the same as that obtained using Cramer's rule.

2.5.1.2 Gauss Elimination with Partial Pivoting

Consider the system of equations from (2.93).

$$[A]\{x\} = \{b\} \tag{2.106}$$

In some cases the coefficients in the system of equations (2.106) may be such that $a_{11} = 0$ even though the system of equations (2.106) does have a unique solution. In this case the naive Gauss elimination method will fail due to the fact that we must divide by the pivot a_{11}. In such situations we can employ *partial pivoting* that helps in avoiding zero pivots. This procedure involves the interchange of rows for a column under consideration during upper triangulation such that the largest element (absolute value) in this column becomes the pivot. This is followed by the upper triangulation for the column under consideration. This procedure is continued for subsequent, columns keeping in mind that the columns (and corresponding rows) that are already in upper triangular form are exempted or are not considered in searching for the next pivot. Consider

$$\begin{bmatrix} a_{11} & a_{12} & a_{13} \\ a_{21} & a_{22} & a_{23} \\ a_{31} & a_{32} & a_{33} \end{bmatrix} \begin{Bmatrix} x_1 \\ x_2 \\ x_3 \end{Bmatrix} = \begin{Bmatrix} b_1 \\ b_2 \\ b_3 \end{Bmatrix} \tag{2.107}$$

We augment coefficient matrix $[A]$ in (2.107) by the column vector $\{b\}$.

$$[A_{ag}] = \left[\begin{array}{ccc|c} a_{11} & a_{12} & a_{13} & b_1 \\ a_{21} & a_{22} & a_{23} & b_2 \\ a_{31} & a_{32} & a_{33} & b_3 \end{array}\right] \tag{2.108}$$

We make column one in (2.108) upper triangular using the largest element in column one of (2.108) as a pivot. Let $|a_{31}|$ (the absolute value of a_{31}) be the largest element in column one, then we interchange row one with row three in (2.108) to obtain the following:

$$\left[\begin{array}{ccc|c} a_{31} & a_{32} & a_{33} & b_3 \\ a_{21} & a_{22} & a_{23} & b_2 \\ a_{11} & a_{12} & a_{13} & b_1 \end{array}\right] \tag{2.109}$$

We make column one in (2.109) upper triangular by using elementary row operations (as discussed in naive Gauss elimination), i.e., we make a_{21} and a_{11} zero.

$$\left[\begin{array}{ccc|c} a_{31} & a_{32} & a_{33} & b_3 \\ 0 & a'_{22} & a'_{23} & b'_2 \\ 0 & a'_{12} & a'_{13} & b'_1 \end{array}\right] \tag{2.110}$$

In the next step we make column two in (2.110) upper triangular using the element with the largest magnitude (absolute value) out of a'_{22} and a'_{12} as the pivot. Let us assume that $|a'_{12}| > |a'_{22}|$, then we interchange row two with row three in (2.110).

$$\left[\begin{array}{ccc|c} a_{31} & a_{32} & a_{33} & b_3 \\ 0 & a'_{12} & a'_{13} & b'_1 \\ 0 & a'_{22} & a'_{23} & b'_2 \end{array}\right] \tag{2.111}$$

Now we can make column two in (2.111) upper triangular by elementary row operation using a'_{12} as the pivot, i.e., we make a'_{22} in (2.111) zero. This gives us:

$$\left[\begin{array}{ccc|c} a_{31} & a_{32} & a_{33} & b_3 \\ 0 & a'_{12} & a'_{13} & b'_1 \\ 0 & 0 & a''_{23} & b''_2 \end{array}\right] \tag{2.112}$$

Using (2.112), we can write the expanded form of the equations.

$$\begin{aligned} a_{31}x_1 + a_{32}x_2 + a_{33}x_3 &= b_3 \\ a'_{12}x_2 + a'_{13}x_3 &= b'_1 \\ a''_{23}x_3 &= b''_2 \end{aligned} \tag{2.113}$$

Using (2.112) or (2.113) we can use back substitution to find x_3, x_2, and x_1 (in this order) beginning with the last equation and then proceeding to the previous equation in succession.

Remarks.

(1) The procedure described above is called *Gauss elimination with partial pivoting.* In this procedure we interchange rows to make sure that in the column to be made upper triangular, the largest element is the pivot. This helps in avoiding divisions by small numbers or zeros during triangulation.

(2) We only consider the diagonal element and the elements below it in the column under consideration to determine the largest element for making the row interchanges.

(3) The partial pivoting procedure is computationally quite efficient even for large systems of algebraic equations.

Example 2.8 (Gauss Elimination with Partial Pivoting). Consider the following system of equations:

$$\begin{aligned} x_1 + x_2 + x_3 &= 6 \\ 8x_1 + 1.6x_2 + 8x_3 &= 35.2 \\ 0.1x_1 + x_2 + 0.2x_3 &= 2.7 \end{aligned}$$

Or in matrix and vector form:

$$[A] = \begin{bmatrix} 1 & 1 & 1 \\ 8 & 1.6 & 8 \\ 0.1 & 1 & 0.2 \end{bmatrix} \qquad \{x\} = \begin{Bmatrix} x_1 \\ x_2 \\ x_3 \end{Bmatrix} \qquad \{b\} = \begin{Bmatrix} 6 \\ 35.2 \\ 2.7 \end{Bmatrix}$$

We augment $[A]$ by $\{b\}$ to obtain $[A_{ag}]$.

$$[A_{ag}] = \left[\begin{array}{ccc|c} 1 & 1 & 1 & 6 \\ 8 & 1.6 & 8 & 35.2 \\ 0.1 & 1 & 0.2 & 2.7 \end{array}\right]$$

We want to make column one upper triangular by using the largest element, i.e., 8, as the pivot. This requires that we interchange rows one and two in the augmented matrix.

$$R1 \rightleftarrows R2 \left[\begin{array}{ccc|c} 8 & 1.6 & 8 & 35.2 \\ 1 & 1 & 1 & 6 \\ 0.1 & 1 & 0.2 & 2.7 \end{array}\right]$$

We then make column one upper triangular by using elementary row operations.

$$\begin{matrix} \\ R2-\left(\frac{1}{8}\right)R1 \\ R3-\left(\frac{0.1}{8}\right)R1 \end{matrix}\left[\begin{array}{ccc|c} 8 & 1.6 & 8 & 35.2 \\ 0 & 0.8 & 0 & 1.6 \\ 0 & 0.98 & 0.1 & 2.26 \end{array}\right]$$

Next we consider column two. The elements on the diagonal and below the diagonal in column two are 0.8 and 0.98. We want to use 0.98 as the pivot (the larger of the two). This requires that we interchange rows two and three.

$$R2 \rightleftarrows R3 \left[\begin{array}{ccc|c} 8 & 1.6 & 8 & 35.2 \\ 0 & 0.98 & 0.1 & 2.26 \\ 0 & 0.8 & 0 & 1.6 \end{array}\right]$$

We now make column two upper triangular by elementary row operations.

$$\begin{matrix} \\ \\ R3-\left(\frac{0.8}{0.98}\right)R2 \end{matrix}\left[\begin{array}{ccc|c} 8 & 1.6 & 8 & 35.2 \\ 0 & 0.98 & 0.1 & 2.26 \\ 0 & 0.8-\left(\frac{0.8}{0.98}\right)(0.98) & 0-\left(\frac{0.8}{0.98}\right)(0.1) & 1.6-\left(\frac{0.8}{0.98}\right)(2.26) \end{array}\right]$$

which after simplification becomes:

$$\left[\begin{array}{ccc|c} 8 & 1.6 & 8 & 35.2 \\ 0 & 0.98 & 0.1 & 2.26 \\ 0 & 0 & 0.0816 & 0.2449 \end{array}\right]$$

This augmented form contains the final upper triangular form of $[A]$. Now we can find x_3, x_2, and x_1 using back substitution. Using the last equation:

$$x_3 = \frac{0.2448979}{0.0816326} = 3$$

Using the second equation with the known value of x_3:

$$x_2 = \frac{(2.26 - 0.1x_3)}{0.98} = \frac{(2.26 - 0.1(3))}{0.98} = \frac{1.96}{0.98} = 2$$

Now, we can find x_1 using the first equation and the known values of x_2 and x_3:

$$x_1 = \frac{1}{8}(35.2 - 1.6x_2 - 8x_3) = \frac{1}{8}(35.2 - 1.6(2) - 8(3)) = \frac{8}{8} = 1$$

Hence we have the solution $\{x\}$:

$$\{x\} = \begin{Bmatrix} x_1 \\ x_2 \\ x_3 \end{Bmatrix} = \begin{Bmatrix} 1 \\ 2 \\ 3 \end{Bmatrix}$$

2.5.1.3 Gauss Elimination with Full Pivoting

In this method also we make matrix $[A]$ upper triangular as in the previous two Gauss elimination methods, except that in this method we choose the largest element of $[A]$ or the largest element of the sub-matrix of upper triangulated $[A]$ during the elimination. This is sometimes necessitated if the elements of the coefficient matrix $[A]$ vary drastically in magnitude, in which case the other two Gauss elimination processes may result in significant roundoff errors or may even fail if a zero pivot is encountered. Consider:

$$[A]\{x\} = \{b\} \tag{2.114}$$

1. Augment matrix $[A]$ by right side $\{b\}$.
2. Search the entire matrix $[A]$ for the element with the largest magnitude (absolute value).
3. Perform simultaneous interchanges of rows and columns to ensure that the element with the largest magnitude is the pivot in the first column.
4. Make column one upper triangular by elementary row operations.
5. Next consider column two of the reduced sub-matrix without row one and column one. Search for the element with largest magnitude (absolute value) in this sub-matrix. Perform row and column interchanges in the augmented matrix (with column one in upper triangular form) so that the element with the largest magnitude is the pivot for column two. Make column two upper triangular by elementary row operations.
6. We continue this procedure for the remaining columns until $[A_{ag}]$ becomes upper triangular.
7. Solution $\{x\}$ is then calculated using the upper triangular matrix in $[A]$ to obtain $\{x\}$ in reverse order ie x_n, x_{n-1}, $\dots$, x_1.

Remarks.

(1) This method is obviously very time consuming as it requires a search for the largest pivot at every step and simultaneous interchanges of rows and columns.

(2) If the $|A| \neq 0$, then this method ensures unique $\{x\}$ if the solution exists.

(3) It is important to note that row interchanges do not effect the order of the variables in vector $\{x\}$, but column interchanges require that we

also interchange the locations of the corresponding variables in the $\{x\}$ vector. This is illustrated in the numerical example presented in the following.

Example 2.9 (Gauss Elimination with Full Pivoting). Consider the following system of equations:

$$\begin{aligned} x_1 + x_2 + x_3 &= 9 \\ 8x_1 + 1.6x_2 + 8x_3 &= 35.2 \\ 0.1x_1 + x_2 + 3x_3 &= 11.1 \end{aligned}$$

in which

$$[A] = \begin{bmatrix} 1 & 1 & 1 \\ 8 & 1.6 & 8 \\ 0.1 & 1 & 3 \end{bmatrix} \qquad \{x\} = \begin{Bmatrix} x_1 \\ x_2 \\ x_3 \end{Bmatrix} \qquad \{b\} = \begin{Bmatrix} 9 \\ 35.2 \\ 11.1 \end{Bmatrix}$$

We augment $[A]$ by $\{b\}$. We label rows and columns as x_1, x_2, x_3.

$$\begin{matrix} & x_1\ x_2\ x_3 & \\ [A_{ag}] = & \left[\begin{array}{ccc|c} 1 & 1 & 1 & 9 \\ 8 & 1.6 & 8 & 35.2 \\ 0.1 & 1 & 3 & 11.1 \end{array}\right] & \begin{matrix} x_1 \\ x_2 \\ x_3 \end{matrix} \end{matrix}$$

In Gauss elimination with full pivoting, in addition to row interchanges, column interchanges may also be required to ensure that the largest element of matrix $[A]$ or the sub-matrix of $[A]$ is the pivot. Since interchanging columns of $[A]$ requires that we also interchange the positions of the corresponding x_i in $\{x\}$, it is prudent in $[A_{ag}]$ to keep x_is with the columns and rows.

Interchange rows one and two in $[A_{ag}]$ so that largest element in column one become the pivot. We choose element a_{21}. There is no incentive to use element a_{23} since $a_{23} = a_{21} = 8$.

$$\begin{matrix} x_1\ x_2\ x_3 & \\ \left[\begin{array}{ccc|c} 8 & 1.6 & 8 & 35.2 \\ 1 & 1 & 1 & 9 \\ 0.1 & 1 & 3 & 11.1 \end{array}\right] & \begin{matrix} x_1 \\ x_2 \\ x_3 \end{matrix} \end{matrix}$$

Make column one upper triangular by using elementary row operations.

$$\begin{matrix} x_1\ \ x_2\ \ x_3 & \\ \left[\begin{array}{ccc|c} 8 & 1.6 & 8 & 35.2 \\ 0 & 0.8 & 0 & 4.6 \\ 0 & 0.98 & 2.9 & 10.66 \end{array}\right] & \begin{matrix} x_1 \\ x_2 \\ x_3 \end{matrix} \end{matrix}$$

Consider the (2×2) sub-matrix (i.e., a'_{22}, a'_{23}, a'_{32}, and a'_{33}). The element with the largest magnitude is 2.9 (a'_{23}). We want 2.9 to be pivot, i.e., at location (2,2). This could be done in two ways:

(i) First interchange row two with row three and then interchange columns two and three.

(ii) Alternatively, first interchange columns two and three and then interchange rows two and three.

Regardless of whether we choose (i) or (ii), the end result is the same. In the following we consider (i).

Interchange rows two and three.

$$\begin{array}{c} \quad x_1 \quad x_2 \quad x_3 \\ \left[\begin{array}{ccc|c} 8 & 1.6 & 8 & 35.2 \\ 0 & 0.98 & 2.9 & 10.66 \\ 0 & 0.8 & 0 & 4.6 \end{array}\right] \begin{array}{c} x_1 \\ x_2 \\ x_3 \end{array} \end{array}$$

Now interchange columns two and three. In doing so, we should also interchange x_2 and x_3 respectively.

$$\begin{array}{c} \quad x_1 \quad x_3 \quad x_2 \\ \left[\begin{array}{ccc|c} 8 & 8 & 1.6 & 35.2 \\ 0 & 2.9 & 0.98 & 10.66 \\ 0 & 0 & 0.8 & 4.6 \end{array}\right] \begin{array}{c} x_1 \\ x_3 \\ x_2 \end{array} \end{array}$$

Since a'_{32} is already zero, column two is already in upper triangular form, hence no elementary row operations are required. This system of equations are in the desired upper triangular form. Now, we can use back substitution to calculate x_2, x_3, and x_1 (in this order).

Consider the last equation from which we can calculate x_2.

$$x_2 = \frac{4.6}{0.8} = 5.75$$

Using the second equation and the known value of x_2 we can calculate x_3.

$$x_3 = \frac{(10.66 - 0.98x_2)}{2.9} = \frac{(10.66 - 0.98(5.75))}{2.9} = 1.73$$

Now using first equation, we can calculate x_1.

$$x_1 = \frac{(35.2 - 8x_3 - 1.6x_2)}{8} = \frac{(35.2 - 8(1.73) - 1.6(5.75))}{8} = 1.52$$

Hence

$$\{x\} = \begin{Bmatrix} x_1 \\ x_2 \\ x_3 \end{Bmatrix} = \begin{Bmatrix} 1.52 \\ 5.75 \\ 1.73 \end{Bmatrix}$$

2.5.2 Gauss-Jordan Elimination

In this method we use the fact that if we premultiply $[A]\{x\} = \{b\}$ by $[A]^{-1}$, then $[I]\{x\} = \{x\} = [A]^{-1}\{b\}$ is the desired solution. Thus, we can augment $[A]$ with $[b]$ and perform elementary row operations on $[A_{ag}]$ such that $[A]$ in $[A_{ag}]$ becomes $[I]$. At this point the locations of $\{b\}$ in $[A_{ag}]$ will contain the solution $\{x\}$. We present details in the following. Consider:

$$[A]\{x\} = \{b\} \tag{2.115}$$

Premultiply (2.115) by $[A]^{-1}$.

$$[A]^{-1}[A]\{x\} = [A]^{-1}\{b\} \tag{2.116}$$

Since $[A]^{-1}[A] = [I]$, (2.116) reduces to:

$$[I]\{x\} = [A]^{-1}\{b\} \tag{2.117}$$

But $[I]\{x\} = \{x\}$, hence (2.117) becomes:

$$\{x\} = [A]^{-1}\{b\} \tag{2.118}$$

Comparing (2.115) with (2.117) suggest that if we augment $[A]$ by $\{b\}$ and then if we can make $[A]$ an identity matrix by elementary row operations, then the modified $\{b\}$ will be the solution $\{x\}$.

Consider a general (3×3) system of linear simultaneous algebraic equations.

$$\begin{aligned} a_{11}x_1 + a_{12}x_2 + a_{13}x_3 &= b_1 \\ a_{21}x_1 + a_{22}x_2 + a_{23}x_3 &= b_2 \\ a_{31}x_1 + a_{32}x_2 + a_{33}x_3 &= b_3 \end{aligned} \tag{2.119}$$

Augment the coefficient matrix $[A]$ of the coefficients a_{ij} by the right side vector $\{b\}$.

$$[A_{ag}] = \left[\begin{array}{ccc|c} a_{11} & a_{12} & a_{13} & b_1 \\ a_{21} & a_{22} & a_{23} & b_2 \\ a_{31} & a_{32} & a_{33} & b_3 \end{array}\right] \tag{2.120}$$

In this first step our goal is to make a_{11} unity and its first column in the upper triangular from using elementary row operations. First, we make a_{11} unity.

$$R1 = \tfrac{R1}{a_{11}} \left[\begin{array}{ccc|c} 1 & a'_{12} & a'_{13} & b'_1 \\ a_{21} & a_{22} & a_{23} & b_2 \\ a_{31} & a_{32} & a_{33} & b_3 \end{array}\right] \tag{2.121}$$

Make column one in (2.121) upper triangular using elementary row operations.

$$\begin{array}{r} \\ R2 - a_{21}R1 \\ R3 - a_{31}R1 \end{array} \left[\begin{array}{ccc|c} 1 & a'_{12} & a'_{13} & b'_1 \\ 0 & a'_{22} & a'_{23} & b'_2 \\ 0 & a'_{32} & a'_{33} & b'_3 \end{array}\right] \tag{2.122}$$

In (2.122), we make a'_{22} unity.

$$\tfrac{R2}{a'_{22}} \left[\begin{array}{ccc|c} 1 & a'_{12} & a'_{13} & b'_1 \\ 0 & 1 & a''_{23} & b''_2 \\ 0 & a'_{32} & a'_{33} & b'_3 \end{array}\right] \tag{2.123}$$

We make the elements of column two below the diagonal zero by using row two and elementary row operations in (2.123).

$$\begin{array}{r} \\ \\ R3 - a'_{32}R2 \end{array} \left[\begin{array}{ccc|c} 1 & a'_{12} & a'_{13} & b'_1 \\ 0 & 1 & a''_{23} & b''_2 \\ 0 & 0 & a''_{33} & b''_3 \end{array}\right] \tag{2.124}$$

Make element a''_{33} in (2.124) unity.

$$\begin{array}{r} \\ \\ \tfrac{R3}{a''_{33}} \end{array} \left[\begin{array}{ccc|c} 1 & a'_{12} & a'_{13} & b'_1 \\ 0 & 1 & a''_{23} & b''_2 \\ 0 & 0 & 1 & b''_3 \end{array}\right] \tag{2.125}$$

Make the elements of column three in (2.125) above the diagonal zero by using row three and elementary row operations.

$$\begin{array}{r} R1 - a'_{13}R3 \\ R2 - a''_{23}R3 \\ \\ \end{array} \left[\begin{array}{ccc|c} 1 & a'_{12} & 0 & b''_1 \\ 0 & 1 & 0 & b'''_2 \\ 0 & 0 & 1 & b'''_3 \end{array}\right] \tag{2.126}$$

Lastly, make the elements of column two in (2.126) above the diagonal zero using row two and elementary row operations.

$$\begin{array}{r} R1 - a''_{12}R2 \\ \\ \\ \end{array} \left[\begin{array}{ccc|c} 1 & 0 & 0 & b'''_1 \\ 0 & 1 & 0 & b'''_2 \\ 0 & 0 & 1 & b'''_3 \end{array}\right] \tag{2.127}$$

In (2.127), the vector $\left[b_1''' \; b_2''' \; b_3'''\right]^T$ is the solution vector $\{x\} = \left[x_1 \; x_2 \; x_3\right]^T$.

Remarks.

(1) If the solution of (2.115) is required for more than one right-hand side, then $[A]$ in (2.121) must be augmented by all right-hand sides before making $[A]$ identity. When $[A]$ becomes identity, the locations of the right-hand side column vectors contain solutions for them.

Example 2.10 (Gauss-Jordan Elimination). Consider the following set of equations:

$$\begin{aligned} x_1 + x_2 + x_3 &= 6 \\ 0.1x_1 + x_2 + 0.2x_3 &= 2.7 \\ x_1 + 0.2x_2 + x_3 &= 4.4 \end{aligned} \tag{2.128}$$

Augment the coefficient matrix $[A]$ in (2.128).

$$[A_{ag}] = \left[\begin{array}{ccc|c} 1 & 1 & 1 & 6 \\ 0.1 & 1 & 0.2 & 2.7 \\ 1 & 0.2 & 1 & 4.4 \end{array}\right] \tag{2.129}$$

The element a_{11} in (2.129) is already unity, hence we can proceed to make column one upper triangular using elementary row operations.

$$\begin{matrix} \\ R2 - 0.1R1 \\ R3 - R1 \end{matrix} \left[\begin{array}{ccc|c} 1 & 1 & 1 & 6 \\ 0 & 0.9 & 0.1 & 2.1 \\ 0 & -0.8 & 0 & -1.6 \end{array}\right] \tag{2.130}$$

Make element a_{22} unity.

$$\left[\begin{array}{ccc|c} 1 & 1 & 1 & 6 \\ 0 & 1 & \frac{1}{9} & \frac{7}{3} \\ 0 & -0.8 & 0 & -1.6 \end{array}\right] \tag{2.131}$$

Make the off-diagonal elements of the second column of (2.131) zero using elementary row operations.

$$\begin{matrix} R1 - R2 \\ \\ R3 - (-0.8)R2 \end{matrix} \left[\begin{array}{ccc|c} 1 & 0 & \frac{8}{9} & \frac{11}{3} \\ 0 & 1 & \frac{1}{9} & \frac{7}{3} \\ 0 & 0 & \frac{0.8}{9} & -1.6 + \frac{7}{3}(0.8) \end{array}\right]$$

or

$$\left[\begin{array}{ccc|c} 1 & 0 & \frac{8}{9} & \frac{11}{3} \\ 0 & 1 & \frac{1}{9} & \frac{7}{3} \\ 0 & 0 & \frac{0.8}{9} & 0.2666 \end{array}\right] \tag{2.132}$$

Make element a_{33} unity in (2.132).

$$\frac{R3}{\frac{0.8}{9}}\left[\begin{array}{ccc|c} 1 & 0 & \frac{8}{9} & \frac{11}{3} \\ 0 & 1 & \frac{1}{9} & \frac{7}{3} \\ 0 & 0 & \frac{0.8}{9} & -1.6+\frac{7}{3}(0.8) \end{array}\right] \tag{2.133}$$

Make the off-diagonal elements of column three zero.

$$\left[\begin{array}{ccc|c} 1 & 0 & 0 & \frac{11}{3}-(\frac{8}{9})3 \\ 0 & 1 & 0 & \frac{7}{3}-(\frac{1}{9})3 \\ 0 & 0 & 1 & -1.6+\frac{7}{3}(0.8) \end{array}\right]$$

or

$$\left[\begin{array}{ccc|c} 1 & 0 & 0 & 1 \\ 0 & 1 & 0 & 2 \\ 0 & 0 & 1 & 3 \end{array}\right] \tag{2.134}$$

The vector in the location of $\{b\}$ in (2.134) is the solution vector $\{x\}$. Thus:

$$\{x\} = \begin{Bmatrix} x_1 \\ x_2 \\ x_3 \end{Bmatrix} = \begin{Bmatrix} 1 \\ 2 \\ 3 \end{Bmatrix} \tag{2.135}$$

2.5.3 Methods Using $[L][U]$ Decomposition

2.5.3.1 Classical $[L][U]$ Decomposition and Solution of $[A]\{x\} = \{b\}$: Cholesky Decomposition

Consider

$$[A]\{x\} = \{b\} \tag{2.136}$$

In this method we express the coefficient matrix $[A]$ as the product of a *unit* lower triangular matrix $[L]$ (that is, a lower triangular matrix with unit diagonal elements) and an upper triangular matrix $[U]$.

$$[A] = [L][U] \tag{2.137}$$

The rules for determining the coefficients of $[L]$ and $[U]$ are established by forming the product $[L][U]$ and equating the elements of the product to the corresponding elements of $[A]$. We present details in the following. Consider

$[A]$ to be a (4×4) matrix. In this case we have:

$$\begin{bmatrix} a_{11} & a_{12} & a_{13} & a_{14} \\ a_{21} & a_{22} & a_{23} & a_{24} \\ a_{31} & a_{32} & a_{33} & a_{34} \\ a_{41} & a_{42} & a_{43} & a_{44} \end{bmatrix} = \begin{bmatrix} 1 & 0 & 0 & 0 \\ L_{21} & 1 & 0 & 0 \\ L_{31} & L_{32} & 1 & 0 \\ L_{41} & L_{42} & L_{43} & 1 \end{bmatrix} \begin{bmatrix} u_{11} & u_{12} & u_{13} & u_{14} \\ 0 & u_{22} & u_{23} & u_{24} \\ 0 & 0 & u_{33} & u_{34} \\ 0 & 0 & 0 & u_{44} \end{bmatrix} \tag{2.138}$$

To determine the coefficients of $[L]$ and $[U]$, we form the product $[L][U]$ in (2.138).

$$\begin{bmatrix} a_{11} & a_{12} & a_{13} & a_{14} \\ a_{21} & a_{22} & a_{23} & a_{24} \\ a_{31} & a_{32} & a_{33} & a_{34} \\ a_{41} & a_{42} & a_{43} & a_{44} \end{bmatrix} =$$

u_{11}	u_{12}	u_{13}	u_{14}
$L_{21}u_{11}$	$L_{21}u_{12} + u_{22}$	$L_{21}u_{13} + u_{23}$	$L_{21}u_{14} + u_{24}$
$L_{31}u_{11}$	$L_{31}u_{12} + L_{32}u_{22}$	$L_{31}u_{13} + L_{32}u_{23} + u_{33}$	$L_{31}u_{14} + L_{32}u_{24} + u_{34}$
$L_{41}u_{11}$	$L_{41}u_{12} + L_{42}u_{22}$	$L_{41}u_{13} + L_{42}u_{23} + L_{43}u_{33}$	$L_{41}u_{14} + L_{42}u_{24} + L_{43}u_{34} + u_{44}$

(2.139)

The elements of $[U]$ and $[L]$ are determined by alternating between a row of $[U]$ and the corresponding column of $[L]$.

First Row of $[U]$:

To determine the first row of $[U]$ we equate coefficients in the first row on both sides of (2.139).

$$u_{11} = a_{11} \qquad u_{12} = a_{12} \qquad u_{13} = a_{13} \qquad u_{14} = a_{14} \tag{2.140}$$

That is, the first row of $[U]$ is the same as the first row of the coefficient matrix $[A]$.

First Column of $[L]$:

If we equate coefficients of the first column on both sides of (2.139), we obtain:

$$L_{21} = \frac{a_{21}}{u_{11}} \qquad L_{31} = \frac{a_{31}}{u_{11}} \qquad L_{41} = \frac{a_{41}}{u_{11}} \tag{2.141}$$

Thus, first column of $[L]$ is determined.

Second Row of $[U]$:

Equate the coefficients of the second row on both sides of (2.139).

$$\begin{aligned} u_{22} &= a_{22} - L_{21}u_{12} \\ u_{23} &= a_{23} - L_{21}u_{13} \\ u_{24} &= a_{24} - L_{21}u_{14} \end{aligned} \tag{2.142}$$

Thus, the second row of $[U]$ is determined.

Second Column of $[L]$:

Equate coefficients of the second column on both sides of (2.139).

$$\begin{aligned} L_{32} &= \frac{(a_{32} - L_{31}u_{12})}{u_{22}} \\ L_{42} &= \frac{(a_{42} - L_{41}u_{12})}{u_{22}} \end{aligned} \tag{2.143}$$

This establishes the elements of the second column of $[L]$.

Third Row of $[U]$:

Equate coefficients of the third row on both sides of (2.139).

$$\begin{aligned} u_{33} &= a_{33} - L_{31}u_{13} - L_{32}u_{23} \\ u_{34} &= a_{34} - L_{31}u_{14} - L_{32}u_{24} \end{aligned} \tag{2.144}$$

Hence, the third row of $[U]$ is known.

Third Column of $[L]$:

Equate coefficients of the third column on both sides of (2.139).

$$L_{43} = \frac{(a_{43} - L_{41}u_{13} - L_{42}u_{23})}{u_{33}} \tag{2.145}$$

Fourth Row of $[U]$:

Equate coefficients of the fourth row on both sides of (2.139).

$$u_{44} = a_{44} - L_{41}u_{14} - L_{42}u_{24} - L_{43}u_{34} \tag{2.146}$$

Thus, the coefficients of $[L]$ and $[U]$ in (2.138) are completely determined. For matrices larger than (4×4) this procedure can be continued for the subsequent rows and columns of $[U]$ and $[L]$.

Remarks.

(1) The coefficients of $[U]$ and $[L]$ can be expressed more compactly as follows:

$$
\begin{aligned}
&u_{1j} = a_{1j} && j = 1, 2, \dots n \quad (i = 1) \\
&L_{i1} = \frac{a_{i1}}{u_{11}} && i = 2, 3, \dots, n \quad (j = 1) \\
&u_{ij} = a_{ij} - \sum_{k=1}^{i-1} L_{ik} u_{kj} && i = 2, 3, \dots, n \quad j = i, i+1, \dots, n \quad \text{(for each value of } i) \\
&L_{ij} = \frac{a_{ij} - \sum\limits_{k=1}^{j-1} L_{ik} u_{kj}}{u_{jj}} && j = 2, 3, \dots, n \quad i = j+1, \dots, n
\end{aligned}
\tag{2.147}
$$

Using $n = 4$ in (2.147) we can obtain (2.140) – (2.146). The form in (2.147) is helpful in programming $[L][U]$ decomposition.

(2) We can economize in the storage of the coefficients of $[L]$ and $[U]$.

(i) There is no need to store zeros in either $[L]$ or $[U]$.

(ii) Ones in the diagonal of $[L]$ do not need to be stored either.

(iii) A closer examination of the expressions for the coefficients of $[L]$ and $[U]$ shows that once the elements of a_{ij} of $[A]$ are used, they do not appear again in the further calculations of the coefficients of $[L]$ and $[U]$.

(iv) Thus we can store coefficients of $[L]$ and $[U]$ in the same storage space for $[A]$.

u_{ij} : stored in the same locations as

$$a_{ij} \qquad i = 1, 2, \dots, n \quad j = i, i+1, \dots, n \text{ (for each } i)$$

L_{ij} : stored in the same locations as

$$a_{ij} \qquad j = 1, 2, \dots, n \quad i = j+1, \dots, n$$

In this scheme of storing coefficients of $[L]$ and $[U]$, the unit diagonal elements of $[L]$ are not stored and the original coefficient matrix $[A]$ is obviously destroyed.

2.5.3.2 Determination of the Solution $\{x\}$ Using $[L][U]$ Decomposition

Consider $[A]\{x\} = \{b\}$, i.e., equation (2.136). Substitute the $[L][U]$ decomposition of $[A]$ from (2.137) into (2.136).

$$[L][U]\{x\} = \{b\} \tag{2.148}$$

Let

$$[U]\{x\} = \{y\} \tag{2.149}$$

Substitute (2.149) in (2.148).

$$[L]\{y\} = \{b\} \tag{2.150}$$

Step 1: We recall that $[L]$ is a unit lower triangular matrix, hence using (2.150) we can determine $y_1, y_2, \dots, y_n$ using the first, second, ..., last equations in (2.150). This is called the *forward pass.*

Step 2: With $\{y\}$ known (right-hand side in (2.149)), we now determine $\{x\}$ using back substitution in (2.149), since $[U]$ is an upper triangular matrix. In this step we determine $x_n, x_{n-1}, \dots, x_1$ (in this order) starting with the last equation (n^{th} equation) and then progressively moving up (i.e., $(n-1)^{th}$ equation, ...).

Remarks.

(1) The $[L][U]$ decomposition does not affect the vector $\{b\}$, hence it is ideally suited for obtaining solutions for more than one right-hand side vector $\{b\}$. For each right side vector we use Steps 1 and 2.

(2) This decomposition is called Cholesky decomposition.

Example 2.11 (Solution of Linear Equations Using Cholesky $[L][U]$ Decomposition). Consider

$$[A]\{x\} = \{b\}$$

in which

$$[A] = \begin{bmatrix} 3 & -0.1 & -0.2 \\ 0.1 & 7 & -0.3 \\ 0.3 & -0.2 & 10 \end{bmatrix} \qquad \{x\} = \begin{Bmatrix} x_1 \\ x_2 \\ x_3 \end{Bmatrix} \qquad \{b\} = \begin{Bmatrix} 7.85 \\ -19.3 \\ 71.4 \end{Bmatrix}$$

$[L][U]$ **Decomposition of** $[A]$:

$$[A] = [L][U] = \begin{bmatrix} 1 & 0 & 0 \\ L_{21} & 1 & 0 \\ L_{31} & L_{32} & 1 \end{bmatrix} \begin{bmatrix} u_{11} & u_{12} & u_{13} \\ 0 & u_{22} & u_{23} \\ 0 & 0 & u_{33} \end{bmatrix}$$

First Row of $[U]$:

The first row of $[U]$ is the same as the first row of $[A]$.

$$u_{11} = a_{11} = 3$$
$$u_{12} = a_{12} = -0.1$$
$$u_{13} = a_{13} = -0.2$$

First Column of $[L]$:

$$L_{21} = \frac{a_{21}}{u_{11}} = \frac{0.1}{3} = 0.033333$$
$$L_{31} = \frac{a_{31}}{u_{11}} = \frac{0.3}{3} = 0.1$$

At this stage $[L]$ and $[U]$ are given by:

$$[L] = \begin{bmatrix} 1 & 0 & 0 \\ 0.033 & 1 & 0 \\ 0.1 & & 1 \end{bmatrix} \qquad [U] = \begin{bmatrix} 3 & -0.1 & -0.2 \\ 0 & & \\ 0 & 0 & \end{bmatrix}$$

Second Row of $[U]$:

$$u_{22} = a_{22} - L_{21}u_{12} = 7 - (0.0333)(-0.1) = 7.00333$$
$$u_{23} = a_{23} - L_{21}u_{13} = -0.3 - (0.0333)(-0.2) = -0.2933$$

Second Column of $[L]$:

$$L_{32} = \frac{(a_{32} - L_{31}u_{12})}{u_{22}} = \frac{-0.2 - (0.1)(-0.1)}{7.00333} = -0.02713$$

At this stage $[L]$ and $[U]$ are:

$$[L] = \begin{bmatrix} 1 & 0 & 0 \\ 0.0333 & 1 & 0 \\ 0.1 & -0.02713 & 1 \end{bmatrix} \qquad [U] = \begin{bmatrix} 3 & -0.1 & -0.2 \\ 0 & 7.00333 & -0.29333 \\ 0 & 0 & \end{bmatrix}$$

Third Row of $[U]$:

$$\begin{aligned} u_{33} &= a_{33} - L_{31}u_{13} - L_{32}u_{23} \\ &= 10 - (0.1)(-0.2) - (-0.02713)(-0.29333) \\ &= 10.012 \end{aligned}$$

This completes the $[L][U]$ decomposition and we have:

$$[L] = \begin{bmatrix} 1 & 0 & 0 \\ 0.0333 & 1 & 0 \\ 0.1 & -0.02713 & 1 \end{bmatrix} \quad [U] = \begin{bmatrix} 3 & -0.1 & -0.2 \\ 0 & 7.00333 & -0.29333 \\ 0 & 0 & 10.012 \end{bmatrix}$$

Solution of $[A]\{x\} = \{b\}$:

In $[A]\{x\} = \{b\}$, we replace $[A]$ by its $[L][U]$ decomposition.

$$\begin{bmatrix} 1 & 0 & 0 \\ 0.0333 & 1 & 0 \\ 0.1 & -0.02713 & 1 \end{bmatrix} \begin{bmatrix} 3 & -0.1 & -0.2 \\ 0 & 7.00333 & -0.29333 \\ 0 & 0 & 10.012 \end{bmatrix} \begin{Bmatrix} x_1 \\ x_2 \\ x_3 \end{Bmatrix} = \begin{Bmatrix} 7.85 \\ -19.3 \\ 71.4 \end{Bmatrix}$$

Let

$$[U]\{x\} = \{y\}$$

$$\therefore \quad [L]\{y\} = \{b\}$$

$$\begin{bmatrix} 1 & 0 & 0 \\ 0.0333 & 1 & 0 \\ 0.1 & -0.02713 & 1 \end{bmatrix} \begin{Bmatrix} y_1 \\ y_2 \\ y_3 \end{Bmatrix} = \begin{Bmatrix} 7.85 \\ -19.3 \\ 71.4 \end{Bmatrix}$$

We can calculate $\{y\}$ using forward pass.

$$y_1 = 7.85$$
$$y_2 = -19.3 - (0.0333)(7.85) = -19.561405$$
$$y_3 = 71.4 - (0.1)(7.85) - (-0.2713)(-19.561405) = 70.0843$$

Now we know $\{y\}$, hence we can use $[U]\{x\} = \{y\}$ to find $\{x\}$ (*backward pass*).

$$\begin{bmatrix} 3 & -0.1 & -0.2 \\ 0 & 7.00333 & -0.29333 \\ 0 & 0 & 10.012 \end{bmatrix} \begin{Bmatrix} x_1 \\ x_2 \\ x_3 \end{Bmatrix} = \begin{Bmatrix} 7.85 \\ -19.56125 \\ 70.0843 \end{Bmatrix}$$

Back substitution of the backward pass gives:

$$
\begin{aligned}
x_3 &= \frac{70.0843}{10.012} = 7.000029872 \\
x_2 &= \frac{(-19.5614 - (-0.29333)(7.000029872))}{7.000333} \\
&= -2.4999652 \\
x_1 &= \frac{(7.85 - (-0.2)(7.000029872) - (-0.1)(-2.499952))}{3.0} \\
&= 3.00003152
\end{aligned}
$$

Therefore the solution $\{x\}$ is:

$$
\{x\} = \begin{Bmatrix} x_1 \\ x_2 \\ x_3 \end{Bmatrix} = \begin{Bmatrix} 3 \\ -2.5 \\ 7 \end{Bmatrix}
$$

2.5.3.3 Crout Decomposition of $[A]$ into $[L][U]$ and Solution of Linear Algebraic Equations

Consider

$$
[A]\{x\} = \{b\} \tag{2.151}
$$

In Crout decomposition, we also express $[A]$ as a product of $[L]$ and $[U]$ as in Cholesky decomposition, except that in this decomposition $[L]$ is a lower triangular matrix and $[U]$ is a *unit* upper triangular matrix. That is, the diagonal elements are $[L]$ are not unity, and instead the diagonal elements of $[U]$ are unity. We begin with:

$$
[A] = [L][U] \tag{2.152}
$$

The rules for determining the elements of $[L]$ and $[U]$ are established by forming the product $[L][U]$ and equating the elements of the product to the corresponding elements of $[A]$. We present details in the following. Consider $[A]$ to be a (4×4) matrix. We equate $[A]$ to the product of $[L]$ and $[U]$.

$$
\begin{bmatrix} a_{11} & a_{12} & a_{13} & a_{14} \\ a_{21} & a_{22} & a_{23} & a_{24} \\ a_{31} & a_{32} & a_{33} & a_{34} \\ a_{41} & a_{42} & a_{43} & a_{44} \end{bmatrix} = \begin{bmatrix} L_{11} & 0 & 0 & 0 \\ L_{21} & L_{22} & 0 & 0 \\ L_{31} & L_{32} & L_{33} & 0 \\ L_{41} & L_{42} & L_{43} & L_{44} \end{bmatrix} \begin{bmatrix} 1 & u_{12} & u_{13} & u_{14} \\ 0 & 1 & u_{23} & u_{24} \\ 0 & 0 & 1 & u_{34} \\ 0 & 0 & 0 & 1 \end{bmatrix} \tag{2.153}
$$

To determine the elements of $[L]$ and $[U]$, we form the product of $[L]$ and $[U]$ in (2.153).

$$\begin{bmatrix} a_{11} & a_{12} & a_{13} & a_{14} \\ a_{21} & a_{22} & a_{23} & a_{24} \\ a_{31} & a_{32} & a_{33} & a_{34} \\ a_{41} & a_{42} & a_{43} & a_{44} \end{bmatrix} =$$

L_{11}	$L_{11}u_{12}$	$L_{11}u_{13}$	$L_{11}u_{14}$
L_{21}	$L_{21}u_{12} + L_{22}$	$L_{21}u_{13} + L_{22}u_{23}$	$L_{21}u_{14} + L_{22}u_{24}$
L_{31}	$L_{31}u_{12} + L_{32}$	$L_{31}u_{13} + L_{32}u_{23} + L_{33}$	$L_{31}u_{14} + L_{32}u_{24} + L_{33}u_{34}$
L_{41}	$L_{41}u_{12} + L_{42}$	$L_{41}u_{13} + L_{42}u_{23} + L_{43}$	$L_{41}u_{14} + L_{42}u_{24} + L_{43}u_{34} + L_{44}$

(2.154)

In the Crout method the procedure for determining the elements of $[L]$ and $[U]$ alternate between a column of $[L]$ and the corresponding row of $[U]$, as opposed to Cholesky decomposition in which we determine a row of $[U]$ first followed by the corresponding column of $[L]$.

First Column of $[L]$:

If we equate the elements of the first column on both sides of (2.154), we obtain:

$$L_{i1} = a_{i1} \qquad i = 1, 2, \ldots, 4 \text{ (or } n \text{ in general)} \tag{2.155}$$

First Row of $[U]$:

Equating elements of the first row of both sides of (2.154):

$$u_{1j} = \frac{a_{1j}}{L_{11}} \qquad j = 1, 2, \ldots, 4 \text{ (or } n \text{ in general)} \tag{2.156}$$

Second Column of $[L]$:

Equating elements of the second column on both sides of (2.154):

$$\begin{aligned} L_{22} &= a_{22} - L_{21}u_{12} \\ L_{32} &= a_{32} - L_{31}u_{12} \\ L_{42} &= a_{42} - L_{41}u_{12} \end{aligned} \tag{2.157}$$

Second Row of $[U]$:

Equating elements of the second row on both sides of (2.154):

$$\begin{aligned} u_{23} &= \frac{(a_{23} - L_{21}u_{13})}{L_{22}} \\ u_{24} &= \frac{(a_{24} - L_{21}u_{14})}{L_{22}} \end{aligned} \tag{2.158}$$

Third Column of $[L]$:

Equating elements of the third column on both sides of (2.154):

$$\begin{aligned} L_{33} &= a_{33} - L_{31}u_{13} - L_{32}u_{23} \\ L_{43} &= a_{43} - L_{41}u_{13} - L_{42}u_{23} \end{aligned} \tag{2.159}$$

Third Row of $[U]$:

Equating the elements of the third row in (2.154):

$$u_{34} = \frac{(a_{34} - L_{31}u_{14} - L_{32}u_{24})}{L_{33}} \tag{2.160}$$

Fourth Column of $[L]$:

Lastly, equating elements of the fourth column on both sides (2.154):

$$L_{44} = a_{44} - L_{41}u_{14} - L_{42}u_{24} - L_{43}u_{34} \tag{2.161}$$

Thus, the elements of $[L]$ and $[U]$ are completely determined.

Remarks.

(1) The elements of $[L]$ and $[U]$ can be expressed more compactly as follows:

$$\begin{aligned}
&L_{ij} = a_{i1} && i = 1, 2, \dots, n \quad (j = 1 \text{ for this case }) \\
&u_{ij} = \frac{a_{1j}}{L_{11}} && j = 2, 3, \dots, n \quad (i = 1 \text{ for this case }) \\
&L_{ij} = a_{ij} - \sum_{k=1}^{j-1} L_{ik}u_{kj} && j = 2, 3, \dots, n \quad i = j, j+1, \dots, n \quad (\text{for each } j) \\
&u_{ij} = \frac{a_{ij} - \sum\limits_{k=1}^{i-1} L_{ik}u_{kj}}{L_{ii}} && i = 2, 3, \dots, n \quad j = i+1, i+2, \dots, n
\end{aligned} \tag{2.162}$$

Using $n = 4$ in (2.162), we can obtain (2.155) - (2.161). The form in (2.162) is helpful in programming $[L][U]$ decomposition based on Crout method.

(2) Just like Cholesky decomposition, $[L]$ and $[U]$ can be stored in the same space that is used for $[A]$, however $[A]$ is obviously destroyed in this case.

Example 2.12 (Decomposition Using Crout Method and Solution of Linear System). Consider

$$[A]\{x\} = \{b\}$$

in which

$$[A] = \begin{bmatrix} 3 & -0.1 & -0.2 \\ 0.1 & 7 & -0.3 \\ 0.3 & -0.2 & 10 \end{bmatrix} \qquad \{x\} = \begin{Bmatrix} x_1 \\ x_2 \\ x_3 \end{Bmatrix} \qquad \{b\} = \begin{Bmatrix} 7.85 \\ -19.3 \\ 71.4 \end{Bmatrix}$$

$[L][U]$ Decomposition of $[A]$:

$$[A] = [L][U] = \begin{bmatrix} L_{11} & 0 & 0 \\ L_{21} & L_{22} & 0 \\ L_{31} & L_{32} & L_{33} \end{bmatrix} \begin{bmatrix} 1 & u_{12} & u_{13} \\ 0 & 1 & u_{23} \\ 0 & 0 & 1 \end{bmatrix}$$

First Column of $[L]$ (same as first column of $[A]$):

$$L_{11} = a_{11} = 3$$
$$L_{21} = a_{21} = 0.1$$
$$L_{31} = a_{31} = 0.3$$

First Row of $[U]$:

$$u_{12} = \frac{a_{12}}{L_{11}} = \frac{-0.1}{3} = -0.033333$$
$$u_{13} = \frac{a_{13}}{L_{11}} = \frac{-0.2}{3} = -0.066667$$

Second Column of $[L]$:

$$L_{22} = a_{22} - L_{21}u_{12} = 7 - 0.1(-0.03333) = 7.003333$$
$$L_{32} = a_{32} - L_{31}u_{12} = -0.2 - 0.3(-0.03333) = -0.19$$

Second Row of $[U]$:

$$u_{23} = \frac{(a_{23} - L_{21}u_{13})}{L_{22}} = \frac{(-0.3 - 0.1(-0.066667))}{7.003333}$$
$$u_{23} = -0.0418848$$

Third Column of $[L]$:

$$L_{33} = a_{33} - L_{31}u_{13} - L_{32}u_{23} = 10 - 0.3(-0.066667) - (-0.19)(-0.0418818)$$
$$L_{33} = 10.012042$$

$$[A] = [L][U] = \begin{bmatrix} 3 & 0 & 0 \\ 0.1 & 7.003333 & 0 \\ 0.3 & -0.19 & 10.012042 \end{bmatrix} \begin{bmatrix} 1 & -0.03333 & -0.66667 \\ 0 & 1 & -0.04188848 \\ 0 & 0 & 1 \end{bmatrix}$$

By taking product of $[L]$ and $[U]$ we recover $[A]$.

Solution $\{x\}$ of $[A]\{x\} = \{b\}$:

Let

$$[U]\{x\} = \{y\}$$

$$\therefore \quad [L]\{y\} = \{b\}$$

Using forward pass, calculate $\{y\}$ using $[L]\{y\} = \{b\}$.

$$\begin{bmatrix} 3 & 0 & 0 \\ 0.1 & 7.003333 & 0 \\ 0.3 & -0.19 & 10.012042 \end{bmatrix} \begin{Bmatrix} y_1 \\ y_2 \\ y_3 \end{Bmatrix} = \begin{Bmatrix} 7.85 \\ -19.3 \\ 71.4 \end{Bmatrix}$$

$$y_1 = \frac{7.85}{3} = 2.616667$$

$$y_2 = \frac{(-19.3 - (0.1)(2.616667))}{7.003333} = -2.7931936$$

$$y_3 = \frac{(71.4 - (0.3)(2.61667) - (-0.19(-2.7931936))}{10.012042} = 7$$

$$\therefore \quad \{y\} = \begin{Bmatrix} y_1 \\ y_2 \\ y_3 \end{Bmatrix} = \begin{Bmatrix} 2.616667 \\ -2.7931936 \\ 7 \end{Bmatrix}$$

Now consider $[U]\{x\} = \{y\}$.

$$\begin{bmatrix} 1 & -0.03333 & -0.66667 \\ 0 & 1 & -0.04188848 \\ 0 & 0 & 1 \end{bmatrix} \begin{Bmatrix} x_1 \\ x_2 \\ x_3 \end{Bmatrix} = \begin{Bmatrix} 2.616667 \\ -2.7931936 \\ 7 \end{Bmatrix}$$

Using the backward pass or back substitution to obtain x_3, x_2, and x_1:

$$x_3 = 7$$

$$x_2 = -2.7931936 - (-0.048848)(7) = -2.5$$

$$x_1 = 2.616667 - (-0.033333)(-2.5) - (-0.066667)(7) = 3$$

$$\therefore \quad \{x\} = \begin{Bmatrix} x_1 \\ x_2 \\ x_3 \end{Bmatrix} = \begin{Bmatrix} 3 \\ -2.5 \\ 7 \end{Bmatrix}$$

This is the same as calculated using Cholesky decomposition.

2.5.3.4 Classical or Cholesky Decomposition of $[A]$ in $[A]\{x\} = \{b\}$ using Gauss Elimination

Consider

$$[A]\{x\} = \{b\} \tag{2.163}$$

In Gauss elimination we make the augmented matrix $[A_{ag}]$ upper triangular by elementary row operations. Consider

$$\begin{bmatrix} a_{11} & a_{12} & a_{13} \\ a_{21} & a_{22} & a_{23} \\ a_{31} & a_{32} & a_{33} \end{bmatrix} \begin{Bmatrix} x_1 \\ x_2 \\ x_3 \end{Bmatrix} = \begin{Bmatrix} b_1 \\ b_2 \\ b_3 \end{Bmatrix} \tag{2.164}$$

When making the first column in (2.164) upper triangular, we need to make a_{21} and a_{31} zero by elementary row operations. In this process we multiply row one by $\frac{a_{21}}{a_{11}} = C_{21}$ and $\frac{a_{31}}{a_{11}} = C_{31}$ and then subtract these from rows two and three of (2.164). This results in

$$\begin{bmatrix} a_{11} & a_{12} & a_{13} \\ 0 & a'_{22} & a'_{23} \\ 0 & a'_{32} & a'_{33} \end{bmatrix} \begin{Bmatrix} x_1 \\ x_2 \\ x_3 \end{Bmatrix} = \begin{Bmatrix} b_1 \\ b'_2 \\ b'_3 \end{Bmatrix} \tag{2.165}$$

To make the second column upper triangular we multiply the second row in (2.165) by $\frac{a'_{32}}{a'_{22}} = C_{32}$ and subtract it from row three of (2.165).

$$\begin{bmatrix} a_{11} & a_{12} & a_{13} \\ 0 & a'_{22} & a'_{23} \\ 0 & 0 & a''_{33} \end{bmatrix} \begin{Bmatrix} x_1 \\ x_2 \\ x_3 \end{Bmatrix} = \begin{Bmatrix} b_1 \\ b'_2 \\ b''_3 \end{Bmatrix} \tag{2.166}$$

This is the upper triangular form, as in Gauss elimination. The coefficients C_{21}, C_{31}, and C_{32} are indeed the elements of $[L]$ and the upper triangular form in (2.166) is $[U]$. Thus, we can write:

$$[A]\{x\} = [L][U]\{x\} = \{b\} \tag{2.167}$$

or

$$\begin{bmatrix} 1 & 0 & 0 \\ C_{21} & 1 & 0 \\ C_{31} & C_{32} & 1 \end{bmatrix} \begin{bmatrix} a_{11} & a_{12} & a_{13} \\ 0 & a'_{22} & a'_{23} \\ 0 & 0 & a''_{33} \end{bmatrix} \begin{Bmatrix} x_1 \\ x_2 \\ x_3 \end{Bmatrix} = \begin{Bmatrix} b_1 \\ b_2 \\ b_3 \end{Bmatrix} \tag{2.168}$$

By using $C_{21} = \frac{a_{21}}{a_{11}}$, $C_{31} = \frac{a_{31}}{a_{11}}$, and $C_{32} = \frac{a'_{32}}{a'_{22}}$ in (2.168) and by carrying the product of $[L]$ and $[U]$ in (2.168), the matrix $[A]$ is recovered.

Example 2.13 (Classical $[L][U]$ Decomposition using Gauss Elimination). Consider $[A]\{x\} = \{b\}$ in which

$$[A] = \begin{bmatrix} 3 & -0.1 & -0.2 \\ 0.1 & 7 & -0.3 \\ 0.3 & -0.2 & 10 \end{bmatrix} \quad \{x\} = \begin{Bmatrix} x_1 \\ x_2 \\ x_3 \end{Bmatrix} \quad \{b\} = \begin{Bmatrix} 7.85 \\ -19.3 \\ 71.4 \end{Bmatrix}$$

In making column one upper triangular we use:

$$\begin{aligned} C_{21} &= \frac{a_{21}}{a_{11}} = \frac{0.1}{3} = 0.033333 \\ C_{31} &= \frac{a_{31}}{a_{11}} = \frac{0.3}{3} = 0.1 \end{aligned} \tag{2.169}$$

and $[A]$ becomes

$$[A] = \begin{bmatrix} 3 & -0.1 & -0.2 \\ 0 & 7.003333 & -0.293333 \\ 0 & -0.19 & 10.012 \end{bmatrix} \tag{2.170}$$

In making column two upper triangular in (2.160), we use:

$$C_{32} = \frac{a'_{32}}{a'_{22}} = \frac{-0.19}{7.003333} = -0.027130 \tag{2.171}$$

The new upper triangular form of $[A]$ is in fact $[U]$ and is given by:

$$\begin{bmatrix} 3 & -0.1 & -0.2 \\ 0 & 7.003333 & -0.293333 \\ 0 & 0 & 10.012 \end{bmatrix} = [U] \tag{2.172}$$

and

$$[L] = \begin{bmatrix} 1 & 0 & 0 \\ L_{21} & 1 & 0 \\ L_{31} & L_{32} & 1 \end{bmatrix} = \begin{bmatrix} 1 & 0 & 0 \\ C_{21} & 1 & 0 \\ C_{31} & C_{32} & 1 \end{bmatrix} = \begin{bmatrix} 1 & 0 & 0 \\ 0.0333 & 1 & 0 \\ 0.1 & -0.02713 & 1 \end{bmatrix} \tag{2.173}$$

We can check that the product of $[L]$ in (2.173) and $[U]$ in (2.172) is in fact $[A]$.

2.5.3.5 Cholesky Decomposition for a Symmetric Matrix $[A]$

If the matrix $[A]$ is symmetric then the following decomposition of $[A]$ is possible. Since $[A] = [A]^T$ we can write:

$$[A] = [\tilde{L}][\tilde{L}]^T \tag{2.174}$$

in which $[\tilde{L}]$ is a lower triangular matrix. If $[A]$ is a (3×3) symmetric matrix, then $[\tilde{L}]$ will have the following form.

$$[\tilde{L}] = \begin{bmatrix} \tilde{L}_{11} & 0 & 0 \\ \tilde{L}_{21} & \tilde{L}_{22} & 0 \\ \tilde{L}_{31} & \tilde{L}_{32} & \tilde{L}_{33} \end{bmatrix} \tag{2.175}$$

Obviously $[\tilde{L}]$ is lower triangular.

The elements of $[\tilde{L}]$ are obtained by substituting $[\tilde{L}]$ from (2.175) in (2.174), carrying out the multiplication of $[\tilde{L}][\tilde{L}]^T$ on the right-hand side of (2.175), and then equating the elements of both sides of (2.174).

$$\begin{bmatrix} a_{11} & a_{12} & a_{13} \\ a_{21} & a_{22} & a_{23} \\ a_{31} & a_{32} & a_{33} \end{bmatrix} = \begin{bmatrix} \tilde{L}_{11} & 0 & 0 \\ \tilde{L}_{21} & \tilde{L}_{22} & 0 \\ \tilde{L}_{31} & \tilde{L}_{32} & \tilde{L}_{33} \end{bmatrix} \begin{bmatrix} \tilde{L}_{11} & \tilde{L}_{21} & \tilde{L}_{31} \\ 0 & \tilde{L}_{22} & \tilde{L}_{32} \\ 0 & 0 & \tilde{L}_{33} \end{bmatrix} \quad a_{ij} = a_{ji} \quad \begin{matrix} i = 1,2,3 \\ j = 1,2,3 \end{matrix}$$

$$= \begin{bmatrix} (\tilde{L}_{11})^2 & (\tilde{L}_{11})(\tilde{L}_{21}) & (\tilde{L}_{11})(\tilde{L}_{31}) \\ (\tilde{L}_{21})(\tilde{L}_{11}) & (\tilde{L}_{21})^2 + (\tilde{L}_{22})^2 & (\tilde{L}_{21})(\tilde{L}_{31}) + (\tilde{L}_{22})(\tilde{L}_{32}) \\ (\tilde{L}_{31})(\tilde{L}_{11}) & (\tilde{L}_{31})(\tilde{L}_{21}) + (\tilde{L}_{32})(\tilde{L}_{22}) & (\tilde{L}_{31})^2 + (\tilde{L}_{32})^2 + (\tilde{L}_{33})^2 \end{bmatrix} \tag{2.176}$$

We note that the $[\tilde{L}][\tilde{L}]^T$ product in (2.176) is symmetric, as expected. Hence, we only need to consider the elements on the diagonal and those above the diagonal in $[\tilde{L}][\tilde{L}]^T$ in (2.176).

Equate the elements of row one on both sides of (2.176).

$$\tilde{L}_{11} = \sqrt{a_{11}}\ ; \quad \tilde{L}_{21} = \frac{a_{21}}{\tilde{L}_{11}}\ ; \quad \tilde{L}_{31} = \frac{a_{31}}{\tilde{L}_{11}} \tag{2.177}$$

Consider row two in (2.176).

$$\tilde{L}_{22} = \sqrt{a_{22} - (\tilde{L}_{21})^2}\ ; \quad \tilde{L}_{32} = \frac{a_{32} - \tilde{L}_{21}\tilde{L}_{31}}{\tilde{L}_{22}} \tag{2.178}$$

Consider row three in (2.176).

$$\tilde{L}_{33} = \sqrt{a_{33} - (\tilde{L}_{31})^2 - (\tilde{L}_{32})^2} \tag{2.179}$$

Hence $[\tilde{L}]$ is completely determined. We can write (2.177) – (2.179) in a more compact form.

$$\begin{aligned} \tilde{L}_{kk}^2 &= a_{kk} - \sum_{j=1}^{k-1} (\tilde{L}_{kj})^2 \qquad k = 1, 2, \dots, n \\ \tilde{L}_{ki} &= \frac{a_{ki} - \sum\limits_{j=1}^{i-1} \tilde{L}_{ij}\tilde{L}_{kj}}{\tilde{L}_{ii}} \qquad \begin{matrix} i = 1, 2, \dots, k-1 \\ k = 1, 2, \dots, n \end{matrix} \end{aligned} \tag{2.180}$$

2.5.3.6 Alternate Derivation of $[L][U]$ Decomposition when $[A]$ is Symmetric

The decomposition in (2.180) can also be derived using the following method when $[A]$ is symmetric. Consider the classical (Cholesky) $[L][U]$ decomposition of a (3×3) matrix $[A]$.

$$[A] = \begin{bmatrix} L_{11} & 0 & 0 \\ L_{21} & L_{22} & 0 \\ L_{31} & L_{32} & L_{33} \end{bmatrix} \begin{bmatrix} 1 & u_{12} & u_{13} \\ 0 & 1 & u_{23} \\ 0 & 0 & 1 \end{bmatrix} \tag{2.181}$$

If we divide columns of $[L]$ by L_{11}, L_{22}, and L_{33} and if we form a diagonal matrix of L_{11}, L_{22}, and L_{33}, then we can write the following:

$$[A] = \begin{bmatrix} 1 & 0 & 0 \\ \frac{L_{21}}{L_{11}} & 1 & 0 \\ \frac{L_{31}}{L_{11}} & \frac{L_{32}}{L_{22}} & 1 \end{bmatrix} \begin{bmatrix} L_{11} & 0 & 0 \\ 0 & L_{22} & 0 \\ 0 & 0 & L_{33} \end{bmatrix} \begin{bmatrix} 1 & u_{12} & u_{13} \\ 0 & 1 & u_{23} \\ 0 & 0 & 1 \end{bmatrix} \tag{2.182}$$

Rewrite the diagonal matrix in (2.182) as a product of two diagonal matrices in which the diagonal elements are the square roots of L_{11}, L_{22}, and L_{33}.

$$[A] = \begin{bmatrix} 1 & 0 & 0 \\ \frac{L_{21}}{L_{11}} & 1 & 0 \\ \frac{L_{31}}{L_{11}} & \frac{L_{32}}{L_{22}} & 1 \end{bmatrix} \begin{bmatrix} \sqrt{L_{11}} & 0 & 0 \\ 0 & \sqrt{L_{22}} & 0 \\ 0 & 0 & \sqrt{L_{33}} \end{bmatrix} \begin{bmatrix} \sqrt{L_{11}} & 0 & 0 \\ 0 & \sqrt{L_{22}} & 0 \\ 0 & 0 & \sqrt{L_{33}} \end{bmatrix} \begin{bmatrix} 1 & u_{12} & u_{13} \\ 0 & 1 & u_{23} \\ 0 & 0 & 1 \end{bmatrix} \tag{2.183}$$

Define

$$[\tilde{L}] = \begin{bmatrix} 1 & 0 & 0 \\ \frac{L_{21}}{L_{11}} & 1 & 0 \\ \frac{L_{31}}{L_{11}} & \frac{L_{32}}{L_{22}} & 1 \end{bmatrix} \begin{bmatrix} \sqrt{L_{11}} & 0 & 0 \\ 0 & \sqrt{L_{22}} & 0 \\ 0 & 0 & \sqrt{L_{33}} \end{bmatrix} \tag{2.184}$$

$$[\tilde{L}]^T = \begin{bmatrix} \sqrt{L_{11}} & 0 & 0 \\ 0 & \sqrt{L_{22}} & 0 \\ 0 & 0 & \sqrt{L_{33}} \end{bmatrix} \begin{bmatrix} 1 & u_{12} & u_{13} \\ 0 & 1 & u_{23} \\ 0 & 0 & 1 \end{bmatrix} \tag{2.185}$$

$$\therefore \quad [A] = [\tilde{L}][\tilde{L}]^T \tag{2.186}$$

This completes the decomposition.

2.6 Solution of Linear Simultaneous Algebraic Equations $[A]\{x\} = \{b\}$ Using the Inverse of $[A]$

Consider

$$[A]\{x\} = \{b\} \tag{2.187}$$

Let $[A]^{-1}$ be inverse of the coefficient matrix $[A]$, then:

$$[A][A]^{-1} = [A]^{-1}[A] = [I] \tag{2.188}$$

Premultiply (2.187) by $[A]^{-1}$.

$$[A]^{-1}[A]\{x\} = [A]^{-1}\{b\} \tag{2.189}$$

Using (2.188):

$$[I]\{x\} = [A]^{-1}\{b\} \tag{2.190}$$

or

$$\{x\} = [A]^{-1}\{b\} \tag{2.191}$$

Thus, if we can find $[A]^{-1}$ then the solution of $\{x\}$ of (2.187) can be obtained using (2.191).

2.6.1 Methods of Finding Inverse of $[A]$

We consider three methods of finding $[A]^{-1}$ in the following sections:

(a) Direct method.

(b) Elementary row transformation as in Gauss-Jordan method.

(c) $[L][U]$ decomposition using Cholesky or Crout method.

2.6.1.1 Direct Method of Finding Inverse of $[A]$

We follow the steps given below.

1. Find the determinant of $[A]$, i.e., $det[A]$ or $|A|$.

2. Find the minors m_{ij}; $i, j = 1, 2, \ldots, n$ of a_{ij}.
The minor m_{ij} of a_{ij} is given by the determinant of $[A]$ after deleting row i and column j.

3. Find the cofactors of a_{ij}, i.e., $\bar{a}_{ij}$; $i,j = 1,2,\dots,n$.

$$\bar{a}_{ij} = (-1)^{i+j} m_{ij} \tag{2.192}$$

4. Find the cofactor matrix of $[A]$, i.e., $[\bar{A}]$, by using cofactors $\bar{a}_{ij}$; $i,j = 1,2,\dots,n$.

$$[\bar{A}] = \begin{bmatrix} \bar{a}_{11} & \bar{a}_{12} & \dots & \bar{a}_{1n} \\ \bar{a}_{21} & \bar{a}_{22} & \dots & \bar{a}_{2n} \\ & \vdots & & \\ \bar{a}_{n1} & \bar{a}_{n2} & \dots & \bar{a}_{nn} \end{bmatrix} \tag{2.193}$$

5. Find the adjoint of $[A]$ $(adj[A])$.

$$adj[A] = [\bar{A}]^T \quad ; \quad \text{transpose of the cofactor matrix of } [A] \tag{2.194}$$

6. Finally, the inverse of $[A]$ is given by:

$$[A]^{-1} = \frac{1}{|A|}(adj[A]) = \frac{1}{|A|}[\bar{A}]^T = \frac{1}{|A|} \begin{bmatrix} \bar{a}_{11} & \bar{a}_{21} & \dots & \bar{a}_{n1} \\ \bar{a}_{12} & \bar{a}_{22} & \dots & \bar{a}_{n2} \\ & \vdots & & \\ \bar{a}_{1n} & \bar{a}_{2n} & \dots & \bar{a}_{nn} \end{bmatrix} \tag{2.195}$$

2.6.1.2 Using Elementary Row Operations and Gauss-Jordan Method to Find the Inverse of $[A]$

Augment the matrix $[A]$ by an identity matrix of the same size (i.e., the same number of rows and columns as in $[A]$).

$$[[A] \,|\, [I]] \tag{2.196}$$

Perform elementary row operations on (2.196) so that $[A]$ becomes an identity matrix (same operations as in Guass-Jordan method).

$$[[A] \,|\, [I]] \xrightarrow[\text{Elementary Row Operations}]{} [[I] \,|\, [B]] \tag{2.197}$$

or

$$\left[\begin{array}{ccc|ccc} a_{11} & a_{12} & a_{13} & 1 & 0 & 0 \\ a_{21} & a_{22} & a_{23} & 0 & 1 & 0 \\ a_{31} & a_{32} & a_{33} & 0 & 0 & 1 \end{array}\right] \xrightarrow[\substack{\text{Elementary} \\ \text{Row} \\ \text{Operations}}]{} \left[\begin{array}{ccc|ccc} 1 & 0 & 0 & b_{11} & b_{12} & b_{13} \\ 0 & 1 & 0 & b_{21} & b_{22} & b_{23} \\ 0 & 0 & 1 & b_{31} & b_{32} & b_{33} \end{array}\right] \tag{2.198}$$

Thus,

$$[B] = [A]^{-1} = \begin{bmatrix} b_{11} & b_{12} & b_{13} \\ b_{21} & b_{22} & b_{23} \\ b_{31} & b_{32} & b_{33} \end{bmatrix} \tag{2.199}$$

Remarks. This procedure is exactly the same as Gauss-Jordan method if we augment $[A]$ by $[I]$ and $\{b\}$. Consider

$$[[A] \,|\, [I] \,|\, \{b\}] \tag{2.200}$$

Using the Gauss-Jordan method, when $[A]$ is made identity using elementary row operations, we have:

$$[[I] \,|\, [A]^{-1} \,|\, \{x\}] \tag{2.201}$$

The location of $[I]$ in (2.200) contains $[A]^{-1}$ and the location of $\{b\}$ in (2.200) contains the solution vector $\{x\}$.

2.6.1.3 Finding the Inverse of $[A]$ by $[L][U]$ Decomposition

Consider the $[L][U]$ decomposition of $[A]$ obtained by any of the methods discussed in the earlier sections.

$$[A] = [L][U] \tag{2.202}$$

Let

$$[B] = [A]^{-1} \tag{2.203}$$

To obtain the first column of $[B]$ we solve the following system of equations:

First column of $[B]$ → ; 1st row

$$[L][U] \begin{Bmatrix} b_{11} \\ b_{21} \\ \vdots \\ b_{i1} \\ \vdots \\ b_{n1} \end{Bmatrix} = \begin{Bmatrix} 1 \\ 0 \\ \vdots \\ \vdots \\ \vdots \\ 0 \end{Bmatrix} \tag{2.204}$$

To obtain the the second column of $[B]$ we consider solution of:

Second column of $[B]$ → ; 2nd row

$$[L][U] \begin{Bmatrix} b_{12} \\ b_{22} \\ \vdots \\ b_{i2} \\ \vdots \\ b_{n2} \end{Bmatrix} = \begin{Bmatrix} 0 \\ 1 \\ \vdots \\ \vdots \\ \vdots \\ 0 \end{Bmatrix} \tag{2.205}$$

For column j of $[B]$, we solve:

Third column of $[B]$

$$[L][U] = \begin{Bmatrix} b_{1j} \\ b_{2j} \\ \vdots \\ b_{ij} \\ \vdots \\ b_{nj} \end{Bmatrix} = \begin{Bmatrix} 0 \\ 0 \\ \vdots \\ 1 \\ \vdots \\ 0 \end{Bmatrix} \longleftarrow j^{th} \text{ row} \tag{2.206}$$

Thus determination of each column of the inverse requires solutions of linear simultaneous algebraic equation in which the right-hand side vector is null except for the row location corresponding to the column number for which we are seeking the solution. This is the only location in the right-hand side vector that contains a nonzero value of one.

2.7 Iterative Methods of Solving Linear Simultaneous Algebraic Equations

Iterative methods of obtaining solutions of linear simultaneous algebraic equations are an alternative to the elimination and other methods we have discussed in earlier sections. In all iterative methods, we begin with an assumed or guess solution, also known as the initial solution, and then use a systematic iterative procedure to obtain successively improved solutions until the solution no longer changes. At this stage we have a *converged* solution that is a close approximation of the true solution. Thus, these are methods of approximation. These methods are generally easier to program. The number of iterations required for convergence is dependent on the coefficient matrix $[A]$ and the choice of the initial solution. Consider the following methods:

(a) Gauss-Seidel method

(b) Jacobi method

(c) Relaxation methods

2.7.1 Gauss-Seidel Method

This is a simple and commonly used iterative method of obtaining solutions of $[A]\{x\} = \{b\}$. We illustrate the details of the method for a system of three linear simultaneous equations.

$$\begin{bmatrix} a_{11} & a_{12} & a_{13} \\ a_{21} & a_{22} & a_{23} \\ a_{31} & a_{32} & a_{33} \end{bmatrix} \begin{Bmatrix} x_1 \\ x_2 \\ x_3 \end{Bmatrix} = \begin{Bmatrix} b_1 \\ b_2 \\ b_3 \end{Bmatrix} \tag{2.207}$$

Solve for x_1, x_2, x_3 using first, second, and third equations in (2.207).

$$x_1 = \frac{b_1 - a_{12}x_2 - a_{13}x_3}{a_{11}} \tag{2.208}$$

$$x_2 = \frac{b_2 - a_{21}x_1 - a_{23}x_3}{a_{22}} \tag{2.209}$$

$$x_3 = \frac{b_3 - a_{31}x_1 - a_{32}x_2}{a_{33}} \tag{2.210}$$

(i) Choose an initial or guess solution:

$$\{x\} = \{\tilde{x}\} = \begin{Bmatrix} \tilde{x}_1 \\ \tilde{x}_2 \\ \tilde{x}_3 \end{Bmatrix} \tag{2.211}$$

$\{\tilde{x}\}$ could be $[0\ 0\ 0]^T$ or $[1\ 1\ 1]^T$ or any other choice.

(ii) Use the $\{\tilde{x}\}$ vector from (2.211) in (2.208) to solve for x_1, say x_1'.

(iii) Update the $\{\tilde{x}\}$ vector in (2.211) by replacing $\tilde{x}_1$ by x_1'. Thus the updated $\{x\}$ is:

$$\{x\} = \begin{Bmatrix} x_1' \\ \tilde{x}_2 \\ \tilde{x}_3 \end{Bmatrix} \tag{2.212}$$

(iv) Use the $\{x\}$ vector from (2.212) in (2.209) to solve for x_2, say x_2''.

(v) Update the $\{x\}$ in (2.212) by replacing $\tilde{x}_2$ with x_2'', hence the updated $\{x\}$ becomes:

$$\{x\} = \begin{Bmatrix} x_1' \\ x_2'' \\ \tilde{x}_3 \end{Bmatrix} \tag{2.213}$$

(vi) Use the vector $\{x\}$ from (2.213) in (2.210) to solve for x_3, say x_3'''.

(vii) Update the $\{x\}$ vector in (2.213) by replacing $\tilde{x}_3$ by x_3''', hence the new, improved $\{x\}$ is:

$$\{x\} = \begin{Bmatrix} x_1' \\ x_2'' \\ x_3''' \end{Bmatrix} \tag{2.214}$$

In (2.214) we have the improved estimate of $\{x\}$. Steps (ii) - (vii) constitute an iteration. The new improved estimate is used to repeat steps (ii) - (vii) until the process is converged, i.e., until two successive estimates of $\{x\}$ do not differ appreciably. More specifically, the process is converged when the solutions from the two successive iterations are within a tolerance based on the desired decimal place accuracy. We discuss the concept of convergence and the convergence criterion in the following section.

Convergence Criterion

Let $\{x\}_{j-1}$ and $\{x\}_j$ be two successive solutions at the end of $(j-1)^{th}$ and j^{th} iterations. We consider the iterative solution procedure converged when the corresponding components of $\{x\}_{j-1}$ and $\{x\}_j$ are within a preset tolerance Δ.

$$(\epsilon_i)_j = \left|\frac{(x_i)_{j-1} - (x_i)_j}{(x_i)_j}\right| \times 100 \leq \Delta \qquad i = 1, 2, \dots, n \tag{2.215}$$

$(\epsilon_i)_j$ is the percentage error in the ith component of $\{x\}$, i.e., x_i, based on the most up to date solution for the ith component of $\{x\}$, $(x_i)_j$. When (2.215) is satisfied, we consider the iterative process converged and we have an approximation $\{x\}_j$ of the true solution $\{x\}$ of (2.207).

Example 2.14 (Gauss-Seidel Method). Consider the following set of linear simultaneous algebraic equations:

$$\begin{aligned} 3x_1 - 0.1x_2 - 0.2x_3 &= 7.85 \\ 0.1x_1 + 7x_2 - 0.3x_3 &= -19.3 \\ 0.3x_1 - 0.2x_2 + 10x_3 &= 71.4 \end{aligned} \tag{2.216}$$

Solve for x_1, x_2, and x_3 using the first, second, and third equations in (2.216).

$$x_1 = \frac{7.85 + 0.1x_2 + 0.2x_3}{3} \tag{2.217}$$

$$x_2 = \frac{-19.3 - 0.1x_1 + 0.3x_3}{7} \tag{2.218}$$

$$x_3 = \frac{71.4 - 0.3x_1 + 0.2x_2}{10} \tag{2.219}$$

(i) Choose

$$\{x\} = \begin{Bmatrix} \tilde{x}_1 \\ \tilde{x}_2 \\ \tilde{x}_3 \end{Bmatrix} = \begin{Bmatrix} 0 \\ 0 \\ 0 \end{Bmatrix} \tag{2.220}$$

as a guess or initial solution.

(ii) Solve for x_1 using (2.217) and (2.220) and denote the new value of x_1 by x_1'.

$$x_1 = \frac{7.85 + 0 + 0}{3} = 2.616667 = x_1' \tag{2.221}$$

(iii) Using the new value of x_1, i.e. x_1', update the starting solution vector (2.220).

$$\{x\} = \begin{Bmatrix} x_1' \\ \tilde{x}_2 \\ \tilde{x}_3 \end{Bmatrix} = \begin{Bmatrix} 2.616667 \\ 0 \\ 0 \end{Bmatrix} \tag{2.222}$$

(iv) Using the most recent $\{x\}$ (2.222), calculate x_2 using (2.218) and denote the new value of x_2 by x_2''.

$$x_2 = \frac{-19.3 - 0.1(2.616667) + 0}{7} = -2.794524 = x_2'' \tag{2.223}$$

(v) Update $\{x\}$ in (2.222) using x_2'' from (2.223).

$$\{x\} = \begin{Bmatrix} x_1' \\ x_2'' \\ \tilde{x}_3 \end{Bmatrix} = \begin{Bmatrix} 2.616667 \\ -2.794524 \\ 0 \end{Bmatrix} \tag{2.224}$$

(vi) Calculate x_3 using (2.219) and (2.224) and denote the new value of x_3 by x_3'''.

$$x_3 = \frac{1}{10}(71.4 - 0.3(2.616667) + 0.2(-2.7974524)) = 7.00561 = x_3'' \tag{2.225}$$

(vii) Update $\{x\}$ in (2.224) using the new value of x_3, i.e. x_3'''.

$$\{x\}_1 = \begin{Bmatrix} x_1' \\ x_2'' \\ x_3''' \end{Bmatrix} = \begin{Bmatrix} 2.616667 \\ -2.794524 \\ 7.00561 \end{Bmatrix} \tag{2.226}$$

Steps (i)-(vii) complete the first iteration. At the end of the first iteration, $\{x\}$ in (2.226) is the most recent estimate of the solution. We denote this by $\{x\}_1$.

Using (2.226) as the initial solution for the second iteration and repeating steps (ii)-(vii), the second iteration would yield the following new estimate of the solution $\{x\}$.

$$\begin{aligned} x_1' &= \frac{7.85 + 0.1(-2.794524) + 0.2(7.005610)}{3} = 2.990557 \\ x_2'' &= \frac{-19.3 - 0.1(2.990557) + 0.3(7.005610)}{7} = -2.499625 \\ x_3''' &= \frac{71.4 - 0.3(2.990557) + 0.2(-2.499625)}{10} = 7.000291 \end{aligned} \tag{2.227}$$

Thus at the end of the second iteration the solution of the vector $\{x\}$ is

$$\{x\}_2 = \begin{Bmatrix} 2.990557 \\ -2.499685 \\ 7.000291 \end{Bmatrix} \tag{2.228}$$

Using $\{x\}_1$ and $\{x\}_2$ in (2.226) and (2.228), we can compute $(\epsilon_i)_2$ (using (2.215)).

$$\begin{aligned} (\epsilon_1)_2 &= \left| \frac{2.990557 - 2.616667}{2.990557} \right| 100 = 12.5\% \\ (\epsilon_2)_2 &= \left| \frac{-2.499625 - (-2.794524)}{-2.4999625} \right| 100 = 11.8\% \\ (\epsilon_3)_2 &= \left| \frac{7.000291 - 7.00561}{7.000291} \right| 100 = 0.076\% \end{aligned} \tag{2.229}$$

More iterations can be performed to reduce $(\epsilon_i)_j$ to obtain desired threshold value Δ. We choose $\Delta = 10^{-7}$ and perform more iterations. When each component of $\{\epsilon\}_j$ for an iteration j becomes lower than 10^{-7}, we consider it to be zero. Details of additional iterations are given in the following.

Table 2.1: Results of Gauss-Seidel method for equations (2.216)

Δ =	0.10000E−07		
iter (j)	$\{x\}_{j-1}$	$\{x\}_j$	$\{\epsilon\}_j$
3	0.299055E+01	0.300003E+01	0.315845E+00
	-0.249962E+01	-0.249998E+01	0.145340E−01
	0.700029E+01	0.699999E+01	0.416891E−02
4	0.300003E+01	0.300000E+01	0.105698E−02
	-0.249998E+01	-0.250000E+01	0.486373E−03
	0.699995E+01	0.700000E+01	0.136239E−04
5	0.300000E+01	0.300000E+01	0.794728E−05
	-0.250000E+01	-0.250000E+01	0.000000E+00
	0.700000E+01	0.700000E+01	0.000000E+00
6	0.300000E+01	0.300000E+01	0.000000E+00
	-0.250000E+01	-0.250000E+01	0.000000E+00
	0.700000E+01	0.700000E+01	0.000000E+00

Thus, at the end of iteration 6, $\{x\}_6$ is the converged solution in which each component of $\{\epsilon\}_6 < 10^{-7}$.

Example 2.15 (Gauss-Seidel Method). Consider the following set of linear simultaneous algebraic equations.

$$\begin{aligned} 10x_1 + x_2 + 2x_3 + 3x_4 &= 10 \\ x_1 + 20x_2 + 2x_3 + 3x_4 &= 20 \\ 2x_1 + 2x_2 + 30x_3 + 4x_4 &= 30 \\ 3x_1 + 3x_2 + 4x_3 + 40x_4 &= 40 \end{aligned} \tag{2.230}$$

Choose $\{\tilde{x}\} = [0\ 0\ 0\ 0]^T$ as the initial guess solution vector and a convergence tolerance of $\Delta = 10^{-7}$ for each component of $\{\epsilon\}_j$, where j is the iteration number. The converged solution is obtained using Gauss-Seidel method in eight iterations. The calculated solution for each iteration is tabulated in the following.

Table 2.2: Results of Gauss-Seidel method for equations (2.230)

Δ =	0.10000E−07		
iter (j)	$\{x\}_{j-1}$	$\{x\}_j$	$\{\epsilon\}_j$
1	0.000000E+00	0.100000E+01	0.100000E+03
	0.000000E+00	0.949999E+00	0.100000E+03
	0.000000E+00	0.870000E+00	0.100000E+03
	0.000000E+00	0.766749E+00	0.100000E+03
2	0.100000E+01	0.500975E+00	0.996107E+02
	0.949999E+00	0.772938E+00	0.229075E+02
	0.870000E+00	0.812839E+00	0.703225E+01
	0.766749E+00	0.823172E+00	0.685428E+01
3	0.500975E+00	0.513186E+00	0.237954E+01
	0.772938E+00	0.769580E+00	0.436318E+00
	0.812839E+00	0.804725E+00	0.100820E+01
	0.823172E+00	0.823319E+00	0.178889E−01
4	0.513186E+00	0.515100E+00	0.371628E+00
	0.769580E+00	0.770274E+00	0.900328E−01
	0.804725E+00	0.804532E+00	0.240483E−01
	0.823319E+00	0.823143E+00	0.214119E−01
5	0.515100E+00	0.515123E+00	0.431586E−02
	0.770274E+00	0.770319E+00	0.579550E−02
	0.804532E+00	0.804551E+00	0.236328E−02
	0.823143E+00	0.823136E+00	0.839974E−03
6	0.515123E+00	0.515116E+00	0.120339E−02
	0.770319E+00	0.770318E+00	0.696389E−04
	0.804551E+00	0.804552E+00	0.170393E−03
	0.823136E+00	0.823137E+00	0.434469E−04
7	0.515116E+00	0.515116E+00	0.694266E−04
	0.770318E+00	0.770318E+00	0.232129E−04
	0.804552E+00	0.804552E+00	0.000000E+00
	0.823137E+00	0.823137E+00	0.724115E−05
8	0.515116E+00	0.515116E+00	0.000000E+00
	0.770318E+00	0.770318E+00	0.000000E+00
	0.804552E+00	0.804552E+00	0.000000E+00
	0.823137E+00	0.823137E+00	0.000000E+00

Thus, at the end of iteration 8, $\{x\}_8$ is the converged solution in which each component of $\{\epsilon\}_8 < 10^{-7}$.

Remarks.

(1) In Gauss-Seidel method, we begin with a starting or assumed solution vector and obtain new values of the components of $\{x\}$ individually. The new computed vector $\{x\}$ is used as the starting vector for the next iteration.

(2) We observe that the coefficient matrix $[A]$ in (2.216), if well-conditioned, generally has the largest elements on the diagonal, i.e., it is a diagonally dominant coefficient matrix. Iterative method have good convergence characteristics for algebraic systems with such coefficient matrices.

(3) The choice of starting vector is crucial. Sometimes the physics from which the algebraic equations are derived provides enough information to prudently select a starting vector. When this information is not available or helpful, null or unity vectors are often useful as initial guess solutions.

2.7.2 Jacobi Method

This method is similar to Gauss-Seidel method, but differs from it in the sense that here we do not continuously update each component of the solution vector $\{x\}$, but rather update all components of $\{x\}$ simultaneously. We consider details in the following.

$$[A]\{x\} = \{b\} \tag{2.231}$$

As in Gauss-Seidel method, here also we solve for $x_1, x_2, \dots, x_n$ using the first, second, $\dots$, n^{th} equations in (2.231).

$$\begin{aligned}
x_1 &= {}^{b_1}/_{a_{11}} - \left(x_1 + {}^{a_{12}}/_{a_{11}}x_2 + {}^{a_{13}}/_{a_{11}}x_3 + \dots + {}^{a_{1n}}/_{a_{11}}x_n\right) + x_1 \\
x_2 &= {}^{b_2}/_{a_{22}} - \left({}^{a_{21}}/_{a_{22}}x_1 + x_2 + {}^{a_{23}}/_{a_{22}}x_3 + \dots + {}^{a_{2n}}/_{a_{22}}x_n\right) + x_2 \\
&\vdots \\
x_n &= {}^{b_n}/_{a_{nn}} - \left({}^{a_{n1}}/_{a_{nn}}x_1 + {}^{a_{n2}}/_{a_{nn}}x_2 + {}^{a_{n3}}/_{a_{nn}}x_3 + \dots + x_n\right) + x_n
\end{aligned} \tag{2.232}$$

In (2.232), we have also added and subtracted $x_1, x_2, \dots, x_n$ in the first, second, $\dots$, n^{th} equations so that right-hand side of each equation in (2.232) contains the complete vector $\{x\}$. Equations (2.232) can be written as:

$$\{x\} = \{\hat{b}\} - [\hat{A}]\{x\} + \{x\}$$

or

$$\{x\} = \{\hat{b}\} - [\hat{A}]\{x\} + [I]\{x\}$$

or

$$\{x\} = \{\hat{b}\} - ([\hat{A}] - [I])\{x\} \tag{2.233}$$

It is more convenient to write (2.233) in the following form for performing iterations.

$$\{x\}_{j+1} = \{\hat{b}\} - ([\hat{A}] - [I])\{x\}_j \tag{2.234}$$

$\{x\}_{j+1}$ is the most recent estimate of $\{x\}$ and $\{x\}_j$ is the immediately preceding estimate of $\{x\}$.

(i) Assume a guess or initial vector $\{x\}_1$ for $\{x\}$ (i.e., $j = 1$ in (2.234)). This could be a null vector, a unit vector, or any other appropriate choice.

(ii) Use (2.234) to solve for $\{x\}_2$. $\{x\}_2$ is the improved estimate of the solution.

(iii) Check for convergence using the criterion defined in the next section or using the same criterion as in the case of Gauss-Seidel method, (2.215). We repeat steps (ii)-(iii) if the most recent estimate of $\{x\}$ is not converged.

2.7.2.1 Condition for Convergence of Jacobi Method

In this section we examine the conditions under which the Jacobi method will converge. Consider:

$$\{x\}_{j+1} = \{\hat{b}\} - ([\hat{A}] - [I])\{x\}_j \tag{2.235}$$

or

$$\{x\}_{j+1} = \{\hat{b}\} - [\hat{B}]\{x\}_j \tag{2.236}$$

For $j = 1$:

$$\{x\}_2 = \{\hat{b}\} - [\hat{B}]\{x\}_1 \tag{2.237}$$

For $j = 2$:

$$\{x\}_3 = \{\hat{b}\} - [\hat{B}]\{x\}_2 \tag{2.238}$$

Substitute $\{x\}_2$ from (2.237) into (2.238).

$$\{x\}_3 = \{\hat{b}\} - [\hat{B}](\{\hat{b}\} - [\hat{B}]\{x\}_1) \tag{2.239}$$

$$= ([I] - [\hat{B}])\{\hat{b}\} + [\hat{B}][\hat{B}]\{x\}_1 \tag{2.240}$$

For $j = 3$:

$$\{x\}_4 = \{\hat{b}\} - [\hat{B}]\{x\}_3 \tag{2.241}$$

Substituting for $\{x\}_3$ from (2.240) in (2.241) and rearranging terms:

$$\{x\}_4 = ([I] - [\hat{B}] + [\hat{B}][\hat{B}])\{\hat{b}\} - [\hat{B}][\hat{B}][\hat{B}]\{x\}_1 \tag{2.242}$$

By induction we can write the following for $i = n$:

$$\{x\}_{n+1} = ([I]+[\hat{B}][\hat{B}]+\cdots+\underbrace{([\hat{B}][\hat{B}]\cdots[\hat{B}])}_{\text{Product of } (n-1)\ [\hat{B}] \text{ matrices}})\{\hat{b}\}+\overbrace{[\hat{B}][\hat{B}]\cdots[\hat{B}]}^{\text{Product of } n\ [\hat{B}] \text{ matrices}}\{x\}_1 \tag{2.243}$$

If we consider the solution $\{x\}_{n+1}$ in the limit $n \to \infty$, then $\{x\}_{n+1}$ must converge to $\{x\}$.

$$\lim_{n\to\infty} \{x\}_{n+1} = \{x\} \tag{2.244}$$

must hold, and (2.244) must be independent of $\{x\}_1$,the starting or initial guess solution. Hence,

$$\lim_{n\to\infty} \underbrace{[\hat{B}][\hat{B}]\cdots[\hat{B}]}_{\text{Product of } n\ [\hat{B}] \text{ matrices}}\{x\}_1 = \{0\} \tag{2.245}$$

for an arbitrary $\{x\}_1$.

Remarks.

(1) The condition in (2.245) is the convergence criterion for Jacobi method. When (2.245) is satisfied, the Jacobi method is assured to converge.

(2) It can be shown that the condition (2.244) is satisfied by the coefficient matrix $[A]$ in $[A]\{x\} = \{b\}$ when $[A]$ is diagonally dominant, i.e., when the elements in the diagonal of $[A]$ are larger in magnitude than the off-diagonal elements.

(3) We can state the convergence criterion (2) in a more concrete form.

   (a) **Row criterion:** Based on each row of $[A]$:

$$\sum_{\substack{j=1\\ j\neq i}}^{n} \left|\frac{a_{ij}}{a_{ii}}\right| \leq 1 \quad ; \quad i = 1, 2, \ldots, n \tag{2.246}$$

   (b) **Column criterion:** Based on each column of $[A]$:

$$\sum_{\substack{i=1\\ i\neq j}}^{n} \left|\frac{a_{ij}}{a_{ii}}\right| \leq 1 \quad ; \quad j = 1, 2, \ldots, n \tag{2.247}$$

   (c) **Normalized off diagonal elements of $[A]$:**

$$\sum_{i=1}^{n}\sum_{\substack{j=1\\ j\neq i}}^{n} \left(\frac{a_{ij}}{a_{ii}}\right)^2 \leq 1 \tag{2.248}$$

Example 2.16 (Jacobi Method). Consider the following set of linear simultaneous algebraic equations:

$$\begin{aligned} 3x_1 - 0.1x_2 - 0.2x_3 &= 7.85 \\ 0.1x_1 + 7x_2 - 0.3x_3 &= -19.3 \\ 0.3x_1 - 0.2x_2 + 10x_3 &= 71.4 \end{aligned} \tag{2.249}$$

Solve for x_1, x_2, and x_3 using the first, second, and third equations in (2.249). Add and subtract x_1, x_2, and x_3 to the right-hand side of the first, second, and third equations, respectively to obtain the appropriate forms of the equations.

$$x_1 = {}^{7.85}/_{3.0} - ({x_1} - {}^{0.1}/_{3.0}x_2 - {}^{0.2}/_{3.0}x_3) + x_1 \tag{2.250}$$

$$x_2 = {}^{-19.3}/_{7.0} - ({}^{0.1}/_{7.0}x_1 + x_2 - {}^{0.3}/_{7.0}x_3) + x_2 \tag{2.251}$$

$$x_3 = {}^{71.4}/_{10.0} - ({}^{0.3}/_{10.0}x_1 - {}^{0.2}/_{10.0}x_2 + x_3) + x_3 \tag{2.252}$$

These equations can be expressed in the iteration form given by (2.234):

$$\{x\}_{j+1} = \{\hat{b}\} - ([\hat{A}] - [I])\{x\}_j \tag{2.253}$$

where

$$[\hat{A}] = \begin{bmatrix} 1.0 & -{}^{0.1}/_{3.0} & -{}^{0.2}/_{3.0} \\ {}^{0.1}/_{7.0} & 1.0 & -{}^{0.3}/_{7.0} \\ {}^{0.3}/_{10.0} & -{}^{0.2}/_{10.0} & 1.0 \end{bmatrix} \; ; \quad \{\hat{b}\} = \begin{Bmatrix} {}^{7.85}/_{3.0} \\ -{}^{19.3}/_{7.0} \\ {}^{71.4}/_{10.0} \end{Bmatrix} \tag{2.254}$$

We choose $\{\tilde{x}\} = [0\ 0\ 0]^T$ as the initial guess solution vector and a convergence tolerance of $\Delta = 10^{-7}$ for each component of the error $\{\epsilon\}$ for each iteration. The converged solution is obtained using Jacobi method in eight iterations. The calculated solution for each iteration is tabulated in the following.

Table 2.3: Results of Jacobi method for equations (2.249)

$\Delta =$ 0.10000E−07

iter (j)	$\{x\}_{j-1}$	$\{x\}_j$	$\{\epsilon\}_j$
1	0.000000E+00	0.261666E+01	0.100000E+03
	0.000000E+00	-0.275714E+01	0.100000E+03
	0.000000E+01	0.714000E+01	0.100000E+03
2	0.261666E+01	0.300076E+01	0.127999E+02
	-0.275714E+01	-0.248852E+01	0.107943E+02
	0.714000E+01	0.700635E+01	0.190744E+01
3	0.300076E+01	0.300080E+01	0.148574E−02
	-0.248852E+01	-0.249973E+01	0.448626E+00
	0.700635E+01	0.700020E+01	0.878648E−01
4	0.300080E+01	0.300002E+01	0.261304E−01
	-0.249973E+01	-0.250000E+01	0.105762E−01
	0.700020E+01	0.699998E+01	0.322206E−02

Table 2.3: Results of Jacobi method for equations (2.249)

5	0.300002E+01	0.299999E+01	0.794728E−03
	-0.250000E+01	-0.250000E+01	0.667571E−04
	0.699998E+01	0.699999E+01	0.258854E−03
6	0.299999E+01	0.299999E+01	0.397364E−04
	-0.250000E+01	-0.250000E+01	0.381469E−04
	0.699999E+01	0.700000E+01	0.136239E−04
7	0.299999E+01	0.300000E+01	0.794728E−05
	-0.250000E+01	-0.250000E+01	0.000000E+00
	0.700000E+01	0.700000E+01	0.000000E+00
8	0.300000E+01	0.300000E+01	0.000000E+00
	-0.250000E+01	-0.250000E+01	0.000000E+00
	0.700000E+01	0.700000E+01	0.000000E+00

Thus, at the end of iteration 8, $\{x\}_8$ is the converged solution in which each component of $\{\epsilon\}_8 < 10^{-7}$.

Example 2.17 (Jacobi Method). Consider the following set of linear simultaneous algebraic equations.

$$\begin{aligned} 10x_1 + x_2 + 2x_3 + 3x_4 &= 10 \\ x_1 + 20x_2 + 2x_3 + 3x_4 &= 20 \\ 2x_1 + 2x_2 + 30x_3 + 4x_4 &= 30 \\ 3x_1 + 3x_2 + 4x_3 + 40x_4 &= 40 \end{aligned} \tag{2.255}$$

Choose $\{\tilde{x}\} = [0\;0\;0\;0]^T$ as the initial guess solution vector and a convergence tolerance of $\Delta = 10^{-7}$ for each component of $\{\epsilon\}_j$, where j is the iteration number. The converged solution is obtained using Jacobi method in seventeen iterations. The calculated solution for each iteration is tabulated in the following.

Table 2.4: Results of Jacobi method for equations (2.255)

Δ =	0.10000E−07		
iter (j)	$\{x\}_{j-1}$	$\{x\}_j$	$\{\epsilon\}_j$
1	0.000000E+00	0.100000E+01	0.100000E+03
	0.000000E+00	0.100000E+01	0.100000E+03
	0.000000E+00	0.100000E+01	0.100000E+03
	0.000000E+00	0.100000E+01	0.100000E+03
2	0.100000E+01	0.399999E+00	0.150000E+03
	0.100000E+01	0.699999E+00	0.428571E+02
	0.100000E+01	0.733333E+00	0.363636E+02
	0.100000E+01	0.750000E+00	0.333333E+02
3	0.399999E+00	0.558333E+00	0.283582E+02
	0.699999E+00	0.794166E+00	0.118572E+02
	0.733333E+00	0.826666E+00	0.112903E+02
	0.750000E+00	0.844166E+00	0.111549E+02
4	0.558333E+00	0.501999E+00	0.112217E+02
	0.794166E+00	0.762791E+00	0.411317E+01
	0.826666E+00	0.797277E+00	0.368615E+01
	0.844166E+00	0.815895E+00	0.346499E+01

Table 2.4: Results of Jacobi method for equations (2.255)

5	0.501999E+00	0.519496E+00	0.336797E+01
	0.762791E+00	0.772787E+00	0.129352E+01
	0.797277E+00	0.806894E+00	0.119181E+01
	0.815895E+00	0.825412E+00	0.346499E+01
6	0.519496E+00	0.513718E+00	0.112475E+01
	0.772787E+00	0.769523E+00	0.424167E+00
	0.806894E+00	0.803792E+00	0.385891E+00
	0.825412E+00	0.822389E+00	0.367663E+00
7	0.513718E+00	0.515572E+00	0.359577E+00
	0.769523E+00	0.770576E+00	0.136601E+00
	0.803792E+00	0.804798E+00	0.124993E+00
	0.822389E+00	0.823377E+00	0.120030E+00
8	0.515572E+00	0.514969E+00	0.117086E+00
	0.770576E+00	0.770234E+00	0.443416E−01
	0.804798E+00	0.804473E+00	0.404687E−01
	0.823377E+00	0.823058E+00	0.387076E−01
9	0.514969E+00	0.515164E+00	0.378224E−01
	0.770234E+00	0.770345E+00	0.143451E−01
	0.804473E+00	0.804578E+00	0.131050E−01
	0.823058E+00	0.823162E+00	0.125630E−01
10	0.515164E+00	0.515101E+00	0.122657E−01
	0.770345E+00	0.770309E+00	0.465038E−02
	0.804578E+00	0.804544E+00	0.423766E−02
	0.823162E+00	0.823128E+00	0.406232E−02
11	0.515101E+00	0.515121E+00	0.396884E−02
	0.770309E+00	0.770321E+00	0.150883E−02
	0.804544E+00	0.804555E+00	0.137055E−02
	0.823128E+00	0.823139E+00	0.131788E−02
12	0.515121E+00	0.515114E+00	0.128439E−02
	0.770321E+00	0.770317E+00	0.487473E−03
	0.804555E+00	0.804551E+00	0.444505E−03
	0.823139E+00	0.823136E+00	0.427228E−03
13	0.515114E+00	0.515116E+00	0.416559E−03
	0.770317E+00	0.770318E+00	0.154753E−03
	0.804551E+00	0.804552E+00	0.140759E−03
	0.823136E+00	0.823137E+00	0.137581E−03
14	0.515116E+00	0.515116E+00	0.127282E−03
	0.770318E+00	0.770318E+00	0.541636E−04
	0.804552E+00	0.804552E+00	0.444505E−04
	0.823137E+00	0.823137E+00	0.434469E−04
15	0.515116E+00	0.515116E+00	0.347132E−04
	0.770318E+00	0.770318E+00	0.154753E−04
	0.804552E+00	0.804552E+00	0.148168E−04
	0.823137E+00	0.823137E+00	0.144823E−04
16	0.515116E+00	0.515116E+00	0.115711E−04
	0.770318E+00	0.770318E+00	0.000000E+00
	0.804552E+00	0.804552E+00	0.000000E+00
	0.823137E+00	0.823137E+00	0.724115E−05
17	0.515116E+00	0.515116E+00	0.000000E+00
	0.770318E+00	0.770318E+00	0.000000E+00
	0.804552E+00	0.804552E+00	0.000000E+00
	0.823137E+00	0.823137E+00	0.000000E+00

Thus, at the end of iteration 17, $\{x\}_{17}$ is the converged solution in which each component of $\{\epsilon\}_{17} < 10^{-7}$.

Remarks.

(1) We note that the convergence characteristics of the Jacobi method for this example are much poorer than the Gauss-Seidel method in terms of the number of iterations.

(2) The observation in Remark (1) is not surprising due to the fact that in

the Gauss-Seidel method, computation of the ith component of $\{x\}$ for each iteration utilizes the most recent updated values of the components $x_1, x_2, \dots, x_{i-1}$ of $\{x\}$, whereas in the Jacobi method all components of $\{x\}$ are updated simultaneously for each iteration.

2.7.3 Relaxation Techniques

The purpose of relaxation techniques is to improve the convergence of the iterative methods of solving the system of linear simultaneous algebraic equations. Let $\{x\}_c$ and $\{x\}_p$ be the currently calculated and the immediately preceding values of the solution vector $\{x\}$. Using these two solution vectors, we construct a new solution vector $\{x\}_{new}$ as follows:

$$\{x\}_{new} = \lambda\{x\}_c + (1-\lambda)\{x\}_p \tag{2.256}$$

In (2.256), λ is a weight factor that is chosen between 0 and 2. Equation (2.256) represents $\{x\}_{new}$ as a weighted average of the two solutions $\{x\}_c$ and $\{x\}_p$. The new vector $\{x\}_{new}$ is used in the next iteration as opposed to $\{x\}_c$. This is continued until the converged solution is obtained.

1. When $\lambda = 1$ in (2.256), then:

$$\{x\}_{new} = \{x\}_c \tag{2.257}$$

   In this case we have the conventional iterative method.

2. If $0 < \lambda < 1$, the vectors $\{x\}_c$ and $\{x\}_p$ get multiplied with factors between 0 and 1. For this choice of λ, the method is called *under-relaxation.*

3. If $1 < \lambda < 2$, then the vector $\{x\}_c$ is multiplied with a factor greater than one and $\{x\}_p$ is multiplied with a factor that is less than zero. The motivation for this is that $\{x\}_c$ is supposedly more accurate than $\{x\}_p$, hence a bigger weight factor is appropriate to assign to $\{x\}_c$ compared to $\{x\}_p$. For this choice of λ, the method is called *successive* or *simultaneous over-relaxation* or SOR.

4. The newly formed vector $\{x\}_{new}$ from either under- or over-relaxation is used for the next iteration in both Gauss-Seidel or Jacobi methods instead of $\{x\}_c$.

5. The choice of λ is unfortunately problem dependent. This is a serious drawback of the method.

6. The relaxation methods are generally helpful in improving the convergence of the iterative methods.

2.8 Condition Number of the Coefficient Matrix

Consider a system of n linear simultaneous equations:

$$[A]\{x\} = \{b\} \tag{2.258}$$

in which $[A]$ is an $(n \times n)$ square matrix that is symmetric and positive-definite, hence its eigenvalues are real and greater than zero. Let λ_i; $i = 1, 2, \dots, n$ be the eigenvalues of $[a]$ arranged in ascending order, i.e., λ_1 is the smallest eigenvalue and λ_n is the greatest (see Chapter 4 for additional details about eigenvalues). The condition number c_n of the matrix $[A]$ is defined as:

$$c_n = {}^{\lambda_n}/_{\lambda_1} \tag{2.259}$$

When the value of c_n is closer to one, the matrix is better conditioned. For a *well-conditioned* matrix $[A]$, the coefficients of $[A]$ (especially diagonal elements) are all roughly of the same order of magnitude. In such cases, when computing $\{x\}$ from (2.258) using, for example, elimination methods, the round off errors are minimal during triangulation (as in Gauss elimination). Higher condition number c_n results in higher round off errors during the elimination process. In extreme cases for very high c_n, the computations may even fail.

With a poorly conditioned coefficient matrix $[A]$, the computed solution $\{x\}$ using (2.258) (if possible to compute) generally does not satisfy (2.258) due to round off errors in the triangulation process. Often, a high conditioning number c_n suggests a large disparity in the magnitudes of the elements of $[A]$, especially the diagonal elements.

2.9 Concluding Remarks

In this chapter, the basic elements of linear algebra are presented first as refresher material. This is followed by the methods of obtaining numerical solutions of a system of linear simultaneous algebraic equations. A list of the methods considered is given is Section 2.3. We remark that groups of methods (B), (C), and Cramer's rule (as listed in Section 2.3) are numerical methods without approximation, whereas graphical methods and methods (D) are methods of approximation. As far as possible, the methods without approximation are meritorious over the methods of approximation. Use of methods of approximation without precise quantification of the solution error is obviously dangerous.

The author's own experience in computational mathematics and finite element computations suggests the Gauss elimination method to be the most efficient and straightforward to program and the method of choice. When the coefficient matrix is symmetric and positive-definite, as is the case in many engineering applications, pivoting is not required. This leads to improved efficiency of computations. When the coefficient matrix is not positive-definite, partial pivoting is worthy of exploiting before using full pivoting as full pivoting is extremely computationally intensive for large systems of equations. Relaxation methods are quite popular in computational fluid dynamics (CFD) due to their simplicity of programming but more importantly due to the fact that coefficient matrices in CFD are rarely positive-definite. Since these are methods of approximation, the computed solutions are always in error.

Problems

2.1 Matrices $[A]$, $[B]$, $[C]$ and the vector $\{x\}$ are given in the following:

$$[A] = \begin{bmatrix} 1 & 2 & -1 & 3 \\ 0 & 4 & -2 & 1 \\ 3 & -1 & 1 & 1 \end{bmatrix} \quad ; \quad [B] = \begin{bmatrix} 0 & 1 & 0 \\ 1 & 2 & -1 \\ 1 & -1 & 3 \\ 2 & -2 & 1 \end{bmatrix}$$

$$[C] = \begin{bmatrix} 0 & 1 & 4 & 2 \\ -1 & -2 & 3 & 1 \\ 0 & 1 & 2 & -1 \end{bmatrix} \quad ; \quad \{x\} = \begin{Bmatrix} -1.2 \\ 3.7 \\ 2.0 \end{Bmatrix}$$

(a) Find $[A]+[B]$, $[A]+[C]$, $[A]-2[C]$

(b) Find $[A][B]$, $[B][A]$, $[A][C]$, $[B]\{x\}$, $[C]\{x\}$

(c) What is the size of $[A][B][C]$ and of $[B]\{x\}$

(d) Show that $([A][B])[C] = [A]([B][C])$

(e) Find $[A]^T$, $[B]^T$, $[C]^T$, and $\{x\}^T$

2.2 Consider the square matrix $[A]$ defined by

$$[A] = \begin{bmatrix} 2 & 2 & -2 \\ -2 & 0 & 2 \\ 2 & -2 & -2 \end{bmatrix}$$

(a) Find $[A]^2$, $[A]^3$

(b) Find $\det[A]$ or $|A|$

(c) Show that $[A][I] = [I][A] = [A]$ where $[I]$ is the (3x3) identity matrix

2.3 Write the following system of equations in the matrix form using x, y, z as vector of unknowns (in this order).

$$\begin{aligned} d_1 &= b_1 y + c_1 z \\ d_2 &= b_2 y + a_2 x \\ d_3 &= a_3 x + b_3 y \end{aligned}$$

Determine the transpose of the coefficient matrix.

2.4 Consider a square matrix $[A]$ given by

$$[A] = \begin{bmatrix} 1 & 3 & 1 \\ 2 & 0 & 1 \\ 0 & -1 & 4 \end{bmatrix}$$

Decompose the matrix $[A]$ as a sum of a symmetric and skew symmetric matrices.

2.5 Consider the following (3×3) matrix

$$[R] = \begin{bmatrix} \cos\theta & \sin\theta & 0 \\ -\sin\theta & \cos\theta & 0 \\ 0 & 0 & 1 \end{bmatrix}$$

in which θ is an angle in radians. Is the matrix $[R]$ orthogonal?

2.6 Consider the following system of equations.

$$\begin{aligned} 3x_1 - x_2 + 2x_3 &= 12 \\ x_1 + 2x_2 + 3x_3 &= 11 \\ 2x_1 - 2x_2 - x_3 &= 2 \end{aligned}$$

or symbolically $[A]\{x\} = \{b\}$.

(a) Obtain the solution $\{x\}$ using naive Gauss elimination

(b) Obtain the solution $\{x\}$ using Gauss-Jordan method. Also obtain the inverse of the coefficient matrix $[A]$ during the Gauss-Jordan elimination. Show that

$$\{x\} = [A]^{-1}\{b\}$$

(c) Obtain the solution $\{x\}$ using Cramer's rule.

(d) Perform Cholesky decomposition of $[A]$ into the product of $[L]$ and $[U]$. Obtain solution $\{x\}$ using Cholesky factors $[L]$ and $[U]$.

2.7 Consider the following system of equations

$$\begin{aligned} 4x_1 + x_2 - x_3 &= -2 \\ 5x_1 + x_2 + 2x_3 &= 4 \\ 6x_1 + x_2 + x_3 &= 6 \end{aligned}$$

(a) Obtain solution $\{x\}$ using Gauss elimination with partial pivoting.

(b) Obtain solution $\{x\}$ using Gauss elimination with full pivoting.

2.8 Consider the following system of equations

$$\begin{aligned} 2x_1 + x_2 - x_3 &= 1 \\ 5x_1 + 2x_2 + 2x_3 &= -4 \\ 3x_1 + x_2 + x_3 &= 5 \end{aligned}$$

(a) Calculate solution $\{x\}$ using Gauss-Jordan method with partial pivoting.

(b) Calculate solution $\{x\}$ using Gauss-Jordan method with full pivoting.

2.9 Consider the following system of equations in which the coefficient matrix $[A]$ is symmetric.

$$\begin{aligned} 2x_1 - x_2 &= 1.5 \\ -x_1 + 2x_2 - x_3 &= -0.25 \\ -x_2 + x_3 &= -0.25 \end{aligned}$$

i,e $[A]\{x\} = \{b\}$

(a) Perform the decomposition

$$[A] = [\tilde{L}][\tilde{L}]^T$$

using Cholesky factors of $[A]$.

(b) Obtain solution $\{x\}$ using $[\tilde{L}][\tilde{L}]^T$ decomposition of $[A]$ in $[A]\{x\} = \{b\}$.

2.10 Write a computer program to solve a system of linear simultaneous algebraic equations using *Gauss-Seidel method*

2.11 Write a computer program to solve a system of linear simultaneous algebraic equations using *Jacobi method*

In both 2.10 and 2.11 consider the following:

For each iteration tabulate the starting solution, computed solution, and percentage error in each component of the computed solution using the calculated solution as the improved solution. Allow maximum of 20 iterations and use a convergence tolerance of 0.1×10^{-6} for the percentage error in each component of the solution vector.

Tabulate starting solution, calculated solution and percentage error as three columns for each iteration. Provide a printed heading showing the iteration number.

Use the following two systems of equations to compute numerical values of the solution using programs in 2.10 and 2.11.

(a)

$$\begin{bmatrix} 3.0 & -0.1 & -0.2 \\ 0.1 & 7.0 & -0.3 \\ 0.3 & -0.2 & 10.0 \end{bmatrix} \begin{Bmatrix} x_1 \\ x_2 \\ x_3 \end{Bmatrix} = \begin{Bmatrix} 7.85 \\ -19.3 \\ 71.4 \end{Bmatrix} \quad ; \quad \text{use } \{x\} = \begin{Bmatrix} 0 \\ 0 \\ 0 \end{Bmatrix} \text{ as initial or starting solution}$$

(b)

$$\begin{bmatrix} 10 & 1 & 2 & 3 \\ 1 & 20 & 2 & 3 \\ 2 & 2 & 30 & 4 \\ 3 & 3 & 4 & 40 \end{bmatrix} \begin{Bmatrix} x_1 \\ x_2 \\ x_3 \\ x_4 \end{Bmatrix} = \begin{Bmatrix} 10 \\ 20 \\ 30 \\ 40 \end{Bmatrix} \quad ; \quad \text{use } \{x\} = \begin{Bmatrix} 0 \\ 0 \\ 0 \\ 0 \end{Bmatrix} \text{ as initial or starting solution}$$

Provide a listing of both computer programs. Document your program with appropriate comments. Provide printed copies of the solutions for (a) and (b) using programs in 2.10 and 2.11. Print all results up to *ten* decimal places.

Based on the numerical studies for (a) and (b) comment on the performance of Gauss-Seidel method and Jacobi method.

2.12 Consider the following system of simultaneous linear algebraic equations

$$\begin{bmatrix} 1 & 2 & 1 \\ 3 & 4 & 0 \\ 2 & 10 & 4 \end{bmatrix} \begin{Bmatrix} x_1 \\ x_2 \\ x_3 \end{Bmatrix} = \begin{Bmatrix} b_1 \\ b_2 \\ b_3 \end{Bmatrix}$$

(a) Use Gauss elimination *only once* to obtain the solutions x_1, x_2, x_3 for

$$\begin{Bmatrix} b_1 \\ b_2 \\ b_3 \end{Bmatrix} = \begin{Bmatrix} 3 \\ 3 \\ 10 \end{Bmatrix} \quad \text{and} \quad \begin{Bmatrix} 1 \\ 1 \\ 3 \end{Bmatrix}$$

2.13 Consider the following system of linear simultaneous algebraic equations

$$\begin{bmatrix} 0.5 & -0.5 & 0 \\ -0.5 & 1 & -0.5 \\ 0 & -0.5 & 0.5 \end{bmatrix} \begin{Bmatrix} x_1 \\ x_2 \\ x_3 \end{Bmatrix} = \begin{Bmatrix} 1 \\ 4 \\ 8 \end{Bmatrix}$$

show whether this system of equations *has a unique solution or not* without computing the solution.

2.14 Consider the following system of equations

$$6x + 4y = 4$$
$$4x + 5y = 1$$

find solution for x, y using Crammer's rule

2.15 Consider the following matrix

$$[A] = \begin{bmatrix} 2 & 4 \\ 6 & 8 \end{bmatrix}$$

(a) Find inverse of $[A]$ using Gauss-Jordan method.
(b) Calculate co-factor matrix of $[A]$.
(c) Perform $[L][U]$ decomposition of $[A]$ where $[L]$ is a unit lower triangular matrix and $[U]$ is upper triangular matrix.
(d) If $[A]\{x\} = \{b\}$, then find $\{x\}$ for $\{b\}^T = [b_1, b_2] = [1, 2]$.

2.16 Obtain solution of the following system of equations using Cholesky decomposition.

$$\begin{bmatrix} 1 & -1 & -1 \\ -1 & 2 & -1 \\ -1 & -1 & 3 \end{bmatrix} \begin{Bmatrix} x_1 \\ x_2 \\ x_3 \end{Bmatrix} = \begin{Bmatrix} -4 \\ 0 \\ 6 \end{Bmatrix}$$

2.17 Consider the following system of linear simultaneous algebraic equations

$$x_1 + 2x_2 = 1$$
$$3x_1 + bx_2 = 2$$

Find condition on b for which the solution exists and is unique.

2.18 Consider the following symmetric matrix $[A]$

$$[A] = \begin{bmatrix} 4 & 2 \\ 2 & 5 \end{bmatrix}$$

The $[L][U]$ decomposition of $[A]$ is given by

$$[A] = [L][U] = \begin{bmatrix} 1 & 0 \\ 0.5 & 1 \end{bmatrix} \begin{bmatrix} 4 & 2 \\ 0 & 4 \end{bmatrix}$$

Using $[L][U]$ decomposition given above, derive the matrix $[\tilde{L}]$, a lower triangular matrix such that, $[A] = [\tilde{L}][\tilde{L}]^T$ where $[\tilde{L}]^T$ is the transpose of $[\tilde{L}]$.

2.19 Consider the following system of equations

$$\begin{aligned} x - y &= 1 \\ -x + 2y &= 2 \end{aligned} \quad \text{or} \quad [A]\begin{Bmatrix} x \\ y \end{Bmatrix} = \begin{Bmatrix} 1 \\ 2 \end{Bmatrix}$$

(a) Find solution x, y and the inverse of A using Gauss-Jordan method.
(b) Find adjoint of $[A]$ where $[A]$ is defined above.

3

Solutions of Nonlinear Simultaneous Equations

3.1 Introduction

Nonlinear simultaneous equations are nonlinear expressions in unknown quantities of interest. These may arise in some physical processes directly due to consideration of the mathematical descriptions of their physics. On the other hand, in many physical processes described by nonlinear differential or partial differential equations, the use of approximation methods such as finite difference, finite volume, and finite element methods for obtaining their approximate numerical solutions naturally results in nonlinear simultaneous equations. Solutions of these nonlinear simultaneous equations provides the solutions of the associated nonlinear differential and partial differential equations.

In this chapter we consider systems of nonlinear simultaneous equations and methods of obtaining their solutions. Consider a system of n nonlinear simultaneous equations:

$$f_i(x_1, x_2, \dots, x_n) = b_i \ ; \qquad i = 1, 2, \dots, n \tag{3.1}$$

In (3.1) some or all $f_i(x_j)$ are nonlinear functions of some or all x_j ; $i, j = 1, 2, \dots, n$. As in the case of linear simultaneous algebraic equations, here also we cannot obtain a solution for any x_j independently of the remaining, i.e., we must solve for all x_j ; $j = 1, 2, \dots, n$ together. Since (3.1) are nonlinear, we must employ iterative methods for obtaining their solution in which we choose a guess or starting solution and improve it iteratively until two successively computed solutions are within a desired tolerance. Thus, generally all methods of obtaining solutions of nonlinear algebraic equations are approximate, i.e., in these methods we only obtain an approximation of the true solution $\{x\}$. Hence, it is appropriate to refer to these methods as *methods of approximation.*

The simplest form of (3.1) is a single nonlinear equation in a single variable.

$$f(x) = f_1(x_1) - b_1 = 0 \tag{3.2}$$

The nonlinear relationship is defined by the function $f_1(\cdot)$ or $f(\cdot)$. Unlike linear equations, nonlinear equations may have multiple solutions. In many cases the methods of obtaining the solutions of (3.1) are extensions or generalizations of the methods employed for (3.2). We first study various numerical methods of approximating the solutions of a single nonlinear equation $f(x) = 0$ in a single independent variable x given by (3.2). *When we have a single nonlinear equation, the values of x that satisfy the equation $f(x) = 0$ are called the roots of $f(x)$. Thus, the methods of obtaining solutions of $f(x) = 0$ are often called root-finding methods.* We consider these in the following section for the nonlinear equation (3.2).

3.2 Root-Finding Methods for (3.2)

Consider $f(x) = 0$ in (3.2), a nonlinear function of x.

(i) If $f(x)$ is a polynomial in x (a linear combination of monomials in x), then for up to third degree polynomials we can solve for the values of x directly using explicit expressions. In this case the solutions are exact. If $f(x)$ is a polynomial in x of degree higher than three, then we must employ numerical methods that are iterative to solve for the roots of $f(x)$.

(ii) If $f(x)$ is not a polynomial in x, then in general we must also use iterative numerical methods to find roots of $f(x)$.

Different Methods of Finding Roots of $f(x)$

In the following sections we consider various methods of finding the roots of $f(x)$; $a \leq x \leq b$, i.e., we find the roots of $f(x)$ that lie in the range $x \in [a, b]$.

(a) Graphical method

(b) Incremental search method

(c) Bisection method or method of half interval

(d) Method of false position

(e) Newton's methods

   (i) Newton-Raphson method, Newton's linear method, or method of tangents

   (ii) Newton's second order method

(f) Secant method

(g) Fixed point iteration method or basic iteration method

3.2.1 Graphical Method

Consider:

$$f(x) = 0 \; ; \quad \forall x \in [a, b] \tag{3.3}$$

We plot a graph of $f(x)$ as a function of x for values of x between a and b. Let Figure 3.1 be a typical graph of $f(x)$ versus x.

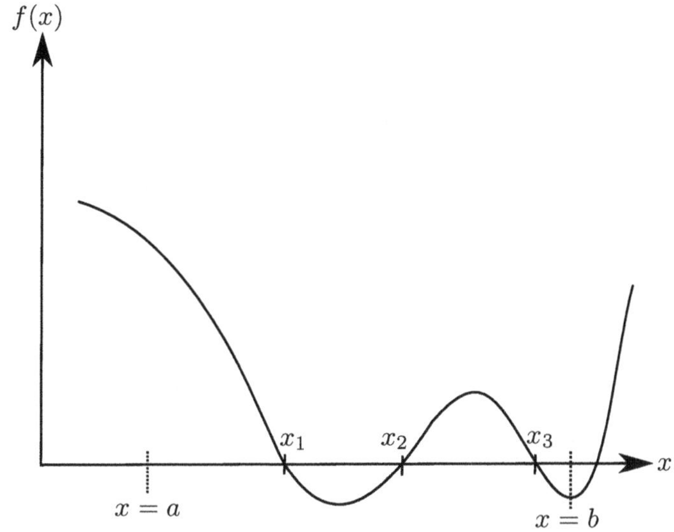

Figure 3.1: Graph of $f(x)$ versus x

From Figure 3.1, we note that $f(x)$ is zero for $x = x_1$, $x = x_2$, and $x = x_3$ $\forall x \in [a, b]$. Thus, x_1, x_2, and x_3 are roots of $f(x)$ in the range $[a, b]$ of x. Figure 3.2 shows exploded view of the behavior of $f(x) = 0$ in the neighborhood of $x = x_1$.

If we choose a value of x slightly less than $x = x_1$, say $x = x_l$, then $f(x_l) > 0$, and if we choose a value of x slightly greather than $x = x_1$, say $x = x_u$, then $f(x_u) < 0$, hence:

$$f(x_l)f(x_u) < 0 \tag{3.4}$$

We note that for the root x_1, this condition holds in the immediate neighborhood of $x = x_1$ as long as $x_l < x_1$ and $x_u > x_1$. For the second root x_2 we have:

$$\begin{aligned} f(x_l) < 0 \; , \; f(x_u) > 0 \\ f(x_l)f(x_u) < 0 \text{ for } x_l < x_2 \; , \; x_u > x_2 \end{aligned} \tag{3.5}$$

For the third root x_3:

$$\begin{aligned} f(x_l) > 0 \; , \; f(x_u) < 0 \\ f(x_l)f(x_u) < 0 \end{aligned} \tag{3.6}$$

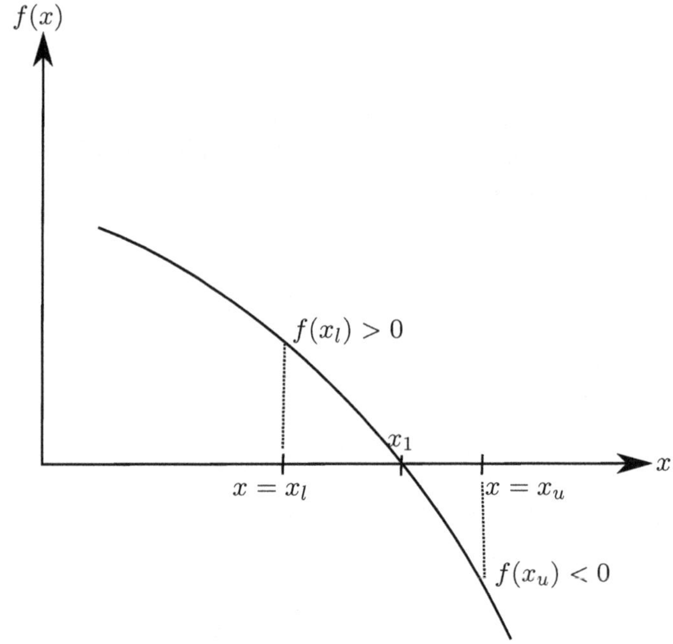

Figure 3.2: Enlarged view of $f(x)$ versus x in the neighborhood of $x = x_1$

Thus, we note that regardless of which root we consider, the condition

$$f(x_l)f(x_u) < 0 \quad ; \quad x_l < x_i \quad x_u > x_i \tag{3.7}$$

in which x_i in the root of $f(x)$, holds for each root. Thus, the condition (3.7) is helpful in the root-finding methods considered in the following sections.

In the graphical method, we simply plot a graph of $f(x)$ versus x and locate values of x for which $f(x) = 0$ in the range $[a, b]$. These of course are the approximate values of the desired roots within the limits of the graphical precision.

3.2.2 Incremental Search Method

Consider

$$f(x) = 0 \quad \forall x \in [a, b] \tag{3.8}$$

In this method we begin with $x = a$ (lower value of x), increment it by Δx, i.e., $x_i = a + i\Delta x$, and find the function values corresponding to each value x_i of x. Let

$$x_i\,,\; f(x_i) \quad ; \quad i = 1, 2, \dots, n \tag{3.9}$$

be the values of x and $f(x)$ at various points between $[a, b]$. Then,

$$f(x_i)f(x_{i+1}) < 0 \quad ; \quad i = 1, 2, \dots, n-1 \tag{3.10}$$

indicates a root between x_i and x_{i+1}. We try progressively reduced values of Δx to ensure that all roots in the range $[a, b]$ are bracketed using (3.10), i.e., no roots are missed.

3.2.2.1 More Accurate Value of a Root

Let

$$f(x_l)f(x_u) < 0 \tag{3.11}$$

hold for values of $x = x_l$ and $x = x_u$ $(x_u > x_l)$. x_l and x_u are typical values of x determined for a root using incremental search. A more accurate value of the root in $[x_l, x_u]$ can be determined by using this range and a smaller Δx (say $\Delta x = {}^{\Delta x}/_{10}$) and by repeating the incremental search. This will yield a yet narrower range x for the root. This process can be continued for progressively reduced values of Δx until the root is determined with desired accuracy.

Remarks.

(1) This method is quite effective in bracketing the roots.

(2) The method is rather 'brute force' and inefficient in determining roots with higher accuracy.

Example 3.1 (Bracketing Roots of (3.12): Incremental Search).

$$f(x) = x^3 + 2.3x^2 - 5.08x - 7.04 = 0 \tag{3.12}$$

Equation (3.12) is a cubic polynomial in x and therefore has three roots. We consider (3.12) and bracket its roots using incremental search. Figure 3.3 shows a graph of $f(x)$ versus x for $x \in [-4, 4]$. From the graph, we note that the roots of $f(x)$ are approximately located at $x = -3, -1, 2$.

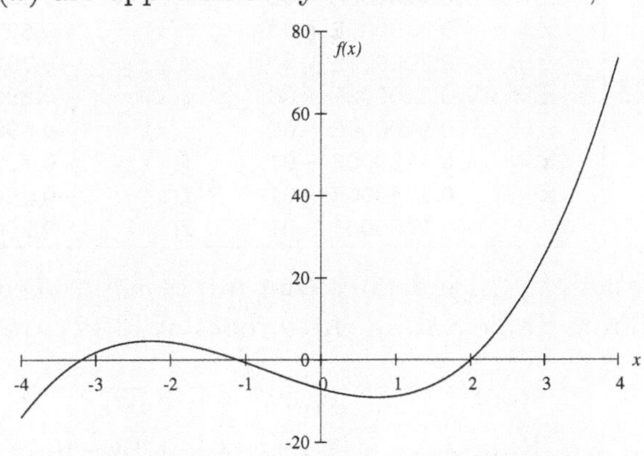

Figure 3.3: Plot of $f(x)$ in (3.12) versus x

We consider incremental search in the range $[x_{min}, x_{max}] = [-4, 4]$ with

$\Delta x = 0.41$ to bracket the roots of $f(x) = 0$ given by (3.12). Let $x_1 = x_{min}$. Calculate $f(x_1)$. Consider

$$x_{i+1} = x_i + \Delta x \ ; \quad i = 1, 2, \dots, n \quad \forall x_{i+1} \in [x_{min}, x_{max}] \tag{3.13}$$

for each i value in (3.13), calculate $f(x_{i+1})$. Using two successive values $f(x_i)$ and $f(x_{i+1})$ of $f(x)$, consider:

$$f(x_i)f(x_{i+1}) \begin{cases} < 0 & \Longrightarrow \text{ a root in } [x_i, x_{i+1}] \\ > 0 & \Longrightarrow \text{ no root in } [x_i, x_{i+1}] \end{cases} \tag{3.14}$$

Details of the calculations are given in the following (Table 3.1).

Table 3.1: Results of incremental search for Example 3.1

Δx =	0.41000E + 00		
x_{min} =	−0.40000E + 01		
x_{max} =	0.40000E + 01		
x =	−0.400000E + 01	f(x) =	−0.139200E + 02
x =	−0.359000E + 01	f(x) =	−0.542845E + 01
x =	−0.318000E + 01	f(x) =	0.215487E + 00
x =	−0.318000E + 01	f(x) =	0.215487E + 00
x =	−0.277000E + 01	f(x) =	0.342534E + 01
x =	−0.236000E + 01	f(x) =	0.461462E + 01
x =	−0.195000E + 01	f(x) =	0.419687E + 01
x =	−0.154000E + 01	f(x) =	0.258562E + 01
x =	−0.113000E + 01	f(x) =	0.194373E + 00
x =	−0.720000E + 00	f(x) =	−0.256333E + 01
x =	−0.720000E + 00	f(x) =	−0.256333E + 01
x =	−0.310000E + 00	f(x) =	−0.527396E + 01
x =	0.100000E + 00	f(x) =	−0.752400E + 01
x =	0.510000E + 00	f(x) =	−0.889992E + 01
x =	0.920000E + 00	f(x) =	−0.898819E + 01
x =	0.133000E + 01	f(x) =	−0.737529E + 01
x =	0.174000E + 01	f(x) =	−0.364770E + 01
x =	0.215000E + 01	f(x) =	0.260812E + 01

A change in sign of $f(x)$ indicates that a root is between the corresponding values of x. From Table 3.1, all three roots of (3.12) are bracketed.

$$\begin{aligned} &\text{Root 1} \quad : \quad [x_l, x_u] = [-3.59, -3.18] \\ &\text{Root 2} \quad : \quad [x_l, x_u] = [-1.13, -0.72] \\ &\text{Root 3} \quad : \quad [x_l, x_u] = [1.74, 2.15] \end{aligned} \tag{3.15}$$

Remarks.

(1) An incremental search with $\Delta x = {}^{\Delta x}/_{2} = 0.205$ undoubtedly will also bracket all three roots, but it will result in smaller range for each root.

(2) A value of Δx larger than 0.41 can be used too, but in this case Δx may be too large, hence we may miss one or more roots.

(3) For each range of the roots in (3.15), we can perform incremental search with progressively reduced Δx to eventually obtain an accurate value of each root. This approach to obtaining accurate values of each root is obviously rather inefficient.

3.2.3 Bisection Method or Method of Half-Interval

When a root has been bracketed using incremental search method, the bisection method, also known as the method of half-interval, can be used more effectively to obtain a more accurate value of the root than continuing incremental search with reduced Δx.

(i) Let $[x_l,x_u]$ be the interval containing a root, then

$$f(x_l)f(x_u) < 0 \tag{3.16}$$

(ii) Divide the interval $[x_l, x_u]$ into two equal intervals, $[x_l, x_k]$ and $[x_k, x_u]$, in which $x_k = {}^{1}/_{2}(x_l + x_u)$.

(iii) Compute $f(x_k)$ and check the products

$$f(x_l)f(x_k) \text{ and } f(x_k)f(x_u) \tag{3.17}$$

for a negative sign. As an example consider the graph of $f(x)$ in Figure 3.4. From Figure 3.4 we note that

$$\begin{aligned} f(x_l)f(x_k) &< 0 \\ f(x_k)f(x_u) &> 0 \end{aligned} \tag{3.18}$$

Since the root lies in $[x_l, x_k]$, the interval $[x_k, x_u]$ can be discarded.

(iv) We now reinitialize the x values, i.e., keep $x = x_l$ the same but set $x_u = x_k$ to create a new, smaller range of $[x_l, x_u]$.

(v) Check if $x_u - x_l < \Delta$, a preset tolerance of accuracy (approximate relative error). If not then repeat steps (ii)-(v). If x_l and x_u are within

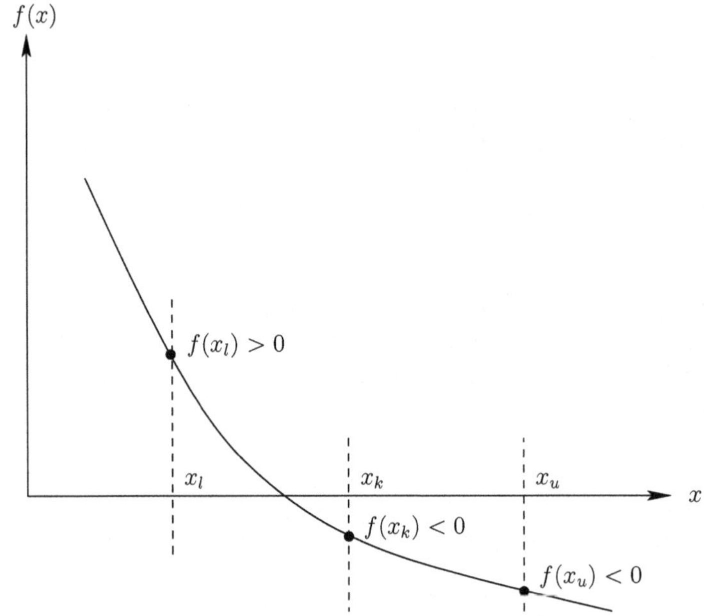

Figure 3.4: Range of a root with half intervals.

the tolerance, then we could use either of x_u, x_l as the desired value of the root.

Remarks.

(1) In each pass we reduce the interval containing the root by half, hence the name half-interval method or bisection method.

(2) This method is more efficient than the incremental search method of finding more accurate value of the root, but is still quite brute force.

Example 3.2 (Bisection or Half-Interval Method). In this example we also consider $f(x) = 0$ given by (3.12) and consider the brackets containing the roots (3.15). We apply bisection method for each root bracket to obtain more accurate values. In this method $[x_l, x_u]$ already contains a root. Consider:

$$x_k = \frac{x_l + x_u}{2} \tag{3.19}$$

and calculate the products:

$$\begin{aligned}
f(x_l)f(x_k) &\begin{cases} < 0 \implies & \text{a root in the range } [x_l, x_k] \\ > 0 \implies & \text{no root in the range } [x_l, x_k] \end{cases} \\
f(x_k)f(x_u) &\begin{cases} < 0 \implies & \text{a root in the range } [x_k, x_u] \\ > 0 \implies & \text{no root in the range } [x_k, x_u] \end{cases}
\end{aligned} \tag{3.20}$$

We discard the range x that does not contain the root and reinitialize the other half-interval to $[x_l, x_u]$. We repeat steps in (3.19) and (3.20) until:

$$\% \text{ Error} = \left| \frac{x_u - x_l}{x_u} \right| \times 100 \leq \Delta, \text{ a preset tolerance} \tag{3.21}$$

We choose a tolerance of $\Delta = 0.0001$ and set the maximum number of iterations $I = 20$. The computations for the three roots are shown in the following.

Root 1: $[x_l, x_u] = [-3.59, -3.18]$, from Example 3.1

Table 3.2: Results of bisection method for the first root of equation (3.12)

x_l =	-0.35900E+01
x_u =	-0.31800E+01
Δ =	0.10000E−03
I =	20

iter	x_l	x_k	x_u	$f(x_l)$	$f(x_k)$	$f(x_u)$	error (%)
1	-0.35900E+01	-0.33850E+01	-0.31800E+01	-0.54284E+01	-0.22764E+01	0.21549E+00	
2	-0.33850E+01	-0.32825E+01	-0.31800E+01	-0.22764E+01	-0.95115E+00	0.21549E+00	0.31226E+01
3	-0.32825E+01	-0.32313E+01	-0.31800E+01	-0.95115E+00	-0.34841E+00	0.21549E+00	0.15861E+01
4	-0.32313E+01	-0.32056E+01	-0.31800E+01	-0.34841E+00	-0.61657E−01	0.21549E+00	0.79938E+00
5	-0.32056E+01	-0.31928E+01	-0.31800E+01	-0.61657E−01	-0.78109E−01	0.21549E+00	0.40129E+00
6	-0.32056E+01	-0.31992E+01	-0.31928E+01	-0.61657E−01	-0.85261E−02	0.78109E−01	0.20024E+00
7	-0.32056E+01	-0.32024E+01	-0.31992E+01	-0.61657E−01	-0.26491E−01	0.85261E−02	0.10002E+00
8	-0.32024E+01	-0.32008E+01	-0.31992E+01	-0.26491E−01	-0.89636E−02	0.85261E−02	0.50036E−01
9	-0.32008E+01	-0.32000E+01	-0.31992E+01	-0.89636E−02	-0.21471E−03	0.85261E−02	0.25023E−01
10	-0.32000E+01	-0.31996E+01	-0.31992E+01	-0.21471E−03	-0.41569E−02	0.85261E−02	0.12515E−01
11	-0.32000E+01	-0.31998E+01	-0.31996E+01	-0.21471E−03	-0.19707E−02	0.41556E−02	0.62588E−02
12	-0.32000E+01	-0.31999E+01	-0.31998E+01	-0.21471E−03	-0.87743E−03	0.19694E−02	0.31293E−02
13	-0.32000E+01	-0.32000E+01	-0.31999E+01	-0.21471E−03	-0.33073E−03	0.87743E−03	0.15646E−02
14	-0.32000E+01	-0.32000E+01	-0.32000E+01	-0.21471E−03	-0.57364E−04	0.33073E−03	0.78231E−03
15	-0.32000E+01	-0.32000E+01	-0.32000E+01	-0.21471E−03	-0.78020E−04	0.57364E−04	0.38743E−03
16	-0.32000E+01	-0.32000E+01	-0.32000E+01	-0.78020E−04	-0.90256E−05	0.57364E−04	0.19744E−03
17	-0.32000E+01	-0.32000E+01	-0.32000E+01	-0.90256E−05	-0.24820E−04	0.57364E−04	0.96858E−04

$\therefore \quad x = -3.2$ is a root of $f(x)$ defined by (3.12) in the range $[x_l, x_u] = [-3.59, 3.18]$.

Root 2: $[x_l, x_u] = [-1.13, -0.72]$, from Example 3.1

Table 3.3: Results of bisection method for the second root of equation (3.12)

x_l =	-0.11300E+01
x_u =	-0.72000E+00
Δ =	0.10000E−03
I =	20

iter	x_l	x_k	x_u	$f(x_l)$	$f(x_k)$	$f(x_u)$	error (%)
1	-0.11300E+01	-0.92500E+00	-0.72000E+00	0.19437E+00	-0.11645E+01	-0.25633E+01	
2	-0.11300E+01	-0.10275E+01	-0.92500E+00	0.19437E+00	-0.47658E+00	-0.11645E+01	0.99757E+01
3	-0.11300E+01	-0.10788E+01	-0.10275E+01	0.19437E+00	-0.13878E+00	-0.47685E+00	0.47509E+01
4	-0.11300E+01	-0.11044E+01	-0.10788E+01	0.19437E+00	0.28462E−01	-0.13878E+00	0.23203E+01
5	-0.11044E+01	-0.10916E+01	-0.10788E+01	0.28462E−01	-0.54999E−01	-0.13878E+00	0.11738E+01
6	-0.11044E+01	-0.10980E+01	-0.10916E+01	0.28462E−01	-0.13228E−01	-0.54999E−01	0.58346E+00
7	-0.11044E+01	-0.11012E+01	-0.10980E+01	0.28462E−01	0.76277E−02	-0.13228E−01	0.29089E+00
8	-0.11012E+01	-0.10996E+01	-0.10980E+01	0.76277E−02	-0.27970E−02	-0.13228E−01	0.14565E+00
9	-0.11012E+01	-0.11004E+01	-0.10996E+01	0.76277E−02	0.24162E−02	-0.27970E−02	0.72774E−01
10	-0.11004E+01	-0.11000E+01	-0.10996E+01	0.24162E−02	-0.19047E−03	-0.27970E−02	0.36403E−01
11	-0.11004E+01	-0.11002E+01	-0.11000E+01	0.24162E−02	0.11129E−02	-0.19047E−03	0.18198E−01
12	-0.11002E+01	-0.11001E+01	-0.11000E+01	0.11129E−02	0.46141E−03	-0.19047E−03	0.90973E−02
13	-0.11001E+01	-0.11000E+01	-0.11000E+01	0.46141E−03	0.13586E−03	-0.19047E−03	0.45461E−02
14	-0.11000E+01	-0.11000E+01	-0.11000E+01	0.13586E−03	-0.27110E−04	-0.19047E−03	0.22758E−02
15	-0.11000E+01	-0.11000E+01	-0.11000E+01	0.13586E−03	0.54375E−04	-0.27110E−04	0.11379E−02
16	-0.11000E+01	-0.11000E+01	-0.11000E+01	0.54375E−04	0.13633E−04	-0.27110E−04	0.56895E−03
17	-0.11000E+01	-0.11000E+01	-0.11000E+01	0.13633E−04	-0.69327E−05	-0.27110E−04	0.28719E−03
18	-0.11000E+01	-0.11000E+01	-0.11000E+01	0.13633E−04	0.31559E−05	-0.69327E−05	0.14088E−03
19	-0.11000E+01	-0.11000E+01	-0.11000E+01	0.31559E−05	-0.18884E−05	-0.69327E−05	0.70442E−04

$\therefore$ $x = -1.1$ is a root of $f(x)$ defined by (3.12) in the range $[x_l, x_u] = [-1.13, -0.72]$.

Root 3: $[x_l, x_u] = [1.74, 2.15]$, from Example 3.1

Table 3.4: Results of bisection method for the third root of equation (3.12)

x_l =	0.17400E+01
x_u =	0.21500E+01
Δ =	0.10000E−03
I =	20

iter	x_l	x_k	x_u	$f(x_l)$	$f(x_k)$	$f(x_u)$	error (%)
1	0.17400E+01	0.19450E+01	0.21500E+01	-0.36477E+01	-0.86166E+00	0.26081E+01	
2	0.19450E+01	0.20475E+01	0.21500E+01	-0.86166E+00	0.78454E+00	0.26081E+01	0.50061E+01
3	0.19450E+01	0.19963E+01	0.20475E+01	-0.86166E+00	-0.60332E-01	0.78454E+00	0.25673E+01
4	0.19963E+01	0.20219E+01	0.20475E+01	-0.60332E−01	0.35661E+00	0.78454E+00	0.12674E+01
5	0.19963E+01	0.20091E+01	0.20219E+01	-0.60332E−01	0.14677E+00	0.35661E+00	0.63773E+00
6	0.19963E+01	0.20027E+01	0.20091E+01	-0.60332E−01	0.42881E−01	0.14677E+00	0.31988E+00
7	0.19963E+01	0.19995E+01	0.20027E+01	-0.60332E−01	-0.88102E−02	0.42881E−01	0.16020E+00
8	0.19995E+01	0.20011E+01	0.20027E+01	-0.88102E−02	0.17014E−01	0.42881E−01	0.80037E−01
9	0.19995E+01	0.20003E+01	0.20011E+01	-0.88102E−02	0.40956E−02	0.17014E−01	0.40037E−01
10	0.19995E+01	0.19999E+01	0.20003E+01	-0.88102E−02	-0.23577E−02	0.40956E−02	0.20017E−01
11	0.19999E+01	0.20001E+01	0.20003E+01	-0.23577E−02	0.86957E−03	0.40956E−02	0.10010E−01
12	0.19999E+01	0.20000E+01	0.20001E+01	-0.23577E−02	-0.74462E−03	0.86957E−03	0.50069E−02
13	0.20000E+01	0.20000E+01	0.20001E+01	-0.74462E−03	0.61493E−04	0.86957E−03	0.25004E−02
14	0.20000E+01	0.20000E+01	0.20000E+01	-0.74462E−03	-0.34205E−03	0.61493E−04	0.12517E−02
15	0.20000E+01	0.20000E+01	0.20000E+01	-0.34205E−03	-0.14028E−03	0.61493E−04	0.62585E−03
16	0.20000E+01	0.20000E+01	0.20000E+01	-0.14028E−03	-0.39394E−04	0.61493E−04	0.31292E−03
17	0.20000E+01	0.20000E+01	0.20000E+01	-0.39394E−04	0.11530E−04	0.61493E−04	0.15795E−03
18	0.20000E+01	0.20000E+01	0.20000E+01	-0.39394E−04	-0.13452E−04	0.11530E−04	0.77486E−04

$\therefore$ $x = 2.0$ is a root of $f(x)$ defined by (3.12) in the range $[x_l, x_u] = [1.74, 2.15]$.

All roots have been determined within $\Delta = 0.0001$ within twenty iterations.

3.2.4 Method of False Position

Let $[x_l, x_u]$ be the range of x containing a root of $f(x)$.

$$f(x_l)f(x_u) < 0 \tag{3.22}$$

(i) Connect the points $(x_l, f(x_l))$ and $(x_u, f(x_u))$ by a straight line (see Figure 3.5) and define the intersection of this line with the x-axis as $x = x_r$. Using the equation of the straight line, similar triangles, or

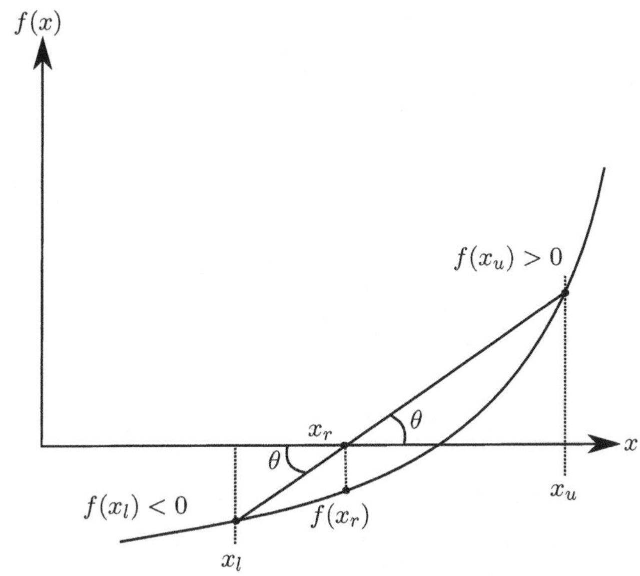

Figure 3.5: Method of false position

simply equating $\tan(\theta)$ from the two triangles shown in Figure 3.5, we obtain:

$$\frac{f(x_l)}{(x_l - x_r)} = \frac{f(x_u)}{(x_u - x_r)} \tag{3.23}$$

Solving for x_r:

$$x_r = \frac{x_u f(x_l) - x_l f(x_u)}{f(x_l) - f(x_u)} \tag{3.24}$$

Equation (3.24) is known as the false position formula. An alternate form of (3.24) can be obtained. From (3.24):

$$x_r = \frac{x_u f(x_l)}{f(x_l) - f(x_u)} - \frac{x_l f(x_u)}{f(x_l) - f(x_u)} \tag{3.25}$$

Add and subtract x_u to the right-hand side of (3.25).

$$x_r = x_u + \frac{x_u f(x_l)}{f(x_l) - f(x_u)} - \frac{x_l f(x_u)}{f(x_l) - f(x_u)} - x_u \tag{3.26}$$

Combine the last three terms on the right side of (3.26).

$$x_r = x_u - \left(\frac{f(x_u)(x_u - x_l)}{f(x_u) - f(x_l)} \right) \tag{3.27}$$

Equation (3.27) is an alternate form of (3.24). In (3.27), x_r is obtained by subtracting the quantity in the bracket on the right-hand side of (3.27) from x_u.

(ii) Calculate $f(x_r)$ and form the products

$$\begin{aligned} &f(x_l)f(x_r) \\ &f(x_r)f(x_u) \end{aligned} \tag{3.28}$$

Check which is less than zero. From Figure 3.5 we note that

$$\begin{aligned} f(x_l)f(x_u) &> 0 \\ f(x_r)f(x_u) &< 0 \end{aligned} \tag{3.29}$$

Hence, the root lies in the interval $[x_r, x_u]$ (for the function shown in Figure 3.5). Therefore, we discard the interval $[x_l, x_r]$.

(iii) Reinitialize the range containing the interval.

$$\begin{aligned} x_l &= x_r \\ \text{and} \qquad x_u &= x_u \qquad \text{unchanged} \end{aligned}$$

(iv) In this method x_r is the new estimate of the root. We check the convergence of the method using the following (approximate percentage relative error):

$$\left| \frac{(x_r)_{i+1} - (x_r)_i}{(x_r)_{i+1}} \right| \times 100 < \Delta \tag{3.30}$$

in which $(x_r)_i$ is the estimate of the root in the ith iteration. When converged, i.e., when (3.30) is satisfied, $(x_r)_{i+1}$ is the final value of the root.

Example 3.3 (False Position Method). In this method, once a root has been bracketed we use the following to obtain an estimate of x_r of the root in the bracketed range:

$$x_r = x_u - \left(\frac{f(x_u)(x_u - x_l)}{f(x_u) - f(x_l)} \right) \tag{3.31}$$

Then, we consider the products $f(x_l)f(x_r)$ and $f(x_r)f(x_u)$ to determine the range containing the root. We discard the range not containing the root and reinitialize the range containing the root to $[x_l, x_u]$. We iterate (3.31) and use the steps following it until:

$$\% \text{ Error} = \left| \frac{(x_r)_{i+1} - (x_r)_i}{(x_r)_{i+1}} \right| \times 100 \leq \Delta \tag{3.32}$$

We consider $f(x) = 0$ defined by (3.12) and the bracketed ranges of the roots determined in Example 3.1 to present details of the false position method for each root. We choose $\Delta = 0.0001$ and maximum of twenty iterations ($I = 20$).

Root 1: $[x_l, x_u] = [-3.59, -3.18]$, from Example 3.1

Table 3.5: Results of false position method for the first root of equation (3.12)

$x_l =$ -0.35900E+01
$x_u =$ -0.31800E+01
$\Delta =$ 0.10000E−03
$I =$ 20

iter	x_l	x_r	x_u	$f(x_l)$	$f(x_r)$	$f(x_u)$	error (%)
1	-0.35900E+01	-0.31957E+01	-0.31800E+01	-0.54284E+01	0.47320E−01	0.21549E+00	
2	-0.35900E+01	-0.31991E+01	-0.31957E+01	-0.54284E+01	0.10238E−01	0.47320E−01	0.10653E+00
3	-0.35900E+01	-0.31998E+01	-0.31991E+01	-0.54284E+01	0.22077E−02	0.10238E−01	0.23000E−01
4	-0.35900E+01	-0.32000E+01	-0.31998E+01	-0.54284E+01	0.47603E−03	0.22077E−02	0.49566E−02
5	-0.35900E+01	-0.32000E+01	-0.32000E+01	-0.54284E+01	0.10240E−03	0.47603E−03	0.10693E−02
6	-0.35900E+01	-0.32000E+01	-0.32000E+01	-0.54284E+01	0.22177E−04	0.10240E−03	0.22957E−03
7	-0.35900E+01	-0.32000E+01	-0.32000E+01	-0.54284E+01	0.47870E−05	0.22177E−04	0.49766E−04

$\therefore$ $x = -3.2$ is a root of $f(x)$ defined by (3.12) in the range $[x_l, x_u] = [-3.59, 3.18]$.

Root 2: $[x_l, x_u] = [-1.13, -0.72]$, from Example 3.1

Table 3.6: Results of false position method for the second root of equation (3.12)

$x_l =$ -0.11700E+01
$x_u =$ -0.72000E+00
$\Delta =$ 0.10000E−03
$I =$ 20

iter	x_l	x_r	x_u	$f(x_l)$	$f(x_r)$	$f(x_u)$	error (%)
1	-0.11700E+01	-0.11027E+01	-0.72000E+00	0.45046E+00	0.17833E−01	-0.25633E+01	
2	-0.11027E+01	-0.11011E+01	-0.72000E+00	0.17833E−01	0.62602E−03	-0.25633E+01	0.24037E+00
3	-0.11001E+01	-0.11000E+01	-0.72000E+00	0.62602E−03	0.21879E−04	-0.25633E+01	0.84366E−02
4	-0.11000E+01	-0.11000E+01	-0.72000E+00	0.21879E−04	0.76075E−06	-0.25633E+01	0.29491E−03
5	-0.11000E+01	-0.11000E+01	-0.72000E+00	0.76075E−06	0.28912E−07	-0.25633E+01	0.10220E−04

$\therefore \quad x = -1.1$ is a root of $f(x)$ defined by (3.12) in the range $[x_l, x_u] = [-1.13, -0.72]$.

Root 3: $[x_l, x_u] = [1.74, 2.15]$, from Example 3.1

Table 3.7: Results of false position method for the third root of equation (3.12)

$x_l =$ 0.17400E+01
$x_u =$ 0.21500E+01
$\Delta =$ 0.10000E−03
I = 20

iter	x_l	x_r	x_u	$f(x_l)$	$f(x_r)$	$f(x_u)$	error (%)
1	0.17400E+01	0.19791E+01	0.21500E+01	-0.36477E+01	-0.33382E+00	0.26081E+01	
2	0.19791E+01	0.19985E+01	0.21500E+01	-0.33382E+00	-0.24770E−01	0.26081E+01	0.97054E+00
3	0.19985E+01	0.19999E+01	0.21500E+01	-0.24770E−01	-0.18080E−02	0.26081E+01	0.71288E−01
4	0.19999E+01	0.20000E+01	0.21500E+01	-0.18080E−02	-0.13182E−03	0.26081E+01	0.51994E−02
5	0.20000E+01	0.20000E+01	0.21500E+01	-0.13182E−03	-0.96658E−05	0.26081E+01	0.37890E−03
6	0.20000E+01	0.20000E+01	0.21500E+01	-0.96658E−05	-0.70042E−06	0.26081E+01	0.27808E−04

$\therefore \quad x = 2.0$ is a root of $f(x)$ defined by (3.12) in the range $[x_l, x_u] = [1.74, 2.15]$.

All roots have been determined within the desired accuracy of $\Delta = 0.0001$ in less than eight iterations, compared to the bisection method which required close to twenty iterations for each root.

3.2.5 Newton-Raphson Method or Newton's Linear Method

This method can be used effectively to converge to a root with very high precision in an efficient manner. Let $[x_l, x_u]$ be the range of x containing a root of $f(x)$. The range $[x_l, x_u]$ can be obtained using the graphical method, incremental search method, or bisection method.

Let $x_i \in [x_l, x_u]$ be the first approximation of the root of $f(x)$ in the range $[x_l, x_u]$. Since x_i is an approximation of the root of $f(x)$, we have:

$$f(x_i) \neq 0 \tag{3.33}$$

Let Δx be a correction to x_i such that $x_{i+1} = x_i + \Delta x$ and

$$f(x_i + \Delta x) = 0 \tag{3.34}$$

If $f(x)$ is analytic (continuous and differentiable) in the neighborhood of x_i, then $f(x_i + \Delta x)$ can be expanded in a Taylor series about x_i.

$$f(x_i + \Delta x) = f(x_i) + \Delta x f'(x_i) + O((\Delta x)^2) = 0 \tag{3.35}$$

If we neglect $O((\Delta x)^2)$ in (3.35) (valid when $(\Delta x)^2 \ll \Delta x$) then (3.35) is approximate, and we can solve for Δx.

$$\Delta x = -\frac{f(x_i)}{f'(x_i)} \tag{3.36}$$

The improved value of the root x_{i+1} is given by,

$$x_{i+1} = x_i + \Delta x \tag{3.37}$$

Substituting for Δx in (3.37) from (3.36).

$$x_{i+1} = x_i - \frac{f(x_i)}{f'(x_i)} \tag{3.38}$$

Remarks.

(1) Since $f(x)$ is given, $f'(x)$ can be obtained.

(2) We use (3.38) for $i = 0, 1, \dots, n$, in which $x_0 \in [x_l, x_u]$ is the initial guess of the root, and iterate until the desired decimal place accuracy is achieved. $x_1, x_2, \dots$ are progressively improved values of the root of $f(x)$ in $[x_l, x_u]$.

3.2.5.1 Alternate Method of Deriving (3.38)

Consider the graph of $f(x)$ versus x shown in Figure 3.6 for $x \in [x_l, x_u]$. Let x_i be an assumed value of x between x_l and x_u. Compute $f(x_i)$ and draw a tangent to the $f(x)$ curve at $x = x_i$. Let x_{i+1} be the intersection of this tangent with the x-axis. Then the slope of the tangent to $f(x)$ at $x = x_i$

$$f'(x_i) = \left.\frac{df}{dx}\right|_{x=x_i} \tag{3.39}$$

can be approximated by:

$$f'(x_i) \approx \frac{f(x_i)}{(x_i - x_{i+1})} \tag{3.40}$$

Using the equality in (3.40) and solving for x_{i+1}:

$$x_{i+1} = x_i - \frac{f(x_i)}{f'(x_i)} \tag{3.41}$$

x_{i+1} is the improved value (i.e., more accurate) of the root of $f(x)$ in $[x_l, x_u]$. Clearly (3.41) is the same as (3.38), hence earlier remarks hold here as well.

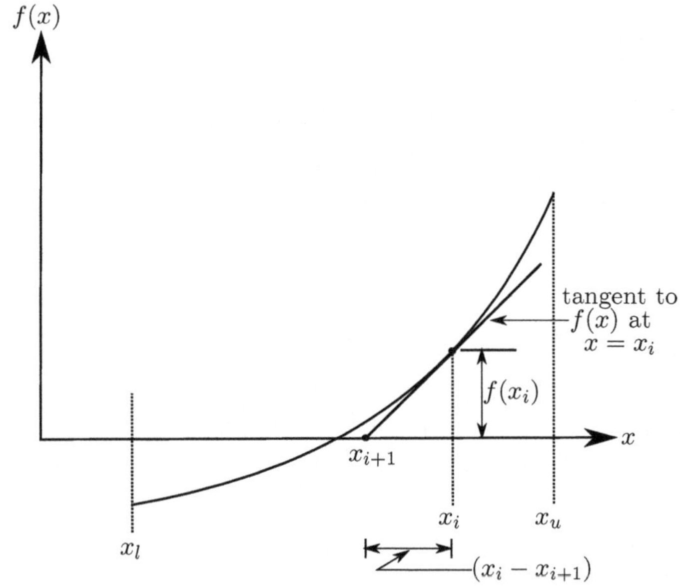

Figure 3.6: Newton Raphson or Newton's linear method

Convergence Criterion

The approximate percentage relative error serves as convergence criteria.

$$\left| \frac{x_{i+1} - x_i}{x_{i+1}} \right| \times 100 \leq \Delta \; ; \qquad \Delta \text{ is a preset value} \tag{3.42}$$

3.2.5.2 General Remarks Regarding Newton-Raphson Method

(1) The method requires a range $[x_l, x_u]$ that brackets the desired root and an initial guess $x_0 \in [x_l, x_u]$ of the root.

(2) The method works extremely well if $f'(x)$ and $f''(x)$ are well-behaved in the range $[x_l, x_u]$.

(3) The method fails if $f'(x_i)$ becomes zero, i.e., $f'(x)$ changes sign in the neighborhood of x_i.

(4) When the initial guess x_i is sufficiently close to the root of $f(x)$, the method has quadratic convergence (shown in a later section), hence only a few iterations are required to obtain a highly accurate value of the root.

(5) Since in this method we construct a tangent to $f(x)$ at $x = x_i$, this method is also called the *method of tangents*, or gradient method.

3.2.5.3 Error Analysis of Newton-Raphson Method

Let the range $[x_l, x_u]$ contain a root of $f(x)$. Let $x_i \in [x_l, x_u]$ be an approximation of the root of $f(x)$. Let Δx be a correction to x_i such that

$$f(x_i + \Delta x) = f(x_{i+1}) = 0 \tag{3.43}$$

where

$$x_{i+1} = x_i + \Delta x \tag{3.44}$$

Expand $f(x_i + \Delta x) = f(x_{i+1})$ in a Taylor series about x_i.

$$f(x_{i+1}) = f(x_i) + f'(x_i)\Delta x + f''(\xi)\frac{\Delta x^2}{2} = 0\ ; \quad \xi \in [x_l, x_u] \tag{3.45}$$

If we neglect $f''(\xi)$ term in (2.138), then we obtain Newton's linear method or Newton-Raphson method.

$$f(x_i) + f'(x_i)\Delta x = 0 \tag{3.46}$$

or

$$f(x_i) + f'(x_i)(x_{i+1} - x_i) = 0 \tag{3.47}$$

$$\therefore \quad x_{i+1} = x_i - \frac{f(x_i)}{f'(x_i)} \tag{3.48}$$

We can use Taylor series expansion to estimate the error in (3.48). We go back to the Taylor series expansion (3.45) and use $\Delta x = x_{i+1} - x_i$.

$$f(x_i) + f'(x_i)(x_{i+1} - x_i) + \frac{f''(\xi)}{2}(x_{i+1} - x_i)^2 = 0 \tag{3.49}$$

Since (3.49) is exact (in the sense that influence of all terms in the Taylor series expansion is accounted for in the last term), x_{i+1} in (3.49) must be the exact root or true root, say $x_{i+1} = x_t$. We substitute x_t for x_{i+1} in (3.49).

$$f(x_i) + f'(x_i)(x_t - x_i) + \frac{f''(\xi)}{2}(x_t - x_i)^2 = 0 \tag{3.50}$$

Subtract (3.47) from (3.50), noting that x_{i+1} in (3.47) is not x_t.

$$f'(x_i)(x_t - x_{i+1}) + \frac{f''(\xi)}{2}(x_t - x_i)^2 = 0 \tag{3.51}$$

Since x_t is the true solution:

$$\begin{aligned} x_t - x_{i+1} &= \varepsilon_{i+1} \quad &&; \text{total error at } x_{i+1} \\ x_t - x_i &= \varepsilon_i \quad &&; \text{total error at } x_i \end{aligned} \tag{3.52}$$

Hence we can write (3.51) as:

$$f'(x_i)\varepsilon_{i+1} + \frac{f''(\xi)}{2}(\varepsilon_i)^2 = 0 \tag{3.53}$$

$$\therefore \quad \varepsilon_{i+1} = \left(-\frac{f''(\xi)}{2f'(x_i)}\right)(\varepsilon_i)^2 \tag{3.54}$$

$$\therefore \quad \varepsilon_{i+1} \propto (\varepsilon_i)^2 \tag{3.55}$$

From (3.55) we conclude that the total error at x_{i+1} is proportional to the square of the total error at x_i. This aspect of the Newton's linear method is referred to as quadratic convergence of the method. In each iteration the total error reduces by two orders of magnitude when the computations are within the radius of convergence. For example, if $\varepsilon_i = O(10^{-2})$, then $\varepsilon_{i+1} = O(10^{-4})$, a reduction of two orders of magnitude.

Example 3.4 (Newton-Raphson or Newton's Linear Method). When a root has been bracketed, this method can be used to find a very accurate value of the root. In this method we use

$$x_{i+1} = x_i - \frac{f(x_i)}{f'(x_i)}\,; \quad i = 0, 1, \dots \tag{3.56}$$

in which x_0 is the initial guess of the root in the bracketed range. We consider $f(x) = 0$ in (3.12) and the bracketed ranges of the roots obtained in Example 3.1 to compute accurate values of each root using (3.56). Using (3.12):

$$\begin{aligned} f(x_i) &= (x_i)^3 + 2.3(x_i)^2 - 5.08x_i - 7.04 \\ f'(x_i) &= 3(x_i)^2 + 4.6x_i - 5.08 \end{aligned} \tag{3.57}$$

We choose $\Delta = 0.1 \times 10^{-4}$, approximate percentage relative error and set a limit of twenty iterations for each root ($I = 20$).

Root 1: $[x_l, x_u] = [-3.59, -3.18]$, from Example 3.1

We choose $x_0 = -3.4$ as the initial guess of the root.

Table 3.8: Results of Newton's linear method for the first root of equation (3.12)

$x_0 =$ -0.340000E+01
$\Delta =$ 0.100000E−04
I = 20

iter	x_i	x_{i+1}	$f(x_i)$	$f(x_{i+1})$	error (%)
1	-0.340000E+01	-0.322206E+01	-0.248400E+01	-0.244493E+00	0.552246E+01
2	-0.322206E+01	-0.320032E+01	-0.244493E+00	-0.347346E−02	0.679467E+00
3	-0.320032E+01	-0.319999E+01	-0.347346E−02	-0.259699E−06	0.993584E−02
4	-0.319999E+01	-0.319999E+01	-0.259699E−06	-0.112885E−05	0.743186E−06

$\therefore \quad x = -3.2$ is a root of $f(x)$ defined by (3.12) in the range $[x_l, x_u] = [-3.59, 3.18]$.

Root 2: $[x_l, x_u] = [-1.13, -0.72]$**, from Example 3.1**

We choose $x_0 = -0.8$ as the initial guess of the root.

Table 3.9: Results of Newton's linear method for the second root of equation (3.12)

x_0 = -0.800000E+00
Δ = 0.100000E−04
I = 20

iter	x_i	x_{i+1}	$f(x_i)$	$f(x_{i+1})$	error (%)
1	-0.800000E+00	-0.109474E+01	-0.201600E+01	-0.342907E−01	0.269231E+02
2	-0.109474E+01	-0.109999E+01	-0.342907E−01	-0.276394E−04	0.478089E+00
3	-0.109999E+01	-0.110000E+01	-0.276394E−04	-0.246789E−06	0.385971E−03
4	-0.110000E+01	-0.110000E+01	-0.246789E−06	-0.298526E−06	0.344629E−05

$\therefore \quad x = -1.1$ is a root of $f(x)$ defined by (3.12) in the range $[x_l, x_u] = [-1.13, -0.72]$.

Root 3: $[x_l, x_u] = [1.74, 2.15]$**, from Example 3.1**

We choose $x_0 = 1.8$ as the initial guess of the root.

Table 3.10: Results of Newton's linear method for the third root of equation (3.12)

x_0 = 0.1800000E+01
Δ = 0.100000E−04
I = 20

iter	x_i	x_{i+1}	$f(x_i)$	$f(x_{i+1})$	error (%)
1	0.180000E+01	0.202446E+01	-0.290000E+01	0.399246E+00	0.110873E+02
2	0.202446E+01	0.200030E+01	0.399246E+00	0.487113E−02	0.120762E+01
3	0.200030E+01	0.200000E+01	0.487113E−02	-0.140689E−06	0.151043E−01
4	0.200000E+01	0.200000E+01	-0.140689E−06	0.140689E−06	0.436379E−06

$\therefore \quad x = 2.0$ is a root of $f(x)$ defined by (3.12) in the range $[x_l, x_u] = [1.74, 2.15]$.

All roots have been determined within the desired accuracy of $\Delta = 0.0001$ in four iterations, compared to up to seven iterations for false position method and almost twenty iterations for the bisection method for each root.

3.2.6 Newton's Second Order Method

Let $[x_l, x_u]$ be a range of x containing a root of $f(x)$. Let $x_i \in [x_l, x_u]$ be the initial guess of the root:

$$f(x_i) \neq 0 \tag{3.58}$$

Let Δx be correction to x_i such that $x_{i+1} = x_i + \Delta x$ and

$$f(x_i + \Delta x) = 0 \tag{3.59}$$

If $f(x)$ is analytic (continuous and differentiable) in the neighborhood of x_i, then $f(x_i + \Delta x)$ can be expanded in a Taylor series about x_i.

$$f(x_i + \Delta x) = f(x_i) + \Delta x f'(x_i) + \frac{(\Delta x)^2}{2!} f''(x_i) + O((\Delta x)^3) = 0 \tag{3.60}$$

If we neglect $O((\Delta x)^3)$, then (3.60) can be written as

$$f(x_i + \Delta x) = f(x_i) + \Delta x f'(x_i) + \frac{(\Delta x)^2}{2!} f''(x_i) = 0 \tag{3.61}$$

In (3.61), $f(x_i)$, $f'(x_i)$, and $f''(x_i)$ are known and have numerical values. Δx is unknown, but in the following we treat it as a known increment in part of the expression, or as completely unknown and to be determined. We consider both cases in the following.

Case I: Treating Δx in (3.61) as unknown

Equation (3.61) is quadratic in Δx, hence there are two values of Δx that satisfy (3.61). Using the expression for the roots of a quadratic equation we can find Δx_1 and Δx_2. Using these two values of Δx, we define improved solutions as:

$$\begin{aligned} x_{i+1} &= x_i + \Delta x_1 \\ x_{i+1} &= x_i + \Delta x_2 \end{aligned} \tag{3.62}$$

Of the two values of x_{i+1} in (3.62), the value that lies in the range (x_l, x_u) is the correct improved or new value of the root. We check for convergence based on approximate percentage relative error given by:

$$\left| \frac{x_{i+1} - x_i}{x_i} \right| 100 \leq \Delta \tag{3.63}$$

If (3.63) is satisfied, then x_{i+1} is the desired value of the root, otherwise use x_{i+1} as the new initial or starting value instead of x_i and repeat the calculations, beginning with (3.61).

Case II: Treating Δx in (3.61) as known

In this case we do not solve for Δx, i.e., we do not calculate Δx_1 and Δx_2 (two values of Δx) using (3.61). Instead we proceed as follows.

Rewrite (3.61) as:

$$f(x_i) + \Delta x \left(f'(x_i) + \left(\frac{\Delta x}{2} \right) f''(x_i) \right) = 0 \tag{3.64}$$

We approximate the $(\Delta x/2)$ term in (3.64) using Newton's linear method, i.e., using equation (3.38) with $\Delta x = x_{i+1} - x_i$.

$$\left(\frac{\Delta x}{2} \right) = -\frac{1}{2} \left(\frac{f(x_i)}{f'(x_i)} \right) \tag{3.65}$$

Substitute for $(\Delta x/2)$ from (3.65) in (3.64):

$$f(x_i) + \Delta x \left(f'(x) - \frac{1}{2} \left(\frac{f(x_i)}{f'(x_i)} \right) f''(x_i) \right) = 0 \tag{3.66}$$

Using (3.66) solve for Δx:

$$\Delta x = -\frac{f(x_i)}{\left(f'(x_i) - \frac{f(x_i) f''(x_i)}{2f'(x_i)} \right)} \tag{3.67}$$

The improved value of the root is:

$$x_{i+1} = x_i + \Delta x \tag{3.68}$$

or

$$x_{i+1} = x_i - \frac{f(x_i)}{\left(f'(x_i) - \frac{f(x_i) f''(x_i)}{2f'(x_i)} \right)} \tag{3.69}$$

We check for convergence:

$$\left| \frac{x_{i+1} - x_i}{x_i} \right| 100 \leq \Delta \tag{3.70}$$

If (3.70) is satisfied then we have a converged value of the root, i.e., x_{i+1} is the desired root of $f(x)$ in the range $[x_l, x_u]$. If not, then using x_{i+1} as the new initial or guess value, we repeat the calculations using (3.69).

Remarks.

(1) Obviously Case I is more accurate than Case II as it requires no further approximation other than that used in the Taylor series expansion.

(2) Case I does require the solution of Δx, i.e., Δx_1 and Δx_2, using the expression for roots of a quadratic equation.

(3) Newton's second order method requires $f''(x)$ to be well-behaved in the neighborhood of x_i.

(4) As in the case of Newton's linear method, Newton's second order method (both Case I and Case II) also have good convergence characteristics as long as the initial or starting solution is in a sufficiently small neighborhood of the correct value of the root.

(5) The convergence rate of Case II is similar to Newton-Raphson method due to the introduction of approximation (3.65).

Example 3.5 (Newton's Second Order Method). When a root has been bracketed, this method can also be used to find a more accurate value of the root in the bracketed range. We consider the same $f(x) = 0$ as in (3.12) for Case I and Case II of Newton's second order method..

Case I:

In this approach we use

$$f(x_i) + f'(x_i) + \frac{(\Delta x)^2}{2} f''(x_i) = 0 \tag{3.71}$$

in which

$$\begin{aligned} f(x) &= x^3 + 2.3x^2 - 5.08x - 7.04 \\ f'(x) &= 3x^2 + 4.6x - 5.08 \\ f''(x) &= 6x + 4.6 \end{aligned} \tag{3.72}$$

and x_i is the current estimate of the root in the bracketed range. Choose $i = 0$, and hence x_0 as the initial guess, and calculate $f(x_i)$, $f'(x_i)$, and $f''(x_i)$ using (3.72). Using these values and (3.71) find two values of Δx, say Δx_1 and Δx_2, using the quadratic formula. Let

$$\begin{aligned} x_{i+1} &= x_i + \Delta x_1 \\ x_{i+1} &= x_i + \Delta x_2 \end{aligned} \quad \text{for } i = 1, 2, \dots \tag{3.73}$$

Choose the value of x_{i+1} that falls within $[x_l, x_u]$, the range that brackets the root. Increment $i = i + 1$ and repeat calculations beginning with (3.71). The convergence criterion is given as (approximate percentage relative error)

$$\left| \frac{x_{i+1} - x_i}{x_{i+1}} \right| \times 100 \leq \Delta \tag{3.74}$$

Root 1: $[x_l, x_u] = [-3.59, -3.18]$, from Example 3.1

We choose $x_0 = -3.4$ as the initial guess of the root.

Table 3.11: Results of Newton's second order method (Case I) for the first root of equation (3.12)

$x_0 =$ -0.340000E+01
$\Delta =$ 0.100000E−04
I = 10

iter	x_{i+1}	$f(x_i)$	$f'(x_i)$	$f''(x_i)$	error (%)
1	-0.319926E+01	-0.248400E+01	0.139600E+02	-0.158000E+02	0.627462E+01
2	-0.320000E+01	0.808915E−02	0.109091E+02	-0.149556E+02	0.231612E−01
3	-0.320000E+01	-0.121498E−05	0.109200E+02	-0.146000E+02	0.418457E−05

$\therefore \quad x = -3.2$ is a root of $f(x)$ defined by (3.12) in the range $[x_l, x_u] = [-3.59, 3.18]$.

Root 2: $[x_l, x_u] = [-1.13, -0.72]$, from Example 3.1

We choose $x_0 = -0.8$ as the initial guess of the root.

Table 3.12: Results of Newton's second order method (Case I) for the second root of equation (3.12)

$x_0 =$ -0.800000E+00
$\Delta =$ 0.100000E−04
I = 10

iter	x_{i+1}	$f(x_i)$	$f'(x_i)$	$f''(x_i)$	error (%)
1	-0.109602E+01	-0.201600E+01	-0.684000E+01	-0.200000E+00	0.270085E+02
2	-0.110000E+01	-0.259336E−01	-0.651791E+01	-0.197611E+01	0.361925E+00
3	-0.110000E+01	-0.517368E−07	-0.651000E+01	-0.200000E+01	0.866976E−06

$\therefore \quad x = -1.1$ is a root of $f(x)$ defined by (3.12) in the range $[x_l, x_u] = [-1.13, -0.72]$.

Root 3: $[x_l, x_u] = [1.74, 2.15]$, from Example 3.1

We choose $x_0 = 1.8$ as the initial guess of the root.

Table 3.13: Results of Newton's second order method (Case I) for the third root of equation (3.12)

$x_0 =$ 0.180000E+01
$\Delta =$ 0.100000E−04
I = 10

iter	x_{i+1}	$f(x_i)$	$f'(x_i)$	$f''(x_i)$	error (%)
1	0.200050E+01	-0.290000E+01	0.129200E+02	0.154000E+02	0.100225E+02
2	0.200000E+01	0.806149E−02	0.161283E+02	0.166030E+02	0.249988E−01
3	0.200000E+01	0.000000E+00	0.161200E+02	0.166000E+02	0.287251E−05

$\therefore \quad x = 2.0$ is a root of $f(x)$ defined by (3.12) in the range $[x_l, x_u] = [1.74, 2.15]$.

All roots have been determined within the desired accuracy of $\Delta = 0.0001$ in three iterations, compared to four iterations for Newton's linear method. Strictly in terms of iteration count, this is an improvement, but this method also requires additional calculation of two values of Δx at each iteration using the quadratic formula, and determination of which choice of Δx is appropriate. This may result in worse overall efficiency compared to Newton's linear method.

Case II

In this approach we approximate $\Delta x/2$ using Newton's linear method to obtain:

$$x_{i+1} = x_i - \frac{x_i}{\left(f'(x_i) - \frac{f(x_i)f''(x_i)}{2f'(x_i)}\right)} \quad ; \quad i = 0, 1, 2, \ldots \tag{3.75}$$

Root 1: $[x_l, x_u] = [-3.59, -3.18]$, from Example 3.1

We choose $x_0 = -3.4$ as the initial guess of the root.

Table 3.14: Results of Newton's second order method (Case II) for the first root of equation (3.12)

x_0 = -0.340000E+01
Δ = 0.100000E−04
I = 10

iter	x_{i+1}	$f(x_i)$	$f'(x_i)$	$f''(x_i)$	error (%)
1	-0.322455E+01	-0.248400E+01	0.139600E+02	-0.158000E+02	0.544108E+01
2	-0.320044E+01	-0.272500E+00	0.112802E+02	-0.147473E+02	0.753168E+00
3	-0.320000E+01	-0.486085E−02	0.109265E+02	-0.146027E+02	0.139016E−01
4	-0.320000E+01	-0.121498E−05	0.109200E+02	-0.146000E+02	0.347694E−05

$\therefore \quad x = -3.2$ is a root of $f(x)$ defined by (3.12) in the range $[x_l, x_u] = [-3.59, 3.18]$.

Root 2: $[x_l, x_u] = [-1.13, -0.72]$, from Example 3.1

We choose $x_0 = -0.8$ as the initial guess of the root.

Table 3.15: Results of Newton's second order method (Case II) for the second root of equation (3.12)

x_0 = -0.800000E+00
Δ = 0.100000E−04
I = 10

iter	x_{i+1}	$f(x_i)$	$f'(x_i)$	$f''(x_i)$	error (%)
1	-0.108251E+01	-0.201600E+01	-0.684000E+01	-0.200000E+00	0.260977E+02
2	-0.109991E+01	-0.114156E+00	-0.654406E+01	-0.189506E+01	0.158174E+01
3	-0.110000E+01	-0.595965E−03	-0.651018E+01	-0.199945E+01	0.832202E−02
4	-0.110000E+01	0.517368E−07	-0.651000E+01	-0.200000E+01	0.722480E−06

$\therefore$ $x = -1.1$ is a root of $f(x)$ defined by (3.12) in the range $[x_l, x_u] = [-1.13, -0.72]$.

Root 3: $[x_l, x_u] = [1.74, 2.15]$, from Example 3.1

We choose $x_0 = 1.8$ as the initial guess of the root.

Table 3.16: Results of Newton's second order method (Case II) for the third root of equation (3.12)

$x_0 =$ 0.180000E+01
$\Delta =$ 0.100000E−04
$I =$ 10

iter	x_{i+1}	$f(x_i)$	$f'(x_i)$	$f''(x_i)$	error (%)
1	0.202107E+01	-0.290000E+01	0.129200E+02	0.154000E+02	0.109383E+02
2	0.200020E+01	0.343354E+00	0.164711E+02	0.167264E+02	0.104352E+01
3	0.200000E+01	0.319411E−02	0.161233E+02	0.166012E+02	0.990540E−02
4	0.200000E+01	0.000000E+00	0.161200E+02	0.166000E+02	0.000000E+00

$\therefore$ $x = 2.0$ is a root of $f(x)$ defined by (3.12) in the range $[x_l, x_u] = [1.74, 2.15]$.

All roots have been determined within the desired accuracy of $\Delta = 0.0001$ in four iterations, the same as when Newton's linear method is used. This is to be expected. By substituting the approximation for $\Delta x/2$ from Newton's linear method, the convergence rate of Newton's second order method is naturally limited to that of Newton's linear method.

3.2.7 Secant Method

In this method we use the same expression for the improved value of the root as in Newton's linear method, i.e., we begin with:

$$x_{i+1} = x_i - \frac{f(x_i)}{f'(x_i)} \tag{3.76}$$

We approximate $f'(x_i)$ using a difference expression. Let x_{i-1}, x_i with $x_{i-1} < x_i$ be such that $x_{i-1},\ x_i \in (x_l, x_u)$. Consider Figure 3.7. The derivative $f'(x_i)$ may be approximated by:

$$f'(x_i) \cong \frac{f(x_i) - f(x_{i-1})}{x_i - x_{i-1}} \tag{3.77}$$

This is a *backwards difference* approximation for $f'(x_i)$. Substituting from (3.77) into (3.76) for $f'(x_i)$:

$$x_{i+1} = x_i - \frac{f(x_i)(x_i - x_{i-1})}{f(x_i) - f(x_{i-1})} \tag{3.78}$$

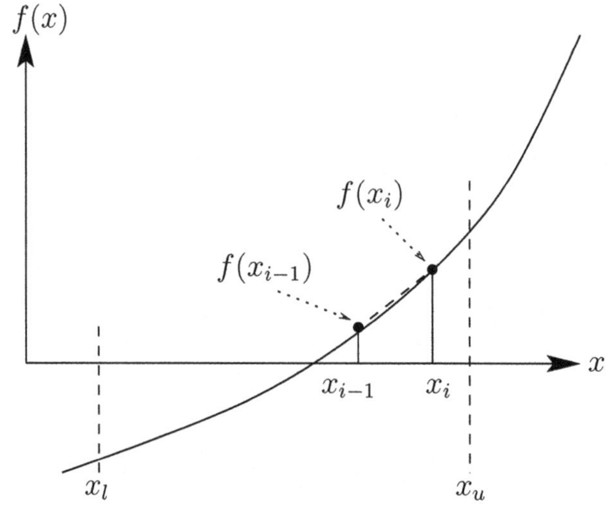

Figure 3.7: Secant Method

This is known as the secant method. This expression for x_{i+1} in (3.78) is the same as that derived for false position method (see Example 3.3 for a numerical example).

Remarks.

(1) We note that in this method, $f'(x_i)$ is approximated compared to Newton's linear method, hence $f'(x_i)$ does not appear in the expression (3.78) used to perform iterations.

(2) This method is helpful when $f(x)$ is complicated, in which case determining $f'(x)$ may be involved, but is avoided in this case.

3.2.8 Fixed Point Method or Basic Iteration Method

In this method we recast $f(x) = 0$ in the form $x = g(x)$. If x^* is a root of $f(x)$ then $f(x^*) = 0$ and hence $x^* = g(x^*)$ holds. We can recast $x = g(x)$ in a different form that is more suitable for iterative computations.

$$x_{i+1} = g(x_i) \; ; \quad i = 1, 2, \ldots \tag{3.79}$$

We begin with $x_i \in [x_l, x_u]$ as the assumed value of the root in the bracketed range $[x_l, x_u]$ and iterate using (3.79) until converged, i.e., until the following holds (approximate percentage relative error):

$$\left| \frac{x_{i+1} - x_i}{x_{i+1}} \right| 100 \leq \Delta \tag{3.80}$$

Remarks. In deriving $x = g(x)$ from $f(x) = 0$ it is possible to use various different approaches.

(1) We can consider $f(x) = 0$ and add x to both sides to obtain $f(x)+x = x$ and then define $f(x) + x = g(x)$.

(2) If possible we can use $f(x) = 0$ to solve for x in terms of quantities that are functions of x. For complex expressions in $f(x) = 0$, this may also lead to more than one possible form of $x = g(x)$.

(3) Example 3.6 illustrates some of these possibilities.

Example 3.6 (Fixed Point Method or Basic Iteration Method). In this example we consider a few problems to demonstrate how to set up the difference form in the basic iteration method and present numerical studies.

Case (a)

Consider

$$f(x) = x^2 - 4x + 3$$

We express $f(x)$ as $x = g(x)$.

Possible choices are

(i) $x = \frac{x^2+3}{4} = g(x)$

$$\therefore \quad x_{i+1} = \frac{x_i^2 + 3}{4} \; ; \quad i = 0, 1, \ldots \; x_0 \text{ is initial guess}$$

(ii) Add x to both sides of $f(x) = 0$

$$x = x^2 - 3x + 3 = g(x)$$

$$\therefore \quad x_{i+1} = x_i^2 - 3x_i + 3 \; ; \quad i = 0, 1, \ldots \; x_0 \text{ is initial guess}$$

(iii) $x = \pm\sqrt{4x - 3} = \pm g(x)$

$$\therefore \quad x_{i+1} = \pm\sqrt{4x_i - 3} \; ; \quad i = 0, 1, \ldots \; x_0 \text{ is initial guess}$$

Case (b)

Consider

$$f(x) = \sin(x) = 0$$

Add x to both sides of $f(x) = 0$.

$$x = \sin(x) + x = g(x)$$

$$\therefore \quad x_{i+1} = \sin(x_i) + x_i \; ; \qquad i = 0, 1, \ldots$$

Case (c)

Consider

$$f(x) = e^{-x} - x$$
$$\therefore \quad x = e^{-x} = g(x)$$

Hence,

$$x_{i+1} = e^{-x_i} \; ; \quad i = 0, 1, \dots \; x_0 \text{ is initial guess}$$

We present a numerical study of Case (c) using $x_0 = 0$. Calculated values are tabulated in the following.

Table 3.17: Results of fixed point method for Case (c) with $x_0 = 0$.

$x_0 =$ 0.00000E+00
$\Delta =$ 0.10000E−03
I = 20

iter	x_i	$x_{i+1} = g(x_i)$	$f(x_i)$	$f(x_{i+1})$	error (%)
1	0.00000E+00	0.10000E+01	0.10000E+01	-0.63212E+00	0.10000E+03
2	0.10000E+01	0.36788E+00	-0.63212E+00	0.32432E+00	0.17183E+03
3	0.36788E+00	0.69220E+00	0.32432E+00	-0.19173E+00	0.46854E+02
4	0.69220E+00	0.50047E+00	-0.19173E+00	0.10577E+00	0.38309E+02
5	0.50047E+00	0.60624E+00	0.10577E+00	-0.60848E−01	0.17447E+02
6	0.60624E+00	0.54540E+00	-0.60848E−01	0.34217E−01	0.11157E+02
7	0.54540E+00	0.57961E+00	0.34217E−01	-0.19497E−01	0.59033E+01
8	0.57961E+00	0.56012E+00	-0.19497E−01	0.11028E−01	0.34809E+01
9	0.56012E+00	0.57114E+00	0.11028E−01	-0.62637E−02	0.19308E+01
10	0.57114E+00	0.56488E+00	-0.62637E−02	0.35493E−02	0.11089E+01
11	0.56488E+00	0.56843E+00	0.35493E−02	-0.20139E−02	0.62441E+00
12	0.56843E+00	0.56641E+00	-0.20139E−02	0.11418E−02	0.35556E+00
13	0.56641E+00	0.56756E+00	0.11418E−02	-0.64772E−03	0.20119E+00
14	0.56756E+00	0.56691E+00	-0.64772E−03	0.36734E−03	0.11426E+00
15	0.56691E+00	0.56728E+00	0.36734E−03	-0.20832E−03	0.64756E−01
16	0.56728E+00	0.56707E+00	-0.20832E−03	0.11814E−03	0.36736E−01
17	0.56707E+00	0.56719E+00	0.11814E−03	-0.66996E−04	0.20829E−01
18	0.56719E+00	0.56712E+00	-0.66996E−04	0.37968E−04	0.11813E−01
19	0.56712E+00	0.56716E+00	0.37968E−04	-0.21517E−04	0.66945E−02
20	0.56716E+00	0.56714E+00	-0.21517E−04	0.12159E−04	0.37940E−02

$\therefore \quad x = 0.56714$ is a root of $f(x)$. The theoretical value of the root is 0.56714329. Percentage error listed above is based on (approximate percentage relative error):

$$\% \text{ error} = \left| \frac{x_{i+1} - x_i}{x_{i+1}} \right| \times 100$$

3.2.9 General Remarks on Root-Finding Methods

(1) Bisection method has the worst performance out of all of the methods except fixed point method (in general). To achieve error $O(10^{-4})$, close to 20 iterations are required in the example problem, more than needed for any other method for the error of the same order of magnitude. Said differently, for a fixed number of iterations, this method has the worst error in the calculated root.

(2) False position method has remarkably improved performance compared to bisection method. For error $O(10^{-4})$, this method required between 5-7 iterations for the numerical example presented here.

(3) Newton's linear method has even better performance than false position method due to the fact that in false position method (or secant method), the function derivative is approximated. Error $O(10^{-5})$ (lower than (1) and (2)) required only four iterations for each root. From third to fourth iteration the error (relative error) reduces from $O(10^{-1})$ or $O(10^{-2})$ to $O(10^{-5})$ or $O(10^{-6})$, better than the theoretical quadratic convergence rate.

(4) Newton's second order method in which Δx is calculated using the quadratic equation has the best performance compared to all of the methods. Only three iterations yield error $O(10^{-5})$ or $O(10^{-6})$. From the second to third iteration relative error changes from $O(10^{-1})$ to $O(10^{-5})$ or $O(10^{-6})$, much better than quadratic convergence rates (established for Newton's linear method). Improved performance of this method is expected as the Taylor series expansion used in deriving this method is more accurate than Newton's linear method.

(5) Newton's second order method in which $(\Delta x/2)$ is approximated using Newton's linear method has performance similar to Newton's linear method. This is not surprising due to the fact that approximating $(\Delta x/2)$ using Newton's linear method will naturally result in the accuracy of this method being comparable to Newton's linear method. Hence, the reason for similar performance of this method to Newton's linear method.

(6) Fixed point method is even worse than bisection method. For the numerical example considered here error $O(10^{-1})$ required 20 iterations.

(7) In all root-finding methods, appropriately bracketed roots are of vital importance. If the initial solution (guess solution) is not close enough to the true value of the root, all root-finding methods will not function properly.

3.3 Solutions of Nonlinear Simultaneous Equations

From the root-finding methods presented in the previous sections, we note that many of the methods can not be extended easily for more than one nonlinear equation. However, Newton's linear method has good mathematical foundation as well as quadratic convergence rate, and can be conveniently extended for obtaining the solution of a system of nonlinear simultaneous equations. We present details in the following.

3.3.1 Newton's Linear Method or Newton-Raphson Method

For the sake of simplicity consider a system of three simultaneous nonlinear equations (not necessarily algebraic):

$$\begin{aligned} f(x,y,z) &= 0 \\ g(x,y,z) &= 0 \\ h(x,y,z) &= 0 \end{aligned} \tag{3.81}$$

Let (x_i, y_i, z_i) be an approximation of the true solution of (3.81) in the small neighborhood of the true solution:

$$\begin{aligned} f(x_i,y_i,z_i) &\neq 0 \\ g(x_i,y_i,z_i) &\neq 0 \\ h(x_i,y_i,z_i) &\neq 0 \end{aligned} \tag{3.82}$$

Let $\Delta x, \Delta y$, and Δz be corrections to the x_i, y_i, z_i such that:

$$\begin{aligned} x_{i+1} &= x_i + \Delta x \\ y_{i+1} &= y_i + \Delta y \\ z_{i+1} &= z_i + \Delta z \end{aligned} \tag{3.83}$$

$$\begin{aligned} f(x_{i+1},y_{i+1},z_{i+1}) &= f(x_i+\Delta x, y_i+\Delta y, z_i+\Delta z) = 0 \\ g(x_{i+1},y_{i+1},z_{i+1}) &= g(x_i+\Delta x, y_i+\Delta y, z_i+\Delta z) = 0 \\ h(x_{i+1},y_{i+1},z_{i+1}) &= h(x_i+\Delta x, y_i+\Delta y, z_i+\Delta z) = 0 \end{aligned} \tag{3.84}$$

Expand $f(x_i+\Delta x, y_i+\Delta y, z_i+\Delta z)$, $g(x_i+\Delta x, y_i+\Delta y, z_i+\Delta z)$ and $h(x_i+\Delta x, y_i+\Delta y, z_i+\Delta z)$ in Taylor series about x_i, y_i, z_i and retaining only up to linear terms in $\Delta x, \Delta y, \Delta z$.

$$\begin{aligned} f(x_i,y_i,z_i) + f_x(x_i,y_i,z_i)\Delta x + f_y(x_i,y_i,z_i)\Delta y + f_z(x_i,y_i,z_i)\Delta z &= 0 \\ g(x_i,y_i,z_i) + g_x(x_i,y_i,z_i)\Delta x + g_y(x_i,y_i,z_i)\Delta y + g_z(x_i,y_i,z_i)\Delta z &= 0 \\ h(x_i,y_i,z_i) + h_x(x_i,y_i,z_i)\Delta x + h_y(x_i,y_i,z_i)\Delta y + h_z(x_i,y_i,z_i)\Delta z &= 0 \end{aligned} \tag{3.85}$$

In (3.85) the subscripts x, y, and z imply partial differentiations with respect to x, y, and z. Equations (3.85) can be arranged in matrix and vector form. Using $\Delta x = x_{i+1} - x_i$, $\Delta y = y_{i+1} - y_i$, and $\Delta z = z_{i+1} - z_i$, we obtain:

$$\begin{Bmatrix} f(x_i, y_i, z_i) \\ g(x_i, y_i, z_i) \\ h(x_i, y_i, z_i) \end{Bmatrix} + \begin{bmatrix} f_x & f_y & f_z \\ g_x & g_y & g_z \\ h_x & h_y & h_z \end{bmatrix}_{x_i, y_i, z_i} \begin{Bmatrix} x_{i+1} - x_i \\ y_{i+1} - y_i \\ z_{i+1} - z_i \end{Bmatrix} = \begin{Bmatrix} 0 \\ 0 \\ 0 \end{Bmatrix} \tag{3.86}$$

From (3.86) we can solve for x_{i+1}, y_{i+1}, and z_{i+1}.

$$\begin{Bmatrix} x_{i+1} - x_i \\ y_{i+1} - y_i \\ z_{i+1} - z_i \end{Bmatrix} = - \begin{bmatrix} f_x & f_y & f_z \\ g_x & g_y & g_z \\ h_x & h_y & h_z \end{bmatrix}^{-1}_{x_i, y_i, z_i} \begin{Bmatrix} f \\ g \\ h \end{Bmatrix}_{x_i, y_i, z_i} \tag{3.87}$$

$$\therefore \quad \begin{Bmatrix} x_{i+1} \\ y_{i+1} \\ z_{i+1} \end{Bmatrix} = \begin{Bmatrix} x_i \\ y_i \\ z_i \end{Bmatrix} - \begin{bmatrix} f_x & f_y & f_z \\ g_x & g_y & g_z \\ h_x & h_y & h_z \end{bmatrix}^{-1}_{x_i, y_i, z_i} \begin{Bmatrix} f \\ g \\ h \end{Bmatrix}_{x_i, y_i, z_i} \tag{3.88}$$

x_{i+1}, y_{i+1}, z_{i+1} from (3.88) are improved values of the solution compared to x_i, y_i, z_i (previous iteration). Now we check for convergence (approximate percentage relative error):

$$\begin{aligned} \left| \frac{x_{i+1} - x_i}{x_{i+1}} \right| 100 &\leq \Delta \\ \left| \frac{y_{i+1} - x_i}{y_{i+1}} \right| 100 &\leq \Delta \\ \left| \frac{z_{i+1} - z_i}{z_{i+1}} \right| 100 &\leq \Delta \end{aligned} \tag{3.89}$$

In which Δ is a preset tolerance for convergence based on the desired accuracy. If (3.89) are satisfied then we have x_{i+1}, y_{i+1}, z_{i+1} as the desired solution of the non-linear equations (3.81). If not, then we set x_i, y_i, z_i to x_{i+1}, y_{i+1}, z_{i+1} and repeat calculations using (3.88) until converged.

Remarks.

(1) As we have seen in the case of root-finding methods, solutions of all nonlinear equations (a single equation or a system of simultaneous equations) are iterative. Thus, a starting guess or initial solution is required to commence the iterative process.

(2) When these nonlinear equations describe a physical process, the physics is generally of help in estimating or guessing a good starting solutions. For example, Stokes flow is a good assumption as a starting solution for a

system of nonlinear equations obtained by discretizing the Navier-Stokes partial differential equations.

(3) Often a null vector or a vector of ones may serve as a crude guess also. Such a choice may result in many more iterations as this choice may be far away from the true solution. In some cases, this choice may also result in lack of convergence.

(4) The most important point to remember is that Newton's method has excellent convergence characteristics, provided the starting solution is in a sufficiently small neighborhood of the correct solution. Thus, a choice of initial solution close to the true solution is necessary, otherwise the method may require too many iterations to converge or may not converge at all.

(5) This is obviously method of tangents or gradient method in $\mathbb{R}^2$.

3.3.1.1 Special Case: Single Equation

If we have only one equation then we only have $f(x) = 0$ in (3.81), and hence (3.88) reduces to:

$$x_{i+1} = x_i - (f_x)^{-1}_{x_i} f(x_i) \tag{3.90}$$

or

$$x_{i+1} = x_i - \frac{f(x_i)}{f'(x_i)} \tag{3.91}$$

which is the same as Newton's linear method or Newton-Raphson method derived in Section 3.2.5 for a single nonlinear equation.

Example 3.7 (Newton's Linear Method for a System of Two Nonlinear Equations). Consider the following system of two nonlinear equations:

$$\begin{aligned} f(x,y) &= x^3 + 3y^2 - 21 = 0 \\ g(x,y) &= x^2 + 2y + 2 = 0 \end{aligned} \tag{3.92}$$

We wish to find all possible values of x,y that satisfy the above equations, i.e., all roots, using Newton's linear method or Newton's first order method. For a system of two equations described by $f(x,y) = 0$ and $g(x,y) = 0$, we have:

$$\begin{Bmatrix} x_{i+1} \\ y_{i+1} \end{Bmatrix} = \begin{Bmatrix} x_i \\ y_i \end{Bmatrix} - \begin{bmatrix} f_x & f_y \\ g_x & g_y \end{bmatrix}^{-1}_{(x_i,y_i)} \begin{Bmatrix} f \\ g \end{Bmatrix}_{(x_i,y_i)} \tag{3.93}$$

Using (3.92), we have:

$$\begin{aligned} f_x &= 3x^2 \\ f_y &= 6y \\ g_x &= 2x \\ g_y &= 2 \end{aligned} \tag{3.94}$$

Hence, (3.93) can be written as:

$$\begin{Bmatrix} x_{i+1} \\ y_{i+1} \end{Bmatrix} = \begin{Bmatrix} x_i \\ y_i \end{Bmatrix} - \begin{bmatrix} 3x_i^2 & 6y_i \\ 2x_i & 2 \end{bmatrix}^{-1} \begin{Bmatrix} x_i^3 + 3y_i^2 - 21 \\ x_i^2 + 2y_i + 2 \end{Bmatrix} \tag{3.95}$$

for $i = 0, 1, \dots$ in which x_0, y_0 is the initial solution of the desired root. We need an initial solution/starting guess x_0, y_0 for each root. In this case we can use a simple graphical procedure to determine these. From (3.92), we can obtain:

$$y = \pm\sqrt{\frac{21 - x^3}{3}} \; ; \qquad y = \left(\frac{-2 - x^2}{3}\right) \tag{3.96}$$

We plot graphs of (3.96) using x as abscissa and y as ordinate (see Figure 3.8). From the graphs in Figure 3.8, we note that the system of equations in (3.92) have two roots and we choose their approximate locations as $(x, y) = (-2, -3)$ and $(x, y) = (1.4, -2.0)$. We use these as initial guess in (3.94) for the two roots. We refer to the root near $(-2, -3)$ as root 1 and the root near $(1.4, -2.0)$ as root 2.

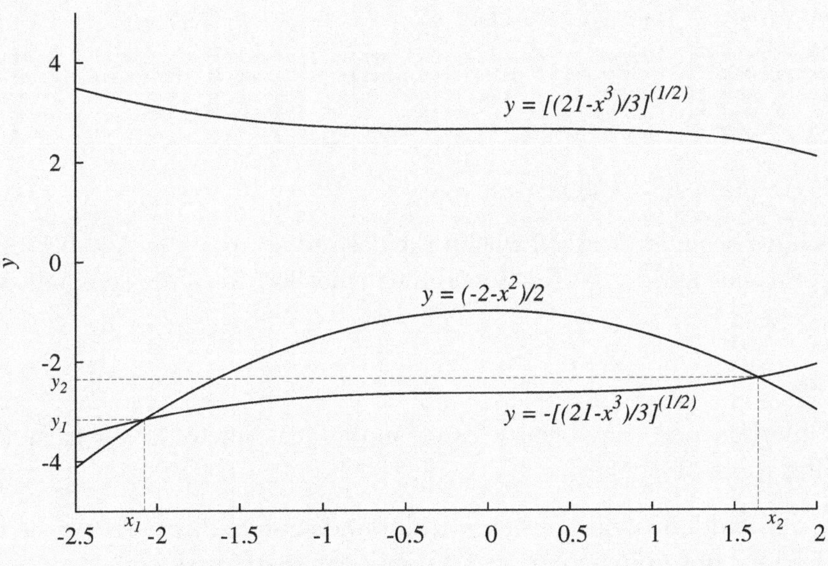

Figure 3.8: Plot of y versus x in (3.96)

Root 1: $(x_0, y_0) = (-2, -3)$

We use the initial solution $(-2, -3)$ and convergence tolerance of $\Delta = 0.1 \times 10^{-5}$ for both x and y. We limit the maximum number of iterations to 10. Calculations are shown below.

Table 3.18: Results of Newton's linear method for the first root

$x_0 =$	-0.20000E+01						
$y_0 =$	-0.30000E+01						
$\Delta =$	0.10000E−05						
I =	10						
x_i	y_i	$f(x_i, y_i)$	$g(x_i, y_i)$	f_x	f_y	g_x	g_y
-0.20000E+01	-0.30000E+01	-0.20000E+01	0.00000E+00	0.12000E+02	-0.18000E+02	-0.40000E+01	0.20000E+01
-0.20833E+01	-0.31667E+01	0.41000E−01	0.69444E−02	0.13021E+02	-0.19000E+02	-0.41667E+01	0.20000E+01
-0.20793E+01	-0.31617E+01	-0.28702E−04	0.16244E−04	0.12971E+02	-0.18970E+02	-0.41586E+01	0.20000E+01
-0.20793E+01	-0.31617E+01	-0.12877E−09	0.22008E−10	0.12970E+02	-0.18970E+02	-0.41586E+01	0.20000E+01
-0.20793E+01	-0.31617E+01	-0.42633E−13	0.62172E−14	0.12970E+02	-0.18970E+02	-0.41586E+01	0.20000E+01

$\therefore \quad (x, y) = (-2.0793, -3.1617)$ is a root.

Root 2: $(x_0, y_0) = (1.4, -2)$

We use the initial solution $(1.4, -2)$ and convergence tolerance of $\Delta = 0.1 \times 10^{-5}$ for both x and y. We limit the maximum number of iterations to 10. Calculations are shown below.

Table 3.19: Results of Newton's linear method for the second root

$x_0 =$	0.14000E+01						
$y_0 =$	-0.20000E+01						
$\Delta =$	0.10000E−05						
I =	10						
x_i	y_i	$f(x_i, y_i)$	$g(x_i, y_i)$	f_x	f_y	g_x	g_y
0.14000E+01	-0.20000E+01	-0.62560E+01	-0.40000E−01	0.58800E+01	-0.12000E+02	0.28000E+01	0.20000E+01
0.16864E+01	-0.23810E+01	0.80350E+00	0.82036E−01	0.85320E+01	-0.14286E+02	0.33728E+01	0.20000E+01
0.16438E+01	-0.23502E+01	0.11947E−01	0.18140E−02	0.81065E+01	-0.14101E+02	0.32877E+01	0.20000E+01
0.16430E+01	-0.23498E+01	0.35454E−05	0.62518E−06	0.80987E+01	-0.14099E+02	0.32861E+01	0.20000E+01
0.16430E+01	-0.23498E+01	0.38014E−12	0.48406E−13	0.80987E+01	-0.14099E+02	0.32861E+01	0.20000E+01

$\therefore \quad (x, y) = (1.6430, -2.3498)$ is a root.

All roots have been determined within the desired accuracy of $\Delta = 0.1 \times 10^{-5}$ in five iterations, similar to Newton's linear method for a single equation.

Remarks.

(1) The method has convergence rates similar to Newton's linear method finding roots of $f(x)$.

(2) This method has good mathematical foundation and is perhaps the best method for obtaining solutions of nonlinear systems of equations.

(3) It is important to note again that Newton's methods (both linear and quadratic) have small radii of convergence. The radius of convergence

is the region or neighborhood about the root such that a choice of the starting solution or guess solution for the root in this region is assured to yield a converged solution. Due to the fact that the radius of convergence is small for Newton's methods, close proximity of the guess or starting solution to the true root or solution is essential for the convergence of the methods.

3.3.2 Concluding Remarks

In this chapter solution methods for a single and a system of nonlinear simultaneous equations are considered. Solution methods for a single nonlinear equation, referred to as root-finding methods, are introduced first. Graphical methods, incremental search method, bisection method, method of false position, secant method, and fixed-point method are presented with numerical examples. This is followed by Newton-Raphson or Newton's linear method as well as Newton's quadratic method. Convergence rates are derived for Newton's method. Newton's linear method is extended to a system of nonlinear simultaneous equations. The solutions of nonlinear equations are always iterative, hence the methods of obtaining solutions of nonlinear equations are always methods of approximation. Due to the iterative nature of the methods, a starting solution is always essential. The physics leading to the nonlinear equations is generally helpful in choosing a starting solution. In case of a single nonlinear equation, the roots can be bracketed to obtain a good starting solution. However in $\mathbb{R}^2$ and $\mathbb{R}^3$ with a system of nonlinear equations this is not possible. In such cases there is no alternative but to resort to the physics described by the non-linear equations. For example, Stokes flow (solution of linear Navier-Stokes equations) is a good starting solution for a system of nonlinear algebraic equations resulting from the discretization of the Navier-Stokes equations.

As a final note, Newton's linear method is the preferred choice for nonlinear equations because of its extremely good convergence rate leading to accurate solutions in just a few iterations. As a word of caution, Newton's linear method has a small radius of convergence, hence generally requires a starting solution in a small neighborhood of the correct solution. Continuity and differentiability of the functions in the neighborhood of the solution sought are obviously essential as the method is based on tangents.

Problems

Consider the following cubic algebraic polynomial.

$$f(x) = x^3 - 6.8x^2 + 5.4x + 19.0 = 0 \tag{1}$$

Consider the range between [-2,6].

3.1 Plot a graph of $f(x)$ versus x to locate the roots of (1), approximately.

Perform incremental search using x between [-2,6] with $\Delta x = 0.61$ to bracket the roots of (1). Tabulate your answers in the same fashion as done in the examples. Clearly show the three brackets of x containing the roots.

3.2 Using the brackets of the roots determined in 3.1, use bisection method to determine more accurate values of each root. Use

$$\left| \frac{(x_m)_{i+1} - (x_m)_i}{(x_m)_{i+1}} \right| \times 100 \leq 0.1 \times 10^{-3}$$

as convergence criterion, where $(x_m)_i$ is the average of x_l and x_u for L^{th} iteration. Tabulate your calculations in the same manner as in the examples. Clearly show the values of the roots. Limit the number of iterations to 20.

3.3 Using brackets of the roots determined in 3.1, use method of false position to determine more accurate values of each root. Use

$$\left| \frac{(x_r)_{i+1} - (x_r)_i}{(x_r)_{i+1}} \right| \times 100 \leq 0.1 \times 10^{-3}$$

as convergence criterion, where $(x_r)_i$ is the value of the root in the ith iteration. Tabulate your calculations in the same manner as in the examples. Clearly show the values of the roots. Limit number of iterations to 20.

3.4 Using the brackets of the roots determined in 3.1, use Newton's linear method (Newton-Raphson method) to determine more accurate values of each root. Use -0.8, 3.3 and 4.8 as starting values (initial guess) of the roots (in ascending order). Use

$$\left| \frac{x_{i+1} - x_i}{x_{i+1}} \right| \times 100 \leq 0.1 \times 10^{-4}$$

as convergence criterion, where x_i is the value of the root in the ith iteration. Tabulate your calculations in the same manner as in the examples. Limit the number of iterations to 10.

3.5 Using the brackets of the roots determined in 3.1, use Newton's second order method:

(a) Case I: Use Newton's linear method as approximation for $(\frac{\Delta x}{2})$. See section 3.2.6.

(b) Case II: Calculate value of Δx using quadratic formula. Use a value of Δx for which the new value of the root lies in the bracketed range.

Use -0.8, 3.3 and 4.8 as initial guess of the roots (considered in ascending order) for Case I. Use -0.8, 2.9, and 4.8 as initial guess of the roots (considered in ascending order) for Case II.

In both cases use

$$\left| \frac{x_{i+1} - x_i}{x_{i+1}} \right| \times 100 \leq 0.1 \times 10^{-4}$$

as convergence criterion, where x_i is the value of the root in the ith iteration. Tabulate your calculations in the same manner as in the examples. Limit the number of iterations to 5.

3.6 Based on the studies performed in problems 3.2 - 3.5, write a short discussion regarding the accuracy of various methods, convergence characteristics, and efficiency.

3.7 Consider

$$f(x) = -x^2 + 5.5x + 11.75$$

(a) Determine roots of $f(x) = 0$ graphically.

(b) Use quadratic formula to determine roots of $f(x) = 0$.

(c) Beginning with $(x_l, x_u) = (5, 10)$, use bisection method to determine the root of $f(x) = 0$. Perform three iterations. At iteration compute relative error and true error using the correct value of root obtained in (b).

3.8 Consider

$$f(x) = 3x^3 - 2.5x^2 + 3.5x - 1$$

(a) Locate roots of $f(x) = 0$ graphically.

(b) Using approximate values of the roots from (a), employ bisection method to determine more accurate values of roots.

3.9 Consider

$$f(x) = x^5 - 13.85x^4 + 69.85x^3 - 135.38x^2 + 126.62x - 40$$

(a) Determine roots of $f(x) = 0$ graphically.

(b) Use bisection method to determine more accurate values of the smallest and the largest roots within relative error of 10%

3.10 Consider

$$f(x) = -x^3 + 6.8x^2 - 8.8x - 4.4$$

(a) Bracket roots of $f(x) = 0$ using graphical method.

(b) Using the brackets of each roots established in (a)

(b1) Use method of false position to determine the roots within four decimal place accuracy.

(b2) Use Newtons linear method to calculate the roots with five decimal place accuracy. Use a starting value for each root within the ranges established in (a).

3.11 Consider

$$f(x) = -x^2 + 1.889x + 2.778$$

(a) Determine roots of $f(x) = 0$ using Newton-Raphson method upto five decimal place accuracy.

(b) Find roots of $f(x) = 0$ using fixed point method upto three decimal place accuracy.

3.12 Consider

$$f(x) = x^3 - 8x^2 + 12x - 4$$

(a) Determine all roots graphically

(b) Determine roots of $f(x) = 0$ using Newton's linear method with five decimal place accuracy.

(c) Also find roots of $f(x) = 0$ using secant method with the same accuracy as in (b).

3.13 Consider

$$f(x) = 0.5x^3 - 3x^2 + 5.5x - 3.05$$

(a) Determine roots of $f(x) = 0$ graphically

(b) Use Newton's linear method to find roots of $f(x) = 0$ accurate up to five decimal places.

(c) Also find roots of $f(x) = 0$ using secant method with the same accuracy as in (b).

3.14 Consider $x^2 = x + 2$. Use Newton's linear method to find root starting with initial guess of $x_0 = 1.0$. Calculate four new estimates. Calculate percentage relative error based on two successive solutions. Comment on the convergence of the method based on relative error.

3.15 Consider

$$x^2 - 3x - 4 = 0 \tag{1}$$

(a) Use basic iteration method or fixed point method to find a root of (1) near $x = 3$. Perform five iterations

(b) Let $(x_l, x_u) = (3.2, 5)$ contain root of (1). Determine a value of the root. Perform four iterations only.

(c) Use method of false position to obtain a root of (1). Perform four iterations only.

Use decimal place accuracy of the computed solutions in (b) and (c) to compare their convergence characteristics.

3.16 Find square of π (3.14159265) using Newton's linear method (but without taking square root) starting with a value of 1.0. Calculate five new estimates.
Hint: Let $x = \sqrt{\pi}$.

3.17 Calculate e^{-1} (inverse of e) without taking its inverse using Newton's linear method with accuracy upto four decimal places. Use $e = 2.7182183$ and a starting value of 0.3.
Hint: $x = e^{-1}$.

3.18 Let $f(x) = \cos(x)$ where x is in radians. Find a root of $f(x)$ starting with $x_0 = 1.0$ using fixed point or basic iteration method. Calculate five new estimates.

3.19 Find cube root of 8 accurate upto three decimal places using Newton's linear method starting with a value of 1.

3.20 Consider system of non-linear algebraic equations

$$\begin{aligned} -2x^2 + 2x - 2y + 1 &= 0 \\ 0.2x^2 - xy - 0.2y &= 0 \end{aligned} \tag{1}$$

Use Newton's linear method to find solutions of (1) using $x_0, y_0 = 1.0, 1.0$ as initial guess with at least five decimal place accuracy. Write a computer program to perform calculations.

3.21 Consider the following system of non-linear algebraic equations

$$\begin{aligned} (x-2)^2 + (y-2)^2 - 3 &= 0 \\ x^2 + y^2 - 4 &= 0 \end{aligned} \tag{1}$$

Plot graphs of the functions in (1) to obtain initial guess of the roots. Use Newton's linear method to obtain values of the roots accurate upto five decimal places. Write a computer program to perform all calculations.

3.22 Consider the following non-linear equations

$$\begin{aligned} x^2 - y + 1 &= 0 \\ 3\cos(x) - y &= 0 \end{aligned} \tag{1}$$

Plot graphs of the functions in (1) in the xy-plane to obtain initial values of the roots. Use Newton's linear method to obtain the values of the roots accurate upto five decimal places. Write a computer program to perform all calculations.

4 Algebraic Eigenvalue Problems

4.1 Introduction

Algebraic eigenvalue problems play a central and crucial role in dynamics, mathematical physics, continuum mechanics, and many areas of engineering. Broadly speaking eigenvalue problems are mathematically classified as standard eigenvalue problems or generalized eigenvalue problems.

Definition 4.1 (Standard Eigenvalue Problem (SEVP)). For a square matrix $[A]$, if there exists a scalar λ and a vector $\{\phi\}$ such that

$$[A]\{\phi\} - \lambda\{\phi\} = [A]\{\phi\} - \lambda[I]\{\phi\} = [[A] - \lambda[I]]\{\phi\} = 0 \tag{4.1}$$

holds, then (4.1) is called the standard eigenvalue problem. The scalar λ (or λs) and the corresponding vector(s) $\{\phi\}$ are called eigenvalue(s) and eigenvector(s). Together we refer to $(\lambda, \{\phi\})$ as an eigenpair of the standard eigenvalue problem (4.1).

Definition 4.2 (Generalized Eigenvalue Problem (GEVP)). For square matrices $[A]$ and $[B]$, if there exists a scalar λ and a vector $\{\phi\}$ such that

$$[A]\{\phi\} - \lambda[B]\{\phi\} = [[A] - \lambda[B]]\{\phi\} = \{0\} \tag{4.2}$$

holds, then (4.2) is called a generalized eigenvalue problem. The scalar λ (or λs) and the corresponding vector(s) $\{\phi\}$ are called eigenvalue(s) and eigenvector(s). Together we refer to $(\lambda, \{\phi\})$ as an eigenpair of the generalized eigenvalue problem (4.2).

Remarks. If $[B]^{-1}$ exists in (4.2), then we can premultiply (4.2) by $[B]^{-1}$ to convert the GEVP (4.2) into a SEVP $[\underset{\sim}{A}]\{\phi\} - \lambda[I]\{\phi\} = \{0\}$, $[\underset{\sim}{A}] = [B]^{-1}[A]$.

4.2 Basic Properties of the Eigenvalue Problems

In the following we list and study some basic properties of the eigenvalue problems.

(i) Consider the standard eigenvalue problem (4.1) in which $[A]$ is $(n \times n)$ and $\{\phi\}$ is $(n \times 1)$. Equation (4.1) represents a system of n homogeneous algebraic equations in n unknowns $\{\phi\}$. We note that λ is unknown as well. For non-zero λ and non-null $\{\phi\}$, the left hand side of (4.2) must yield a null vector, hence the rank of $[[A] - \lambda[I]]$ is less than its size n, i.e., $[[A] - \lambda[I]]$ is rank deficient or singular. Hence,

$$\det\left([A] - \lambda[I]\right) = 0 \tag{4.3}$$

must hold. Likewise, in the case of a GEVP the following most hold:

$$\det\left([A] - \lambda[B]\right) = 0 \tag{4.4}$$

(ii) Expansion of (4.3) and (4.4) using Laplace expansion will result in a nth degree polynomial in λ called the *characteristic polynomial* $p(\lambda)$ corresponding to the eigenvalue problems (4.1) or (4.2).

(iii) The nth degree characteristic polynomial $p(\lambda)$ has n roots $\lambda_1, \lambda_2, \dots, \lambda_n$ called eigenvalues of the eigenvalue problem (4.1) or (4.2). We generally arrange λ_is in ascending order.

$$\lambda_1 < \lambda_2 < \cdots < \lambda_n \tag{4.5}$$

(iv) For each eigenvalue λ_i; $i = 1, 2, \dots, n$ there exists a unique eigenvector $\{\phi\}_i$; $i = 1, 2, \dots, n$ such that each eigenpair $(\lambda_i, \{\phi\}_i)$; $i = 1, 2, \dots, n$ satisfies (4.1) or (4.2).

$$[[A] - \lambda_i[I]]\,\{\phi\}_i = \{0\} \quad ; \quad \text{for SEVP} \tag{4.6}$$

$$[[A] - \lambda_i[B]]\,\{\phi\}_i = \{0\} \quad ; \quad \text{for GEVP} \tag{4.7}$$

For each eigenvalue λ_i ; $i = 1, 2, \dots, n$ we can use (4.6) or (4.7) to find the corresponding eigenvectors $\{\phi\}_i$; $i = 1, 2, \dots, n$. Since for each eigenvalue

$$\det([A] - \lambda_i[I]) = 0 \tag{4.8}$$

$$\det([A] - \lambda_i[B]) = 0 \tag{4.9}$$

hold, the coefficient matrices in (4.6) and (4.7) are rank deficient, i.e., if $[A]$ is $(n \times n)$ then rank of the coefficient matrices in (4.6) and (4.7) is less than n. Thus, the only way we can determine $\{\phi\}_i$ corresponding to λ_i is to assume a value (say one) for one component of $\{\phi\}_i$, ϕ_k (k^{th} row) and then use (4.6) or (4.7) to obtain a reduced $(n - 1 \times n - 1)$ system with a nonzero right side to solve for the remaining components of $\{\phi\}$. Thus, in fact we have calculated

$$\{\phi\}_i^T = \left[\frac{\phi_1}{\phi_k}, \frac{\phi_2}{\phi_k}, \dots, \underset{k^{th}\text{ location}}{1}, \dots, \frac{\phi_n}{\phi_k}\right]_i \tag{4.10}$$

Thus, we note that the magnitude of the components of the eigenvector $\{\phi\}_i$ depends upon the choice of the magnitude of ϕ_k, k^{th} component of $\{\phi\}_i$, which is arbitrary. Hence, the $(n \times 1)$ eigenvector $\{\phi\}_i$ represents a direction in the n-dimensional space. Its magnitude can not be determined in the absolute sense but can be scaled as desired.

(v) Consider the SEVP (4.1). When $[A]$ is symmetric its eigenvalues λ_i ; $i = 1, 2, \dots, n$ are real. When $[A]$ is positive-definite, then λ_i are real and positive, i.e., $\lambda_i > 0$; $i = 1, 2, \dots, n$. When $[A]$ is positive semi-definite then all eigenvalues of $[A]$ are real, but the smallest eigenvalue λ_i (or more) can be zero. When $[A]$ is non-symmetric then its eigenvalues can be real, real and complex, or all of them can be complex. In this course as far as possible we only consider $[A]$ to be symmetric. In case of GEVP the same rules hold for both $[A]$ and $[B]$ together, i.e., either symmetric or non-symmetric.

4.2.1 Orthogonality of Eigenvectors

In this section we consider SEVP and GEVP to show that eigenvectors in both cases are orthogonal and can be orthonormalized.

4.2.1.1 Orthogonality of Eigenvectors in SEVP

Consider the standard eigenvalue problem in which $[A]$ is symmetric:

$$[A]\{\phi\} - \lambda[I]\{\phi\} = \{0\} \tag{4.11}$$

Let $(\lambda_i, \{\phi\}_i)$ and $(\lambda_j, \{\phi\}_j)$ be two distinct eigenpairs of (4.11), i.e., $\lambda_i \neq \lambda_j$. Then we have:

$$[A]\{\phi\}_i - \lambda_i[I]\{\phi\}_i = \{0\} \tag{4.12}$$

$$[A]\{\phi\}_j - \lambda_j[I]\{\phi\}_j = \{0\} \tag{4.13}$$

Premultiply (4.12) by $\{\phi\}_j^T$ and (4.13) by $\{\phi\}_i^T$.

$$\{\phi\}_j^T[A]\{\phi\}_i - \lambda_i\{\phi\}_j^T[I]\{\phi\}_i = 0 \tag{4.14}$$

$$\{\phi\}_i^T[A]\{\phi\}_j - \lambda_j\{\phi\}_i^T[I]\{\phi\}_j = 0 \tag{4.15}$$

Take the transpose of (4.15) (since $[A]$ is symmetric, $[A]^T = [A]$).

$$\{\phi\}_j^T[A]\{\phi\}_i - \lambda_j\{\phi\}_j^T[I]\{\phi\}_i = 0 \tag{4.16}$$

Subtract (4.16) from (4.14):

$$\{\phi\}_j^T[A]\{\phi\}_i - \{\phi\}_j^T[A]\{\phi\}_i - (\lambda_i - \lambda_j)\{\phi\}_j^T[I]\{\phi\}_i = 0 \tag{4.17}$$

or

$$(\lambda_i - \lambda_j)\{\phi\}_j^T[I]\{\phi\}_i = 0 \tag{4.18}$$

Since $\lambda_i \neq \lambda_j$, $\lambda_i - \lambda_j \neq 0$, hence equation (4.18) implies:

$$\begin{aligned} &\{\phi\}_j^T[I]\{\phi\}_i = 0 \implies \{\phi\}_j^T\{\phi\}_i = 0 \\ \text{or} \quad &\{\phi\}_i^T[I]\{\phi\}_j = 0 \implies \{\phi\}_i^T\{\phi\}_j = 0 \end{aligned} \tag{4.19}$$

The property (4.19) is known as the *orthogonal property of the eigenvectors* $\{\phi\}_k$; $k = 1, 2, \dots, n$ of the standard eigenvalue problem (4.11). That is, when $\lambda_i \neq \lambda_j$, the eigenvectors $\{\phi\}_i$ and $\{\phi\}_j$ are orthogonal to each other with respect to identity matrix or simply orthogonal to each other.

4.2.1.2 Normalizing an Eigenvector of SEVP

We note that

$$\{\phi\}_i^T\{\phi\}_i > 0 \tag{4.20}$$

and is equal to zero if and only if $\{\phi\}_i = \{0\}$, a null vector. Since the eigenvectors only represent a direction, we can normalize them such that their length is unity (in this case). Let $||\{\phi\}_i||$ be the *euclidean norm* or the length of the eigenvector $\{\phi\}_i$.

$$||\{\phi\}_i|| = \sqrt{\{\phi\}_i^T\{\phi\}_i} \tag{4.21}$$

Consider:

$$\{\widetilde{\phi}\}_i = \frac{1}{||\{\phi\}_i||}\{\phi\}_i \tag{4.22}$$

$$\left|\left|\{\widetilde{\phi}\}_i\right|\right| = \sqrt{\frac{1}{||\{\phi\}_i||}\{\phi\}_i^T\{\phi\}_i\frac{1}{||\{\phi\}_i||}} = \sqrt{\frac{||\{\phi\}_i||^2}{||\{\phi\}_i||\,||\{\phi\}_i||}} = 1 \tag{4.23}$$

Thus, $\{\widetilde{\phi}\}_i$ is the normalized $\{\phi\}_i$ such that the norm of $\{\widetilde{\phi}\}_i$ is one. With this normalization (4.19) reduces to:

$$\{\widetilde{\phi}\}_i^T[I]\{\widetilde{\phi}\}_j = \{\widetilde{\phi}\}_i^T\{\widetilde{\phi}\}_j = \delta_{ij} = \begin{cases} 1 & \text{if} \quad j = i \\ 0 & \text{if} \quad j \neq i \end{cases} \tag{4.24}$$

The quantity δ_{ij} is called the *Kronecker delta*. The condition (4.24) is called the *orthonormality condition* of the normalized eigenvectors of SEVP.

4.2.1.3 Orthogonality of Eigenvectors in GEVP

Consider the GEVP given by:

$$[A]\{\phi\} - \lambda[B]\{\phi\} = \{0\} \tag{4.25}$$

Let $(\lambda_i, \{\phi\}_i)$ and $(\lambda_j, \{\phi\}_j)$ be two eigenpairs of (4.25) in which λ_i and λ_j are distinct, i.e., $\lambda_i \neq \lambda_j$. Then we have:

$$[A]\{\phi\}_i - \lambda_i[B]\{\phi\}_i = \{0\} \tag{4.26}$$

$$[A]\{\phi\}_j - \lambda_j[B]\{\phi\}_j = \{0\} \tag{4.27}$$

Premultiply (4.26) by $\{\phi\}_j^T$ and (4.27) by $\{\phi\}_i^T$.

$$\{\phi\}_j^T[A]\{\phi\}_i - \lambda_i\{\phi\}_j^T[B]\{\phi\}_i = 0 \tag{4.28}$$

$$\{\phi\}_i^T[A]\{\phi\}_j - \lambda_j\{\phi\}_i^T[B]\{\phi\}_j = 0 \tag{4.29}$$

Take the transpose of (4.29) (since $[A]$ and $[B]$ are symmetric $[A]^T = [A]$ and $[B]^T = [B]$).

$$\{\phi\}_j^T[A]\{\phi\}_i - \lambda_j\{\phi\}_j^T[B]\{\phi\}_i = 0 \tag{4.30}$$

Subtract (4.30) from (4.28).

$$\{\phi\}_j^T[A]\{\phi\}_i - \{\phi\}_j^T[A]\{\phi\}_i - (\lambda_i - \lambda_j)\{\phi\}_j^T[B]\{\phi\}_i = 0 \tag{4.31}$$

Which reduces to:

$$(\lambda_i - \lambda_j)\{\phi\}_j^T[B]\{\phi\}_i = 0 \tag{4.32}$$

Since $\lambda_i \neq \lambda_j$, $\lambda_i - \lambda_j \neq 0$ and the following holds:

$$\{\phi\}_j^T[B]\{\phi\}_i = \{\phi\}_i^T[B]\{\phi\}_j = 0 \tag{4.33}$$

This is known as the *$[B]$-orthogonal property of the eigenvectors* $\{\phi\}_i$ and $\{\phi\}_j$ of the GEVP (4.25).

4.2.1.4 Normalizing an Eigenvector of GEVP

We note that

$$\{\phi\}_i^T[B]\{\phi\}_i \tag{4.34}$$

and is equal to zero if and only if $\{\phi\}_i = \{0\}$, a null vector. Since the eigenvectors only represent a direction, we can $[B]$-normalize them, i.e., their norm with respect to $[B]$ becomes unity. Let the $[B]$-norm of $\{\phi\}_i$, denoted by $||\{\phi\}_i||_B$, be defined as:

$$||\{\phi\}_i||_B = \sqrt{\{\phi\}_i^T[B]\{\phi\}_i} \tag{4.35}$$

Consider:

$$\{\widetilde{\phi}\}_i = \frac{\{\phi\}_i}{||\{\phi\}_i||_B} \tag{4.36}$$

Taking the $[B]$-norm of $\{\widetilde{\phi}\}_i$:

$$||\{\widetilde{\phi}\}_i||_B = \sqrt{\{\widetilde{\phi}\}_i^T[B]\{\widetilde{\phi}\}_i} \tag{4.37}$$

Substitute from (4.36) into (4.37).

$$||\{\widetilde{\phi}\}_i||_B = \sqrt{\frac{\{\phi\}_i^T[B]\{\phi\}_i}{||\{\phi\}_i||_B^2}} = \frac{||\{\phi\}_i||_B}{||\{\phi\}_i||_B} = 1 \tag{4.38}$$

Thus, the $[B]$-norm of $\{\widetilde{\phi}\}_i$, $||\{\widetilde{\phi}\}_i||_B$, is one, i.e., $\{\widetilde{\phi}\}_i$ is the $[B]$-normalized $\{\phi\}_i$. Using the condition (4.38) for the eigenvectors $\{\widetilde{\phi}\}_i$, (4.33) can be written as:

$$\{\widetilde{\phi}\}_i^T[B]\{\widetilde{\phi}\}_j = \delta_{ij} = \begin{cases} 1 & \text{if} \quad j = i \\ 0 & \text{if} \quad j \neq i \end{cases} \tag{4.39}$$

Condition (4.39) is called the $[B]$-*orthonormality condition* of the eigenvectors of GEVP. The eigenvectors $\{\widetilde{\phi}\}_i$ are orthogonal and normalized, hence orthonormal.

4.2.2 Scalar Multiples of Eigenvectors

From the material in Section 4.2.1 it is straightforward to conclude that eigenvectors are only determined within a scalar multiple, i.e., if $(\lambda_i, \{\phi\}_i)$ is an eigenpair for SEVP or GEVP then $(\lambda_i, \beta\{\phi\}_i)$ is also an eigenpair of SEVP or GEVP, β being a nonzero scalar.

4.2.2.1 SEVP

If $(\lambda_i, \{\phi\}_i)$ is an eigenpair of the SEVP, then:

$$[A]\{\phi\}_i - \lambda[I]\{\phi\}_i = \{0\} \tag{4.40}$$

To establish if $(\lambda_i, \beta\{\phi\}_i)$ is an eigenpair of the SEVP:

$$[A]\beta\{\phi\}_i - \lambda_i[I]\beta\{\phi\}_i = 0 \qquad \text{must hold} \tag{4.41}$$

or

$$\beta([A]\{\phi\}_i - \lambda_i[I]\{\phi\}_i) = \{0\} \tag{4.42}$$

Since $\beta \neq 0$:

$$[A]\{\phi\}_i - \lambda_i[I]\{\phi\}_i = \{0\} \qquad \text{must hold} \tag{4.43}$$

Hence, $(\lambda_i, \beta\{\phi\}_i)$ is an eigenpair of SEVP as (4.43) is identical to the SEVP.

GEVP

If $(\lambda_i, \{\phi\}_i)$ is an eigenpair of the GEVP, then:

$$[A]\{\phi\}_i - \lambda_i[B]\{\phi\}_i = \{0\} \tag{4.44}$$

To establish if $(\lambda_i, \beta\{\phi\}_i)$ is an eigenpair of the GEVP:

$$[A]\beta\{\phi\}_i - \lambda_i[B]\beta\{\phi\}_i = \{0\} \qquad \text{must hold} \tag{4.45}$$

or

$$\beta([A]\{\phi\}_i - \lambda_i[B]\{\phi\}_i) = \{0\} \tag{4.46}$$

Since $\beta \neq 0$:

$$[A]\{\phi\}_i - \lambda_i[B]\{\phi\}_i - \{0\} \qquad \text{must hold} \tag{4.47}$$

Hence, $(\lambda_i, \beta\{\phi\}_i)$ is an eigenpair of the GEVP as (4.47) is identical to the GEVP.

4.2.3 Consequences of Orthonormality of $\{\widetilde{\phi}\}$

4.2.3.1 Orthonormality of $\{\widetilde{\phi}\}$ in SEVP

Consider:

$$[A]\{\phi\} - \lambda[I]\{\phi\} = \{0\} \tag{4.48}$$

For an eigenpair $(\lambda_i, \{\widetilde{\phi}\}_i)$:

$$[A]\{\widetilde{\phi}\}_i - \lambda_i[I]\{\widetilde{\phi}\}_i = \{0\} \tag{4.49}$$

Premultiply (4.49) by $\{\widetilde{\phi}\}_i^T$.

$$\{\widetilde{\phi}\}_i^T[A]\{\widetilde{\phi}\}_i - \lambda_i\{\widetilde{\phi}\}_i^T[I]\{\widetilde{\phi}\}_i = 0 \tag{4.50}$$

Since $\{\widetilde{\phi}\}_i$ is normalized with respect to $[I]$:

$$\{\widetilde{\phi}\}_i^T[I]\{\widetilde{\phi}\}_i = 1 \tag{4.51}$$

Hence, (4.50) reduces to:

$$\{\widetilde{\phi}\}_i^T[A]\{\widetilde{\phi}\}_i = \lambda_i \tag{4.52}$$

The property (4.52) is known as the $[A]$-*orthonormal property* of $[I]$-normalized eigenvectors of the SEVP.

4.2.3.2 Orthonormality of $\{\widetilde{\phi}\}$ in GEVP

Consider:

$$[A]\{\phi\} - \lambda[B]\{\phi\} = \{0\} \tag{4.53}$$

For an eigenpair $(\lambda_i, \{\widetilde{\phi}\}_i)$:

$$[A]\{\widetilde{\phi}\}_i - \lambda_i[B]\{\widetilde{\phi}\}_i = \{0\} \tag{4.54}$$

Premultiply (4.54) by $\{\widetilde{\phi}\}_i^T$.

$$\{\widetilde{\phi}\}_i^T[A]\{\widetilde{\phi}\}_i - \lambda_i\{\widetilde{\phi}\}_i^T[B]\{\widetilde{\phi}\}_i = 0 \tag{4.55}$$

Since $\{\widetilde{\phi}\}_i$ is normalized with respect to $[B]$:

$$\{\widetilde{\phi}\}_i^T[B]\{\widetilde{\phi}\}_i = 1 \tag{4.56}$$

Hence, (4.55) reduces to:

$$\{\widetilde{\phi}\}_i^T[A]\{\widetilde{\phi}\}_i = \lambda_i \tag{4.57}$$

The property (4.57) is known as the $[A]$-*orthonormal property* of $[B]$-normalized eigenvectors of the GEVP.

4.3 Methods of Determining Eigenpairs of SEVPs and GEVPs

Since the eigenvalues λ_i of an eigenvalue problem are roots of the characteristic polynomial $p(\lambda)$ obtained by setting the determinant of the coefficient matrix in the homogeneous system of equations to zero, the methods of finding eigenvalues λ_i are similar to root-finding methods, hence are iterative (when the degree of the characteristic polynomial is greater than three). Since the characteristic polynomial of degree up to three is rather academic, in general the methods of determining the eigenvalues and eigenvectors are iterative, hence are methods of approximation. We consider the following methods for both SEVPs and GEVPs.

(I) Using characteristic polynomial $p(\lambda) = 0$

(II) Vector iteration methods

  (a) Inverse iteration method

  (b) Forward iteration method

  (c) Gram-Schmidt orthogonalization or iteration vector deflation technique for calculating intermediate or subsequent eigenpairs

(III) Transformation methods

(a) Jacobi method (for SEVP)

(b) Generalized Jacobi method (for GEVP)

(c) Householder method with QR iterations for SEVP

(d) Subspace iteration method and other

4.3.1 Characteristic Polynomial Method

In this method we construct the characteristic polynomial $p(\lambda)$ corresponding to the eigenvalue problem (either SEVP or GEVP). The roots of the characteristic polynomial are the eigenvalues. For each eigenvalue we determine the eigenvector using the eigenvalue problem.

Basic Steps:

(i) Find characteristic polynomial $p(\lambda)$ using:

$$\begin{aligned} \det[[A] - \lambda[I]] = 0 \quad &; \quad \text{for SEVP} \\ \det[[A] - \lambda[B]] = 0 \quad &; \quad \text{for GEVP} \end{aligned} \tag{4.58}$$

(ii) Find roots of the characteristic polynomial using root-finding methods. This gives us eigenvalues λ_i; $i = 1, 2, \dots, n$. Arrange λ_i in ascending order.

$$\lambda_1 < \lambda_2 < \dots < \lambda_n \tag{4.59}$$

(iii) Corresponding to each λ_i we find the corresponding eigenvector $\{\phi\}_i$ using:

$$\begin{aligned} [[A] - \lambda[I]]\{\phi\}_i = 0 \quad &; \quad \text{for SEVP} \\ [[A] - \lambda[B]]\{\phi\}_i = 0 \quad &; \quad \text{for GEVP} \end{aligned} \tag{4.60}$$

Thus now we have all pairs $(\lambda_i, \{\phi\}_i)$; $i = 1, 2, \dots, n$.

The characteristic polynomial $p(\lambda)$ can be obtained by using Laplace expansion when $[A]$ and $[B]$ are not too large. But, when $[A]$ and $[B]$ are large, Laplace expansion is algebraically too cumbersome. In such cases we can use a more efficient method of obtaining characteristic polynomial, the Faddeev-Leverrier method presented in the following.

4.3.1.1 Faddeev-Leverrier Method of Obtaining the Characteristic Polynomial $p(\lambda)$

We present details of the method (without proof) for the SEVP keeping in mind that the GEVP can be converted to the SEVP. Consider the SEVP:

$$\big[[A]-\lambda[I]\big]\{\phi\}=\{0\} \tag{4.61}$$

Let

$$\begin{aligned}
&[B_1]=[A] &&;\quad p_1=\text{tr}[B_1]=\sum_{i=1}^{n}(b_1)_{ii}\\
&[B_2]=[A]\big[[B_1]-p_1[I]\big] &&;\quad p_2=\tfrac{1}{2}\,\text{tr}[B_2]=\tfrac{1}{2}\sum_{i=1}^{n}(b_2)_{ii}\\
&[B_3]=[A]\big[[B_2]-p_2[I]\big] &&;\quad p_3=\tfrac{1}{3}\,\text{tr}[B_3]=\tfrac{1}{3}\sum_{i=1}^{n}(b_3)_{ii}\\
&\quad\vdots && \qquad\quad\vdots\\
&[B_n]=[A]\big[[B_{n-1}-p_{n-1}[I]\big] &&;\quad p_n=\tfrac{1}{n}\,\text{tr}[B_n]=\tfrac{1}{n}\sum_{i=1}^{n}(b_n)_{ii}
\end{aligned} \tag{4.62}$$

Then, the characteristic polynomial $p(\lambda)$ is given by:

$$(-1)^n(\lambda^n-p_1\lambda^{n-1}-\lambda_2\lambda^{n-2}\cdots p_n) \tag{4.63}$$

The inverse of $[A]$ can be obtained using:

$$[A]^{-1}=\frac{1}{p_n}\big[[B_{n-1}-p_{n-1}[I]\big] \tag{4.64}$$

From (4.63), we have (if we premultiply by $[A]$):

$$[A][A]^{-1}=\frac{1}{p_n}[A]\big[[B_{n-1}]-p_{n-1}[I]\big] \tag{4.65}$$

or

$$[I]=\frac{1}{p_n}[A]\big[[B_{n-1}]-p_{n-1}[I]\big] \tag{4.66}$$

$$\therefore\quad \big[[B_{n-1}]-p_{n-1}[I]\big]=p_n[I] \tag{4.67}$$

In (4.67), $p_n[I]$ is a diagonal matrix with p_n as diagonal elements.

Example 4.1 (Characteristic Polynomial $p(\lambda)$). Consider the following SEVP:

$$\left[\begin{bmatrix}2&-1&0\\-1&2&-1\\0&-1&1\end{bmatrix}-\lambda\begin{bmatrix}1&0&0\\0&1&0\\0&0&1\end{bmatrix}\right]\begin{Bmatrix}\phi_1\\\phi_2\\\phi_3\end{Bmatrix}=\begin{Bmatrix}0\\0\\0\end{Bmatrix} \tag{4.68}$$

(a) *Characteristic polynomial $p(\lambda)$ using the determinant of the coefficient matrix*
Consider :

$$\det\left[\begin{bmatrix} 2 & -1 & 0 \\ -1 & 2 & -1 \\ 0 & -1 & 1 \end{bmatrix} - \lambda \begin{bmatrix} 1 & 0 & 0 \\ 0 & 1 & 0 \\ 0 & 0 & 1 \end{bmatrix}\right] \begin{Bmatrix} \phi_1 \\ \phi_2 \\ \phi_3 \end{Bmatrix} = 0$$

or

$$\begin{vmatrix} 2-\lambda & -1 & 0 \\ -1 & 2-\lambda & -1 \\ 0 & -1 & 1-\lambda \end{vmatrix} = 0$$

Laplace expansion using the first row:

$$(2-\lambda)\begin{vmatrix} 2-\lambda & -1 \\ -1 & 1-\lambda \end{vmatrix} + (-1)\begin{vmatrix} -1 & -1 \\ 1-\lambda & 0 \end{vmatrix} + (0)\begin{vmatrix} -1 & 2-\lambda \\ 0 & -1 \end{vmatrix} = 0$$

$$(2-\lambda)\big((2-\lambda)(1-\lambda) - (-1)(-1)\big) + (-1)\big((-1)(0) - (-1)(1-\lambda)\big) = 0$$

or

$$p(\lambda) = -\lambda^3 + 5\lambda^2 - 6\lambda + 1 = 0$$

(b) *Faddeev-Leverrier method*

$$[A] = \begin{bmatrix} 2 & -1 & 0 \\ -1 & 2 & -1 \\ 0 & -1 & 1 \end{bmatrix}$$

$$[B_1] = [A] = \begin{bmatrix} 2 & -1 & 0 \\ -1 & 2 & -1 \\ 0 & -1 & 1 \end{bmatrix} \quad ; \quad p_1 = \text{tr}[B_1] = 2+2+1 = 5$$

$$[B_2] = [A]\big[[B_1] - p_1[I]\big]$$

$$= \begin{bmatrix} 2 & -1 & 0 \\ -1 & 2 & -1 \\ 0 & -1 & 1 \end{bmatrix}\left[\begin{bmatrix} 2 & -1 & 0 \\ -1 & 2 & -1 \\ 0 & -1 & 1 \end{bmatrix} - (5)\begin{bmatrix} 1 & 0 & 0 \\ 0 & 1 & 0 \\ 0 & 0 & 1 \end{bmatrix}\right]$$

$$= \begin{bmatrix} 2 & -1 & 0 \\ -1 & 2 & -1 \\ 0 & -1 & 1 \end{bmatrix}\begin{bmatrix} -3 & -1 & 0 \\ -1 & -3 & -1 \\ 0 & -1 & -4 \end{bmatrix}$$

$$\therefore \quad [B_2] = \begin{bmatrix} -5 & 1 & 1 \\ 1 & -4 & 2 \\ 1 & 2 & -3 \end{bmatrix} \quad ; \quad p_2 = \frac{1}{2}\text{tr}[B_2] = \frac{1}{2}(-5-4-3) = -6$$

$$
\begin{aligned}
[B_3] &= [A]\big[[B_2] - p_2[I]\big] \\
&= \begin{bmatrix} 2 & -1 & 0 \\ -1 & 2 & -1 \\ 0 & -1 & 1 \end{bmatrix} \left[\begin{bmatrix} 2 & -1 & 0 \\ -1 & 2 & -1 \\ 0 & -1 & 1 \end{bmatrix} - (-6) \begin{bmatrix} 1 & 0 & 0 \\ 0 & 1 & 0 \\ 0 & 0 & 1 \end{bmatrix} \right] \\
&= \begin{bmatrix} 2 & -1 & 0 \\ -1 & 2 & -1 \\ 0 & -1 & 1 \end{bmatrix} \begin{bmatrix} 1 & 1 & 1 \\ 1 & 2 & 2 \\ 1 & 2 & 3 \end{bmatrix}
\end{aligned}
$$

$$
\therefore \quad [B_3] = \begin{bmatrix} 1 & 0 & 0 \\ 0 & 1 & 0 \\ 0 & 0 & 1 \end{bmatrix} \quad ; \qquad p_3 = \frac{1}{3}\,\mathrm{tr}[B_3] = \frac{1}{3}(1+1+1) = 1
$$

The characteristic polynomial $p(\lambda)$ is given by:

$$
p(\lambda) = (-1)^3(\lambda^3 - p_1\lambda^2 - p_2\lambda - p_3)
$$

or

$$
p(\lambda) = (-1)^3(\lambda^3 - 5\lambda^2 - (-6)\lambda - 1)
$$

or

$$
p(\lambda) = -\lambda^3 + 5\lambda^2 - 6\lambda + 1
$$

which is the same as obtained using the determinant method. We note that:

$$
[A]^{-1} = \frac{1}{p_3}\big[[B_2] - p_2[I]\big] = \frac{1}{1} \begin{bmatrix} 1 & 1 & 1 \\ 1 & 2 & 2 \\ 1 & 2 & 3 \end{bmatrix}
$$

$$
\therefore \quad [A]^{-1} = \begin{bmatrix} 1 & 1 & 1 \\ 1 & 2 & 2 \\ 1 & 2 & 3 \end{bmatrix}
$$

We could verify that $[A]^{-1}[A] = [A][A]^{-1} = [I]$ holds.

Example 4.2 (Determination of Eigenpairs Using Characteristic Polynomial). Consider the following SEVP:

$$
[A]\{x\} - \lambda[I]\{\phi\} = \{0\} \tag{4.69}
$$

in which

$$
[A] = \begin{bmatrix} 2 & -1 \\ -1 & 2 \end{bmatrix}
$$

We determine the eigenpairs of (4.69) using characteristic polynomial.

Characteristic Polynomial $p(\lambda)$

From (4.69):

$$[[A]-\lambda[I]]\{\phi\}=\{0\} \tag{4.70}$$

Hence

$$\det\left[[A]-\lambda[I]\right]=0$$

or

$$\det\left[\begin{bmatrix}2 & -1\\ -1 & 2\end{bmatrix}-\lambda\begin{bmatrix}1 & 0\\ 0 & 1\end{bmatrix}\right]=0$$

or

$$\det\begin{bmatrix}2-\lambda & -1\\ -1 & 2-\lambda\end{bmatrix}=0$$

or

$$(2-\lambda)^2-1=0$$

Therefore

$$p(\lambda)=\lambda^2-4\lambda+3=0 \tag{4.71}$$

Roots of the Characteristic Polynomial $p(\lambda)$

$$\lambda^2-4\lambda+3=0 \implies (\lambda-1)(\lambda-3)=0 \tag{4.72}$$

$$\therefore \quad \lambda=1 \text{ and } \lambda=3$$

Hence, the eigenvalues λ_1 and λ_2 are (in ascending order):

$$\lambda_1=1 \quad , \quad \lambda_2=3 \tag{4.73}$$

Eigenvectors

Corresponding to eigenvalues $\lambda_1=1$, $\lambda_2=3$, we calculate eigenvectors in the following. Each eigenvector must satisfy the eigenvalue problem (4.69).

(a) Eigenvector Corresponding to $\lambda_1=1$

Using $\lambda=\lambda_1=1$ in (4.69):

$$\left[\begin{bmatrix}2 & -1\\ -1 & 2\end{bmatrix}-(1)\begin{bmatrix}1 & 0\\ 0 & 1\end{bmatrix}\right]\begin{Bmatrix}\phi_1\\ \phi_2\end{Bmatrix}=\begin{Bmatrix}0\\ 0\end{Bmatrix} \tag{4.74}$$

$\{\phi\}_1 = \begin{Bmatrix} \phi_1 \\ \phi_2 \end{Bmatrix}_1$ is the desired eigenvector corresponding to $\lambda_1 = 1$. From equation (4.74):

$$\begin{bmatrix} 1 & -1 \\ -1 & 1 \end{bmatrix} \begin{Bmatrix} \phi_1 \\ \phi_2 \end{Bmatrix} = \begin{Bmatrix} 0 \\ 0 \end{Bmatrix} \tag{4.75}$$

Obviously $\det \begin{bmatrix} 1 & -1 \\ -1 & 1 \end{bmatrix} = 0$ as expected.

To determine $\{\phi\}_1$, we must choose a value for either ϕ_1 or ϕ_2 in (4.75) and then solve for the other. Let $\phi_1 = 1$, then using:

$$\phi_1 - \phi_2 = 0 \quad \text{or} \quad -\phi_1 + \phi_2 = 0$$

from (4.75), we obtain:

$$\phi_2 = 1$$

Hence

$$\{\phi\}_1 = \begin{Bmatrix} \phi_1 \\ \phi_2 \end{Bmatrix} = \begin{Bmatrix} 1 \\ 1 \end{Bmatrix} \tag{4.76}$$

We now have the first eigenpair corresponding to the lowest eigenvalue $\lambda_1 = 1$:

$$(\lambda_1, \{\phi\}_1) = \left(1, \begin{Bmatrix} 1 \\ 1 \end{Bmatrix}\right) \tag{4.77}$$

(b) Eigenvector Corresponding to $\lambda_2 = 3$

Using $\lambda = \lambda_2 = 3$ in (4.69):

$$\left[\begin{bmatrix} 2 & -1 \\ -1 & 2 \end{bmatrix} - (3)\begin{bmatrix} 1 & 0 \\ 0 & 1 \end{bmatrix}\right] \begin{Bmatrix} \phi_1 \\ \phi_2 \end{Bmatrix} = \begin{Bmatrix} 0 \\ 0 \end{Bmatrix} \tag{4.78}$$

$\{\phi\}_2 = \begin{Bmatrix} \phi_1 \\ \phi_2 \end{Bmatrix}_2$ is the desired eigenvector corresponding to $\lambda_2 = 3$. From equation (4.78):

$$\begin{bmatrix} -1 & -1 \\ -1 & -1 \end{bmatrix} \begin{Bmatrix} \phi_1 \\ \phi_2 \end{Bmatrix} = \begin{Bmatrix} 0 \\ 0 \end{Bmatrix} \tag{4.79}$$

In this case we also note that the determinant of the coefficient matrix in (4.79) is zero (as expected). To determine $\{\phi\}_2$ we must choose a value of ϕ_1 or ϕ_2 in (4.79) and then solve for the other. Let $\phi_1 = 1$, then using:

$$-\phi_1 - \phi_2 = 0 \tag{4.80}$$

we obtain:

$$\phi_2 = -1$$

Hence

$$\{\phi\}_2 = \begin{Bmatrix} \phi_1 \\ \phi_2 \end{Bmatrix} = \begin{Bmatrix} 1 \\ -1 \end{Bmatrix} \tag{4.81}$$

We now have the second eigenpair corresponding to the second eigenvalue $\lambda_2 = 3$:

$$(\lambda_1, \{\phi\}_2) = \left(3, \begin{Bmatrix} 1 \\ -1 \end{Bmatrix}\right) \tag{4.82}$$

Thus, the two eigenpairs in acending order of the eigenvalues are:

$$\left(1, \begin{Bmatrix} 1 \\ 1 \end{Bmatrix}\right) \quad \text{and} \quad \left(3, \begin{Bmatrix} 1 \\ -1 \end{Bmatrix}\right) \tag{4.83}$$

Orthogonality of Eigenvectors

We note that

$$\begin{aligned} \{\phi\}_1^T[I]\{\phi\}_2 &= \{\phi\}_1^T\{\phi\}_2 = \begin{Bmatrix} 1 \\ 1 \end{Bmatrix}^T \begin{Bmatrix} 1 \\ -1 \end{Bmatrix} = [1\ 1]\begin{Bmatrix} 1 \\ -1 \end{Bmatrix} = 0 \\ \{\phi\}_2^T[I]\{\phi\}_1 &= \{\phi\}_2^T\{\phi\}_1 = \begin{Bmatrix} 1 \\ -1 \end{Bmatrix}^T \begin{Bmatrix} 1 \\ 1 \end{Bmatrix} = [1\ -1]\begin{Bmatrix} 1 \\ 1 \end{Bmatrix} = 0 \end{aligned} \tag{4.84}$$

That is, $\{\phi\}_1$ and $\{\phi\}_2$ are orthogonal to each other or with respect to $[I]$. Since

$$\begin{aligned} \{\phi\}_1^T\{\phi\}_1 &= \begin{Bmatrix} 1 \\ 1 \end{Bmatrix}^T \begin{Bmatrix} 1 \\ 1 \end{Bmatrix} = [1\ 1]\begin{Bmatrix} 1 \\ 1 \end{Bmatrix} = 2 \\ \text{and} \quad \{\phi\}_2^T\{\phi\}_2 &= \begin{Bmatrix} 1 \\ -1 \end{Bmatrix}^T \begin{Bmatrix} 1 \\ -1 \end{Bmatrix} = [1\ -1]\begin{Bmatrix} 1 \\ -1 \end{Bmatrix} = 2 \end{aligned} \tag{4.85}$$

Therefore

$$\{\phi\}_i^T\{\phi\}_j \neq \delta_{ij} \quad ; \quad i,j = 1,2 \tag{4.86}$$

Hence, $\{\phi\}_1$ and $\{\phi\}_2$ are not orthonormal.

Normalized Eigenvectors and their Orthonormality

Since this is a SEVP we normalize $\{\phi\}_1$ and $\{\phi\}_2$ with respect to $[I]$.

$$||\{\phi\}_1|| = \sqrt{\{\phi\}_1^T[I]\{\phi\}_1} = \sqrt{\{\phi\}_1^T\{\phi\}_1} = \sqrt{\begin{Bmatrix} 1 \\ 1 \end{Bmatrix}^T \begin{Bmatrix} 1 \\ 1 \end{Bmatrix}} = \sqrt{2} \tag{4.87}$$

$$\therefore \quad \{\widetilde{\phi}\}_1 = \frac{1}{||\{\phi\}_1||}\{\phi\}_1 = \frac{1}{\sqrt{2}}\begin{Bmatrix} 1 \\ 1 \end{Bmatrix} = \begin{Bmatrix} 1/\sqrt{2} \\ 1/\sqrt{2} \end{Bmatrix} \tag{4.88}$$

and

$$||\{\phi\}_2|| = \sqrt{\{\phi\}_2^T[I]\{\phi\}_2} = \sqrt{\{\phi\}_2^T\{\phi\}_2} = \sqrt{\begin{Bmatrix} 1 \\ -1 \end{Bmatrix}^T \begin{Bmatrix} 1 \\ -1 \end{Bmatrix}} = \sqrt{2} \tag{4.89}$$

$$\therefore \quad \{\widetilde{\phi}\}_2 = \frac{1}{||\{\phi\}_2||}\{\phi\}_2 = \frac{1}{\sqrt{2}}\begin{Bmatrix} 1 \\ -1 \end{Bmatrix} = \begin{Bmatrix} 1/\sqrt{2} \\ -1/\sqrt{2} \end{Bmatrix} \tag{4.90}$$

We note that $\{\widetilde{\phi}\}_i$; $i = 1, 2$ satisfy the following orthonormal property:

$$\{\widetilde{\phi}\}_i^T\{\widetilde{\phi}\}_j = \delta_{ij} = \begin{cases} 1 & ; \quad j = i \\ 0 & ; \quad j \neq i \end{cases} \qquad i, j = 1, 2 \tag{4.91}$$

Thus, $\{\widetilde{\phi}\}_1$ and $\{\widetilde{\phi}\}_2$ are orthogonal and normalized (with respect to $[I]$), hence these eigenvectors are orthonormal.

4.3.2 Vector Iteration Method of Finding Eigenpairs

It can be shown that the vector iteration method always yields the smallest eigenvalue and the corresponding eigenvector regardless of whether it is a SEVP or GEVP or regardless of the specific form in which SEVP and GEVP are recast.

4.3.2.1 Inverse Iteration Method: Setting Up an Eigenvalue Problem for Determining Smallest Eigenpair

Consider the following SEVP and GEVP:

$$[A]\{x\} = \lambda[I]\{x\} \qquad ; \qquad \text{SEVP} \tag{4.92}$$

$$[A]\{x\} = \lambda[B]\{x\} \qquad ; \qquad \text{GEVP} \tag{4.93}$$

The only difference between (4.92) and (4.93) is that in the right side of (4.92) we have $[I]$ instead of $[B]$. Thus (4.92) can be obtained from (4.93) by redefining $[B]$ as $[I]$. Hence, instead of (4.92) and (4.93) we could define a new eigenvalue problem:

$$[\underset{\sim}{A}]\{x\} = \lambda[\underset{\sim}{B}]\{x\} \tag{4.94}$$

If we choose $[\underset{\sim}{A}] = [A]$ and $[\underset{\sim}{B}] = [I]$,we recover (4.92) and if we set $[\underset{\sim}{A}] = [A]$ and $[\underset{\sim}{B}] = [B]$, then we obtain (4.93). Eigenvalue problem (4.94) is the preferred form of defining the SEVP as well as the GEVP given by (4.92) and (4.93). Other forms of (4.92) and (4.93) are possible as well which can also be recast as (4.94). For example, we could premultiply (4.92) by $[A]^{-1}$ (provided $[A]$ is invertible):

$$[A]^{-1}[A]\{x\} = \lambda[A]^{-1}\{x\}$$

or

$$[I]\{x\} = \lambda[A]^{-1}\{x\} \tag{4.95}$$

If we define $[\underset{\sim}{A}] = [I]$ and $[\underset{\sim}{B}] = [A]^{-1}$ in (4.95) then we obtain (4.94). In the case of (4.93), we could premultiply it by $[A]^{-1}$ (provided $[A]^{-1}$ exists) to obtain:

$$[A]^{-1}[A]\{x\} = \lambda[A]^{-1}[B]\{x\}$$

or

$$[I]\{x\} = \lambda([A]^{-1}[B])\{x\} \tag{4.96}$$

If we define $[\underset{\sim}{A}] = [I]$ and $[\underset{\sim}{B}] = [A]^{-1}[B]$ in (4.96), then again we obtain (4.94). Alternatively, in the case of (4.93), we can also premultiply by $[B]^{-1}$ (provided $[B]^{-1}$ exists) to obtain:

$$[B]^{-1}[A]\{x\} = \lambda[B]^{-1}[B]\{x\}$$

or

$$([B]^{-1}[A])\{x\} = \lambda[I]\{x\} \tag{4.97}$$

If we define $[\underset{\sim}{A}] = [B]^{-1}[A]$ and $[\underset{\sim}{B}] = [I]$ in (4.97), then we also obtain (4.94).

Thus, the eigenvalue problem in the form (4.94) is the most general representation of any one of the five eigenvalue problem forms defined by (4.92), (4.93), (4.95), (4.96), and (4.97). Regardless of the form we use, the eigenvalues remain unaffected due to the fact that in all these various forms λ has never been changed. However, a $[\underset{\sim}{B}]$-normalized eigenvector may be different if $[\underset{\sim}{B}]$ is not the same.

It then suffices to consider the eigenvalue problem (4.94) for presenting details of the vector iteration method. The specific choices of $[\underset{\sim}{A}]$ and $[\underset{\sim}{B}]$ may be application dependent, but these choices do not influence the eigenvalue. As mentioned earlier, when vector iteration method is applied to an eigenvalue problem such as (4.94), it always yields lowest eigenpair (proof omitted) $(\lambda_1, \{\phi\}_1)$. Calculation of lowest eigenpair by vector iteration method is also called the *Inverse Iteration Method.*

4.3.2.2 Inverse Iteration Method: Determination of Smallest Eigenpair $(\lambda_1, \{\phi\}_1)$

Consider the eigenvalue problem (4.94):

$$[\underset{\sim}{A}]\{x\} = \lambda[\underset{\sim}{B}]\{x\} \tag{4.98}$$

We present details of the inverse iteration method in the following. We want to calculate $(\lambda_1, \{\phi\}_1)$ in which λ_1 is the lowest eigenvalue and $\{\phi\}_1$ is the corresponding eigenvector.

1. Choose $\lambda = 1$ and rewrite (4.98) in the difference equation form as follows:

$$[\underset{\sim}{A}]\{\bar{x}\}_{k+1} = [\underset{\sim}{B}]\{x\}_k \quad ; \quad k = 1, 2, \dots \tag{4.99}$$

   For $k = 1$, choose $\{x\}_1$, a starting vector or initial guess of the eigenvector as a vector whose components are unity. Thus $(1, \{x\}_1)$ is the initial guess of $(\lambda_1, \{\phi\}_1)$. $\{x\}_1$ *should be such that it is not orthogonal to* $\{\phi\}_1$.

2. Use (4.99) to solve for $\{\bar{x}\}_{k+1}$ (solution of linear simultaneous algebraic equations). This is a new estimate of the non-normalized eigenvector.

3. Calculate a new estimate of the eigenvalue, say $P(\{\bar{x}\}_{k+1})$ using (4.98) and $\{\bar{x}\}_{k+1}$, i.e., replace $\{x\}$ in (4.98) by $\{\bar{x}\}_{k+1}$ and λ by $P(\{\bar{x}\}_{k+1})$.

$$[\underset{\sim}{A}]\{\bar{x}\}_{k+1} = P(\{\bar{x}\}_{k+1})[\underset{\sim}{B}]\{\bar{x}\}_{k+1} \tag{4.100}$$

   Premultiply (4.100) by $\{\bar{x}\}_{k+1}^T$ and solve for $P(\{\bar{x}\}_{k+1})$.

$$\{\bar{x}\}_{k+1}^T[\underset{\sim}{A}]\{\bar{x}\}_{k+1} = P(\{\bar{x}\}_{k+1})\{\bar{x}\}_{k+1}^T[\underset{\sim}{B}]\{\bar{x}\}_{k+1}$$

$$\therefore \quad P(\{\bar{x}\}) = \frac{\{\bar{x}\}_{k+1}^T[\underset{\sim}{A}]\{\bar{x}\}_{k+1}}{\{\bar{x}\}_{k+1}^T[\underset{\sim}{B}]\{\bar{x}\}_{k+1}} \tag{4.101}$$

4. Normalize eigenvector $\{\bar{x}\}_{k+1}$ with respect to $[\underset{\sim}{B}]$ using $\underset{\sim}{B}$-norm of $\{\bar{x}\}_{k+1}$.

$$||\{\bar{x}\}_{k+1}|| = \sqrt{\{\bar{x}\}_{k+1}^T[\underset{\sim}{B}]\{\bar{x}\}_{k+1}} \tag{4.102}$$

$$\therefore \quad \{x\}_{k+1} = \frac{1}{||\{\bar{x}\}_{k+1}||}\{\bar{x}\}_{k+1} \tag{4.103}$$

   The vector $\{x\}_{k+1}$ is the $[\underset{\sim}{B}]$-normalized new estimate of $\{\phi\}_1$. Thus, the new estimate of the eigenpair is:

$$(P(\{\bar{x}\}_{k+1}, \{x\}_{k+1})) \tag{4.104}$$

   Using (4.99) and Steps 2 – 4 for $k = 1, 2, \dots$ we obtain a sequence of approximations (4.104) for the first eigenpair $(\lambda_1, \{\phi\}_1)$.

5. For each value of k we check for the convergence of the calculated eigenpair.

$$\text{For eigenvalue:} \quad \left|\frac{P(\{\bar{x}\}_{k+1}) - P(\{\bar{x}\}_k)}{P(\{\bar{x}\}_{k+1})}\right| \le \Delta_1 \tag{4.105}$$

$$\text{For eigenvector:} \quad ||\{x\}_{k+1} - \{x\}_k|| \le \Delta_2 \tag{4.106}$$

The scalars Δ_1 and Δ_2 are preset tolerances. In (4.106), we check for the norm of the difference between $\{x\}_{k+1}$ and $\{x\}_k$, i.e., the norm of the relative error in the eigenvector.

If converged, then $(P(\{\bar{x}\}_{k+1}, \{x\}_{k+1}))$ is the lowest eigenpair, i.e., $(\lambda_1, \{\phi\}_1)$. If not converged, then k is incremented by one and Steps 2 – 5 are repeated using (4.99) until (4.105) and (4.106) both are satisfied.

Remarks.

(1) The method described above to calculate the smallest eigenpair is called the inverse iteration method, in which we iterate for an eigenvector and then for an eigenvalue, hence the method is sometimes also called the vector iteration method.

(2) In the vector iteration method described above for $k = 1, 2, \dots,$ the following holds.

$$\begin{aligned} \lim_{k\to\infty} P(\{\bar{x}\}_k) &= \lambda_1 \ ; \qquad \text{lowest eigenvalue} \\ \lim_{k\to\infty} \{x\}_k &= \{\phi\}_1 \ ; \qquad \text{eigenvector corresponding to } \lambda_1 \end{aligned} \tag{4.107}$$

(3) Using the method presented here it is only possible to determine $(\lambda_1, \{\phi\}_1)$ (proof omitted), the smallest eigenvalue and the corresponding eigenvector.

(4) Use of (4.98) in the development of the computational procedure permits treatments of SEVP as well as GEVP in any one of the desired forms shown earlier by choosing appropriate definitions of $[\underset{\sim}{A}]$ and $[\underset{\sim}{B}]$.

4.3.2.3 Forward Iteration Method: Setting Up an Eigenvalue Problem for Determining Largest Eigenpair

Consider the SEVP and GEVP given by:

$$[A]\{x\} = \lambda[I]\{x\} \tag{4.108}$$

$$[A]\{x\} = \lambda[B]\{x\} \tag{4.109}$$

Since the vector iteration technique described in the previous section only determines the lowest eigenpair, we must recast (4.108) and (4.109) in alternate forms that will allow us to determine the largest eigenpair. Divide both (4.108) and (4.109) by λ and switch sides.

$$[I]\{x\} = \frac{1}{\lambda}[A]\{x\} \tag{4.110}$$

$$[B]\{x\} = \frac{1}{\lambda}[A]\{x\} \tag{4.111}$$

Let

$$\frac{1}{\lambda} = \widetilde{\lambda} \tag{4.112}$$

Then, (4.110) and (4.111) become:

$$[I]\{x\} = \widetilde{\lambda}[A]\{x\} \tag{4.113}$$

$$[B]\{x\} = \widetilde{\lambda}[A]\{x\} \tag{4.114}$$

We can represent (4.113) or (4.114) by :

$$[\widetilde{A}]\{x\} = \widetilde{\lambda}[\widetilde{B}]\{x\} \tag{4.115}$$

The alternate forms of (4.113) and (4.114) described in Section 4.3.2.1 are possible to define here too. If we premultiply (4.113) by $[A]^{-1}$ (provided $[A]^{-1}$ exists), we obtain:

$$[A]^{-1}\{x\} = \widetilde{\lambda}[I]\{x\} \tag{4.116}$$

If we define $[\widetilde{A}] = [A]^{-1}$ and $[\widetilde{B}] = [I]$ in (4.116), we obtain (4.115).

If we premultiply (4.114) by $[B]^{-1}$ (provided $[B]^{-1}$ exists), we obtain

$$[I]\{x\} = \widetilde{\lambda}([B]^{-1}[A])\{x\} \tag{4.117}$$

If we define $[\widetilde{A}] = [I]$ and $[\widetilde{B}] = [B]^{-1}[A]$ in (4.117), then we obtain (4.115). If we premultiply (4.114) by $[A]^{-1}$ (provided $[A]^{-1}$ exists), we obtain:

$$([A]^{-1}[B])\{x\} = \widetilde{\lambda}[I]\{x\} \tag{4.118}$$

If we define $[\widetilde{A}] = [A]^{-1}[B]$ and $[\widetilde{B}] = [I]$ in (4.118), then we obtain (4.115). Thus, the eigenvalue problem in the form (4.115) is the most general representation of any one of the five eigenvalue problems defined by (4.113), (4.115), (4.116), (4.117) and (4.118). Regardless of the form we use, the eigenvalues remain unaffected due to the fact that in the various forms $\widetilde{\lambda}$ has never been changed. But the $[\widetilde{B}]$-normalized eigenvector may differ if the definition of $[\widetilde{B}]$ changes. Thus, instead of considering five different forms of the eigenvalue problem, it suffices to consider the eigenvalue problem (4.115).

The vector iteration method when applied to (4.115) will yield the lowest eigenpair $(\widetilde{\lambda}_1, \{\widetilde{\phi}\}_1)$ but $\lambda = {}^1\!/\widetilde{\lambda}$, hence, when $\widetilde{\lambda}$ is $\widetilde{\lambda}_1$, the lowest eigenvalue, ${}^1\!/\widetilde{\lambda}_1$ is the largest eigenvalue.

$$(\widetilde{\lambda}_1, \{\widetilde{\phi}\}_1) \text{ gives us } ({}^1\!/\widetilde{\lambda}, \{\widetilde{\phi}\}_1) = (\lambda_n, \{\phi\}_n)$$

Thus, it is the largest eigenpair. Details of determining $(\widetilde{\lambda}_1, \{\widetilde{\phi}\}_1)$ using (4.115) are exactly the same as described earlier for $(\lambda_1, \{\phi\}_1)$ using (4.94) except the fact that in (4.94) we have λ and in (4.115) we have $\widetilde{\lambda}$ instead of λ. This procedure of calculating the largest eigenpair is called *Forward Iteration Method.* Here we calculate smallest $\widetilde{\lambda}$, i.e., $\widetilde{\lambda}_1$, using (4.116) and vector iteration method and determine the largest eigenvalue λ_n by taking the reciprocal of $\widetilde{\lambda}_1$.

4.3.2.4 Forward Iteration Method: Determination of Largest Eigenpair $(\lambda_n, \{\phi\}_n)$

Consider the eigenvaue problem (4.115):

$$[\widetilde{A}]\{x\} = \widetilde{\lambda}[\widetilde{B}]\{x\} \tag{4.119}$$

The details are exactly the same as presented for calculating $(\lambda_1, \{\phi\}_1)$, but we repeat these in the following for the sake of completeness. We want to calculate $(\widetilde{\lambda}, \{\widetilde{\phi}\}_1)$ in which $\widetilde{\lambda}_1$ is the lowest eigenvalue and $\{\widetilde{\phi}\}$ is the associated eigenvector.

1. Choose $\widetilde{\lambda} = 1$ and rewrite (4.119) in the difference equation form as follows:

   $$[\widetilde{A}]\{\bar{x}\}_{k+1} = [\widetilde{B}]\{x\}_k \quad ; \quad k = 1, 2, \dots \tag{4.120}$$

   For $k = 1$, choose $\{x\}_1$, a starting vector or initial guess of the eigenvector as a vector whose components are unity. Thus $(1, \{x\}_1)$ is the initial guess of $(\widetilde{\lambda}_1, \{\widetilde{\phi}\}_1)$. $\{x\}_1$ *should be such that it is not orthogonal to* $\{\widetilde{\phi}\}_1$.

2. Use (4.120) to solve for $\{\bar{x}\}_{k+1}$ (solution of linear simultaneous algebraic equations). This is a new estimate of the non-normalized eigenvector.

3. Calculate a new estimate of the eigenvalue, say $\widetilde{P}(\{\bar{x}\}_{k+1})$, using (4.119) and $\{\bar{x}\}_{k+1}$, i.e., replace $\{x\}$ in (4.119) by $\{\bar{x}\}_{k+1}$ and $\widetilde{\lambda}$ by $\widetilde{P}(\{\bar{x}\}_{k+1})$.

   $$[\widetilde{A}]\{\bar{x}\}_{k+1} = \widetilde{P}(\{\bar{x}\}_{k+1})[\widetilde{B}]\{\bar{x}\}_{k+1} \tag{4.121}$$

   Premultiply (4.121) by $\{\bar{x}\}_{k+1}^T$ and solve for $\widetilde{P}(\{\bar{x}\}_{k+1})$.

   $$\{\bar{x}\}_{k+1}^T[\widetilde{A}]\{\bar{x}\}_{k+1} = \widetilde{P}(\{\bar{x}\}_{k+1})\{\bar{x}\}_{k+1}^T[\widetilde{B}]\{\bar{x}\}_{k+1}$$

$$\therefore \quad \tilde{P}(\{\bar{x}\}) = \frac{\{\bar{x}\}_{k+1}^T[\widetilde{A}]\{\bar{x}\}_{k+1}}{\{\bar{x}\}_{k+1}^T[\widetilde{B}]\{\bar{x}\}_{k+1}} \tag{4.122}$$

4. Normalize the eigenvector $\{\bar{x}\}_{k+1}$ with respect to $[\widetilde{B}]$ using $\widetilde{B}$-norm of $\{\bar{x}\}_{k+1}$.

$$||\{\bar{x}\}_{k+1}|| = \sqrt{\{\bar{x}\}_{k+1}^T[\widetilde{B}]\{\bar{x}\}_{k+1}} \tag{4.123}$$

$$\therefore \quad \{x\}_{k+1} = \frac{1}{||\{\bar{x}\}_{k+1}||}\{\bar{x}\}_{k+1} \tag{4.124}$$

The vector $\{x\}_{k+1}$ is the $[\widetilde{B}]$-normalized new estimate of $\{\phi\}_1$. Thus, the new estimate of the eigenpair is:

$$(\widetilde{P}(\{\bar{x}\}_{k+1}, \{x\}_{k+1})) \tag{4.125}$$

Using (4.120) and Steps 2 – 4 for $k = 1, 2, \ldots,$ we obtain a sequence of approximations (4.125) for the first eigenpair $(\widetilde{\lambda}_1, \{\widetilde{\phi}\}_1)$.

5. For each value of k we check for the convergence of the calculated eigenpair.

$$\text{For eigenvalue:} \quad \left|\frac{\widetilde{P}(\{\bar{x}\}_{k+1}) - \widetilde{P}(\{\bar{x}\}_k)}{\widetilde{P}(\{\bar{x}\}_{k+1})}\right| \leq \Delta_1 \tag{4.126}$$

$$\text{For eigenvector:} \quad ||\{x\}_{k+1} - \{x\}_k|| \leq \Delta_2 \tag{4.127}$$

The scalars Δ_1 and Δ_2 are preset tolerances. In (4.126), we check for the norm of the difference between $\{x\}_{k+1}$ and $\{x\}_k$, i.e., the norm of the relative error in the eigenvector. If converged then $(\widetilde{P}(\{\bar{x}\}_{k+1}, \{x\}_{k+1}))$ is the lowest eigenpair, i.e., $(\widetilde{\lambda}_1, \{\widetilde{\phi}\}_1)$. If not converged, then k is incremented by one and Steps 2 – 5 are repeated using (4.120) until (4.126) and (4.127) both are satisfied.

Remarks.

(1) In the vector iteration method described above for $k = 1, 2, \ldots$ the following holds.

$$\lim_{k\to\infty} P(\{\bar{x}\}_k) = \widetilde{\lambda}_1 \ ; \quad \text{lowest eigenvalue} \tag{4.128}$$

$$\frac{1}{\widetilde{\widetilde{\lambda}}} = \lambda_n \ ; \quad \text{largest eigenvalue} \tag{4.129}$$

$$\lim_{k\to\infty} \{x\}_k = \{\widetilde{\phi}\}_1; \text{ eigenvector corresponding to } \lambda_n \tag{4.130}$$

(2) Using the method presented it is only possible to determine one eigenpair.

(3) Use of (4.115) in the development of the computational procedure permits treatment of the SEVP as well as the GEVP in any one of the desired forms shown earlier by choosing appropriate definitions of $[\widetilde{A}]$ and $[\widetilde{B}]$.

Example 4.3 (Determining the Smallest Eigenpair: Inverse Iteration Method). Consider the following SEVP $[A]\{x\} = \lambda[I]\{x\}$.

$$\begin{bmatrix} 2 & -1 \\ -1 & 4 \end{bmatrix} \begin{Bmatrix} x_1 \\ x_2 \end{Bmatrix} = \lambda \begin{bmatrix} 1 & 0 \\ 0 & 1 \end{bmatrix} \begin{Bmatrix} x_1 \\ x_2 \end{Bmatrix} \tag{4.131}$$

When we compare (4.131) with (4.94) we find that:

$$[\underset{\sim}{A}] = \begin{bmatrix} 2 & -1 \\ -1 & 4 \end{bmatrix} \quad \text{and} \quad [\underset{\sim}{B}] = \begin{bmatrix} 1 & 0 \\ 0 & 1 \end{bmatrix} \tag{4.132}$$

We will calculate the lowest eigenvalue λ_1 of (4.131) and the corresponding eigenvector using the inverse iteration method.

Characteristic Polynomial and its Roots

First we calculate the eigenvalues of (4.131) using the characteristic polynomial. This will serve as a check for the inverse iteration method.

$$p(\lambda) = \det \left[\begin{bmatrix} 2 & -1 \\ -1 & 4 \end{bmatrix} - \lambda \begin{bmatrix} 1 & 0 \\ 0 & 1 \end{bmatrix} \right] = 0 \tag{4.133}$$

or

$$p(\lambda) = (2 - \lambda)(4 - \lambda) - (-1)(-1) = \lambda^2 - 6\lambda + 7 = 0 \tag{4.134}$$

We can find the roots of (4.134) using the quadratic formula.

$$\lambda = \frac{-(-6) \pm \sqrt{(6)^2 - 4(1)(7)}}{2} \tag{4.135}$$

or

$$\lambda = \frac{6 \pm \sqrt{8}}{2} = 3 \pm \sqrt{2} \tag{4.136}$$

Hence

$$\begin{aligned} \lambda_1 &= 3 - \sqrt{2} \quad \text{(smallest eigenvalue)} \\ \lambda_2 &= 3 + \sqrt{2} \quad \text{(largest eigenvalue)} \end{aligned} \tag{4.137}$$

Determination of eigenvectors $\{\phi\}_1$ and $\{\phi\}_2$ is straightforward.

Inverse Iteration Method to Determine $(\lambda_1, \{\phi\}_1)$

Let $\lambda = 1 = \lambda_1$ and $\{x\}_1 = \begin{Bmatrix} x_1 \\ x_2 \end{Bmatrix}_1 = \begin{Bmatrix} 1 \\ 1 \end{Bmatrix}$ for $k = 1$ be the initial guess of the lowest eigenpair. We use:

$$[\underset{\sim}{A}]\{x\} = \lambda[\underset{\sim}{B}]\{x\}$$

$$[\underset{\sim}{A}] = \begin{bmatrix} 2 & -1 \\ -1 & 4 \end{bmatrix} \qquad [\underset{\sim}{B}] = \begin{bmatrix} 1 & 0 \\ 0 & 1 \end{bmatrix}$$

The difference equation form of the eigenvalue problems becomes:

$$\begin{bmatrix} 2 & -1 \\ -1 & 4 \end{bmatrix} \begin{Bmatrix} \bar{x}_1 \\ \bar{x}_2 \end{Bmatrix}_{k+1} = \begin{bmatrix} 1 & 0 \\ 0 & 1 \end{bmatrix} \{x\}_k = [I]\{x\}_k \tag{4.138}$$

We use (4.138) for $k = 1, 2, \dots$

For $k = 1$

We have $\left(1, \begin{Bmatrix} 1 \\ 1 \end{Bmatrix}\right)$ for the initial guess of the eigenpair (not orthogonal to $\{\phi\}_1$). Using (4.138), we have:

$$\begin{bmatrix} 2 & -1 \\ -1 & 4 \end{bmatrix} \begin{Bmatrix} \bar{x}_1 \\ \bar{x}_2 \end{Bmatrix}_2 = \begin{bmatrix} 1 & 0 \\ 0 & 1 \end{bmatrix} \begin{Bmatrix} 1 \\ 1 \end{Bmatrix}$$

Hence

$$\begin{Bmatrix} \bar{x}_1 \\ \bar{x}_2 \end{Bmatrix}_2 = \begin{Bmatrix} 0.71429 \\ 0.42857 \end{Bmatrix} = \{\bar{x}\}_2$$

$$P(\{\bar{x}\}_2) = \frac{\{\bar{x}\}_2^T[\underset{\sim}{A}]\{\bar{x}\}_2}{\{\bar{x}\}_2^T[I]\{\bar{x}\}_2} = \frac{\begin{Bmatrix} 0.71429 \\ 0.42857 \end{Bmatrix}^T \begin{bmatrix} 2 & -1 \\ -1 & 4 \end{bmatrix} \begin{Bmatrix} 0.71429 \\ 0.42857 \end{Bmatrix}}{\begin{Bmatrix} 0.71429 \\ 0.42857 \end{Bmatrix}^T [I] \begin{Bmatrix} 0.71429 \\ 0.42857 \end{Bmatrix}} = 1.6471$$

$$\{x\}_2 = \frac{1}{\sqrt{\{\bar{x}\}_2^T[I]\{\bar{x}\}_2}}\{\bar{x}\}_2$$

$$= \frac{1}{\sqrt{\begin{Bmatrix} 0.71429 \\ 0.42857 \end{Bmatrix}^T \begin{Bmatrix} 0.71429 \\ 0.42857 \end{Bmatrix}}} \begin{Bmatrix} 0.71429 \\ 0.42857 \end{Bmatrix} = \begin{Bmatrix} 0.85749 \\ 0.51450 \end{Bmatrix}$$

$\therefore$ The new estimate of the first eigenpair is

$$\left(1.6471, \begin{Bmatrix} 0.85749 \\ 0.51450 \end{Bmatrix}\right) = \left(P(\{\bar{x}\}_2), \begin{Bmatrix} x_1 \\ x_2 \end{Bmatrix}_2\right)$$

For $k = 2$

$$\begin{bmatrix} 2 & -1 \\ -1 & 4 \end{bmatrix} \begin{Bmatrix} \bar{x}_1 \\ \bar{x}_2 \end{Bmatrix}_3 = [I] \begin{Bmatrix} x_1 \\ x_2 \end{Bmatrix}_2$$

Hence

$$\begin{Bmatrix} \bar{x}_1 \\ \bar{x}_2 \end{Bmatrix}_3 = \begin{Bmatrix} 0.56350 \\ 0.26950 \end{Bmatrix} = \{\bar{x}\}_3$$

$$P(\{\bar{x}\}_3) = \frac{\{\bar{x}\}_3^T [\underset{\sim}{A}] \{\bar{x}\}_3}{\{\bar{x}\}_3^T [I] \{\bar{x}\}_3} = \frac{\begin{Bmatrix} 0.56350 \\ 0.26950 \end{Bmatrix}^T \begin{bmatrix} 2 & -1 \\ -1 & 4 \end{bmatrix} \begin{Bmatrix} 0.56350 \\ 0.26950 \end{Bmatrix}}{\begin{Bmatrix} 0.56350 \\ 0.26950 \end{Bmatrix}^T [I] \begin{Bmatrix} 0.56350 \\ 0.26950 \end{Bmatrix}} = 1.5938$$

$$\{x\}_3 = \frac{1}{\sqrt{\{\bar{x}\}_3^T [I] \{\bar{x}\}_3}} \{\bar{x}\}_3$$

$$= \frac{1}{\sqrt{\begin{Bmatrix} 0.56350 \\ 0.26950 \end{Bmatrix}^T [I] \begin{Bmatrix} 0.56350 \\ 0.26950 \end{Bmatrix}}} \begin{Bmatrix} 0.56350 \\ 0.26950 \end{Bmatrix} = \begin{Bmatrix} 0.90213 \\ 0.43146 \end{Bmatrix}$$

$\therefore$ The new estimate of the eigenpair is

$$\left(1.5938, \begin{Bmatrix} 0.90213 \\ 0.43146 \end{Bmatrix}\right) = \left(P(\{\bar{x}\}_3), \begin{Bmatrix} x_1 \\ x_2 \end{Bmatrix}_3\right)$$

We proceed in a similar fashion for $k = 3, 4, \ldots$ until converged. A summary of the calculations is given in the following.

Table 4.1: Results of the inverse iteration method for Example 4.3

k	$P(\{\bar{x}\}_{k+1})$	$\left\lvert\frac{P(\{\bar{x}\}_{k+1}) - P(\{\bar{x}\}_k)}{P(\{\bar{x}\}_{k+1})}\right\rvert$	Normalized Eigenvector x_1	x_2
1	0.16470588E+01	0.00000E+00	0.85649E+00	0.51450E+00
2	0.15938462E+01	0.33386E−01	0.90213E+00	0.43146E+00
3	0.15868292E+01	0.44220E−02	0.91636E+00	0.40035E+00
4	0.15859211E+01	0.57262E−03	0.92122E+00	0.38905E+00
5	0.15858038E+01	0.73933E−04	0.92293E+00	0.38497E+00
6	0.15857887E+01	0.95422E−05	0.92354E+00	0.38351E+00
7	0.15857867E+01	0.12315E−05	0.92376E+00	0.38298E+00

The converged first eigenvalue is $1.5857867 \approx 3 - \sqrt{2}$ (theoretical value) and the corresponding eigenvector is

$$\{\phi\}_1 = \begin{Bmatrix} 0.92376 \\ 0.38298 \end{Bmatrix}$$

The convergence tolerance $\Delta_1 = O(10^{-5})$ has been used for the eigenvalue in the results listed in Table 4.1.

Example 4.4 (Determining the Largest Eigenpair: Forward Iteration Method). Consider the following SEVP:

$$\begin{bmatrix} 2 & -1 \\ -1 & 4 \end{bmatrix} \begin{Bmatrix} x_1 \\ x_2 \end{Bmatrix} = \lambda \begin{bmatrix} 1 & 0 \\ 0 & 1 \end{bmatrix} \begin{Bmatrix} x_1 \\ x_2 \end{Bmatrix} \tag{4.139}$$

To set up (4.139) for determining the largest eigenpair, divide (4.139) by λ and let $\frac{1}{\lambda} = \widetilde{\lambda}$.

$$\begin{bmatrix} 1 & 0 \\ 0 & 1 \end{bmatrix} \begin{Bmatrix} x_1 \\ x_2 \end{Bmatrix} = \widetilde{\lambda} \begin{bmatrix} 2 & -1 \\ -1 & 4 \end{bmatrix} \begin{Bmatrix} x_1 \\ x_2 \end{Bmatrix} \tag{4.140}$$

Comparing (4.140) with (4.119) we find that:

$$[\widetilde{A}] = \begin{bmatrix} 1 & 0 \\ 0 & 1 \end{bmatrix} \qquad \text{and} \qquad [\widetilde{B}] = \begin{bmatrix} 2 & -1 \\ -1 & 4 \end{bmatrix}$$

The basic steps to be used in this method have already been presented, so we do not repeat these here. Instead we present calculation details and numerical results.

Choose $\widetilde{\lambda} = 1.0$ and $\begin{Bmatrix} x_1 \\ x_2 \end{Bmatrix} = \begin{Bmatrix} 1.0 \\ 1.0 \end{Bmatrix}$ and rewrite (4.140) in the difference equation form with $\widetilde{\lambda} = 1$. The vector $\{x\}_1$ must not be orthogonal to $\{\widetilde{\phi}\}_1$.

$$\begin{bmatrix} 1 & 0 \\ 0 & 1 \end{bmatrix} \begin{Bmatrix} \bar{x}_1 \\ \bar{x}_2 \end{Bmatrix}_{k+1} = \begin{bmatrix} 2 & -1 \\ -1 & 4 \end{bmatrix} \begin{Bmatrix} x_1 \\ x_2 \end{Bmatrix}_k \quad ; \quad k = 1, 2, \dots \tag{4.141}$$

For $k = 1$

We have $\left(1, \begin{Bmatrix} 1 \\ 1 \end{Bmatrix}\right)$ as the initial guess of the largest eigenvalue and the corresponding eigenvector, hence:

$$\begin{bmatrix} 1 & 0 \\ 0 & 1 \end{bmatrix} \begin{Bmatrix} \bar{x}_1 \\ \bar{x}_2 \end{Bmatrix} = \begin{bmatrix} 2 & -1 \\ -1 & 4 \end{bmatrix} \begin{Bmatrix} 1 \\ 1 \end{Bmatrix} \begin{bmatrix} 1 & 0 \\ 0 & 1 \end{bmatrix}$$

$$\therefore \quad \begin{Bmatrix} \bar{x}_1 \\ \bar{x}_2 \end{Bmatrix} = \begin{Bmatrix} 1.0 \\ 3.0 \end{Bmatrix} = \{\bar{x}\}_2 \quad ; \quad \text{new estimate of eigenvector}$$

Using $\{\bar{x}\}_2$ and (4.140), obtain a new estimate of $\widetilde{\lambda}$.

$$P(\{\bar{x}\}_2) = \frac{\{\bar{x}\}_2^T[\widetilde{A}]\{\bar{x}\}_2}{\{\bar{x}\}_2^T[\widetilde{B}]\{\bar{x}\}_2} = \frac{\begin{Bmatrix}1\\3\end{Bmatrix}^T \begin{bmatrix}1 & 0\\0 & 1\end{bmatrix}\begin{Bmatrix}1\\3\end{Bmatrix}}{\begin{Bmatrix}1\\3\end{Bmatrix}^T \begin{bmatrix}2 & -1\\-1 & 4\end{bmatrix}\begin{Bmatrix}1\\3\end{Bmatrix}} = 0.31250$$

Normalize $\{\bar{x}\}_2$ to obtain $\{x\}_2$.

$$\{x\}_2 = \frac{1}{\sqrt{\{\bar{x}\}_2^T[\widetilde{B}]\{\bar{x}\}_2}}\{\bar{x}\}_2 = \frac{1}{\sqrt{\begin{Bmatrix}1\\3\end{Bmatrix}^T \begin{bmatrix}2 & -1\\-1 & 4\end{bmatrix}\begin{Bmatrix}1\\3\end{Bmatrix}}}\begin{Bmatrix}1\\3\end{Bmatrix} = \begin{Bmatrix}0.17678\\0.53033\end{Bmatrix}$$

$\therefore$ The new estimate of the eigenpair is

$$\left(0.31250, \begin{Bmatrix}0.17678\\0.53033\end{Bmatrix}\right)$$

For $k = 2$

$$\begin{bmatrix}1 & 0\\0 & 1\end{bmatrix}\begin{Bmatrix}\bar{x}_1\\\bar{x}_2\end{Bmatrix}_3 = \begin{bmatrix}2 & -1\\-1 & 4\end{bmatrix}\begin{Bmatrix}x_1\\x_2\end{Bmatrix}_2$$

$$\begin{bmatrix}1 & 0\\0 & 1\end{bmatrix}\begin{Bmatrix}\bar{x}_1\\\bar{x}_2\end{Bmatrix}_3 = \begin{bmatrix}2 & -1\\-1 & 4\end{bmatrix}\begin{Bmatrix}0.17678\\0.53033\end{Bmatrix}$$

$$\begin{Bmatrix}\bar{x}_1\\\bar{x}_2\end{Bmatrix}_3 = \begin{Bmatrix}-0.17678\\1.9445\end{Bmatrix} = \{\bar{x}\}_3$$

Using $\{\bar{x}\}_3$ and (4.140), we obtain a new estimate of $\widetilde{\lambda}$.

$$P(\{\bar{x}\}_3) = \frac{\{\bar{x}\}_3^T[\widetilde{A}]\{\bar{x}\}_3}{\{\bar{x}\}_3^T[\widetilde{B}]\{\bar{x}\}_3} = \frac{\begin{Bmatrix}-0.17678\\1.9445\end{Bmatrix}^T \begin{bmatrix}1 & 0\\0 & 1\end{bmatrix}\begin{Bmatrix}-0.17678\\1.9445\end{Bmatrix}}{\begin{Bmatrix}-0.17678\\1.9445\end{Bmatrix}^T \begin{bmatrix}2 & -1\\-1 & 4\end{bmatrix}\begin{Bmatrix}-0.17678\\1.9445\end{Bmatrix}} = 0.24016$$

Normalize $\{\bar{x}\}_3$ to obtain $\{x\}_3$.

$$\{x\}_3 = \frac{1}{\sqrt{\{\bar{x}\}_3^T[\widetilde{B}]\{\bar{x}\}_3}}\{\bar{x}\}_3$$

$$= \frac{1}{\sqrt{\begin{Bmatrix}-0.17678\\1.9445\end{Bmatrix}^T \begin{bmatrix}2 & -1\\-1 & 4\end{bmatrix}\begin{Bmatrix}-0.17678\\1.9445\end{Bmatrix}}}\begin{Bmatrix}-0.17678\\1.9445\end{Bmatrix} = \begin{Bmatrix}-0.44368\\0.48805\end{Bmatrix}$$

$\therefore$ The new estimate of eigenpair is

$$\left(0.24016, \begin{Bmatrix} -0.44368 \\ 0.48805 \end{Bmatrix}\right)$$

We proceed in a similar fashion for $k = 3, 4, \ldots$ until converged. A summary of the calculations is given in the following.

Table 4.2: Results of the forward iteration method for Example 4.4

k	$P(\{\bar{x}\}_{k+1})$	$\left\lvert\frac{P(\{\bar{x}\}_{k+1})-P(\{\bar{x}\}_k)}{P(\{\bar{x}\}_{k+1})}\right\rvert$	Normalized Eigenvector x_1	x_2
1	0.31250000E+00	0.00000E+00	0.17678E+00	0.53033E+00
2	0.24015748E+00	0.30123E+00	-0.44368E−01	0.48805E+00
3	0.22835137E+00	0.51701E−01	-0.13263E+00	0.45909E+00
4	0.22677549E+00	0.69491E−02	-0.16441E+00	0.44693E+00
5	0.22657121E+00	0.90161E−03	-0.17578E+00	0.44235E+00
6	0.22654483E+00	0.11644E−03	-0.17986E+00	0.44068E+00
7	0.22654142E+00	0.15029E−04	-0.18132E+00	0.44007E+00
8	0.22654098E+00	0.19396E−05	-0.18185E+00	0.43985E+00

Thus, in this process we obtain $\widetilde{\lambda} = 0.22653547$, the smallest eigenvalue of $[I]\{x\} = \widetilde{\lambda}[A]$. Hence, the largest eigenvalue (which in this case is λ_2) will be

$$\lambda_2 = \frac{1}{\widetilde{\lambda}} = \frac{1}{0.22653547} = 4.414320 \approx 3 + \sqrt{2}$$

which matches the theoretical value of the largest eigenvalue. Hence, the largest eigenvalue and the corresponding eigenvector are

$$\left(4.414320, \begin{Bmatrix} -0.38246 \\ 0.92397 \end{Bmatrix}\right)$$

Example 4.5 (Forward Iteration Method by Converting a GEVP to a SEVP). Consider the following eigenvalue problem:

$$\begin{bmatrix} 2 & -1 \\ -1 & 4 \end{bmatrix} \begin{Bmatrix} x_1 \\ x_2 \end{Bmatrix} = \lambda \begin{bmatrix} 1 & 0 \\ 0 & 1 \end{bmatrix} \begin{Bmatrix} x_1 \\ x_2 \end{Bmatrix} \tag{4.142}$$

To find the largest eigenvalue and the corresponding eigenvector we convert (4.142) to the following:

$$\begin{bmatrix} 1 & 0 \\ 0 & 1 \end{bmatrix} \begin{Bmatrix} x_1 \\ x_2 \end{Bmatrix} = \widetilde{\lambda} \begin{bmatrix} 2 & -1 \\ -1 & 4 \end{bmatrix} \begin{Bmatrix} x_1 \\ x_2 \end{Bmatrix} \quad ; \quad \widetilde{\lambda} = \frac{1}{\lambda} \tag{4.143}$$

As shown in the previous example, we can use (4.143) and the forward iteration method to find the smallest $\widetilde{\lambda}$, i.e., $\widetilde{\lambda}_1$, and hence the largest eigenvalue would be $\lambda_2 = \frac{1}{\tilde{\lambda}}$. The eigenvector remains unchanged. The eigenvalue problem (4.143) is a GEVP in which:

$$[A] = \begin{bmatrix} 1 & 0 \\ 0 & 1 \end{bmatrix} \quad \text{and} \quad [B] = \begin{bmatrix} 2 & -1 \\ -1 & 4 \end{bmatrix} \tag{4.144}$$

We can also take another approach. We can convert (4.143) to a SEVP by premultiplying (4.143) by:

$$[B]^{-1} = \begin{bmatrix} 2 & -1 \\ -1 & 4 \end{bmatrix}^{-1} \tag{4.145}$$

$$\begin{bmatrix} 2 & -1 \\ -1 & 4 \end{bmatrix}^{-1} \begin{bmatrix} 1 & 0 \\ 0 & 1 \end{bmatrix} \begin{Bmatrix} x_1 \\ x_2 \end{Bmatrix} = \tilde{\lambda} \begin{bmatrix} 1 & 0 \\ 0 & 1 \end{bmatrix} \begin{Bmatrix} x_1 \\ x_2 \end{Bmatrix} \tag{4.146}$$

or

$$\begin{bmatrix} 0.57143 & 0.14286 \\ 0.14286 & 0.28571 \end{bmatrix} \begin{Bmatrix} x_1 \\ x_2 \end{Bmatrix} = \tilde{\lambda} \begin{bmatrix} 1 & 0 \\ 0 & 1 \end{bmatrix} \begin{Bmatrix} x_1 \\ x_2 \end{Bmatrix} \tag{4.147}$$

In this case in (4.147), we have $[\widetilde{A}]\{x\} = \tilde{\lambda}[\widetilde{B}]\{x\}$. Equations (4.147) define a SEVP. We can now use (4.147) and the vector iteration method to find the smallest eigenvalue $\tilde{\lambda}_1$ and the corresponding eigenvector $\{\tilde{\phi}\}_1$. This is obviously the forward iteration method. Hence, $\frac{1}{\tilde{\lambda}} = \lambda_2$ is the largest eigenvalue and we have the desired eigenpair $(\lambda_2, \{\phi\}_2)$. $\{\phi\}_2$ is the same as $\{\tilde{\phi}\}_1$, of the eigenvalue problem (4.142). Calculations are summarized in Table 4.3.

Table 4.3: Results of the forward iteration method for Example 4.5

k	$P(\{\bar{x}\}_{k+1})$	$\left\lvert\frac{P(\{\bar{x}\}_{k+1})-P(\{\bar{x}\}_k)}{P(\{\bar{x}\}_{k+1})}\right\rvert$	Normalized Eigenvector x_1	x_2
1	0.39999240E+00	0.00000E+00	0.31621E+00	0.94869E+00
2	0.26228664E+00	0.52502E+00	-0.90552E−01	0.99589E+00
3	0.23153436E+00	0.13282E+00	-0.27755E+00	0.96071E+00
4	0.22718759E+00	0.19133E−01	-0.34526E+00	0.93851E+00
5	0.22661973E+00	0.25058E−02	-0.36930E+00	0.92931E+00
6	0.22654633E+00	0.32399E−03	-0.37788E+00	0.92585E+00
7	0.22653685E+00	0.41821E−04	-0.38096E+00	0.92459E+00
8	0.22653563E+00	0.53397E−05	-0.38206E+00	0.92414E+00
9	0.22653547E+00	0.69651E−06	-0.38246E+00	0.92397E+00

Thus, $\tilde{\lambda}_1 = 0.22653547$ (same as previous example) and we have:

$$\lambda_2 = \frac{1}{\tilde{\lambda}_1} = \frac{1}{0.22653547} = 4.414320 \quad \text{(largest eigenvalue)}$$

The eigenvector in this case is different than previous example due to the fact that it is normalized differently. Hence, we have

$$(\lambda_2, \{\phi\}_2) = \left(4.414320, \begin{Bmatrix} -0.38246 \\ 0.92397 \end{Bmatrix}\right)$$

4.3.3 Gram-Schmidt Orthogonalization or Iteration Vector Deflation to Calculate Intermediate or Subsequent Eigenpairs

We recall that the inverse iteration method only yields the lowest eigenpair while the forward iteration method gives the largest eigenpair. These two methods do not have a mechanism for determining intermediate or subsequent eigenpairs. For this purpose we utilize *Gram-Schmidt orthogonalization* or the *iteration vector deflation method* in conjunction with the inverse or forward iteration method.

The basis for Gram-Schmidt orthogonalization or iteration vector deflation is that in order for an assumed eigenvector (iteration vector) to converge to the desired eigenvector in the inverse or forward iteration method, the iteration vector must not be orthogonal to the desired eigenvector. In other words, if the iteration vector is orthogonalized to the eigenvectors that have already been calculated, then we can eliminate the possibility of convergence of iteration vector to any one of them and hence convergence will occur to the next eigenvector. A particular orthogonalization procedure used extensively is called Gram-Schmidt orthogonalization process or iteration vector deflation method. This procedure can be used for the SEVP as well as the GEVP in the inverse or forward iteration methods.

Based on the material presented for the inverse and forward iteration methods it suffices to consider:

$$\begin{aligned} & [\underset{\sim}{A}]\{x\} = \lambda[\underset{\sim}{B}]\{x\} \; ; \quad \text{Inverse iteration} \\ \text{or} \quad & [\widetilde{A}]\{x\} = \widetilde{\lambda}[\widetilde{B}]\{x\} \; ; \quad \text{Forward iteration} \end{aligned}$$

By choosing a specific definition of $[\underset{\sim}{A}]$, $[\underset{\sim}{B}]$ and $[\widetilde{A}]$, $[\widetilde{B}]$ we can have these forms yield what we need. Further, the vector iteration method for both forms yields the lowest eigenpair hence, for discussing iteration vector deflation method we can consider either of the two forms without over or under tilde ($\sim$).

4.3.3.1 Gram-Schmidt Orthogonalization or Iteration Vector Deflation

Consider the eigenvalue problem:

$$[A]\{x\} = \lambda[B]\{x\} \tag{4.148}$$

Let $\{\phi\}_1$, $\{\phi\}_2$, ..., $\{\phi\}_m$ corresponding to m eigenpairs be the eigenvectors that have already been determined or calculated and are $[B]$-orthogonal, i.e., normalized with respect to $[B]$. We wish to calculate the $(m+1)^{th}$ eigenpair $(\lambda_{m+1}, \{\phi\}_{m+1})$. Let $\{x\}_1$ be the initial guess (or starting) eigenvector (iteration vector). We subtract a linear combination of $\{\phi\}_i$, $i = 1, 2, \dots, m$ from $\{x\}_1$ to obtain a new starting or guess iteration vector $\{\underset{\sim}{x}\}_1$ as follows.

$$\{\underset{\sim}{x}\}_1 = \{x\}_1 - \sum_{i=1}^{m} \alpha_i \{\phi\}_i \tag{4.149}$$

Where $\alpha_i \in \mathbb{R}$ are scalars yet to be determined. We refer to $\{\underset{\sim}{x}\}_1$ as the deflated starting iteration vector due to the fact that it is obtained from $\{x\}_1$ by removing the influence of $\{\phi\}_i$, $i = 1, 2, \dots, m$ from it. We determine α_i, $i = 1, 2, \dots, m$ by orthogonalizing $\{\underset{\sim}{x}\}_1$ to $\{\phi\}_i$; $i = 1, 2, \dots, m$ with respect to $[B]$. By doing so we ensure that $\{\underset{\sim}{x}\}_1$ as a starting iteration vector will not converge to any one of the $\{\phi\}_i$; $i = 1, 2, \dots, m$. If $\{\underset{\sim}{x}\}_1$ is $[B]$-orthogonal to $\{\phi\}_i$; $i = 1, 2, \dots, m$, then:

$$\{\phi\}_i^T[B]\{\underset{\sim}{x}\}_1 = 0 \quad ; \quad i = 1, 2, \dots, m \tag{4.150}$$

Premultiply (4.149) by $\{\phi\}_k^T[B]$; $k = 1, 2, \dots, m$.

$$\{\phi\}_k^T[B]\{\underset{\sim}{x}\}_1 = \{\phi\}_k^T[B]\{x\}_1 - \sum_{i=1}^{m} \alpha_i \{\phi\}_k^T[B]\{\phi\}_i \; ; \quad k = 1, 2, \dots m \tag{4.151}$$

Since $\{\phi\}_i$ are $[B]$-orthogonal, we have:

$$\{\phi\}_k[B]\{\phi\}_i = \begin{cases} 1 & ; \quad k = i \\ 0 & ; \quad k \neq i \end{cases} \tag{4.152}$$

and since $\{\underset{\sim}{x}\}_1$ is also $[B]$-orthogonal to $\{\phi\}_i$; $i = 1, 2, \dots, m$, we also have:

$$\{\phi\}_k[B]\{\underset{\sim}{x}\}_1 = 0 \quad ; \quad k = 1, 2, \dots, m \tag{4.153}$$

Using (4.152) and (4.153) in (4.151), we obtain:

$$\alpha_k = \{\phi\}_k^T[B]\{x\}_1 \quad ; \quad k = 1, 2, \dots, m \tag{4.154}$$

In (4.154), $\{\phi\}_k$; $k = 1, 2, \dots, m$, $[B]$, and $\{x\}_1$ are known, therefore we can determine α_k ; $k = 1, 2, \dots m$. Knowing α_i ; $i = 1, 2, \dots, m$, $\{\underset{\sim}{x}\}_1$ can be calculated using (4.149). The deflated vector $\{\underset{\sim}{x}\}_1$ is used in inverse or forward iteration methods to extract the eigenpair $(\lambda_{m+1}, \{\phi\}_{m+1})$.

Remarks.

(1) In the inverse iteration method we find the lowest eigenpair $(\lambda_1, \{\phi\}_1)$ using the usual procedure and then use vector deflation to find $(\lambda_2, \{\phi\}_2)$, $(\lambda_3, \{\phi\}_3), \dots$, in ascending order.

(2) In the forward iteration method we find the largest eigenpair $(\lambda_n, \{\phi\}_n)$ using usual procedure and then use iteration vector deflation to find $(\lambda_{n-1}, \{\phi\}_{n-1})$, $(\lambda_{n-2}, \{\phi\}_{n-2})$, $\dots$, in descending order.

(3) Consider the GEVP:

$$[A]\{x\} = \lambda[B]\{x\} \tag{4.155}$$

For determining the eigenpairs in ascending order, we use (4.155) in inverse iteration method with iteration vector deflation. Using (4.155) we consider:

$$[B]\{x\} = \widetilde{\lambda}[A]\{x\} \; ; \quad \widetilde{\lambda} = \frac{1}{\lambda} \tag{4.156}$$

$$\text{or} \qquad [\widetilde{A}]\{x\} = \widetilde{\lambda}[\widetilde{B}]\{x\} \tag{4.157}$$

$$\text{where} \qquad [\widetilde{A}] = [B] \quad ; \quad [\widetilde{B}] = [A] \tag{4.158}$$

The eigenvalue problem (4.157) is identical to (4.155) except with new definitions of $[A]$ and $[B]$, hence we can also use (4.157) in inverse iteration with iteration vector deflation to extract:

$$(\widetilde{\lambda}_1, \{\widetilde{\phi}\}_1), \quad (\widetilde{\lambda}_2, \{\widetilde{\phi}\}_2), \quad \dots \tag{4.159}$$

in ascending order. Since $\lambda_i = 1/\widetilde{\lambda}_i$ we have the folowing from (4.159).

$$(\lambda_n, \{\phi\}_n), \quad (\lambda_{n-1}, \{\phi\}_{n-1}), \quad \dots \tag{4.160}$$

In (4.160) we have eigenpairs of (4.155) in descending order.

4.3.3.2 Basic Steps in Iteration Vector Deflation

Consider the GEVP:

$$[A]\{x\} = \lambda[B]\{x\} \tag{4.161}$$

We present details for extracting the eigenpairs of (4.161) in ascending order, i.e., we use inverse iteration with iteration vector deflation. Let $(\lambda_i, \{\phi\}_i)$; $i = 1, 2, \dots, m$ be the eigenpairs that have already been calculated.

1. Choose $\lambda = 1$ in (4.161) and let $\{x\}_k^T = [1, 1, \dots, 1]$ be the initial guess for the eigenvector. λ_1 is the initial estimate of λ_{m+1}.

2. Calculate scalars α_i ; $i = 1, 2, \dots, m$ using:

$$\alpha_i = \{\phi\}_i^T [B]\{x\}_k \tag{4.162}$$

3. Calculate deflated starting iteration vector $\{\underset{\sim}{x}\}_k$ using:

$$\{\underset{\sim}{x}\}_k = \{x\}_k - \sum_{i=1}^{m} \alpha_i \{\phi\}_i \tag{4.163}$$

   $\{\underset{\sim}{x}\}_k$ is the initial estimate of $\{\phi\}_{m+1}$.

4. Using (4.161), we set up difference form with $\lambda = 1$.

$$[A]\{\bar{x}\}_{k+1} = [B]\{\underset{\sim}{x}\}_k \tag{4.164}$$

   Solve for $\{\bar{x}\}_{k+1}$ (solution of linear simultaneous algebraic equations).

5. Calculate a new estimate of λ_{m+1}, say $P_{m+1}(\{\bar{x}\}_{k+1})$, using (4.161) and $\{\bar{x}\}_{k+1}$.

$$P_{m+1}(\{\bar{x}\}_{k+1}) = \frac{\{\bar{x}\}_{k+1}^T [A]\{\bar{x}\}_{k+1}}{\{\bar{x}\}_{k+1}^T [B]\{\bar{x}\}_{k+1}} \tag{4.165}$$

6. $[B]$-normalize the new estimate $\{\bar{x}\}_{k+1}$ of $\{\phi\}_{m+1}$.

$$||\{\bar{x}\}_{k+1}|| = \sqrt{\{\bar{x}\}_{k+1}^T [B]\{\bar{x}\}_{k+1}} \tag{4.166}$$

$$\therefore \quad \{x\}_{k+1} = \frac{1}{||\{\bar{x}\}_{k+1}||}\{\bar{x}\}_{k+1} \tag{4.167}$$

   Hence, $(P_{m+1}(\{\bar{x}\}_{k+1}), \{x\}_{k+1})$ is the new estimate of $(\lambda_{m+1}, \{\phi\}_{m+1})$.

7. Convergence check:

$$\text{Eigenvalue :} \quad \left| \frac{P_{m+1}(\{\bar{x}\}_{k+1}) - P_{m+1}(\{\bar{x}\}_k)}{P_{m+1}(\{\bar{x}\}_{k+1})} \right| \leq \Delta_1 \tag{4.168}$$

$$\text{Eigenvector :} \quad ||\{x\}_{k+1} - \{x\}_k|| \leq \Delta_2 \tag{4.169}$$

   If converged then $(P_{m+1}(\{\bar{x}\}_{k+1}), \{x\}_{k+1}) \approx (\lambda_{m+1}, \{\phi\}_{m+1})$ and we stop, otherwise increment k by 1, i.e., set $k = k + 1$ and repeat steps 2 to 7.

Remarks.

(1) In the iteration process, we note that:

$$\lim_{k \to \infty} P_{m+1}(\{\bar{x}\}_{k+1}) = \lambda_{m+1} \tag{4.170}$$

$$\lim_{k \to \infty} \{x\}_{k+1} = \{\phi\}_{m+1} \tag{4.171}$$

(2) Once $(\lambda_{m+1}, \{\phi\}_{m+1})$ has been determined, we change m to $m+1$ and repeat steps 1 to 7 to extract $(\lambda_{m+2}, \{\phi\}_{m+2})$. This process can be continued until all eigenpairs have been determined.

(3) If eigenpairs are desired to be extracted in descending order, then we use (4.157) instead of (4.155) and follow exactly the same procedure as described above.

Example 4.6 (Gram-Schmidt Orthogonalization or Iteration Vector Deflation). Consider the following GEVP:

$$[A]\{x\} = \lambda[B]\{x\}$$

in which

$$[A] = \begin{bmatrix} 1 & -1 \\ -1 & 2 \end{bmatrix} \quad \text{and} \quad [B] = \begin{bmatrix} 2 & 1 \\ 1 & 2 \end{bmatrix}$$

We want to calculate both eigenpairs of this GEVP by using inverse iteration method with vector deflation.

First Eigenpair $(\lambda_1, \{\phi\}_1)$

First calculate $(\lambda_1, \{\phi\}_1)$ using the standard inverse iteration method with $\{x\}_1^T = [1 \;\; 1]$ as a guess of the eigenvector and $\lambda = 1$ as the corresponding eigenvalue and using a tolerance of 0.00001 for the relative error in the eigenvalue. This is similar to what has already been described in detail in Example 4.3. A summary of the calculations is given below.

Table 4.4: Results of inverse iteration method for first eigenpair of Example 4.6

k	$P(\{\bar{x}\}_{k+1})$	$\left\lvert\frac{P(\{\bar{x}\}_{k+1})-P(\{\bar{x}\}_k)}{P(\{\bar{x}\}_{k+1})}\right\rvert$	Normalized Eigenvector x_1	Normalized Eigenvector x_2
1	0.13157895E+00	0.00000E+00	0.48666E+00	0.32444E+00
2	0.13148317E+00	0.72846E−03	0.49058E+00	0.31995E+00
3	0.13148291E+00	0.19595E−05	0.49079E+00	0.31971E+00

Second Eigenpair $(\lambda_2, \{\phi\}_2)$

We already have:

$$\{\phi\}_1 = \begin{Bmatrix} 0.49079 \\ 0.31971 \end{Bmatrix}$$

hence m, the number of known eigenpairs, is one.

For $k = 1$

1. Choose $\lambda = 1$ and $\{x\}_1^T = [1 \;\; 1]$ (corresponding to $k = 1$).

2. Calculate scalars α_i ; $i = 1, 2, \dots, m$ using:

$$\alpha_i = \{\phi\}_i^T [B] \{x\}_1$$

In this case $m = 1$, hence we have:

$$\alpha_1 = \{\phi\}_1^T [B] \{x\}_1 = \begin{Bmatrix} 0.49079 \\ 0.31971 \end{Bmatrix}^T \begin{bmatrix} 2 & 1 \\ 1 & 2 \end{bmatrix} \begin{Bmatrix} 1 \\ 1 \end{Bmatrix} = 2.4315$$

3. Calculate the deflated starting iteration vector $\{\underset{\sim}{x}\}_1$ using:

$$\{\underset{\sim}{x}\}_1 = \{x\}_1 - \sum_{i=1}^{m} \alpha_i \{\phi\}_i = \{x\}_1 - \alpha_1 \{\phi\}_1$$

$$\text{or} \quad \{\underset{\sim}{x}\}_1 = \begin{Bmatrix} 1 \\ 1 \end{Bmatrix} - 2.4315 \begin{Bmatrix} 0.49079 \\ 0.31971 \end{Bmatrix} = \begin{Bmatrix} -0.19335 \\ 0.22262 \end{Bmatrix}$$

4. Construct the difference form of the eigenvalue problem for $\lambda = 1$.

$$[A]\{\bar{x}\}_2 = [B]\{\underset{\sim}{x}\}_1$$

Calculate $\{\bar{x}\}_2$ using $\{\underset{\sim}{x}\}_1$ from Step 3.

$$\{\bar{x}\}_2 = \begin{bmatrix} 1 & -1 \\ -1 & 2 \end{bmatrix}^{-1} \begin{bmatrix} 2 & 1 \\ 1 & 2 \end{bmatrix} \begin{Bmatrix} -0.19335 \\ 0.22262 \end{Bmatrix} = \begin{Bmatrix} -0.076285 \\ 0.087799 \end{Bmatrix}$$

5. Calculate the new estimate of λ_2, say $P_2(\{\bar{x}\}_2)$, using $\{\bar{x}\}_2$ in Step 4.

$$P_2(\{\bar{x}\}_2) = \frac{\{\bar{x}\}_2^T [A] \{\bar{x}\}_2}{\{\bar{x}\}_2^T [B] \{\bar{x}\}_2} = \frac{\begin{Bmatrix} -0.076285 \\ 0.087799 \end{Bmatrix}^T \begin{bmatrix} 1 & -1 \\ -1 & 2 \end{bmatrix} \begin{Bmatrix} -0.076285 \\ 0.087799 \end{Bmatrix}}{\begin{Bmatrix} -0.076285 \\ 0.087799 \end{Bmatrix}^T \begin{bmatrix} 2 & 1 \\ 1 & 2 \end{bmatrix} \begin{Bmatrix} -0.076285 \\ 0.087799 \end{Bmatrix}}$$

$$\therefore \quad P_2(\{\bar{x}\}_2) = 2.5352 \quad \text{(new estimate of } \lambda_2\text{)}$$

6. $[B]$-normalize the new estimate $\{\bar{x}\}_2$ of $\{\phi\}_2$.

$$||\{\bar{x}\}_2|| = \sqrt{\{\bar{x}\}_2^T[B]\{\bar{x}\}_2} = \sqrt{\begin{Bmatrix}-0.076285\\0.087799\end{Bmatrix}^T \begin{bmatrix}2&1\\1&2\end{bmatrix}\begin{Bmatrix}-0.076285\\0.087799\end{Bmatrix}}$$

or

$$||\{\bar{x}\}_2|| = 0.11688$$

$$\therefore \quad \{x\}_2 = \frac{1}{||\{\bar{x}\}_2||}\{\bar{x}\}_2 = \frac{1}{0.11688}\begin{Bmatrix}-0.076285\\0.087799\end{Bmatrix} = \begin{Bmatrix}0.65268\\0.75120\end{Bmatrix}$$

Thus at the end of the calculations for $k = 1$, we have the following estimate of the second eigenpair $(\lambda_2, \{\phi\}_2)$.

$$\left(2.5352, \begin{Bmatrix}0.65268\\0.75120\end{Bmatrix}\right) = (P_2(\{\bar{x}\}_2, \{x\}_2))$$

For $k = 2$

1. Choose $\lambda = 1$ and $\{x\}_2 = \begin{Bmatrix}0.65268\\0.75120\end{Bmatrix}$, from Step 6 for $k = 1$.

2. Calculate the scalar α_1.

$$\alpha_i = \{\phi\}_i^T[B]\{x\}_2 = \begin{Bmatrix}0.49079\\0.31971\end{Bmatrix}\begin{bmatrix}2&1\\1&2\end{bmatrix}\begin{Bmatrix}0.65268\\0.75120\end{Bmatrix}$$

or

$$\alpha_1 = -0.31083 \times 10^{-3}$$

3. Calculate the deflated iteration vector $\{\underset{\sim}{x}\}_2$ using:

$$\{\underset{\sim}{x}\}_2 = \{x\}_2 - \sum_{i=1}^{m} \alpha_i\{\phi\}_i = \{x\}_1 - \alpha_1\{\phi\}_1$$

$$\{\underset{\sim}{x}\}_2 = \begin{Bmatrix}0.65268\\0.75120\end{Bmatrix} - (-0.31083 \times 10^{-3})\begin{Bmatrix}0.49079\\0.31971\end{Bmatrix}$$

$$\text{or} \quad \{\underset{\sim}{x}\}_2 = \begin{Bmatrix}-0.65253\\0.75130\end{Bmatrix}$$

4. Construct the difference form of the eigenvalue problem for $\lambda = 1$.

$$[A]\{\bar{x}\}_3 = [B]\{\underset{\sim}{x}\}_2$$

Calculate $\{\bar{x}\}_3$ using $\{\underset{\sim}{x}\}_2$ from Step 3.

$$\{\bar{x}\}_3 = \begin{bmatrix}1&-1\\-1&2\end{bmatrix}^{-1}\begin{bmatrix}2&1\\1&2\end{bmatrix}\begin{Bmatrix}-0.65253\\0.75130\end{Bmatrix} = \begin{Bmatrix}-0.25745\\0.29631\end{Bmatrix}$$

5. Calculate the new estimate of λ_2, $P_2(\{\bar{x}\}_3)$, using $\{\bar{x}\}_3$ in Step 4.

$$P_2(\{\bar{x}\}_3) = \frac{\{\bar{x}\}_3^T[A]\{\bar{x}\}_3}{\{\bar{x}\}_3^T[B]\{\bar{x}\}_3} = 2.5352 \quad \begin{array}{l}\text{same as for } k = 1 \text{ up} \\ \text{to four decimal places}\end{array}$$

6. $[B]$-normalize the new estimate $\{\bar{x}\}_3$ of $\{\phi\}_2$.

$$||\{\bar{x}\}_3|| = \sqrt{\{\bar{x}\}_3^T[B]\{\bar{x}\}_3} = 0.39445$$

$$\therefore \quad \frac{1}{||\{\bar{x}\}_3||}\{\bar{x}\}_3 = \begin{Bmatrix} -0.65268 \\ 0.75120 \end{Bmatrix} \quad ; \quad \begin{array}{l}\text{same as } \{x\}_2 \text{ for } k = 1 \\ \text{up to 5 decimal places}\end{array}$$

Thus, at the end of the calculations for $k = 2$, we have the following estimate of the second eigenpair $(\lambda_2, \{x\}_2)$.

$$\left(2.5352, \begin{Bmatrix} -0.65268 \\ 0.75120 \end{Bmatrix}\right) = (P_2(\{\bar{x}\}_3), \{x\}_3)$$

The calculations are summarized in the following, including relative error.

Table 4.5: Results of vector deflation for the second eigenpair of Example 4.6

k	$P(\{\bar{x}\}_{k+1})$	$\left\lvert\frac{P(\{\bar{x}\}_{k+1})-P(\{\bar{x}\}_k)}{P(\{\bar{x}\}_{k+1})}\right\rvert$	Normalized Eigenvector x_1	x_2
1	0.25351835E+01	0.00000E+00	-0.65268E+00	0.75120E+00
2	0.25351835E+01	0.17517E−15	-0.65268E+00	0.75120E+00

From the relative error, we note that for $k = 2$, we have converged values of the second eigenpair, hence:

$$(\lambda_2, \{\phi\}_2) = \left(2.5352, \begin{Bmatrix} -0.65268 \\ 0.75120 \end{Bmatrix}\right)$$

Thus, the second eigenpair is determined using $(\lambda_1, \{\phi\}_1)$ and iteration vector deflation method. Just in case the estimates of the second eigenpair are not accurate enough for $k = 2$, the process can be continued for $k = 3, 4, \ldots$ until desired accuracy is achieved.

4.3.4 Shifting in Eigenpair Calculations

Shifting is a technique in eigenpair calculations that can be used to achieve many desired features.

(i) Shifting may be used to avoid calculations of zero eigenvalues.

(ii) Shifting may be used to improve convergence of the inverse or forward iteration methods.

(iii) Shifting can be used to calculate eigenpairs other than $(\lambda_1, \{\phi\}_1)$ and $(\lambda_n, \{\phi\}_n)$ in inverse and forward iteration methods.

(iv) In inverse iteration method, if $[A]$ is singular or positive-semidefinite, then shift can be used to make it positive-definite without influencing the eigenvector.

(v) In forward iteration method, if $[B]$ is singular or positive-semidefinite, then shift can be used to make it positive-definite without influencing the eigenvector.

4.3.4.1 What is a Shift?

Consider the GEVP:

$$[A]\{x\} = \lambda[B]\{x\} \tag{4.172}$$

Consider $\mu \in \mathbb{R}$, $\mu \neq 0$. Consider the new GEVP defined by:

$$[[A] - \mu[B]]\{y\} = \eta[B]\{y\} \tag{4.173}$$

in which η and $\{y\}$ are an eigenvalue and eigenvector of the GEVP (4.173). μ is called the shift and the GEVP defined by (4.173) is called the shifted GEVP.

4.3.4.2 Consequences of Shifting

Using (4.172) and (4.173), we determine relationship(s) between $(\lambda, \{x\})$ and $(\eta, \{y\})$, the eigenpairs of (4.172) and (4.173). From (4.173) we can write:

$$[A]\{y\} = (\eta + \mu)[B]\{y\} \tag{4.174}$$

Comparing (4.172) and (4.174) we note that in both GEVPs we have the same $[A]$ and $[B]$, hence the following must hold.

$$\begin{gathered} \lambda = \eta + \mu \qquad \text{or} \qquad \eta = \lambda - \mu \\ \{y\} = \{x\} \end{gathered} \tag{4.175}$$

Thus,

(i) The eigenvectors of the original GEVP and shifted GEVP are the same.

(ii) The eigenvalues η of the shifted GEVP are shifted by μ compared to the eigenvalues λ of the original GEVP.

Remarks.

(1) Shifting also holds for the SEVP as in this case the only difference compared to GEVP is that $[B] = [I]$.

(2) If $\lambda = 0$ is the smallest eigenvalue of the GEVP or SEVP, then by using shifting (a negative value of μ) we can construct shifted eigenvalue problem (4.173) such that the smallest eigenvalue of the shifted eigenvalue problem will be greater than zero.

(3) When $[A]$ or $[B]$ is singular or positive-semidefinite, then shifting can be used in inverse or forward iteration methods to make a new $[A]$ or $[B]$ in the shifted eigenvalue problem positive-definite, thereby avoiding difficulties in their inverses.

4.4 Transformation Methods for Eigenvalue Problems

In transformation methods the treatment of the SEVP and GEVP differs somewhat but the basic principle is the same. In the following we present the basic ideas employed for the SEVP and GEVP in designing transformation methods.

4.4.1 SEVP: Orthogonal Transformation, Change of Basis

Consider the following SEVP:

$$[A]\{x\} = \lambda[I]\{x\} \tag{4.176}$$

in which $[A]$ is a symmetric matrix. First, we show that an orthogonal transformation on (4.176) does not alter its eigenvalues, but the eigenvectors do change. In an orthogonal transformation we perform a change of basis, i.e., we replace $\{x\}$ by $\{x\}_1$ through an orthogonal transformation of the type:

$$\{x\} = [P_1]\{x\}_1 \tag{4.177}$$

in which $[P_1]$ is orthogonal, i.e., $[P_1]^{-1} = [P_1]^T$. Substituting from (4.177) into (4.176) and premultiplying by $[P_1]^T$, we obtain:

$$[P_1]^T[A][P_1]\{x\}_1 - \lambda[P_1]^T[I][P_1]\{x\} = \{0\}$$

or

$$\left[[P_1]^T\left[[A] - \lambda[I]\right][P_1]\right]\{x\}_1 \tag{4.178}$$

In the eigenvalue problem (4.178), $\{x\}_1$ is the eigenvector (and not $\{x\}$), thus a change of basis alters eigenvectors. To determine the eigenvalues of

(4.178), we determine the characteristic polynomial associated with (4.178), i.e., we set the determinant of the coefficient matrix in (4.178) to zero.

$$\det\left[[P_1]^T\left[[A]-\lambda[I]\right][P_1]\right]=0 \tag{4.179}$$

or

$$\det[P_1]^T\det[[A]-\lambda[I]]\det[P_1]=0 \tag{4.180}$$

Since $[P_1]$ is orthogonal:

$$\det[P_1]=\det[P_1]^T=1 \tag{4.181}$$

Hence, (4.180) reduces to:

$$\det[[A]-\lambda[I]]=0=p(\lambda) \tag{4.182}$$

Which is the same as the characteristic polynomial of the original eigenvalue problem (4.176). Thus, a change of basis in the SEVP through an orthogonal transformation does not alter its eigenvalues. The eigenvectors of the original SEVP (4.176) and the transformed SEVP (4.178), $\{x\}$ and $\{x\}_1$, are naturally related through $[P_1]$ as shown in (4.177).

4.4.2 GEVP: Orthogonal Transformation, Change of Basis

Consider the following GEVP:

$$[A]\{x\}=\lambda[B]\{x\} \tag{4.183}$$

in which $[A]$ and $[B]$ are symmetric matrices. Here also, we show that an orthogonal transformation on (4.183) does not alter its eigenvalues but the eigenvectors change. As in the case of the SEVP, we replace $\{x\}$ by $\{x\}_1$ through an orthogonal transformation of the type:

$$\{x\}=[P_1]\{x\}_1 \tag{4.184}$$

in which

$$[P_1]^{-1}=[P_1]^T \quad \text{and} \quad \det[P_1]=\det[P_1]^T=1 \tag{4.185}$$

Substituting from (4.184) into (4.183) and premultiplying (4.183) by $[P_1]^T$:

$$[P_1]^T[A][P_1]\{x\}_1-\lambda[P_1]^T[B][P_1]\{x\}_1-\{0\} \tag{4.186}$$

or

$$\left[[P_1]^T\left[[A]-\lambda[B]\right][P_1]\right]\{x\}_1 \tag{4.187}$$

In the eigenvalue problem (4.187), $\{x\}_1$ is the eigenvector (and not $\{x\}$), thus change of basis alters eigenvectors. To determine the eigenvalues of (4.187), we determine the characteristic polynomial associated with the eigenvalue problem (4.187), i.e., we set the determinant of the coefficient matrix in (4.187) to zero.

$$\det\left[[P_1]^T\left[[A]-\lambda[B]\right][P_1]\right]=0 \tag{4.188}$$

or

$$\det[P_1]^T\det\left[[A]-\lambda[B]\right]\det[P_1]=0 \tag{4.189}$$

Using (4.185), (4.189) reduces to:

$$\det\left[[A]-\lambda[B]\right]=0=p(\lambda) \tag{4.190}$$

which is the same as the characteristic polynomial of the original GEVP (4.183). Thus, a change of basis in the GEVP through an orthogonal transformation does not alter its eigenvalues. The eigenvectors of the original GEVP (4.183) and the transformed GEVP (4.187), $\{x\}$ and $\{x\}_1$, are naturally related through $[P_1]$ as shown in (4.184).

Remarks.

(1) In all transformation methods we perform a series of orthogonal transformations on the original eigenvalue problem such that:

(a) In the case of the SEVP, $[A]$ becomes a diagonal matrix but $[I]$ matrix remains unaltered. Then, the diagonals of this transformed $[A]$ matrix are the eigenvalues and columns of the products of the transformation matrices contain the eigenvectors.

(b) In the case of the GEVP, we make both $[A]$ and $[B]$ diagonal matrices through orthogonal transformations. Then, the ratios of the corresponding diagonals of transformed $[A]$ and $[B]$ are the eigenvalues and the columns of the products of transformation matrices contain the eigenvectors.

(2) In transformation methods all eigenpairs are extracted simultaneously.

(3) The eigenpairs are not in any particular order, hence these must be arranged in ascending order.

(4) Just like the root-finding methods used for the characteristic polynomial, the transformation methods are also iterative. Thus, the eigenpairs are determined only within the accuracy of preset thresholds for the eigenvalues and eigenvectors. The transformation methods are indeed methods of approximation.

(5) In the following sections we present details of the Jacobi method for the SEVP, the Generalized Jacobi method for the GEVP, and only provide

basic concepts of Householder QR and subspace iteration methods as these methods are quite involved, hence detailed presentations of these methods are beyond the scope of study in this book.

4.4.3 Jacobi Method for SEVP

Consider the SEVP:

$$[A]\{x\} = \lambda[I]\{x\} \tag{4.191}$$

in which $[A]$ is a symmetric matrix. Our aim is to perform a series of orthogonal transformations (change of basis) on (4.191) such that all off-diagonal elements of $[A]$ become zero, i.e., $[A]$ becomes a diagonal matrix while $[I]$ on right side of (4.191) remains unaltered.

Consider the change of basis using an orthogonal matrix $[P_1]$.

$$\{x\} = [P_1]\{x\}_1 \tag{4.192}$$

Substitute (4.192) in (4.191) and premultiply (4.191) by $[P_1]^T$.

$$[P_1]^T[A][P_1]\{x\}_1 = \lambda[P_1]^T[I][P_1]\{x\}_1 \tag{4.193}$$

Since $[P_1]$ is orthogonal:

$$[P_1]^T[I][P_1] = [I] \tag{4.194}$$

Hence, (4.193) becomes:

$$[P_1]^T[A][P_1]\{x\}_1 = \lambda[I]\{x\}_1 \tag{4.195}$$

We construct $[P_1]$ in such a way that it makes an off-diagonal element of $[A]$ zero in $[P_1]^T[A][P_1]$. Perform another change of basis using an orthogonal matrix $[P_2]$.

$$\{x\}_1 = [P_2]\{x\}_2 \tag{4.196}$$

Substitute from (4.196) in (4.195) and premultiply (4.195) by $[P_2]^T$.

$$[P_2]^T[P_1]^T[A][P_1][P_2]\{x\}_2 = \lambda[P_2]^T[I][P_2]\{x\}_2 \tag{4.197}$$

Since $[P_2]$ is orthogonal:

$$[P_2]^T[I][P_2] = [I] \tag{4.198}$$

Hence, (4.197) reduces to:

$$[P_2]^T[P_1]^T[A][P_1][P_2]\{x\}_2 = \lambda[I]\{x\}_2 \tag{4.199}$$

and

$$\{x\} = [P_1][P_2]\{x\}_2 \tag{4.200}$$

Equations (4.200) describe how the eigenvectors $\{x\}_2$ of (4.199) are related to the eigenvectors $\{x\}$ of the original SEVP (4.191). We construct $[P_2]$ such that the transformation (4.199) makes an off-diagonal element zero in $[P_2]^T([P_1]^T[A][P_1])[P_2]$. This process is continued by choosing off-diagonal elements of the progressively transformed $[A]$ in sequence until all off-diagonal elements have been considered. In this process it is possible that when we zero out a specific element of the transformed $[A]$, the element that was made zero in the immediately preceding transformation may not remain zero, but may be of a lower magnitude than its original value. Thus, to make $[A]$ diagonal it may be necessary to make more than one pass through the transformed off-diagonal elements of $[A]$. We discuss details in the following sections.

Thus, after k transformations, we obtain:

$$[P_k]^T[P_{k-1}]\dots[P_2]^T[P_1]^T[A][P_1][P_2]\dots[P_{k-1}][P_k]\{x\}_k = \lambda[I]\{x\}_k \quad (4.201)$$

And we have the following:

$$\lim_{k\to\infty}\left[[P_k]^T[P_{k-1}]\dots[P_2]^T[P_1]^T[A][P_1][P_2]\dots[P_{k-1}][P_k]\right] = [\Lambda] \quad (4.202)$$

$$\lim_{k\to\infty}\left[[P_1][P_2]\dots[P_{k-1}][P_k]\right] = [\Phi] \quad (4.203)$$

in which $[\Lambda]$ is a diagonal matrix containing the eigenvalues and the columns of the square matrix $[\Phi]$ are the corresponding eigenvectors. Thus in the end, when $[A]$ becomes the diagonal matrix $[\Lambda]$ we have all eigenvalues in $[\Lambda]$ and $[\Phi]$ contains all eigenvectors. In (4.202) and (4.203), each value of k corresponds to a complete pass through all of the off-diagonal elements of the progressively transformed $[A]$ that are not zero.

4.4.3.1 Constructing $[P_l]$; $l = 1, 2, \dots, k$ Matrices

$[P_l]$ orthogonal matrices in the Jacobi method are called rotation matrices as these represent rigid rotations of the coordinate axes. A specific $[P_l]$ is designed to make a specific off-diagonal term of the transformed $[A]$ (beginning with $[P_1]$ for $[A]$) zero. Since $[A]$ is symmetric, $[P_l]$ can be designed to make an off-diagonal term of $[A^l]$ (transformed $[A]$) as well as its transposed term zero at the same time.

Let us say that we have already performed some transformations and the new transformed $[A]$ is $[A^l]$. We wish to make a^l_{ij} and a^l_{ji} of $[A^l]$ zero. We design $[P_l]$ as follows to accomplish this.

$$
[P_l] = \begin{bmatrix}
1 & & & & & & & \\
 & 1 & & & & & & \\
 & & \cos\theta & & & -\sin\theta & & \\
 & & & 1 & & & & \\
 & & & & 1 & & & \\
 & & \sin\theta & & & \cos\theta & & \\
 & & & & & & 1 & \\
 & & & & & & & 1
\end{bmatrix}
\begin{matrix} \\ \\ \leftarrow \text{Row } i \\ \\ \\ \leftarrow \text{Row } j \\ \\ \\ \end{matrix}
\quad \text{(Column } i \text{ and Column } j \text{ marked above the entries } \cos\theta, \sin\theta \text{)} \tag{4.204}
$$

In $[P_l]$, the elements corresponding to rows i, j and columns i, j are non-zero. The remaining diagonal elements are unity, and all other elements are zero. θ is chosen such that in the following transformed matrix:

$$[P_l]^T[A^l][P_l] \tag{4.205}$$

The elements at the locations i, j and j, i, i.e., $a^l_{ij} = a^l_{ji}$, have become zero. This allows us to determine θ.

$$
\begin{aligned}
\tan(2\theta) &= \frac{2a^l_{ij}}{a^l_{ii} - a^l_{jj}} \quad ; \quad a^l_{jj} \neq a^l_{ii} \\
\text{and} \quad \theta &= \frac{\pi}{4} \quad \text{when} \quad a^l_{jj} = a^l_{ii}
\end{aligned} \tag{4.206}
$$

4.4.3.2 Using Jacobi Method

Consider the SEVP:

$$[A]\{x\} = \lambda[I]\{x\} \tag{4.207}$$

Since $[A]$ is symmetric, we only need to consider making the elements above the diagonal zero as the transformations (4.204) ensure that the corresponding element below the diagonal will automatically become zero. We follow the steps outlined below.

1. Consider the off-diagonal elements row-wise, and each element of the row in sequence. That is first consider row one of the original matrix $[A]$ and the off-diagonal element a_{12} ($a_{21} = a_{12}$). We use a_{11}, a_{22}, and a_{12} to determine θ using (4.206) and then use this value of θ to construct $[P_1]$ using (4.204) and perform the orthogonal transformation on $[A]$ to obtain $[A^1]$.

   $$[P_1]^T[A][P_1] = [A^1] \tag{4.208}$$

   In $[A^1]$, a^1_{12} and a^1_{21} have become zero.

2. Next consider $[A^1]$ and the next element in row one, i.e., a^1_{13} ($a^1_{31} = a^1_{13}$). Using a^1_{11}, a^1_{33}, and a^1_{13} and (4.206) determine θ and then use this value

of θ in (4.204) to determine $[P_2]$. Perform the orthogonal transformation on $[A^1]$ to obtain $[A^2]$.

$$[P_2]^T[A^1][P_2] = [A^2] \tag{4.209}$$

In $[A^2]$, a^2_{13} and a^2_{31} have become zero but the elements at the locations 1,2 and 2,1 made zero in (1) may have become non-zero, however its magnitude may be less than a_{12} in the original $[A]$. This is a drawback of this transformation. Also accumulate the product $[P_1][P_2]$ as it is needed to recover the eigenvector.

3. Next consider a^2_{14} ($a^2_{41} = a^2_{14}$) of $[A^2]$ in row one. Construct $[P_3]$ using a^2_{11}, a^2_{44}, and a^2_{14} in (4.206) and then θ in (4.204). Perform the orthogonal transformation on $[A^2]$ to obtain $[A^3]$.

$$[P_3]^T[A^2][P_3] = [A^3] \tag{4.210}$$

In $[A^3]$, a^3_{14} and a^3_{41} have become zero but the elements at locations 1,3 and 3,1 made zero in (2) may have become non-zero, however their magnitudes may be smaller than a_{13} and a_{31} in the original $[A]$. Also accumulate the product $[P_1][P_3][P_3]$ as it is needed for recovering the eigenvectors of the original eigenvalue problem.

4. We continue this process for all elements of row one using progressively transformed $[A]$. When row one is exhausted, we begin with the off-diagonal elements of row two of the most recently transformed $[A]$. This process is continued until all off-diagonal elements of all rows of the progressively transformed $[A]$ have been considered once. This constitutes a 'sweep', i.e., we have swept all off-diagonal elements of progressively transformed $[A]$ matrices once. At the end of the sweep only the last off-diagonal element made zero remains zero. All other off-diagonal elements may have become non-zero, but their magnitudes are generally smaller than those in the original $[A]$.

5. Using the more recently transformed $[A]$ in (4), we begin another sweep starting with row one. At the end of sweep two the off-diagonal elements in the transformed $[A]$ will even be smaller than those after sweep one. We also continue accumulating the products of the transformation matrices continuously in the sweeps. We make as many sweeps as necessary to ensure that each off-diagonal element is the most recently transformed $[A]$ is below a preset tolerance Δ of numerically computed zero.

6. The procedure described above is called *Cyclic Jacobi* method due to the fact we have cycles of sweeps of the off-diagonal elements until all off-diagonal elements are below a preset tolerance of numerically computed zero.

7. *Threshold Jacobi*
   In threshold Jacobi we perform an orthogonal transformation to zero out an off-diagonal element of $[A]$ (or of the most recently transformed $[A]$) and then check the magnitude of the next off-diagonal element to be made zero. If it is below the threshold we skip the orthogonal transformation for it and move to the next off-diagonal element in sequence, keeping in mind that the same rule applies to this current off-diagonal elements as well as those to come. It is clear that in this procedure we avoid unnecessary transformation for the elements that are already within the threshold of zero. Thus, the threshold Jacobi clearly is more efficient than cyclic Jacobi.

8. A sweep is generally quite fast if $[A]$ is not too large.

Example 4.7 (Jacobi Method for SEVP). Consider the SEVP:

$$[A]\{x\} = \lambda[I]\{x\}$$

in which

$$[A] = \begin{bmatrix} 2 & -1 \\ -1 & 4 \end{bmatrix}$$

In this case there is only one off-diagonal term in $[A]$, $a_{12} = a_{21} = -1$ (row 1, column 2 ; $i = 1$, $j = 2$). We construct $[P_1]$ or $[P_{12}]$ to make $a_{12} = a_{21} = -1$ zero in $[A]$. The subscript 12 in $[P_{12}]$ implies the matrix $[P]$ corresponding to the element of $[A]$ located at row 1, column 2.

$$[P_{12}] = \begin{bmatrix} \cos\theta & -\sin\theta \\ \sin\theta & \cos\theta \end{bmatrix}$$

$$\tan 2\theta = \frac{2a_{12}}{a_{11} - a_{22}} = \frac{2(-1)}{2-4} = \frac{-2}{-2} = 1$$

$$\therefore \quad 2\theta = \frac{\pi}{4} \quad ; \quad \theta = \frac{\pi}{8}$$

$$\cos\left(\frac{\pi}{8}\right) = 0.92388$$

$$\sin\left(\frac{\pi}{8}\right) = 0.38268$$

$$\therefore \quad [P_{12}] = \begin{bmatrix} 0.92388 & -0.38268 \\ 0.38268 & 0.92388 \end{bmatrix}$$

$$\therefore \quad [A^1] = [P_{12}]^T[A][P_{12}]$$

or

$$
\begin{aligned}
[A^1] &= \begin{bmatrix} 0.92388 & 0.38268 \\ -0.38268 & 0.92388 \end{bmatrix} \begin{bmatrix} 2 & -1 \\ -1 & 4 \end{bmatrix} \begin{bmatrix} 0.92388 & -0.38268 \\ 0.38268 & 0.92388 \end{bmatrix} \\
&= \begin{bmatrix} 0.92388 & 0.38268 \\ -0.38268 & 0.92388 \end{bmatrix} \begin{bmatrix} 1.46508 & -1.68924 \\ 0.60684 & 4.0782 \end{bmatrix} \\
&= \begin{bmatrix} 1.58578 & 0 \\ 0 & 4.41421 \end{bmatrix} = \begin{bmatrix} \lambda_1 & 0 \\ 0 & \lambda_2 \end{bmatrix}
\end{aligned}
$$

The eigenvectors corresponding to λ_1 and λ_2 are the columns of $[P_{12}]$.

$$
[\Phi] = [P_{12}] = \begin{bmatrix} 0.92388 & -0.38268 \\ 0.38268 & 0.92388 \end{bmatrix} = [\{\phi\}_1, \{\phi\}_2]
$$

Hence, we have:

$$
(\lambda_1, \{\phi\}_1) = \left(1.52578, \begin{Bmatrix} 0.92388 \\ 0.38268 \end{Bmatrix}\right)
$$

$$
\text{and} \qquad (\lambda_2, \{\phi\}_2) = \left(4.41421, \begin{Bmatrix} -0.38268 \\ 0.92388 \end{Bmatrix}\right)
$$

as the two eigenpairs of the SEVP considered here.

4.4.4 Generalized Jacobi Method for GEVP

Consider the GEVP:

$$
[A]\{x\} = \lambda[B]\{x\} \tag{4.211}
$$

in which $[A]$ and $[B]$ are symmetric matrices and $[B] \neq [I]$. Our aim in the generalized Jacobi method is to perform a series of orthogonal transformation on (4.211) such that:

(i) $[A]$ becomes a diagonal matrix and $[B]$ becomes $[I]$. The diagonal elements of the final transformed $[A]$, i.e., $[A^k]$ (after k transformations), will be the eigenvalues of (4.211) and the columns of the product of the transformation matrices will contain the corresponding eigenvectors.

(ii) $[A]$ and $[B]$ both become diagonal, but $[B]$ is not an identity matrix. In this approach the ratios of the diagonal elements of the transformed $[A]$ and $[B]$, i.e., $[A^k]$ and $[B^k]$ (after k transformations), will be the eigenvalues and the columns of the product of the transformation matrices will contain the corresponding eigenvectors.

(iii) Using either (i) or (ii), the results remain unaffected. In designing transformation matrices, (ii) is easier.

4.4.4.1 Basic Theory of Generalized Jacobi Method

Consider the change of basis in (4.211) using:

$$\{x\} = [P_1]\{x_1\} \tag{4.212}$$

in which $[P_1]$ is orthogonal, i.e., $[P_1]^T = [P_1]^{-1}$. Substituting from (4.212) in (4.211) and premultiplying by $[P_1]^T$:

$$[P_1]^T[A][P_1]\{x\}_1 = \lambda[P_1]^T[B][P_1]\{x\}_1 \tag{4.213}$$

We choose $[P_1]$ such that an off-diagonal element of $[A]$ and the corresponding off-diagonal element of $[B]$ become zero. If we define:

$$\begin{aligned}[A^1] &= [P_1]^T[A][P_1] \\ [B^1] &= [P_1]^T[B][P_1]\end{aligned} \tag{4.214}$$

Then we can write (4.213) as:

$$[A^1]\{x\}_1 = \lambda[B_1]\{x\}_1 \tag{4.215}$$

$[A^1]$ and $[B^1]$ are the transformed $[A]$ and $[B]$ after the first orthogonal transformation. Perform another change of basis on (4.215).

$$\{x\}_2 = [P_2]\{x\}_1 \tag{4.216}$$

Substituting from (4.216) in (4.215) and premultiplying by $[P_2]^T$:

$$[P_2]^T[A^1][P_2]\{x\}_2 = \lambda[P_2]^T[B^1][P_2]\{x\}_2 \tag{4.217}$$

We choose $[P_2]$ such that it makes an off-diagonal element of $[A^1]$ and the corresponding element of $[B^1]$ zero and

$$\{x\} = [P_1][P_2]\{x\}_k \tag{4.218}$$

Continuing this process we obtain after k transformations:

$$\begin{aligned}&[P_k]^T[P_{k-1}]^T \dots [P_2]^T[P_1]^T[A][P_1][P_2] \dots [P_{k-1}][P_k] \\ &\qquad = \lambda[P_k]^T[P_{k-1}]^T \dots [P_2]^T[P_1]^T[B][P_1][P_2] \dots [P_{k-1}][P_k]\end{aligned} \tag{4.219}$$

$$\text{or} \quad [P_k]^T[A^{k-1}][P^k]\{x\}_k = \lambda[P_k]^T[B_{k-1}][P_k]\{x\}_k \tag{4.220}$$

$$\text{or} \quad [A^k]\{x\} = \lambda[B_k]\{x\}_k \tag{4.221}$$

Definitions of $[A^k]$ and $[B^k]$ are clear from (4.219), and

$$\{x\} = [P_1][P_2]\dots[P_{k-1}][P_k]\{x\}_k \tag{4.222}$$

And we have the following.

$$
\begin{aligned}
&\lim_{k\to\infty}[A^k] = [\tau] \quad ; \quad \text{a diagonal matrix} \\
&\lim_{k\to\infty}[B^k] = [\Lambda] \quad ; \quad \text{a diagonal matrix} \\
\text{and} \quad &\lim_{k\to\infty}[P_1][P_2]\dots[P_{k-1}][P_k] = [\Phi]
\end{aligned} \tag{4.223}
$$

The eigenvalues are given by (not in any particular order):

$$\lambda_i = \frac{\tau_{ii}}{\Lambda_{ii}} \quad ; \quad i = 1, 2, \dots, n \tag{4.224}$$

The columns of $[\Phi]$ are the corresponding eigenvectors. What remains in this method is to consider the details of constructing $[P_l]$; $l = 1, 2, \dots, k$ matrices.

4.4.4.2 Construction of $[P_l]$ Matrices

The $[P_l]$ matrices are called rotation matrices, same as in the case of the Jacobi method for the SEVP. In the design of $[P_l]$ we take into account that $[A]$ and $[B]$ are symmetric. To be general, consider $[A^l]$ and $[B^l]$ after l transformations. Let us say that we want to make a^l_{ij} and b^l_{ij} zero (a^l_{ji} and b^l_{ji} are automatically made zero as we consider symmetry of $[A]$ and $[B]$ in designing $[P_{l+1}]$), then $[P_{l+1}]$ can have the following form.

$$
[P_{l+1}] =
\begin{array}{c}
\begin{array}{cccccccc} & & i & & & j & & \end{array} \\
\left[\begin{array}{cccccccc}
1 & & & & & & & \\
 & 1 & & & & & & \\
 & & 1 & & & \alpha & & \\
 & & & 1 & & & & \\
 & & & & 1 & & & \\
 & & \beta & & & 1 & & \\
 & & & & & & 1 & \\
 & & & & & & & 1
\end{array}\right]
\end{array}
\begin{array}{c} \\ \\ \\ i \\ \\ \\ j \\ \\ \\ \end{array}
\tag{4.225}
$$

The parameters α and β are determined such that in the transformed $[A^l]$ and $[B^l]$, i.e., in

$$[P_{l+1}]^T[A^l][P_{l+1}] \quad \text{and} \quad [P_{l+1}]^T[B^l][P_{l+1}] \tag{4.226}$$

a^l_{ij} and b^l_{ij} are zero (hence, a^l_{ji} and b^l_{ji} are zero). Using $[P_{l+1}]$ in (4.226) and setting $a^l_{ij} = b^l_{ij} = 0$ gives us the following two equations.

$$\begin{aligned} \alpha a^l_{ii} + (1+\alpha\beta)a^l_{ii} + \beta a^l_{jj} &= 0 \\ \alpha b^l_{ii} + (1+\alpha\beta)b^l_{ij} + \beta b^l_{jj} &= 0 \end{aligned} \tag{4.227}$$

These are nonlinear equations in α and β and have the following solution.

$$\alpha = \frac{\bar{a}^l_{jj}}{X} \quad , \quad \beta = -\frac{a^l_{ii}}{X}$$

where

$$\begin{aligned} X &= \frac{\bar{a}^l}{2} + \text{sign}(\bar{a}^1)\sqrt{\left(\frac{\bar{a}^{-l}}{2}\right)^2 + \bar{a}^l_{ii}\bar{a}^l_{jj}} \\ \bar{a}^l_{ii} &= a^l_{ii}b^l_{ij} - b^l_{ii}a^l_{ij} \\ \bar{a}^l_{jj} &= a^l_{jj}b^l_{ij} - b^l_{jj}a^l_{ij} \\ \bar{a}^l &= a^l_{ii}b^l_{jj} - a^l_{jj}b^l_{ii} \end{aligned} \tag{4.228}$$

• The basic steps in this method are identical to the Jacobi method described for the SEVP.

• Thus, we have the cyclic generalized Jacobi and threshold generalized Jacobi methods.

Example 4.8 (Generalized Jacobi Method for GEVP). Consider the GEVP:

$$[A]\{x\} = \lambda[B]\{x\}$$

in which

$$[A] = \begin{bmatrix} 1 & -1 \\ -1 & 1 \end{bmatrix} \quad \text{and} \quad [B] = \begin{bmatrix} 2 & 1 \\ 1 & 2 \end{bmatrix}$$

The off-diagonal elements of $[A]$ and $[B]$ at location (1,2) are to be made zero. We perform the change of basis, i.e., orthogonal transformation, on the eigenvalue problem to diagonalize $[A]$ and $[B]$. Since $[A]$ and $[B]$ have only one off-diagonal element to be made zero (due to symmetry), one orthogonal transformation is needed to accomplish this. In this case:

$$\begin{aligned} a_{12} &= a_{21} = -1 \\ b_{12} &= b_{21} = 1 \end{aligned}$$

$$[P_{12}] = \begin{bmatrix} 1 & \alpha \\ \beta & 1 \end{bmatrix}$$

α and β are calculated as

$$\bar{a}_{11} = a_{11}b_{12} - b_{11}a_{12} = 1(1) - 2(-1) = 3$$
$$\bar{a}_{22} = a_{22}b_{12} - b_{22}a_{12} = 1(1) - 2(-1) = 3$$
$$\bar{a} = a_{11}b_{22} - a_{22}b_{11} = 1(2) - 1(2) = 0 \quad \text{(sign is positive)}$$
$$\therefore \quad X = \frac{\bar{a}}{2} + \text{sign}(\bar{a})\sqrt{\left(\frac{\bar{a}}{2}\right)^2 + \bar{a}_{11}\bar{a}_{22}}$$
$$\therefore \quad X = \frac{0}{2} + \text{sign}(\bar{a})\sqrt{\left(\frac{0}{2}\right)^2 + (3)(3)} = 0 + 3 = 3$$
$$\therefore \quad \alpha = \frac{\bar{a}_{22}}{X} = \frac{3}{3} = 1\,; \qquad \beta = -\frac{\bar{a}_{11}}{X} = -\frac{3}{3} = -1$$
$$\therefore \quad [P_{12}] = \begin{bmatrix} 1 & 1 \\ -1 & 1 \end{bmatrix}$$

Hence

$$[P_{12}]^T[A][P_{12}]\{x\}_1 = \lambda[P_{12}]^T[B][P_{12}]\{x\}_1$$

or

$$\begin{bmatrix} 1 & -1 \\ 1 & 1 \end{bmatrix} \begin{bmatrix} 1 & -1 \\ -1 & 1 \end{bmatrix} \begin{bmatrix} 1 & 1 \\ -1 & 1 \end{bmatrix} \begin{Bmatrix} x_1 \\ x_1 \end{Bmatrix}_1 = \lambda \begin{bmatrix} 1 & -1 \\ 1 & 1 \end{bmatrix} \begin{bmatrix} 2 & 1 \\ 1 & 2 \end{bmatrix} \begin{bmatrix} 1 & 1 \\ -1 & 1 \end{bmatrix} \begin{Bmatrix} x_1 \\ x_1 \end{Bmatrix}_1$$

or

$$\begin{bmatrix} 1 & -1 \\ 1 & 1 \end{bmatrix} \begin{bmatrix} 2 & 0 \\ -2 & 0 \end{bmatrix} \begin{Bmatrix} x_1 \\ x_1 \end{Bmatrix}_1 = \lambda \begin{bmatrix} 1 & -1 \\ 1 & 1 \end{bmatrix} \begin{bmatrix} 1 & 3 \\ -1 & 3 \end{bmatrix} \begin{Bmatrix} x_1 \\ x_1 \end{Bmatrix}_1$$

or

$$\begin{bmatrix} 4 & 0 \\ 0 & 0 \end{bmatrix} \begin{Bmatrix} x_1 \\ x_1 \end{Bmatrix}_1 = \lambda \begin{bmatrix} 2 & 0 \\ 0 & 6 \end{bmatrix} \begin{Bmatrix} x_1 \\ x_1 \end{Bmatrix}_1$$
$$\therefore \quad [\tau] = \begin{bmatrix} 4 & 0 \\ 0 & 4 \end{bmatrix} \quad \text{and} \quad [\Lambda] = \begin{bmatrix} 2 & 0 \\ 0 & 6 \end{bmatrix}$$
$$\lambda_i = \frac{\tau_{ii}}{\Lambda_{ii}}$$
$$\lambda_1 = \frac{\tau_{11}}{\Lambda_{11}} = \frac{4}{2} = 2$$
$$\lambda_2 = \frac{\tau_{22}}{\Lambda_{22}} = \frac{0}{6} = 0$$
$$\text{and} \quad [\Phi] = [P_{12}] = \begin{bmatrix} 1 & 1 \\ -1 & 1 \end{bmatrix} = \left[\begin{Bmatrix} 1 \\ -1 \end{Bmatrix}, \begin{Bmatrix} 1 \\ 1 \end{Bmatrix}\right]$$
$$\therefore \quad (\lambda_1, \{\phi\}_1) = \left(2, \begin{Bmatrix} 1 \\ -1 \end{Bmatrix}\right)$$
$$(\lambda_2, \{\phi\}_2) = \left(0, \begin{Bmatrix} 1 \\ 1 \end{Bmatrix}\right)$$

We note that λ's are not in ascending order, but can be arranged in ascending or descending order.

4.4.5 Householder Method with QR Iterations

The Householder method can only be used for the SEVP. Thus, to use this method for the GEVP, we must first transform it into the SEVP (only possible if $[B]$ in the GEVP is invertible). Consider the SEVP:

$$[A]\{x\} = \lambda[I]\{x\} \tag{4.229}$$

This method consists of two steps:

1. Perform a series of orthogonal transformations (change of basis) on (4.229) such that $[A]$ becomes tridiagonal but $[I]$ on right side of (4.229) remains unaffected. These transformations are similarity transformations called Householder transformations.

2. Using the tridiagonal form of $[A]$ and $[I]$ in the transformed (4.229), we perform QR iterations to extract the eigenvalues and eigenvectors of (4.229).

4.4.5.1 Step 1: Householder Transformations to Tridiagonalize $[A]$

Consider (4.229) and perform a series of transformations such that:

(i) Each transformation is designed to transform a row and the corresponding column into tridiagonal form.

(ii) Unlike the Jacobi method, in the Householder method once a row and the corresponding column are in tridiagonal form, subsequent transformations for other rows and columns do not affect them.

(iii) Thus, for an $(n \times n)$ matrix, only $(n-2)$ Householder transformations are needed, i.e., this process is not iterative (as in the Jacobi method).

Details of the change of bases on (4.229) follow the usual procedure used for Jacobi method, except that the orthogonal transformation matrix $[P_l]$ in this method is not the same as the transformation matrix in the Jacobi method.

$$\{x\}_1 = [P_1]\{x\} \quad , \quad \{x\}_2 = [P_2]\{x\}_1 \quad \dots \tag{4.230}$$

After k transformations we have:

$$[A^k]\{x\}_k = \lambda[I]\{x\}_k \tag{4.231}$$

in which $[P_1]$; $i = 1, 2, \dots, k$ are orthogonal and:

$$\{x\} = [P_1][P_2]\dots.[P_k] \tag{4.232}$$

$$[A^k] = [P_k]^T[P_{k-1}]^T \dots [P_2]^T[P_1][A][P_1][P_2]\dots[P_{k-1}][P_k] \tag{4.233}$$

After $(n-2)$ transformations, the transformed $[A]$ will be tridiagonal.

$$[A^{n-2}] = [P_{n-2}]^T[P_{n-3}]^T \dots [P_2]^T[P_1][A][P_1][P_2]\dots[P_{n-3}][P_{n-3}] \tag{4.234}$$

$$\{x\} = [P_1][P_2]\dots.[P_{n-3}][P_{n-2}] \tag{4.235}$$

$[A^{n-2}]$ is the final tridiagonal form of $[A]$. We note that $[I]$ remains unaffected.

4.4.5.2 Using Householder Transformations

Let $[P_l]$ be the Householder transformation matrix that makes row l and column l of $[A^{l-1}]$ tridiagonal (i.e., only one element above and below the diagonal are non-zero) without affecting the tridiagonal forms of rows and columns 1 through $l-1$. $[P_l]$ is given by:

$$[P_l] = [I] - \theta\{w_l\}\{w_l\}^T \tag{4.236}$$

$$\theta = \frac{2}{\{w_l\}^T\{w_l\}} \tag{4.237}$$

Thus, $[P_l]$ is completely defined once $\{w_l\}$ is defined.

It is perhaps easier to understand $[P_l]$ matrices if we begin with $[P_1]$ that operates on row one and column one and then consider $[P_2]$, $[P_3]$, etc. subsequently. Consider $[P_1]$ $(l=1)$ in (4.236) and (4.237). We partition $[P_1]$, $[A]$, and $\{w_l\}$ as follows:

$$[P_1] = \left[\begin{array}{c|c} 1 & [0] \\ \hline \{0\} & [\bar{P}_1] \end{array}\right] \; ; \quad [A] = \left[\begin{array}{c|c} a_{11} & \{a_1\}^T \\ \hline \{a_1\} & [A_{11}] \end{array}\right] \; ; \quad \{w_1\} = \left\{\begin{array}{c} 0 \\ \hline \{\bar{w}_1\} \end{array}\right\} \tag{4.238}$$

where $[\bar{P}_1]$, $\{\bar{w}_1\}$ and $[A_{11}]$ are of order $(n-1)$. Premultiply $[A]$ by $[P_1]^T$ and post-multiply by $[P_1]$ to obtain $[A^1]$.

$$[A^1] = \left[\begin{array}{c|c} 1 & [0] \\ \hline \{0\} & [\bar{P}_1]^T \end{array}\right] \left[\begin{array}{c|c} a_{11} & \{a_1\}^T \\ \hline \{a_1\} & [A_{11}] \end{array}\right] \left[\begin{array}{c|c} 1 & [0] \\ \hline \{0\} & [\bar{P}_1] \end{array}\right] \tag{4.239}$$

or

$$[A^1] = \left[\begin{array}{c|c} a_{11} & \{a_1\}^T[\bar{P}_1] \\ \hline [\bar{P}_1]^T\{a_1\} & [\bar{P}_1]^T[A_{11}][\bar{P}_1] \end{array}\right] \tag{4.240}$$

In $[A^1]$ first row and the first column should be tridiagonal, i.e. $[A^1]$ should have the following form:

$$[A^1] = \left[\begin{array}{c|c} a_{11} & x\ 0\ 0 \dots 0 \\ \hline x & \\ 0 & \\ 0 & [\bar{A}^1] \\ \vdots & \\ 0 & \end{array}\right] \tag{4.241}$$

where x is a nonzero element and

$$[\bar{A}^1] = [\bar{P}_1]^T[A_{11}][\bar{P}_1] \tag{4.242}$$

$[\bar{P}_1]$ is called the reflection matrix. We are using $[\bar{P}_1]$ to reflect $\{a_1\}$ of $[A]$ into a vector such that only its first component is non-zero (obvious by comparing (4.240) and (4.241)). Since the length of the vector corresponding to row one or column one (excluding a_{11}) must be the same as the length of $\{a_1\}$, we can use this condition to determine $\{w_1\}$ (i.e., first $\{\bar{w}_1\}$ and then $\{w_1\}$).

$$\begin{gathered} [[I] - \theta\{\bar{w}_1\}\{w_1\}]\{a_1\} = \pm\,||\{a_{11}\}||_{L2}\,\{e\}_1 \\ \text{where} \quad \{e_1\} = \begin{Bmatrix} 1 \\ 0 \\ \vdots \\ 0 \end{Bmatrix} \end{gathered} \tag{4.243}$$

A positive or negative sign is selected for numerical stability. From (4.243), we can solve for $\{\bar{w}_1\}$.

$$\{\bar{w}_1\} = \{a_1\} + \text{sign}(a_{21})\,||\{a_1\}||_{L2}\,\{e_1\} \tag{4.244}$$

where a_{21} is the element (2,1) of matrix $[A]$. The vector $\{w_1\}$ is obtained using $\{\bar{w}_1\}$ from (4.244) in (4.238). Thus, $[P_1]$ is defined and we can perform the Householder transformation for column one and row one to obtain $[A^1]$, in which column one and row one are in tridiagonal form.

$$[A^1] = [P_1]^T[A][P_1] \tag{4.245}$$

Next we consider column two and row two to obtain $[P_2]$ and then use:

$$[A^2] = [P_2]^T[A^1][P_2] \tag{4.246}$$

In $[A^2]$ the first two columns and rows are in tridiagonal form. We continue this $(n-2)$ times to finally obtain $[A^{n-2}]$ in tridiagonal form, and we can write:

$$[A^{n-2}]\{x\}_{n-2} = \lambda[I]\{x\}_{n-2} \tag{4.247}$$

$$\text{and} \quad \{x\} = [P_1][P_2]\dots[P_{n-3}][P_{n-2}]\{x\}_{n-2} \tag{4.248}$$

Equation (4.248) is essential to recover the original eigenvector $\{x\}$.

4.4.5.3 Step 2: QR Iterations to Extract Eigenpairs

We apply QR iterations to the tridiagonal form (4.247) to extract the eigenvalues of the original SEVP (same as the eigenvalues of (4.247)). QR iterations can also be applied to the original $[A]$ but QR iterations are more efficient with tridiagonal form of $[A]$. The purpose of QR iterations is to decompose $[A^{n-2}]$ into a product of $[Q]$ and $[R]$.

$$[A^{n-2}] = [Q][R] \tag{4.249}$$

The matrix $[Q]$ is orthogonal and $[R]$ is upper triangular. Since $[Q]$ is orthogonal we perform a change of basis on (4.247).

$$\{x\}_{n-2} = [Q_1]\{x\}^1_{n-2} \tag{4.250}$$

$$\text{with} \quad [A^{n-2}] = [Q_1][R_1] \tag{4.251}$$

Substitute (4.250) in (4.247) and premultiply (4.247) by $[Q_1]^T$.

$$[Q_1]^T[A^{n-2}][Q_1]\{x\}^1_{n-2} = \lambda[Q_1]^T[I][Q_1]\{x\}^1_{n-2} \tag{4.252}$$

or

$$[Q_1]^T[A^{n-2}][Q_1]\{x\}^1_{n-2} = \lambda[I]\{x\}^1_{n-2} \tag{4.253}$$

Using (4.251) in the left side of (4.253) for $[A^{n-2}]$:

$$[Q_1]^T[A^{n-2}][Q_1] = [Q_1]^T[Q_1][R_1][Q_1] = [R_1][Q_1] \tag{4.254}$$

That is, performing an orthogonal transformation on $[A^{n-2}]$ using $[Q_1]$ is the same as taking the product of $[R_1][Q_1]$. Thus, once we have the decomposition (4.251), the orthogonal transformation on $[A^{n-2}]$ is simply the product $[R_1][Q_1]$.

4.4.5.4 Determining $[Q]$ and $[R]$

We wish to perform the decomposition:

$$[A^{n-2}] = [Q][R] \tag{4.255}$$

$[R]$ can be obtained by premultiplying $[A^{n-2}]$ and successively transformed $[A^{n-2}]$ by a series of rotation matrices designed to make the elements of $[A^{n-2}]$ below the diagonal zero, so that the transformed $[A^{n-2}]$ will be upper triangular. Thus,

$$[R] = [P]_{n,n-1} \dots [P]_{3,3} \dots [P]_{1,1}[P]_{3,1}[P]_{2,1}[A^{n-1}] \tag{4.256}$$

where $[P]_{i,j}$ corresponding to a row i and column j that makes a_{ji}^{n-2} zero is given by:

$$[P]_{i,j} = \begin{array}{c} \text{Column } i \quad \text{Column } j \\ \left[\begin{array}{cccccccc} 1 & & & & & & & \\ & 1 & & & & & & \\ & & \cos\theta & & & -\sin\theta & & \\ & & & 1 & & & & \\ & & & & 1 & & & \\ & & \sin\theta & & & \cos\theta & & \\ & & & & & & 1 & \\ & & & & & & & 1 \end{array}\right] \begin{array}{l} \\ \\ \text{Row } i \\ \\ \\ \text{Row } j \\ \\ \\ \end{array} \end{array} \tag{4.257}$$

The diagonals of (4.256) are unity and:

$$\sin\theta = \frac{a_{ji}^{n-2}}{\sqrt{(a_{ii}^{n-2})^2 + (a_{jj}^{n-2})^2}} \quad ; \quad \cos\theta = \frac{a_{ii}^{n-2}}{\sqrt{(a_{ii}^{n-2})^2 + (a_{ji}^{n-2})^2}} \tag{4.258}$$

If $a_{ii}^2 + a_{ji}^2 = 0$, no transformation is required. Therefore:

$$[Q] = [P]_{2,1}[P]_{3,1}[P]_{1,1} \dots [P]_{3,2}[P]_{3,3} \dots [P]_{n,n-1} \tag{4.259}$$

4.4.5.5 Using QR Iteration

Given $[A^{n-2}]$, we want to obtain $[Q]$ and $[R]$ and take the product of $[R][Q]$, which is the same as an orthogonal transformation on $[A^{n-2}]$ to get $[A^{n-2}]_1$. The process of obtaining $[Q]$ and $[R]$ follows the previous section. Now we take $[A^{n-2}]_1$ and repeat QR iterations and continue doing so until $[A^{n-2}]$ has become a diagonal matrix. Keeping in mind that each QR iteration changes the eigenvector the columns of

$$[[Q_1][Q_2] \dots [Q_n]] = [\Phi] \tag{4.260}$$

contain the eigenvectors of the SEVP and the diagonal elements of the final transformed $[A^{n-2}]$ are the eigenvalues. We also note that these eigenvalues are not in any particular order.

Example 4.9 (Householder Transformation). Consider matrix $[A]$ appearing in a standard eigenvalue problem.

$$[A] = \begin{bmatrix} 5 & -4 & 1 & 0 \\ -4 & 6 & -4 & 1 \\ 1 & -4 & 6 & -4 \\ 0 & 1 & -4 & 5 \end{bmatrix}$$

We use Householder transformations to reduce $[A]$ to tridiagonal form.

1. *Reducing column one to tridiagonal form: making a_{11} zero*

$$\begin{aligned} &\{a_1\}^T = [-4 \;\; 1 \;\; 0] \\ &\text{sign}(a_{21}) = \text{sign}(-4) = - \\ &||a_1|| = \sqrt{(-4)^2 + (1)^2 + (0)^2} = \sqrt{17} = 4.123 \end{aligned}$$

$$\therefore \quad \{\bar{w}\} = \begin{Bmatrix} -4 \\ 1 \\ 0 \end{Bmatrix} - 4.123 \begin{Bmatrix} 1 \\ 0 \\ 0 \end{Bmatrix} = \begin{Bmatrix} -8.1231 \\ 1 \\ 0 \end{Bmatrix}$$

$$\{w_1\} = \begin{Bmatrix} 0 \\ \dots \\ \{\bar{w}_1\} \end{Bmatrix} = \begin{Bmatrix} 0 \\ -8.1231 \\ 1 \\ 0 \end{Bmatrix}$$

$$\theta = \frac{2}{\{w_1\}^T\{w_1\}} = \frac{2}{0 + (-8.1231)(-8.1231) + (1)(1) + 0} = 0.029857$$

$$\therefore \quad [P_1] = [I] - \theta\{w_1\}\{w_1\}^T = \begin{bmatrix} 1 & 0 & 0 & 0 \\ 0 & -0.9701 & 0.2425 & 0 \\ 0 & 0.2425 & 0.9701 & 0 \\ 0 & 0 & 0 & 1 \end{bmatrix}$$

$$[A^1] = [P_1]^T[A][P_1] = \begin{bmatrix} 5 & 4.1231 & 0 & 0 \\ 4.1231 & 7.8823 & 3.5294 & -1.9403 \\ 0 & 3.5294 & 4.1177 & -3.6380 \\ 0 & -1.9403 & -3.6380 & 5 \end{bmatrix}$$

2. *Reducing column two of $[A^1]$ to tridiagonal form: making a^1_{41} zero*

$$\begin{aligned} &\{a_2\}^T = [3.5294 \;\; -1.9403] \\ &\text{sign}(a^2_{21}) = \text{sign}(3.5294) = + \\ &||a_2|| = \sqrt{(3.5294)^2 + (-1.9403)^2} = 4.0276 \end{aligned}$$

$$\therefore \quad \{\bar{w}_2\} = \begin{Bmatrix} 3.5294 \\ -1.9403 \end{Bmatrix} + 4.0276 \begin{Bmatrix} 1 \\ 0 \end{Bmatrix} = \begin{Bmatrix} 7.5570 \\ -1.9403 \end{Bmatrix}$$

$$\{w_2\} = \begin{Bmatrix} 0 \\ 0 \\ 7.5570 \\ -1.9403 \end{Bmatrix} \quad ; \quad \theta = \frac{2}{\{w_2\}^T\{w_2\}} = 0.032855$$

$$\therefore \quad [P_2] = [I] - \theta\{w_2\}\{w_2\}^T = \begin{bmatrix} 1 & 0 & 0 & 0 \\ 0 & 1 & 0 & 0 \\ 0 & 0 & -0.8763 & 0.4817 \\ 0 & 0 & 0.4817 & 0.8763 \end{bmatrix}$$

$$\therefore \quad [P_2][A^1][P_2] = [A^2] \begin{bmatrix} 5 & 4.1231 & 0 & 0 \\ 4.1231 & 7.8823 & -4.0276 & 0 \\ 0 & -4.0276 & 7.3941 & 2.3219 \\ 0 & 0 & 2.3218 & 1.7236 \end{bmatrix}$$

$[A^2]$ is the final tridiagonal form of $[A]$. The tridiagonal form is symmetric as expected.

Remarks. The QR iteration process is computationally intense, hence not practical to illustrate in this example. QR iterations need to be programmed.

4.4.6 Subspace Iteration Method

In this method a large eigenvalue problem

$$[K]\{\phi\} - \lambda[M]\{\phi\} = \{0\} \tag{4.261}$$

in which $[K]$ and $[M]$ are $(n \times n)$ is reduced to a much smaller eigenvalue problem based on the desired number of eigenpairs to be extracted. The steps of this method are given in the following.

If p $(p \ll n)$ is the desired number of eigenpairs, then choose a $n \times q$ $(q > p)$ starting matrix $[x_1]_{n\times q}$ whose columns are initial guesses for the eigenvectors.

(1) Consider the following inverse iteration problem $(k = 1)$:

$$[K]_{n\times n}[\bar{x}_{k+1}]_{n\times q} = [M]_{n\times n}[x_k]_{n\times q} \tag{4.262}$$

We calculate $[\bar{x}_{k+1}]_{n\times q}$ matrix using (4.262).

(2) We find the projections of $[K]$ and $[M]$ in the space spanned by the q matrices using:

$$\begin{aligned}[K_{k+1}]_{q\times q} &= [\bar{x}_{k+1}]^T_{q\times n}[K]_{n\times n}[\bar{x}_{k+1}]_{n\times q}\\ [M_{k+1}]_{q\times q} &= [\bar{x}_{k+1}]^T_{q\times n}[M]_{n\times n}[\bar{x}_{k+1}]_{n\times q}\end{aligned} \tag{4.263}$$

(3) We solve the eigenvalue problem constructed using $[K_{k+1}]$ and $[M_{k+1}]$.

$$[K_{k+1}][\Phi^q_{k+1}] = [\Lambda_{k+1}][M_{k+1}][\Phi^q_{k+1}] \tag{4.264}$$

$[\Lambda_{k+1}]$ is a diagonal matrix containing approximated eigenvalues and $[\Phi^q_{k+1}]$ is a matrix whose columns are eigenvectors of the reduced-dimension problem corresponding to the eigenvalues in $[\Lambda_{k+1}]$.

(4) Construct new starting matrices:

$$[x_{k+1}]_{n\times q} = [\bar{x}_{k+1}]_{n\times q}[\Phi^q_{k+1}]_{q\times q} \tag{4.265}$$

(5) Use $k = k+1$ and repeat steps (1) – (4).

Provided the vectors in the columns of the starting matrix $[x_1]$ are not orthogonal to one of the required eigenvectors, we have:

$$[\Lambda_{k+1}]_{q\times q} \rightarrow [\Lambda]_{q\times q} \qquad \text{and} \qquad [x_{k+1}]_{n\times q} \rightarrow [\Phi^q]_{n\times q} \qquad \text{as } k \rightarrow \infty \tag{4.266}$$

where $[\Lambda]$ is a diagonal (square) matrix containing q eigenvalues and $[\Phi^q]$ is a (rectangular) matrix whose columns are eigenvectors of the original eigenvalue problem corresponding to the eigenvalues in $[\Lambda]$.

Remarks.

(1) The choice of starting vectors in $[x_1]$ is obviously critical. First, they should be orthogonal to each other. Secondly, none of these should be orthogonal to any of the desired eigenvectors.

(2) Based on the diagonal elements of $[K]$ and $[M]$, some guidelines can be used.

   (a) Construct $r_i = {}^{k_{ii}}/_{m_{ii}}$. If $\{e_1\}, \{e_2\}, \dots, \{e_n\}$ are unit vectors containing one at rows $1, 2, \dots, n$, respectively, and zeroes everywhere else and if r_j, r_k, r_l are progressively increasing values of r_i, then $[x]_{n\times 3}$ consists of $[\{e_j\}, \{e_k\}, \{e_l\}]$.

(3) The eigenvalue problem (4.264) can be solved effectively using Generalized Jacobi or QR-Householder method in which all eigenpairs are determined.

Example 4.10 (Starting Vectors in Subspace Iteration Method). Consider the following eigenvalue problem:

$$[K]\{\phi\} - \lambda[M]\{\phi\} = \{0\}$$

in which

$$[K] = \begin{bmatrix} 2 & -1 & 0 & 0 \\ -1 & 2 & -1 & 0 \\ 0 & -1 & 2 & -1 \\ 0 & 0 & -1 & 1 \end{bmatrix} \quad ; \quad [M] = \begin{bmatrix} 1 & 0 & 0 & 0 \\ 0 & 2 & 0 & 0 \\ 0 & 0 & 4 & 0 \\ 0 & 0 & 0 & 3 \end{bmatrix}$$

In this case:

$$\underset{i=1,2,3,4}{r_i} = \underset{i=1,2,3,4}{k_{ii}/m_{ii}} = 2, 1, 0.5, 1/3$$

Let $q = 2$, i.e., we choose two starting vectors. We see that $i = 4$ and $i = 3$ correspond to the lowest values of r_i. Then, based on r_i; $i = 1, 2, 3, 4$ we choose the following.

$$[x_1]_{4\times 2} = [\{e_4\}, \{e_3\}] = \begin{bmatrix} 0 & 0 \\ 0 & 0 \\ 0 & 1 \\ 1 & 0 \end{bmatrix}$$

4.5 Concluding Remarks

(1) The characteristic polynomial method is rather primitive due to the fact that (i) it requires determination of the coefficients of the polynomial and (ii) it uses root-finding methods that may be computationally inefficient or even ineffective when large numbers of eigenpairs are required. Generally for eigensystems smaller than (10×10), this method may be employed.

(2) The vector iteration method is quite effective in determining the smallest and the largest eigenpairs. The vector iteration method with iteration vector deflation is quite effective in calculating a few eigenpairs. For a large number of eigenpairs, the orthogonalization process becomes error prone due to inaccuracies in the numerically computed eigenvectors (hence, not ensuring their orthogonal property). This can cause the computations to become erroneous or even to break down completely.

(3) The Jacobi and generalized Jacobi methods yield all eigenpairs, hence

are not practical for eigensystems larger than (50×50) or at the most (100×100). In these methods, the off-diagonal terms made zero in an orthogonal transformation become non-zero in the next orthogonal transformation, thus these methods may require a larger number of cycles or sweeps before convergence is achieved.

(4) In the Householder method with QR iterations for the SEVP (only), we tridiagonalize the $[A]$ matrix by Householder transformations and then use QR iterations on the tridiagonal form to extract the eigenpairs. The tridiagonalization process is not iterative, but QR iterations are (as the name suggests). This method also yields all eigenpairs and hence is only efficient for eigensystems that are smaller than (50×50) or at the most (100×100). Extracting eigenpairs using QR iterations is more efficient with the tridiagonal form of $[A]$ than the original matrix $[A]$. This is the main motivation for converting (transforming) $[A]$ to the tridiagonal form before extracting eigenpairs.

(5) The subspace iteration method is perhaps the most practical method for larger eigensystems as in this method a large eigenvalue problem is reduced to a very small eigenvalue problem. Computation of the eigenpairs only requires working with an eigensystem that is $(q \times q)$, $q > p$, where p is the desired number of eigenpairs. Generally we choose $q = 2p$.

Problems

4.1 Use minors to expand and compute the determinant of

$$\begin{bmatrix} 2-\lambda & 2 & 10 \\ 8 & 3-\lambda & 4 \\ 10 & 4 & 5-\lambda \end{bmatrix}$$

Use Faddeev-Leverrier method to perform the same computations. Also compute the matrix inverse and verify that it is correct.

Write a computer program to perform computations for problems 4.2 and 4.3.

4.2 Consider the eigenvalue problem

$$[A]\{x\} = \lambda[B]\{x\} \tag{1}$$

where

$$[A] = \begin{bmatrix} 1 & 0 & 0 \\ 0 & 2 & 0 \\ 0 & 0 & 3 \end{bmatrix} \quad ; \quad [B] = \begin{bmatrix} 1 & 0 & 0 \\ 0 & 1 & 0 \\ 0 & 0 & 1 \end{bmatrix}$$

(a) Use inverse iteration method to compute the lowest eigenvalue and the corresponding eigenvector.

(b) Transform (1) into new SEVP such that the transformed eigenvalue problem can be used to determine the largest eigenvalue and the corresponding eigenvector of (1). Compute the largest eigenvalue and the corresponding eigenvector using this form.

(c) For eigenvalue problem (1), use inverse iteration with iteration vector deflation technique to compute all its eigenpairs.

(d) Consider the transformed SEVP in (b). Apply inverse iteration with iteration vector deflation to compute all eigenpairs.

(e) Tabulate and compare the eigenpairs computed in (c) and (d). Discuss and comment on the results.

4.3 Consider the eigenvalue problem

$$[A]\{x\} = \lambda[B]\{x\} \tag{1}$$

where

$$[A] = \begin{bmatrix} 2 & -1 & 0 & 0 \\ -1 & 2 & -1 & 0 \\ 0 & -1 & 2 & -1 \\ 0 & 0 & -1 & 1 \end{bmatrix} \quad ; \quad [B] = \begin{bmatrix} 1 & 0 & 0 & 0 \\ 0 & 1 & 0 & 0 \\ 0 & 0 & 1 & 0 \\ 0 & 0 & 0 & 1 \end{bmatrix}$$

(a) Use inverse iteration method with iteration vector deflation to compute all eigenpairs of (1).

(b) Transform (1) into SEVP such that the transformed eigenvalue problem can be used to find the largest eigenvalue and the corresponding eigenvector of (1). Apply inverse iteration method to this transformed eigenvalue problem with iteration vector deflection to compute all eigenpairs.

(c) Tabulate and compare the eigenpairs computed in (a) and (b). Discuss and comment on the results.

4.4 (a) Show that $\lambda = 1$ and $\lambda = 4$ are the eigenvalues of the following eigenvalue problem without calculating them.

$$\begin{bmatrix} 5 & 3 \\ 3 & 5 \end{bmatrix} \begin{Bmatrix} x \\ y \end{Bmatrix} = \lambda \begin{bmatrix} 2 & 0 \\ 0 & 2 \end{bmatrix} \begin{Bmatrix} x \\ y \end{Bmatrix}$$

(b) For what values of a, b and c, the following vectors constitute a system of eigenvectors.

$$\begin{Bmatrix} 1 \\ 1 \\ c \end{Bmatrix} \; ; \; \begin{Bmatrix} a \\ -1 \\ 1 \end{Bmatrix} \; ; \; \begin{Bmatrix} 1 \\ b \\ -1 \end{Bmatrix}$$

4.5 Consider the following eigenvalue problem

$$\begin{bmatrix} 4 & -4 \\ -4 & 8 \end{bmatrix} \begin{Bmatrix} x_1 \\ x_2 \end{Bmatrix} = \lambda \begin{bmatrix} 1 & 0 \\ 0 & 1 \end{bmatrix} \begin{Bmatrix} x_1 \\ x_2 \end{Bmatrix} \tag{1}$$

Let

$$(\lambda_1, \{\phi\}_1) = (1.52786, \begin{Bmatrix} 0.85064 \\ 0.52574 \end{Bmatrix})$$

be the first eigenpair of (1).
Use inverse iteration method with iteration vector deflation to calculate second eigenpair of (1) using $\{x\} = \begin{Bmatrix} 1 \\ 1 \end{Bmatrix}$ as initial starting vector. Write a computer program to perform the calculations.

4.6 Consider the following SEVP

$$\begin{bmatrix} 4 & -4 \\ -4 & 8 \end{bmatrix} \begin{Bmatrix} x_1 \\ x_2 \end{Bmatrix} = \lambda \begin{bmatrix} 1 & 0 \\ 0 & 1 \end{bmatrix} \begin{Bmatrix} x_1 \\ x_2 \end{Bmatrix} \tag{1}$$

Calculate both eigenpairs of (1) using standard Jacobi method. Discuss and show the orthogonality of the eigenvectors calculated in this method.

4.7 Consider the following GEVP

$$\begin{bmatrix} 2 & 1 \\ 1 & 2 \end{bmatrix} \begin{Bmatrix} x_1 \\ x_2 \end{Bmatrix} = \lambda \begin{bmatrix} 2 & 0 \\ 0 & 1 \end{bmatrix} \begin{Bmatrix} x_1 \\ x_2 \end{Bmatrix} \tag{1}$$

Calculate both eigenvalues of (1) using generalized Jacobi method.

4.8 Consider the following GEVP

$$\begin{bmatrix} 4 & 1 \\ 1 & 4 \end{bmatrix} \begin{Bmatrix} x_1 \\ x_2 \end{Bmatrix} = \lambda \begin{bmatrix} 2 & -1 \\ -1 & 2 \end{bmatrix} \begin{Bmatrix} x_1 \\ x_2 \end{Bmatrix} \tag{1}$$

(a) Using basic properties of the eigenvalue problem, show if $\lambda = 1$ and $\lambda = 5$ are the eigenvalues of (1). If yes, then calculate the corresponding eigenvectors and show if they are orthogonal or not.

(b) Convert the eigenvalue problem in (1) to a SEVP in the form $[A]\{x\} = \lambda[I]\{x\}$ and then compute its eigenpairs using standard Jacobi method.

4.9 Consider the following GEVP

$$\begin{bmatrix} 3 & 1 \\ 1 & 3 \end{bmatrix} \begin{Bmatrix} x_1 \\ x_2 \end{Bmatrix} = \lambda \begin{bmatrix} 2 & -1 \\ -1 & 2 \end{bmatrix} \begin{Bmatrix} x_1 \\ x_2 \end{Bmatrix} \tag{1}$$

(a) Compute eigenvalues and the corresponding eigenvectors and show that the eigenvectors are orthogonal.

(b) Transform the eigenvalue problem in (1) to the standard form $[A]\{x\} = \lambda[I]\{x\}$ and then compute its eigenpairs using Jacobi method.

4.10 Consider the following eigenvalue problem $[K]\{x\} = \lambda[M]\{x\}$ in which

$$[K] = \begin{bmatrix} 2 & -1 & 0 \\ -1 & 4 & -1 \\ 0 & -1 & 2 \end{bmatrix} \quad \text{and} \quad [M] = \begin{bmatrix} 1/2 & 0 & 0 \\ 0 & 1 & 0 \\ 0 & 0 & 1/2 \end{bmatrix} \tag{1}$$

(a) Perform $[L][D][L]^T$ factorization of $\big[[K] - \lambda[M]\big]$ at $\lambda = 5$.

(b) Describe the significance of this factorization.

(c) What does this factorization for this eigenvalue problem indicate.

4.11 Consider the same eigenvalue problem 4.10 in problem 4.5. If $\lambda_1 = 2$, $\lambda_2 = 4$ and $\lambda_3 = 6$ are its eigenvalues and

$$\{\phi\}_1 = \begin{Bmatrix} 0.707 \\ 1.0 \\ 0.707 \end{Bmatrix} \; ; \; \{\phi\}_2 = \begin{Bmatrix} -1 \\ 0 \\ 1 \end{Bmatrix} \; \text{and} \; \{\phi\}_3 = \begin{Bmatrix} 0.707 \\ -1.0 \\ 0.707 \end{Bmatrix}$$

are its eigenvectors, then how are the eigenvalues and the eigenvectors of this problem related to the eigenvalue problem $[\hat{K}]\{x\} = \mu[M]\{x\}$? Where

$$[\hat{K}] = [K] + 1.5[M]$$

$[K]$ and $[M]$ are same as in problem 4.10

4.12 Consider the following eigenvalue problem $\left([A]\{x\} = \lambda[B]\{x\}\right)$

$$\begin{bmatrix} 1 & -1 \\ -1 & 1 \end{bmatrix} \begin{Bmatrix} x_1 \\ x_2 \end{Bmatrix} = \lambda \begin{bmatrix} 1 & 0 \\ 0 & 1 \end{bmatrix} \begin{Bmatrix} x_1 \\ x_2 \end{Bmatrix} \tag{1}$$

(a) Use Faddeev-Leverrier method to determine the characteristic polynomial.

(b) What can you conclude about the inverse of $[A]$ in (a).

(c) Based on the inverse in (b) what can you conclude about the nature of $[A]$. Can the same conclusion be arrived at independent of the method used here.

4.13 Consider the following eigenvalue problem $[A]\{x\} = \lambda[I]\{x\}$

$$[A] = \begin{bmatrix} 3 & -1 & 0 \\ -1 & 2 & -1 \\ 0 & -1 & 3 \end{bmatrix}$$

If $\lambda_1 = 4$ is an eigenvalue and $\{\phi\}_1 = \begin{Bmatrix} -1 \\ -1 \\ -1 \end{Bmatrix}$ is the corresponding eigenvector of the above eigenvalue problem, then determine one subsequent eigenpair using inverse iteration with iteration vector deflation. Use $\{x\} = \begin{Bmatrix} 1 \\ 1 \\ 1 \end{Bmatrix}$ as starting vector.

5
Interpolation and Mapping

5.1 Introduction

When taking measurements in experiments, we often collect discrete data that may describe the behavior of a desired quantity of interest at selected discrete locations. These data in general may be over irregular domains in $\mathbb{R}^1$, $\mathbb{R}^2$, and $\mathbb{R}^3$. Constructing a mathematical description of these data is helpful and sometimes necessary if we desire to perform operations of integration, differentiation, etc. for the physics described by these data. One of the techniques or methods of constructing a mathematical description for discrete data is called interpolation. Interpolation yields an analytical expression for the discrete data. This expression then can be integrated or differentiated, thus permitting operations of integration or differentiation on discrete data. The interpolation technique ensures that the mathematical expression so generated will yield precise function values at the discrete locations that are used in generating it.

When the discrete data belong to irregular domains in $\mathbb{R}^1$, $\mathbb{R}^2$, and $\mathbb{R}^3$, the interpolations may be quite difficult to construct. To facilitate the interpolations over irregular domains, the data in the irregular domains are mapped into regular domains of known shape and size in $\mathbb{R}^1$, $\mathbb{R}^2$, and $\mathbb{R}^3$. The desired operations of integration and differentiation are also performed in the mapped domain and then mapped back to the original (physical) irregular domain. Details of the mapping theories and interpolations are considered in this chapter. First, we introduce the concepts of interpolation theory in $\mathbb{R}^1$ in the physical coordinate space (say x). This is followed by mapping theory in $\mathbb{R}^1$ that maps data in the physical coordinate space to the natural coordinate space ξ in a domain of two unit length with the origin located at the center of the two unit length. The concepts of piecewise mapping in $\mathbb{R}^1$, $\mathbb{R}^2$, and $\mathbb{R}^3$ as well as interpolations over the mapped domains are presented with illustrative examples.

5.2 Interpolation Theory in $\mathbb{R}^1$

First, we consider basic elements of interpolation theory in $\mathbb{R}^1$.

Definition 5.1 (Interpolation). Given a set of values (x_i, f_i) ; $i = 1, 2, \ldots, n+$

1, if we can establish an analytical expression $f(x)$ such that $f(x_i) = f_i$, then $f(x)$ is called the *interpolation* associated with the data set (x_i, f_i) ; $i = 1, 2, \dots, n+1$.

Important properties of $f(x)$ are that (i) it is an analytical expression and (ii) at each x_i in the given data set the function $f(x)$ has a value $f(x_i)$ that agrees with f_i, i.e., $f(x_i) = f_i$. There are many approaches one could take that would satisfy the requirement in the definition.

5.2.1 Piecewise Linear Interpolation

Let (x_i, f_i) ; $i = 1, 2, \dots, n+1$ be a set of given data points. In this method we assume that the interpolation $f(x)$ associated with the data set is linear between each pair of data points and hence $f(x)$ for $x_1 \le x \le x_{n+1}$ is a piecewise linear function (Figure 5.1).

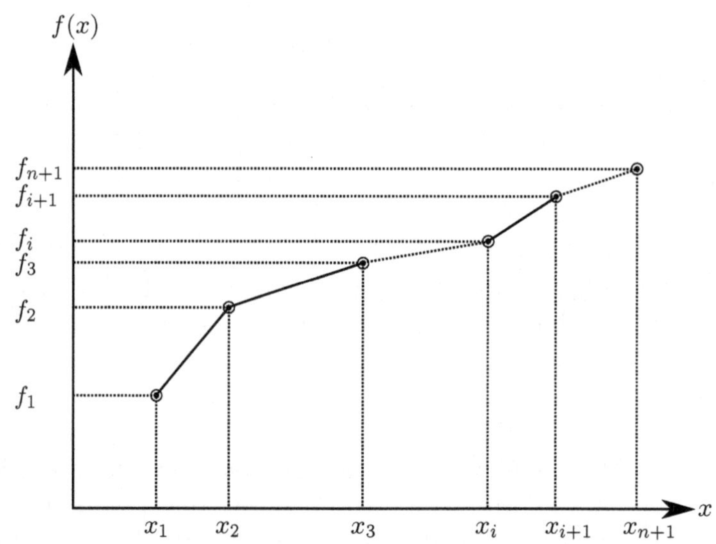

Figure 5.1: Piecewise linear interpolation

Consider a pair of points (x_i, f_i) and (x_{i+1}, f_{i+1}). The equation of a straight line describing linear interpolation between these two points is given by

$$\frac{f(x) - f_i}{x - x_i} = \frac{f(x) - f_{i+1}}{x - x_{i+1}} \tag{5.1}$$

or

$$f(x) = f_i + \left(\frac{f_{i+1} - f_i}{x_{i+1} - x_i}\right)(x - x_i) \quad ; \quad i = 1, 2, \dots, n \tag{5.2}$$

The function $f(x)$ in equation (5.2) is the desired interpolation function for the data set (x_i, f_i) ; $i = 1, 2, \dots, n+1$. For $i = 1, 2, \dots, n$ we obtain piecewise linear interpolations between each pair of successive data points. It is obvious from Figure 5.1 that $f(x)$ is continuous for $x_1 \leq x \leq x_{n+1}$ but has non-unique $\frac{df}{dx}$ at $x = x_i$; $i = 2, 3, \dots, n$. This may be an undesirable feature in some instances.

5.2.2 Polynomial Interpolation

Consider data points (x_i, f_i) ; $i = 1, 2, \dots, n+1$. In this approach we consider the interpolation $f(x)$ for this data set to be a linear combination of the monomials x^j ; $j = 0, 1, \dots, n$, i.e. $1, x, x^2, \dots, x^n$. Using the constants a_i ; $i = 0, 1, \dots, n$ (to be determined) we can write

$$f(x) = a_0 + a_1 x + a_2 x^2 + \dots + a_n x^n \tag{5.3}$$

Based on the given data set (x_i, f_i); $i = 1, 2, \dots, n+1$, $f(x)$ in (5.3) must satisfy the following conditions:

$$f(x)|_{x=x_i} = f_i \quad ; \quad i = 1, 2, \dots, n+1 \tag{5.4}$$

If we substitute $x = x_i$; $i = 1, 2, \dots, n+1$ in (5.3), we must obtain $f(x_i) = f_i$; $i = 1, 2, \dots, n+1$. Using (5.4) in (5.3), we obtain $n+1$ linear simultaneous algebraic equations in the unknowns a_i ; $i = 0, 1, \dots, n$. These can be written in the matrix and vector form.

$$\begin{bmatrix} 1 & x_1 & x_1^2 & \dots & x_1^n \\ 1 & x_2 & x_2^2 & \dots & x_2^n \\ \vdots & \vdots & \vdots & \dots & \vdots \\ 1 & x_{n+1} & x_{n+1}^2 & \dots & x_{n+1}^n \end{bmatrix} \begin{Bmatrix} a_0 \\ a_1 \\ \vdots \\ a_n \end{Bmatrix} = \begin{Bmatrix} f_1 \\ f_2 \\ \vdots \\ f_{n+1} \end{Bmatrix} \tag{5.5}$$

In equations (5.5), the first equation corresponds to (x_1, f_1), the second equation corresponds to (x_2, f_2) and so on. We can write (5.5) in compact notation as:

$$[A]\{a\} = \{F\} \tag{5.6}$$

From (5.6) we can calculate the unknown coefficients $\{a\}$ by solving the system of linear simultaneous algebraic equations. When we substitute a_i; $i = 0, 1, \dots, n$ in (5.3), we obtain the desired polynomial interpolations $f(x)$ for the data set (x_i, f_i) ; $i = 1, 2, \dots, n+1$ $\forall x \in [x_1, x_{n+1}]$.

Remarks.

(1) In order for $\{a\}$ to be unique $\det[A] \neq 0$ must hold. This is ensured if each x_i location is distinct or unique.

(2) When two data point locations (i.e., x_i values) are extremely close to each other the coefficient matrix $[A]$ may become ill-conditioned.

(3) For large data sets (large values of n), this method requires solutions of a large system of linear simultaneous algebraic equations. This obviously leads to inefficiency in its computations.

5.2.3 Lagrange Interpolating Polynomials

Let (x_i, f_i) ; $i = 1, 2, \dots, n$ be the given data points.

Theorem 5.1. *There exists a unique polynomial $\psi(x)$ of degree not exceeding n called the Lagrange interpolating polynomial such that*

$$\psi(x_i) = f_i \quad ; \quad i = 1, 2, \dots, n \tag{5.7}$$

Proof. The existence of the polynomial $\psi(x)$ can be proven if we can establish the existence of polynomials $L_k(x)$; $k = 1, 2, \dots, n$ with the following properties:

$$\begin{aligned}
&(i) \quad \text{Each } L_k(x) \text{ is a polynomial of degree less than or equal to } n \\
&(ii) \quad L_i(x_j) = \begin{cases} 1 & ; \quad j = i \\ 0 & ; \quad j \neq i \end{cases} \\
&(iii) \quad \sum_{k=1}^{n} L_k(x) = 1
\end{aligned} \tag{5.8}$$

Assuming the existence of the polynomials $L_k(x)$ we can can write:

$$\psi(x) = \sum_{k=1}^{n} f_k L_k(x) \tag{5.9}$$

We note that for $x = x_i$ in (5.9):

$$\psi(x_i) = \sum_{k=1}^{n} f_k L_k(x_i) = f_i$$

Hence $\psi(x)$ has the desired properties of $f(x)$, an interpolation for the data set (x_i, f_i) ; $i = 1, 2, \dots, n$.

$$f(x) = \psi(x) = \sum_{i=1}^{n} f_i L_i(x) \tag{5.10}$$

□

Remarks.

(1) $L_k(x)$; $k = 1, 2, \dots, n$ are polynomials of degree less than or equal to n.

(2) $\psi(x)$ is a linear combination of f_k and $L_k(x)$, hence $\psi(x)$ is also a polynomial of degree less than or equal to n.

(3) $\psi(x_k) = f_k = f(x_k)$ because $L_k(x_i) = 0$ for $i \neq k$ and $L_k(x_k) = 1$.

(4) $L_k(x)$ are called Lagrange interpolating polynomials or Lagrange interpolation functions.

(5) The property $\sum\limits_{i=1}^{n} L_i(x) = 1$ is essential due to the fact that if $f_i = f^*$; $i = 1, 2, \dots, n$, then $f(x)$ from (5.10) must be f^* for all values of x, which is only possible if $\sum\limits_{i=1}^{n} L_i(x) = 1$.

5.2.3.1 Construction of $L_k(x)$: Lagrange Interpolating Polynomials

The Lagrange interpolating polynomials $L_k(x)$ can be constructed using:

$$L_k(x) = \prod_{\substack{m=1 \\ m \neq k}}^{n} \frac{x - x_m}{x_k - x_m} \quad ; \quad k = 1, 2, \dots, n \tag{5.11}$$

The functions $L_k(x)$ defined in (5.11) have the desired properties (5.8). Hence we can write

$$f(x) = \psi(x) = \sum_{i=1}^{n} f_i L_i(x) \quad ; \quad f(x_i) = f_i \tag{5.12}$$

Example 5.1 (Quadratic Lagrange Interpolating Polynomials). Consider the set of points (x_1, f_1), (x_2, f_2) and (x_3, f_3) where (x, x_2, x_3) are (-1,0,1). Then we can express these data by $f(x)$ using the Lagrange interpolating polynomials.

$$f(x) = f_1 L_1(x) + f_2 L_2(x) + f_3 L_3(x) \qquad \forall x \in [x_1, x_3] = [-1, 1] \tag{5.13}$$

We establish $L_k(x)$; $k = 1, 2, 3$ in the following. In this case $x_1 = -1$, $x_2 = 0$, $x_3 = -1$ and

$$L_k(x) = \prod_{\substack{m=1 \\ m \neq k}}^{3} \frac{x - x_m}{x_k - x_m} \qquad k = 1, 2, 3 \tag{5.14}$$

Therefore we have:

$$\begin{aligned}
\underset{(k=1)}{L_1(x)} &= \prod_{\substack{m=1\\ m\neq 1}}^{3} \frac{x-x_m}{x_1-x_m} = \frac{(x-x_2)(x-x_3)}{(x_1-x_2)(x_1-x_3)} = \frac{(x-0)(x-1)}{(-1-0)(-1-1)} = \frac{x(x-1)}{2} \\
\underset{(k=2)}{L_2(x)} &= \prod_{\substack{m=1\\ m\neq 2}}^{3} \frac{x-x_m}{x_2-x_m} = \frac{(x-x_1)(x-x_3)}{(x_2-x_1)(x_2-x_3)} = \frac{(x-(-1))(x-1)}{(0-(-1))(0-1)} = 1-x^2 \\
\underset{(k=3)}{L_3(x)} &= \prod_{\substack{m=1\\ m\neq 3}}^{3} \frac{x-x_m}{x_3-x_m} = \frac{(x-x_1)(x-x_2)}{(x_3-x_1)(x_3-x_2)} = \frac{(x-(-1))(x-0)}{(1-(-1))(1-0)} = \frac{x(x+1)}{2}
\end{aligned} \tag{5.15}$$

$L_1(x)$, $L_2(x)$, and $L_3(x)$ defined in (5.15) are the desired Lagrange polynomials in (5.13), hence $f(x)$ is defined.

Remarks. $L_k(x)$; $k = 1, 2, 3$ in (5.15) have the desired properties.

(i)

$$L_i(x_j) = \begin{cases} 1 & ; \quad j = i \\ 0 & ; \quad j \neq i \end{cases} \tag{5.16}$$

(ii)

$$\sum_{i=1}^{3} L_i(x) = \frac{x(x-1)}{2} + (1-x^2) + \frac{x(x+1)}{2} = 1 \tag{5.17}$$

(iii) Plots of $L_i(x)$; $i = 1, 2, 3$ versus x for $x \in [-1, 1]$ are shown in Figure 5.2.

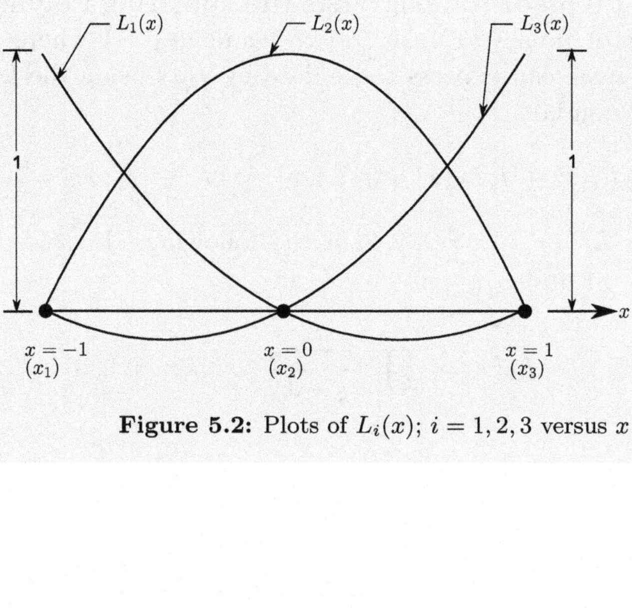

Figure 5.2: Plots of $L_i(x)$; $i = 1, 2, 3$ versus x

(iv) We have

$$f(x) = f_1 L_1(x) + f_2 L_2(x) + f_3 L_3(x) \tag{5.18}$$

Substituting for $L_1(x)$, $L_2(x)$, and $L_3(x)$ from (5.15) in (5.18):

$$f(x) = f_1 \frac{x(x-1)}{2} + f_2(1-x^2) + f_3 \frac{x(x+1)}{2} \tag{5.19}$$

where f_1, f_2, f_3 are given numerical values. The function $f(x)$ in (5.19) is the desired interpolating polynomial for the three data points. We note that $f(x)$ is a quadratic polynomial in x (i.e., a polynomial of degree two).

Example 5.2. Let (x_i, f_i) ; $i = 1, 2, 3, 4$ be the given data set in which

$$(x_1, x_2, x_3, x_4) = (-1, -1/3, 1/3, 1)$$

The Lagrange interpolating polynomial $f(x)$ for this data set can be written as:

$$f(x) = f_1 L_1(x) + f_2 L_2(x) + f_3 L_3(x) + f_4 L_4(x) \qquad \forall x \in [x_1, x_4] = [-1, 1] \tag{5.20}$$

We need to determine $L_i(x)$; $i = 1, 2, \dots, 4$ in (5.20). In this case $x_1 = -1$, $x_2 = -1/3$, $x_3 = 1/3$, $x_4 = 1$.

$$L_k(x) = \prod_{\substack{m=1 \\ m \neq k}}^{4} \frac{x - x_m}{x_k - x_m} \qquad k = 1, 2, \dots, 4 \tag{5.21}$$

We have the following:

$$\begin{aligned}
\underset{(k=1)}{L_1(x)} &= \prod_{\substack{m=1 \\ m \neq 1}}^{4} \frac{x - x_m}{x_1 - x_m} = \frac{(x - x_2)(x - x_3)(x - x_4)}{(x_1 - x_2)(x_1 - x_3)(x_1 - x_4)} \\
&= -\frac{9}{16}(1 - x)\,(1/3 + x)\,(1/3 - x) \\
\underset{(k=2)}{L_2(x)} &= \prod_{\substack{m=1 \\ m \neq 2}}^{4} \frac{x - x_m}{x_2 - x_m} = \frac{(x - x_1)(x - x_3)(x - x_4)}{(x_2 - x_1)(x_2 - x_3)(x_2 - x_4)} \\
&= \frac{27}{16}(1 + x)(1 - x)\,(1/3 - x)
\end{aligned} \tag{5.22}$$

$$
\begin{aligned}
\underset{(k=3)}{L_3(x)} &= \prod_{\substack{m=1\\m\neq 3}}^{4} \frac{x-x_m}{x_3-x_m} = \frac{(x-x_1)(x-x_2)(x-x_4)}{(x_3-x_1)(x_3-x_2)(x_3-x_4)} \\
&= \frac{27}{16}(1+x)(1-x)\,(1/3+x) \\
\underset{(k=4)}{L_4(x)} &= \prod_{\substack{m=1\\m\neq 4}}^{4} \frac{x-x_m}{x_4-x_m} = \frac{(x-x_1)(x-x_2)(x-x_3)}{(x_4-x_1)(x_4-x_2)(x_4-x_3)} \\
&= -\frac{9}{16}\,(1/3+x)\,(1/3-x)\,(1+x)
\end{aligned}
\tag{5.22}
$$

$L_i(x)$; $i = 1, 2, \dots, 4$ in (5.22) are the desired Lagrange interpolating polynomials in (5.20).

Remarks. The usual properties of $L_i(x)$; $i = 1, 2, \dots, 4$ in (5.22) hold.

(i)

$$
L_i(x_j) = \begin{cases} 1 & ; \quad j = i \\ 0 & ; \quad j \neq i \end{cases} \tag{5.23}
$$

(ii)

$$
\sum_{i=1}^{n} L_i(x) = 1 \tag{5.24}
$$

These can be verified by using $L_i(x)$; $i = 1, 2, \dots, 4$ defined by (5.22).

5.3 Mapping in $\mathbb{R}^1$

Consider a line segment in one-dimensional space x with equally spaced coordinates x_i ; $i = 1, 2, \dots, n$ (Figure 5.3). We want to map this line segment in another coordinate space ξ in which its length becomes two units and x_i; $i = 1, 2, \dots, n$ are mapped into locations ξ_i; $i = 1, 2, \dots, n$ respectively. Thus x_1 maps into location $\xi_1 = -1$, and x_n into location $\xi_n = +1$ and so on. This can be done rather easily if we recall that for the data set (x_i, f_i), the Lagrange interpolation $f(x)$ is given by:

$$
f(x) = \sum_{i=1}^{n} f_i L_i(x) \tag{5.25}
$$

If we replace x_i by ξ_i in (5.25) (and thereby replacing x with ξ) and f_i

(a) A line segment with equally spaced n points in $\mathbb{R}^1$

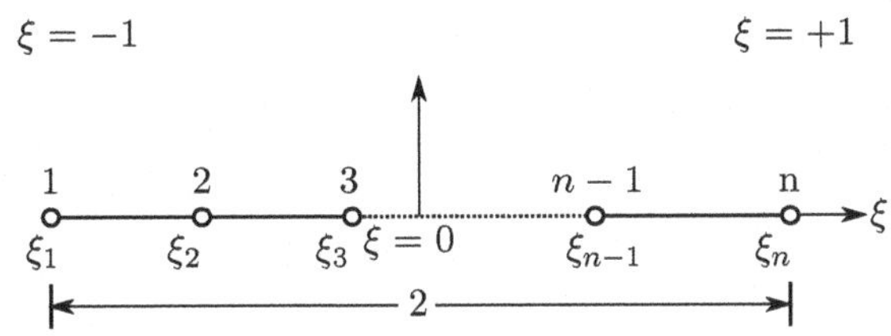

(b) Map of line segment of (a) in natural coordinate ξ

Figure 5.3: A line segment in x-space and its map in ξ-space

by x_i, then the data set (x_i, f_i) becomes (ξ_i, x_i) and (5.25) becomes:

$$x(\xi) = \sum_{i=1}^{n} x_i L_i(\xi) \tag{5.26}$$

The ξ-coordinate space is called natural coordinate space. The origin of the ξ-coordinate space is considered at the middle of the map of two unit length (Figure 5.3). Equation (5.26) indeed is the desired equation that describes the mapping of points in x- and ξ-coordinate spaces.

Remarks.

(1) The Lagrange interpolation functions $L_i(\xi)$ are constructed using the configuration of Figure 5.3 in the ξ-coordinate space.

(2) x_i; $i = 1, 2, \dots, n$ are the Cartesian coordinates of the points on the line segment in the Cartesian coordinate space.

(3) In equation (5.26), $x(\xi)$ is expressed as a linear combination of the Lagrange polynomials $L_i(\xi)$ using the Cartesian coordinates of the points in the x-space.

(4) If we choose a point $-1 \leq \xi^* \leq 1$, then (5.26) gives its corresponding location in x-space.

$$x^* = x(\xi^*) = \sum_{i=1}^{n} x_i L_i(\xi^*) \tag{5.27}$$

Thus given a location $-1 \leq \xi^* \leq 1$, the mapping (5.26) explicitly gives the corresponding location x^* in x-space. But given x^* we need to find the correct root of the polynomial in ξ to determine ξ^*. Thus, the mapping (5.26) is explicit in ξ but implicit in x, i.e., we do not have an explicit expression for $\xi = \xi(x)$.

(5) We generally consider x_i; $i = 1, 2, \dots, n$ to be equally spaced but this is not a strict requirement. However, regardless of the spacing of x_i; $i = 1, 2, \dots, n$ in the x-space, the points ξ_i; $i = 1, 2, \dots, n$ are always taken to be equally spaced in constructing the Lagrange interpolation functions $L_i(\xi)$; $i = 1, 2, \dots, n$.

(6) The Lagrange interpolation functions $L_k(\xi)$ are given by replacing x and x_i; $i = 1, 2, .., n$ with ξ and ξ_i; $i = 1, 2, \dots, n$ in (5.11).

$$L_k(\xi) = \prod_{\substack{m=1 \\ m \neq k}}^{n} \left(\frac{\xi - \xi_m}{\xi_k - \xi_m} \right) \quad ; \quad k = 1, 2, \dots, n \tag{5.28}$$

We consider some examples in the following.

Example 5.3 (1D Mapping: Two Points). Consider a line segment in $\mathbb{R}^1$ consisting of two end points $(x_1, x_2) = (2, 6)$. Derive mapping to map this line segment in two unit length in ξ-space with the origin of ξ-coordinate system at the center of $[-1, 1]$. Mapping of (a) into (b) is given by:

$$x(\xi) = L_1(\xi)x_1 + L_2(\xi)x_2 \tag{5.29}$$

$L_1(\xi)$ and $L_2(\xi)$ are derived using:

$$L_k(\xi) = \prod_{\substack{m=1 \\ m \neq k}}^{2} \left(\frac{\xi - \xi_m}{\xi_k - \xi_m} \right) \quad ; \quad k = 1, 2$$

$$\begin{aligned} \therefore \quad L_1(\xi) &= \left(\frac{\xi - \xi_2}{\xi_1 - \xi_2} \right) = \frac{\xi - 1}{-1 - 1} = \left(\frac{1 - \xi}{2} \right) \\ L_2(\xi) &= \left(\frac{\xi - \xi_1}{\xi_2 - \xi_1} \right) = \frac{\xi - (-1)}{1 - (-1)} = \left(\frac{1 + \xi}{2} \right) \end{aligned} \tag{5.30}$$

$$
\begin{aligned}
\therefore \quad x(\xi) &= \left(\frac{1-\xi}{2}\right)x_1 + \left(\frac{1+\xi}{2}\right)x_2 \\
\text{or} \quad x(\xi) &= \left(\frac{1-\xi}{2}\right)(2) + \left(\frac{1+\xi}{2}\right)(6) \\
\text{or} \quad x(\xi) &= (1-\xi) + 3(1-\xi) \\
\text{or} \quad x(\xi) &= 4 + 2\xi
\end{aligned} \tag{5.31}
$$

We note that when $\xi = -1$, then $x(-1) = 2 = x_1$ and when $\xi = 1$, $x(1) = 6 = x_2$, i.e., with $\xi = -1, 1$ we recover the x-coordinates of points 1 and 2 in the x-space. In this case the mapping is a stretch mapping. The line segment of length 4 units in x-space is uniformly compressed into a line segment of length 2 units in ξ-space.

Example 5.4 (1D Mapping: Three Points). Consider a line segment in $\mathbb{R}^1$ containing three equally spaced points $(x_1, x_2, x_3) = (2, 4, 6)$. Derive the mapping to map this line segment in two unit length ξ-space with the origin of the ξ-coordinate system at the center of $[-1, 1]$. Mapping of (a) into (b) is given by:

$$
x(\xi) = L_1(\xi)x_1 + L_2(\xi)x_2 + L_3(\xi)x_3 \tag{5.32}
$$

$L_1(\xi)$, $L_2(\xi)$ and $L_3(\xi)$ are derived using (with $\xi_1 = -1$, $\xi_2 = 0$, $\xi_3 = -1$):

$$
L_k(\xi) = \prod_{\substack{m=1 \\ m \neq k}}^{3} \left(\frac{\xi - \xi_m}{\xi_k - \xi_m}\right) \quad ; \quad k = 1, 2, 3
$$

$$
\begin{aligned}
\therefore \quad L_1(\xi) &= \frac{(\xi - \xi_2)(\xi - \xi_3)}{(\xi_1 - \xi_2)(\xi_1 - \xi_3)} = \frac{\xi(\xi - 1)}{2} \\
L_2(\xi) &= \frac{(\xi - \xi_1)(\xi - \xi_3)}{(\xi_2 - \xi_1)(\xi_2 - \xi_3)} = 1 - \xi^2 \\
L_3(\xi) &= \frac{(\xi - \xi_1)(\xi - \xi_2)}{(\xi_3 - \xi_1)(\xi_3 - \xi_2)} = \frac{\xi(\xi + 1)}{2}
\end{aligned} \tag{5.33}
$$

$$
\begin{aligned}
\text{and} \quad x(\xi) &= \frac{\xi(\xi - 1)}{2}x_1 + (1 - \xi^2)x_2 + \frac{\xi(\xi + 1)}{2}x_3 \\
x(\xi) &= \frac{\xi(\xi - 1)}{2}(2) + (1 - \xi^2)(4) + \frac{\xi(\xi + 1)}{2}(6) \\
x(\xi) &= 4 + 2\xi
\end{aligned} \tag{5.34}
$$

Remarks.

(1) We note that the mapping (5.34) is the same as (5.31). This is not a surprise as the mapping in this case is also a linear stretch mapping due to the fact that points in x-space are equally spaced. Hence, in this case we could have used points 1 and 3 with coordinates x_1 and x_3 in x-space and linear Lagrange interpolation functions corresponding to points at $\xi = -1$ and $\xi = 1$, i.e., $\left(\frac{1-\xi}{2}\right)$ and $\left(\frac{1+\xi}{2}\right)$, to derive the mapping:

$$x(\xi) = \frac{1-\xi}{2}(2) + \frac{1+\xi}{2}(6) = 4 + 2\xi \tag{5.35}$$

which is the same as (5.31) and (5.34).

(2) The conclusion in (1) also holds for more than three equally spaced points in the x-space.

(3) From (1) and (2) we conclude that when the points in x-space are equally spaced it is only necessary to use the coordinates of the two end points with Lagrange linear polynomials $\left(\frac{1-\xi}{2}\right)$ and $\left(\frac{1+\xi}{2}\right)$ for the mapping between x- and ξ-spaces. Thus, if we have n equally spaced points in x-space that are mapped in ξ-space, the following $x(\xi)$ can be used for defining the mapping.

$$x(\xi) = \left(\frac{1-\xi}{2}\right)x_1 + \left(\frac{1+\xi}{2}\right)x_n \tag{5.36}$$

Example 5.5 (1D Mapping: Three Unequally Spaced Points). Consider a line segment in $\mathbb{R}^1$ containing three unequally spaced points $(x_1, x_2, x_3) = (2, 3, 6)$. Derive the mapping to map this line segment in two unit length ξ-space with origin of the ξ-coordinate system at the center of $[-1, 1]$. We recall that points in the ξ-space are always equally spaced. Mapping of (a) into (b) is given by (using $L_k(\xi)$; $k = 1, 2, 3$ derived in Example 5.4):

$$\begin{aligned} x(\xi) &= \frac{1-\xi}{2}(2) + (1-\xi^2)(3) + \frac{1+\xi}{2}(6) \\ \text{or} \quad x(\xi) &= 3 + 2\xi + \xi^2 \end{aligned} \tag{5.37}$$

On the other hand if we used linear mapping, i.e,. x_1, x_3 and $\left(\frac{1-\xi}{2}\right)$, $\left(\frac{1+\xi}{2}\right)$, we obtain:

$$x(\xi) = \left(\frac{1-\xi}{2}\right)(2) + \left(\frac{1+\xi}{2}\right)(6)$$

$$\text{or} \quad x(\xi) = 4 + 2\xi \qquad \text{(a linear stretch mapping)} \tag{5.38}$$

When

$$\begin{aligned} \xi &= -1 \quad &; \quad x &= 2 = x_1 \\ \xi &= 1 \quad &; \quad x &= 6 = x_3 \\ \xi &= 0 \quad &; \quad x &= 4 \neq x_2 \end{aligned}$$

Thus, mapping (5.38) is not valid in this case.

Remarks.

(1) Mapping (5.37) is not a stretch mapping due to the fact that points in x-space are not equally spaced. In mapping (5.37) the length between points 2 and 1 in x-space ($x_2 - x_1 = 3 - 2 = 1$) is mapping into unit length ($\xi_2 - \xi_1 = 1$) in the ξ-space. On the other hand the length between points 3 and 2 ($x_3 - x_2 = 6 - 3 = 3$) is also mapped in the unit length ($\xi_3 - \xi_2 = 1$) in the ξ-space. Hence, this mapping is not a linear stretch mapping for the entire domain in x-space.

(2) From this example we conclude that when the points in the x-space are not equally spaced, we must utilize all points in the x-space in deriving in the mapping. This is necessitated due to the fact that in this case the mapping is not a linear stretch mapping.

5.4 Lagrange Interpolation in $\mathbb{R}^1$ using Mapping

In this section we present details of Lagrange interpolation using mapping, i.e., using the mapped domain in ξ-space. Let (x_i, f_i) ; $i = 1, 2, \dots, n$ be given data points. Let x_i be equally spaced points in the x-space. Then the mapping of points x_i ; $i = 1, 2, \dots, n$ from x-space to ξ-space in a two unit length is given by (using only the two end points in x-space)

$$x(\xi) = \left(\frac{1-\xi}{2}\right)x_1 + \left(\frac{1+\xi}{2}\right)x_n \tag{5.39}$$

The mapping defined by (5.39) maps x_i; $i = 1, 2, \dots, n$ equally spaced points in x-space into ξ_i; $i = 1, 2, \dots, n$ equally spaced points in a two unit length in ξ-space. Let $L_i(\xi)$; $i = 1, 2, \dots, n$ be Lagrange interpolating polynomials corresponding to ξ_i ; $i = 1, 2, \dots, n$ in the ξ-space. We note that even though x_i are equally spaced, the function values f_i; $i = 1, 2, \dots, n$ may not have a constant increment between two successive values. This necessitates that we use all values of f_i; $i = 1, 2, \dots, n$ in the interpolation $f(\xi)$. Now we can construct Lagrange interpolating polynomial $f(\xi)$ in the ξ-space as follows:

$$f(\xi) = \sum_{i=1}^{n} f_i L_i(\xi) \tag{5.40}$$

For a given ξ^*, we obtain $f(\xi^*)$ from (5.40) that correspond to x^* in x-space obtained using (5.39) i.e.

$$x^* = x(\xi^*) = \left(\frac{1-\xi^*}{2}\right) x_1 + \left(\frac{1+\xi^*}{2}\right) x_n$$

Thus, (5.40) suffices as interpolation for data points (x_i, f_i) ; $i = 1, 2, \dots, n$.

The mapping (5.39) together with Lagrange interpolation $f(\xi)$ given by (5.40) in ξ-space completes the interpolation for the data (x_i, f_i) ; $i = 1, 2, \dots, n$. We note that the data are interpolated in ξ-space and the mapping of geometry (lengths) between the x- and ξ-spaces establishes the correspondence in x-space for a location in ξ-space.

Remarks.

(1) In the process of interpolation described here, the interpolating polynomials are constructed in ξ-space. The correspondence of a value of $f(\xi)$ at ξ^*, i.e. $f(\xi^*)$ to x-space, is established by the mapping $x^* = x(\xi^*)$.

(2) The Lagrange polynomials for mapping can be chosen suitably (as done above) depending upon the spacing of the coordinates x_i. These can be independent of the Lagrange polynomials used to interpolate f_i ; $i = 1, 2, \dots, n$.

(3) Given a set of data (x_i, f_i), all x_i must be mapped in ξ-space and for each ξ_i ; $i = 1, 2, \dots, n$ we must construct Lagrange polynomials so that $f(\xi)$ can be expressed as a linear combination of $L_i(\xi)$; $i = 1, 2, \dots, n$ using f_i ; $i = 1, 2, \dots, n$.

Example 5.6 (1D Lagrange Interpolation in ξ-Space). Consider (x_i, f_i) ; $i = 1, 2, 3$ given below:

i	1	2	3
x_i	2	4	6
f_i	0	10	0

Derive Lagrange interpolating polynomial for this data set using the map of x_i ; $i = 1, 2, 3$ in ξ-space in two unit length.

Mapping from x-space to ξ-space
In this case the points x_i are equally spaced in the x-space hence the mapping $x(\xi)$ can be defined by

$$x(\xi) = \left(\frac{1-\xi}{2}\right) x_1 + \left(\frac{1+\xi}{2}\right) x_3 = \left(\frac{1-\xi}{2}\right)(2) + \left(\frac{1+\xi}{2}\right)(6)$$

$$\text{or} \quad x(\xi) = 4 + 2\xi \tag{5.41}$$

Lagrange Interpolation in ξ-space

$$f(\xi) = \sum_{i=1}^{3} f_i L_i(\xi) \tag{5.42}$$

In which

$$L_1(\xi) = \left(\frac{1-\xi}{2}\right) \quad ; \quad L_2(\xi) = (1-\xi^2) \quad ; \quad L_3(\xi) = \left(\frac{1+\xi}{2}\right)$$

$$\therefore \quad f(\xi) = \left(\frac{1-\xi}{2}\right)(0) + (1-\xi^2)(10) + \left(\frac{1+\xi}{2}\right)(0)$$

$$\therefore \quad f(\xi) = 10(1-\xi^2) \tag{5.43}$$

Hence, (5.41) and (5.43) complete the interpolation of the data in the table. For a given ξ^*, we obtain $f(\xi^*)$ from (5.43) that corresponds to x^* (in x-space) obtained using (5.41), i.e., $x^* = x(\xi^*)$.

5.5 Piecewise Mapping and Lagrange Interpolation in $\mathbb{R}^1$

When a large number of data (x_i, f_i) ; $i = 1, 2, \dots, n$ are given, it may not be practical to construct a single Lagrange interpolating polynomial for all the data in the data set. In such cases we can perform piecewise interpolation and mapping, keeping in mind that this approach will undoubtedly yield different results than a single interpolating polynomial for the entire data set, but may be necessitated due to practical considerations.

Let $f(\xi)$ be the single interpolating polynomial for the entire data set (x_i, f_i) ; $i = 1, 2, \dots, n$. We divide the domain $[x_1, x_n] = \bar{\Omega}$ into subdomains

$\bar{\Omega}^{(e)}$; $e = 1, 2, \dots, M$ such that

$$\bar{\Omega} = \bigcup_{e=1}^{M} \bar{\Omega}^{(e)} \tag{5.44}$$

Each subdomain $\bar{\Omega}^{(e)}$ consists of suitably chosen consecutive points x_i. A subdomain is connected to the adjacent subdomains through their end points. The choice of the number of points for each subdomain $\bar{\Omega}^{(e)}$ depends upon the desired degree of interpolation of the Lagrange interpolating polynomial for the subdomain and the regularity of data (x_i, f_i). A subdomain containing m points will permit a Lagrange interpolating polynomial of degree $m-1$. For each subdomain consider:

(i) Mapping of $\bar{\Omega}^{(e)}$ into $\bar{\Omega}^{(\xi)} = [-1, 1]$ in the ξ-space.

(ii) Lagrange interpolating polynomial $f^{(e)}(\xi)$ for the subdomain $\bar{\Omega}^{(e)}$ using its map $\bar{\Omega}^{(\xi)}$.

Then $f(\xi)$ for the entire domain $[x_1, x_n]$ is given by

$$f(\xi) = \bigcup_{e=1}^{M} f^{(e)}(\xi) \tag{5.45}$$

We can illustrate this in the following:

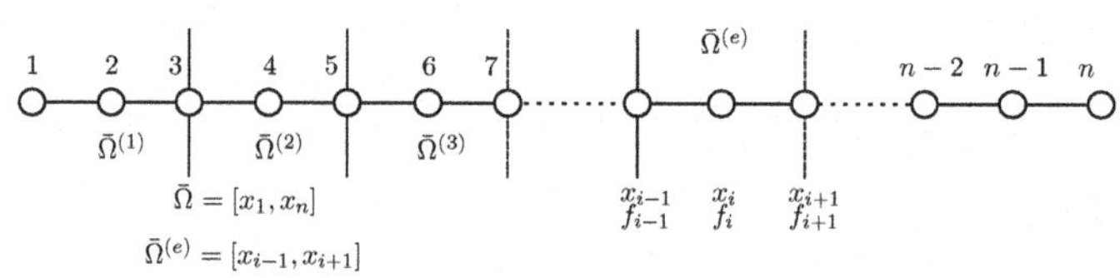

Figure 5.4: Subdomain of $\bar{\Omega}$, each consisting of three points

Let us consider $\bar{\Omega}^{(e)}$, a subdomain of $\bar{\Omega}$ consisting of three points (Figure 5.4).

(i) We map each $\bar{\Omega}^{(e)}$ in the x-space (Figure 5.5(a)) into a two unit length $\bar{\Omega}^{(\xi)}$ in ξ-space (Figure 5.5(b)), and then construct Lagrange interpolation over $\bar{\Omega}^{(\xi)}$.

(ii) Then

$$\bar{\Omega} = [x_1, x_n] = \bigcup_{e=1}^{M} \bar{\Omega}^{(e)}$$

$$\text{and} \quad f(\xi) = \bigcup_{e=1}^{M} f^{(e)}(\xi)$$

(a) Subdomain $\bar{\Omega}^{(e)}$ (b) $\bar{\Omega}^{\xi}$ map of $\bar{\Omega}^{(e)}$

Figure 5.5: Mapping of x into ξ

This completes the interpolation of the data (x_i, f_i) ; $i = 1, 2, \dots, n$.

Remarks.

(1) Rather than choosing three points for a subdomain $\bar{\Omega}^{(e)}$, we could have chosen four points in which case $f^{(e)}(\xi)$ over $\bar{\Omega}^{(\xi)}$ would be a polynomial of degree three.

(2) Choice of the number of points for a subdomain depends upon the degree $p^{(e)}$ of the Lagrange interpolation $f^{(e)}(\xi)$ desired.

(3) We note that when the entire data set (x_i, f_i) ; $i = 1, 2, \dots, n$ is interpolated using a single Lagrange interpolating polynomial $f(\xi)$, then $f(\xi)$ is a polynomial of degree $(n-1)$, hence it is of class C^{n-1}, i.e., derivatives of $f(\xi)$ of up to order $(n-1)$ are continuous.

(4) When we use piecewise mapping and interpolation, then $f^{(e)}(\xi)$ is of class C^{p^e} ; $p^e \leq n$ but $f(\xi)$ given by $\bigcup_{e=1}^{M} f^{(e)}(\xi)$ is only of class C^0. This is due to the fact that at the mating boundaries between the subdomains $\bar{\Omega}^{(e)}$, only the function f is continuous. This is a major and fundamental difference between piecewise interpolation and using a single Lagrange polynomial describing the interpolation of the data.

(5) It is also possible to design piecewise interpolations that would yield higher order differentiability for the whole domain $\bar{\Omega}$. Spline interpolation is an example. Other approaches are possible too.

Example 5.7 (1D Piecewise Lagrange Interpolation). Consider the same problem in Example 5.6 in which (x_i, f_i) ; $i = 1, 2, 3$ are given by:

i	1	2	3
x_i	2	4	6
f_i	0	10	0

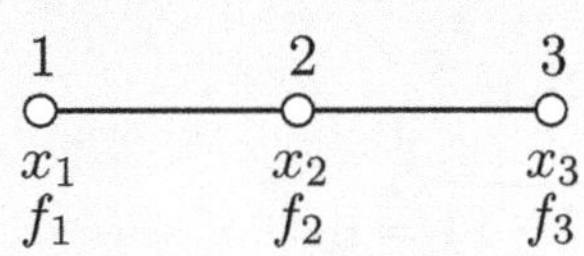

In the first case we construct a single Lagrange interpolating polynomial using all three data points. The result is the same as in Example 5.6, and we have:

$$x(\xi) = 4 + 2\xi \tag{5.46}$$
$$f(\xi) = 10(1 - \xi^2) \tag{5.47}$$

In the second case, we consider piecewise mapping and interpolations using the subdomains $\bar{\Omega}^{(1)} = [x_1, x_2]$, $\bar{\Omega}^{(2)} = [x_2, x_3]$.

Consider $\bar{\Omega}^{(1)}$

$$x_1 = 2 \quad , \quad x_2 = 4 \quad ; \quad f_1 = 0 \quad , \quad f_2 = 10$$

(a) x-space (b) ξ-space

$$\therefore \quad x^{(1)}(\xi) = \left(\frac{1-\xi}{2}\right) x_1 + \left(\frac{1+\xi}{2}\right) x_2$$
$$x^{(1)}(\xi) = \left(\frac{1-\xi}{2}\right)(2) + \left(\frac{1+\xi}{2}\right)(4)$$
$$\therefore \quad x^{(1)}(\xi) = 3 + \xi \tag{5.48}$$

and

$$f^{(1)}(\xi) = \left(\frac{1-\xi}{2}\right) f_1 + \left(\frac{1+\xi}{2}\right) f_2$$

or

$$f^{(1)}(\xi) = \left(\frac{1-\xi}{2}\right)(0) + \left(\frac{1+\xi}{2}\right)(10)$$
$$\therefore \quad f^{(1)}(\xi) = 5(1+\xi) \tag{5.49}$$

Consider $\bar{\Omega}^{(2)}$

$$x_2 = 4 \quad , \quad x_3 = 6 \quad ; \quad f_2 = 10 \quad , \quad f_3 = 0$$

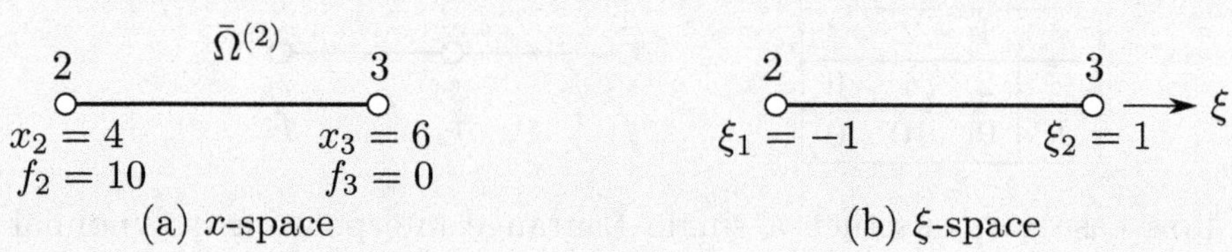

(a) x-space (b) ξ-space

$$
\begin{aligned}
\therefore \quad x^{(2)}(\xi) &= \left(\frac{1-\xi}{2}\right)x_2 + \left(\frac{1+\xi}{2}\right)x_3 \\
\text{or} \quad x^{(2)}(\xi) &= \left(\frac{1-\xi}{2}\right)(4) + \left(\frac{1+\xi}{2}\right)(6) \\
x^{(2)}(\xi) &= 5+\xi
\end{aligned}
\tag{5.50}
$$

and

$$
\begin{aligned}
f^{(2)}(\xi) &= \left(\frac{1-\xi}{2}\right)f_2 + \left(\frac{1+\xi}{2}\right)f_3 \\
\text{or} \quad f^{(2)}(\xi) &= \left(\frac{1-\xi}{2}\right)(10) + \left(\frac{1+\xi}{2}\right)(0) \\
\therefore \quad f^{(2)}(\xi) &= 5(1-\xi)
\end{aligned}
\tag{5.51}
$$

Summary

(i) Single Lagrange polynomial for the whole domain $\Omega = [x_1, x_3]$

$$
\begin{aligned}
x(\xi) &= 4 + 2\xi \\
f(\xi) &= 10(1-\xi^2)
\end{aligned}
\tag{5.52}
$$

(ii) Piecewise interpolation

(a) Subdomain $\bar{\Omega}^{(1)}$:

$$
\begin{aligned}
x^{(1)}(\xi) &= 3 + \xi \\
f^{(1)}(\xi) &= 5(1+\xi)
\end{aligned}
\tag{5.53}
$$

(b) Subdomain $\bar{\Omega}^{(2)}$:

$$
\begin{aligned}
x^{(2)}(\xi) &= 5 + \xi \\
f^{(2)}(\xi) &= 5(1-\xi)
\end{aligned}
\tag{5.54}
$$

Figure 5.6 shows plots of $f(\xi)$, $f^{(1)}(\xi)$, and $f^{(2)}(\xi)$ over $\bar{\Omega} = [x_1, x_3] = [2, 6]$.

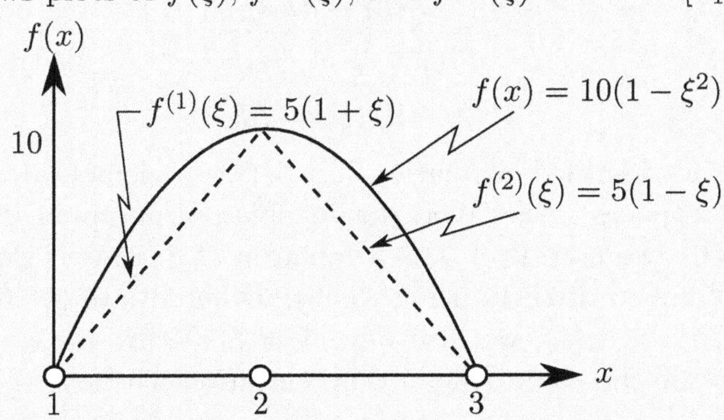

Figure 5.6: Plots of $f(\xi)$, $f^{(1)}(\xi)$, and $f^{(2)}(\xi)$

It is clear that $f(\xi) = 10(1-\xi^2)$ is of class $C^2(\bar{\Omega})$ where as $f(\xi) = \bigcup_{e=1}^{2} f^{(e)}(\xi)$ is of class $C^0(\bar{\Omega})$, due to the fact that $f(x)$ is continuous $\forall x \in [x_1, x_3]$ but $\frac{df}{dx}$ is discontinuous at point 2 ($x_2 = 4$).

5.6 Mapping of Length and Derivatives of $f(\cdot)$ in x- and ξ-spaces ($\mathbb{R}^1$)

The concepts presented in the following can be applied to $x(\xi)$, $f(\xi)$ for $\bar{\Omega}$ as well as to $x^{(e)}(\xi)$, $f^{(e)}(\xi)$ for a subdomain $\bar{\Omega}^{(e)}$. Consider

$$x(\xi) = \sum_{i=1}^{\tilde{n}} x_i \widetilde{L}_i(\xi) \tag{5.55}$$

and

$$f(\xi) = \sum_{i=1}^{n} f_i L_i(\xi) \tag{5.56}$$

$\widetilde{L}_i(\xi)$ and $L_i(\xi)$ are suitable Lagrange polynomials in ξ for mapping of points and interpolation. We note that (5.55) only describes mapping of points, i.e., given ξ^* we can obtain x^* using (5.55), $x^* = x(\xi^*)$. Mapping of length in x- and ξ-spaces requires a different relationship than (5.55). Consider the differential of (5.55):

$$dx(\xi) = \sum_{i=1}^{\tilde{n}} x_i \frac{d\widetilde{L}_i(\xi)}{d\xi} d\xi = \left(\sum_{i=1}^{\tilde{n}} \left(x_i \frac{d\widetilde{L}_i(\xi)}{d\xi} \right) \right) d\xi \tag{5.57}$$

Let

$$J = \sum_{i=1}^{\tilde{n}} \left(x_i \frac{d\widetilde{L}_i(\xi)}{d\xi} \right) \tag{5.58}$$

$$\therefore \quad dx(\xi) = J d\xi \tag{5.59}$$

Equation (5.59) describes a relationship between elemental lengths $d\xi$ and dx in ξ- and x-spaces. J is called the Jacobian of mapping.

From (5.56), we note that f is a function of ξ, thus if we require $\frac{df}{dx}$, it can not be obtained directly using (5.56). Differentiate (5.56) with respect to ξ and since $x = x(\xi)$, we also have $\xi = \xi(x)$ (inverse of the mapping), hence we can use the chain rule of differentiation whenever needed.

$$\frac{df(\xi)}{dx} = \sum_{i=1}^{n} f_i \frac{dL_i(\xi)}{dx} \tag{5.60}$$

Thus, determination of $\frac{df(\xi)}{dx}$ requires $\frac{dL_i(\xi)}{dx}$. We can differentiate $L_i(\xi)$ with respect to ξ using the chain rule of differentiation.

$$\frac{dL_i(\xi)}{d\xi} = \frac{dL_i(\xi)}{dx}\frac{dx}{d\xi} = \left(\frac{dx}{d\xi}\right)\frac{dL_i(\xi)}{dx}$$

or

$$\frac{dL_i(\xi)}{d\xi} = J\frac{dL_i(\xi)}{dx}$$

$$\therefore \quad \frac{dL_i(\xi)}{dx} = \frac{1}{J}\frac{dL_i(\xi)}{d\xi} \tag{5.61}$$

Similarly

$$\frac{d^2L_i(\xi)}{d\xi^2} = \frac{d}{d\xi}\left(\frac{dL_i(\xi)}{d\xi}\right) = \frac{d}{d\xi}\left(J\frac{dL_i(\xi)}{dx}\right) \tag{5.62}$$

If we assume that the mapping $x(\xi)$ is a linear stretch, then $x(\xi)$ is a linear function of ξ and hence $J = \frac{dx}{d\xi}$ is not a function of ξ. Thus, we can write (5.62) as:

$$\frac{d^2L_i(\xi)}{d\xi^2} = J\left(\frac{d}{d\xi}\left(\frac{dL_i(\xi)}{dx}\right)\right) = J\left(\frac{d^2L_i(\xi)}{dx^2}\frac{dx}{d\xi}\right) = J\left(\frac{d^2L_i(\xi)}{dx^2}J\right)$$

$$\text{or} \quad \frac{d^2L_i(\xi)}{d\xi^2} = J^2\frac{d^2L_i(\xi)}{dx^2}$$

Hence,

$$\frac{d^2L_i(\xi)}{dx^2} = \frac{1}{J^2}\frac{d^2L_i(\xi)}{d\xi^2} \tag{5.63}$$

In general, for the derivative of order k we can write:

$$\frac{d^kL_i(\xi)}{dx^k} = \frac{1}{J^k}\frac{d^kL_i(\xi)}{d\xi^k} \tag{5.64}$$

If we substitute from (5.61) in (5.60), then we obtain:

$$\frac{df(\xi)}{dx} = \frac{1}{J}\left(\sum_{i=1}^{n} f_i\frac{dL_i(\xi)}{d\xi}\right) = \frac{1}{J}\frac{df(\xi)}{d\xi} \tag{5.65}$$

Likewise, using (5.63):

$$\frac{d^2f(\xi)}{dx^2} = \sum_{i=1}^{n} f_i\frac{d^2L_i(\xi)}{dx^2} = \frac{1}{J^2}\left(\sum_{i=1}^{n} f_i\frac{d^2L_i(\xi)}{d\xi^2}\right) = \frac{1}{J^2}\frac{d^2f(\xi)}{d\xi^2} \tag{5.66}$$

In general for the derivative of $f(\xi)$ of order k, we have:

$$\frac{d^kf(\xi)}{dx^k} = \frac{1}{J^k}\left(\sum_{i=1}^{n} f_i\frac{d^kL_i(\xi)}{d\xi^k}\right) = \frac{1}{J^k}\frac{d^kf(\xi)}{d\xi^k} \tag{5.67}$$

The derivatives $\frac{dL_i(\xi)}{d\xi}$, $\frac{d^2L_i(\xi)}{d\xi^2}$, ..., $\frac{d^kL_i(\xi)}{d\xi^k}$; $k = 1, 2, \dots, n$ can be determined by differentiating $L_i(\xi)$; $i = 1, 2, \dots, n$ with respect to ξ. Hence, the derivatives of $f(\xi)$ with respect to x of any desired order can be determined.

The mapping of length and derivatives of f in the two spaces (x and ξ) is quite important in many other instances than just obtaining derivatives of $f(\cdot)$ with respect to x. We illustrate this in the following.

(i) Suppose we require the integral of $f(x)$ (interpolation of data (x_i, f_i) ; $i = 1, 2, \dots, n$) over $\bar{\Omega} = [x_1, x_n]$.

$$I = \int_{x_1}^{x_n} f(x)dx \tag{5.68}$$

If $[x_1, x_n] \to \Omega^{(\xi)} = [-1, 1]$, then (5.68) can be written as

$$I = \int_{-1}^{1} f(\xi)Jd\xi \tag{5.69}$$

$f(\xi)$ is Lagrange interpolating polynomial in ξ-space corresponding to the data set (x_i, f_i) ; $i = 1, 2, \dots, n$. The integrand in (5.69) is an algebraic polynomial in ξ, hence can be easily integrated. See Chapter 6 for Gauss quadrature to obtain values of the integral I.

(ii) Suppose we require the integral of $\frac{d^kf}{dx^k}$; $k = 1, 2, \dots, n$ over the domain $\bar{\Omega}$, then

$$I = \int_{x_1}^{x_n} \frac{d^kf}{dx^k}dx = \int_{-1}^{1} \frac{1}{J^k}\left(\frac{d^kf}{d\xi^k}\right) J\,d\xi \tag{5.70}$$

Thus various differentiation and integration processes are now possible using $f(\xi)$, its derivatives with respect to ξ, J, and the mapping.

Example 5.8 (Mapping of Derivatives in 1D). Consider the following sets of data.

i	1	2	3
x_i	2	4	6
f_i	0	10	0

Use all three data points to derive the mapping $x(\xi)$. Also derive the mapping using points x_1 and x_3. Construct Lagrange polynomial $f(\xi)$ using all three points. Determine the Jacobian of mapping J using both approaches of mapping. Determine an expression for $\frac{df}{dx}$ and $\frac{d^2f}{dx^2}$.

As seen in previous examples, in this case the mapping $x(\xi)$ is a linear stretch mapping, hence using x_1, x_2, x_3 or x_1, x_3 we obtain the same mapping. In the following we present a slightly different exercise.

Using points 1 and 3:

$$x(\xi) = \left(\frac{1-\xi}{2}\right)x_1 + \left(\frac{1+\xi}{2}\right)x_3 \tag{5.71}$$

$$\therefore \quad J = \frac{dx}{d\xi} = \left(\frac{x_3 - x_1}{2}\right) = \frac{h}{2}\,; \quad h = \text{length of the domain} \tag{5.72}$$

On the other hand if we use points 1,2, and 3, then

$$x(\xi) = \left(\frac{1-\xi}{2}\right)x_1 + (1-\xi^2)x_2\left(\frac{1+\xi}{2}\right)x_3 \tag{5.73}$$

But since the points are equally spaced $x_2 = \frac{x_1+x_3}{2}$, hence we obtain the following (by substituting $x_2 = \frac{x_1+x_3}{2}$ into (5.73)).

$$x(\xi) = \left(\frac{1-\xi}{2}\right)x_1 + \left(\frac{1+\xi}{2}\right)x_3$$

which is the same mapping as (5.71), hence in this case also $J = \frac{h}{2}$. Thus for linear stretch mapping between x and ξ, mapping is always linear in ξ and, hence $J = \frac{h}{2}$, h being the length of the domain in x-space.

As derived earlier, $f(\xi)$ is given by

$$f(\xi) = 10(1-\xi^2) \qquad , \qquad h = \frac{x_3 - x_1}{2} = \frac{6-1}{2} = 2$$

$$\frac{df}{dx} = \frac{1}{J}\frac{df}{d\xi} = \frac{1}{J}(-20\xi) = \frac{1}{2}(-20\xi) = -10\xi$$

$$\frac{d^2f}{dx^2} = \frac{1}{J^2}\frac{d^2f}{d\xi^2} = \frac{1}{J^2}(-20) = \frac{1}{2^2}(-20) = -5$$

5.7 Mapping and Interpolation Theory in $\mathbb{R}^2$

Consider a domain $\bar{\Omega} \subset \mathbb{R}^2$ and let $((x_i, y_i), f_i)\,;\, i = 1, 2, \dots, n$ be the data points in $\bar{\Omega}$. Our objective is to construct $f(x, y)$ that interpolates this data set, i.e., establish an analytical expression $f(x,y)$ such that $f(x_i,y_i) = f_i$; $i = 1,2,\dots,n$.

This problem may appear simple but in reality it is quite complex. In many applications, the complexity of $\bar{\Omega}$ adds to the complexity of the problem of interpolation. In such cases rather than constructing a single interpolation function $f(x,y)$ for $\bar{\Omega}$, we may consider the possibility of piecewise mapping and the interpolation using the mapping. Even though from mapping in $\mathbb{R}^1$ we know that these two approaches are not the same, it may be necessary to use the second approach due to complexity of $\bar{\Omega}$.

5.7.1 Division of $\bar{\Omega}$ into Subdivisions $\bar{\Omega}^{(e)}$

Consider data points $((x_i,y_i),f_i)$; $i = 1,2,\dots,n$ shown in Figure 5.7(a). We regularize this data into four sided quadrilateral subdomains shown in Figure 5.7(c) such that each data point is the vertex of the quadrilaterals. In this case we say that $\bar{\Omega}$ is discretized into $\bar{\Omega}^T = \bigcup_e \bar{\Omega}^{(e)}$ in which $\bar{\Omega}^{(e)}$ is a typical quadrilateral subdomain containing data only at the vertices (Figure 5.8(a)).

The numbers (not shown) at the vertices are local numbers assigned to each vertex of the quadrilateral $\bar{\Omega}^{(e)}$ of $\bar{\Omega}^T$. Figure 5.7(b) shows another set of data $((x_i,y_i),f_i)$; $i = 1,2,\dots,n$ which are regularized into subdomains containing nine data points (Figure 5.7(d)). A typical subdomain of Figure 5.7(d) is shown in Figure 5.8(b).

Now the problem of interpolating data in Figure 5.7(a) reduces to constructing piecewise interpolation for each $\bar{\Omega}^{(e)}$ (containing four data points) of $\bar{\Omega}^T$. Likewise the interpolation of data in Figure 5.7(b) reduces to piecewise interpolation for $\bar{\Omega}^{(e)}$ containing nine data points (Figure 5.8(b)). If $f^{(e)}(x,y)$ is the piecewise interpolation of data for $\bar{\Omega}^{(e)}$ of either Figure 5.8(a) or Figure 5.8(b), then the interpolation $f(x,y)$ for the entire data set of $\bar{\Omega}$ (approximated by $\bar{\Omega}^T$) is given by:

$$f(x,y) = \bigcup_e f^{(e)}(x,y) \tag{5.74}$$

Choice of data points for subdomains (i.e., Figure 5.8(a) or 5.8(b)) is not arbitrary but depends upon the degree of interpolation desired for $\bar{\Omega}^{(e)}$ as shown later.

Instead of choosing quadrilateral subdomains we could have chosen triangular or any other desired shape of subdomain. We illustrate the details using quadrilateral subdomains. Constructing interpolations of data over subdomains of Figures 5.8(a) and 5.8(b) is quite a difficult task due to irregular geometry. This task can be simplified by using mapping of the domains of Figures 5.8(a) and 5.8(b) into regular shapes such as two unit squares, and then constructing interpolation in the mapped domain.

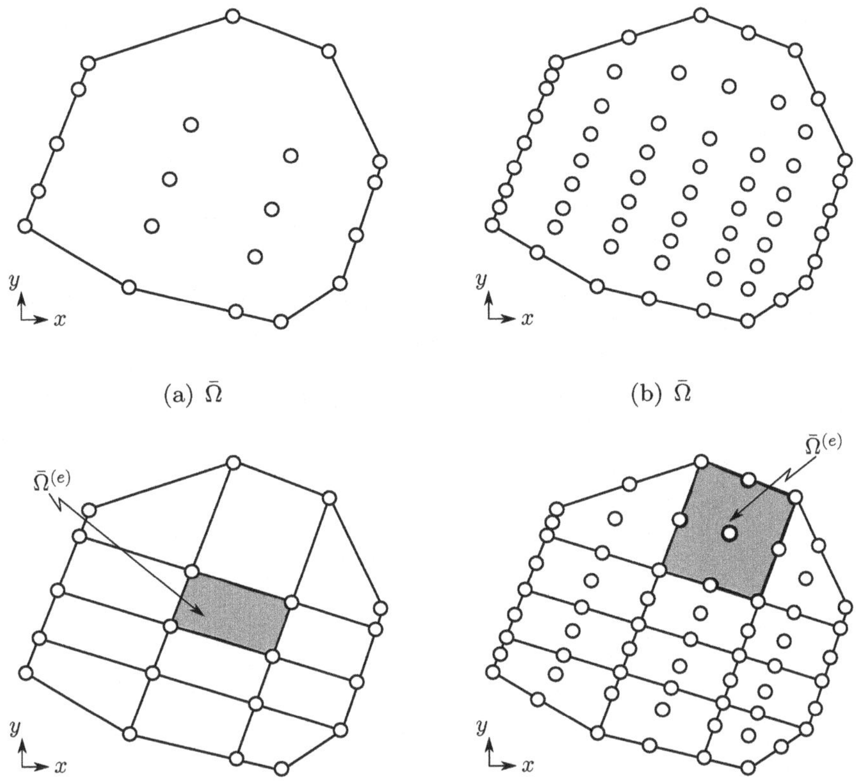

(a) $\bar{\Omega}$

(b) $\bar{\Omega}$

(c) Subdivision of $\bar{\Omega}$ of (a) in $\bar{\Omega}^T = \bigcup_e \bar{\Omega}^{(e)}$

(d) Subdivision of $\bar{\Omega}$ of (b) in $\bar{\Omega}^T = \bigcup_e \bar{\Omega}^{(e)}$

Figure 5.7: Discrete data points in $\mathbb{R}^2$ and the subdivision

5.7.2 Mapping of $\bar{\Omega}^{(e)} \subset \mathbb{R}^2$ into $\bar{\Omega}^{(\xi\eta)} \subset \mathbb{R}^2$

Figure 5.9(a) shows a four-node quadrilateral subdomain in xy-space and Figure 5.9(b) shows its map in a two unit square $\bar{\Omega}^{(\xi\eta)}$ in the $\xi\eta$ natural coordinate space with the origin of the coordinate system at the center of the subdomain $\bar{\Omega}^{(\xi\eta)}$.

Figure 5.9(c) shows a nine-node distorted quadrilateral subdomain with curved faces. Figure 5.9(d) shows its map in the natural coordinate space $\xi\eta$ in a two unit square $\bar{\Omega}^{(\xi\eta)}$ with the origin of the coordinate system at the center of $\bar{\Omega}^{(\xi\eta)}$. For convenience we assign local numbers to the data points (node numbers).

Consider a four-node quadrilateral of Figure 5.9(a) and its map in $\xi\eta$-space shown in Figure 5.9(b). Let $L_i(\xi, \eta)$; $i = 1, 2, \dots, 4$ be Lagrange polynomials corresponding to nodes $1, 2, 3, 4$ of Figure 5.9(b) with the following

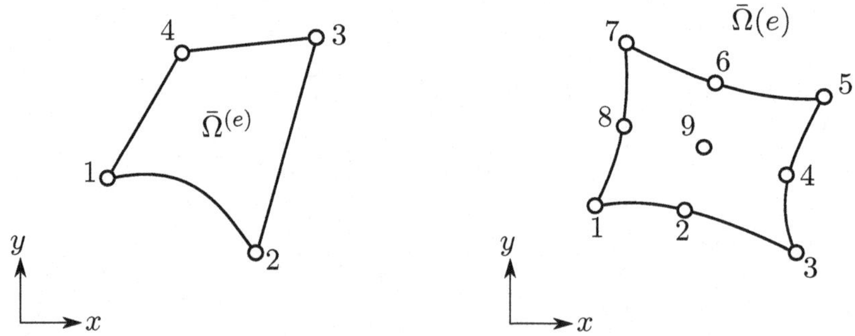

(a) A typical four-node quadrilateral subdomain of $\bar{\Omega}^T$

(b) A typical nine-node quadrilateral subdomain of $\bar{\Omega}^T$

Figure 5.8: Sample subdomains of $\bar{\Omega}^T$

properties.

$$
\begin{aligned}
&1. \quad L_i(\xi_j,\eta_j) = \begin{cases} 1 \; ; & j = i \\ 0 \; ; & j \neq i \end{cases} \qquad ; \qquad i,j = 1,2,\dots,4 \\
&2. \quad \sum_{i=1}^{4} L_i(\xi,\eta) = 1 \\
&3. \quad L_i(\xi,\eta) \; ; \; i = 1,2,\dots,4 \text{ are polynomials of degree} \\
&\qquad\qquad \text{less than or equal to 2 in } \xi \text{ and } \eta
\end{aligned}
\tag{5.75}
$$

Using $L_i(\xi,\eta)$, we can define $x(\xi,\eta)$ and $y(\xi,\eta)$

$$
\begin{aligned}
x(\xi,\eta) &= \sum_{i=1}^{4} L_i(\xi,\eta)x_i \\
y(\xi,\eta) &= \sum_{i=1}^{4} L_i(\xi,\eta)y_i
\end{aligned}
\tag{5.76}
$$

Equations (5.76) are the desired equations for mapping of points between $\bar{\Omega}^{(e)}$ and $\bar{\Omega}^{(\xi\eta)}$.

Remarks.

(1) Equations (5.76) are explicit in ξ and η, i.e., given values of ξ and η (ξ^*,η^*), we can use (5.76) to determine their map (x^*,y^*) using (5.76), $(x^*,y^*) = (x(\xi^*,\eta^*), y(\xi^*,\eta^*))$.

(2) Equations (5.76) are implicit in x and y. Given (x^*,y^*) in xy-space, determination of its map (ξ^*,η^*) in $\xi\eta$-space requires solution of simultaneous

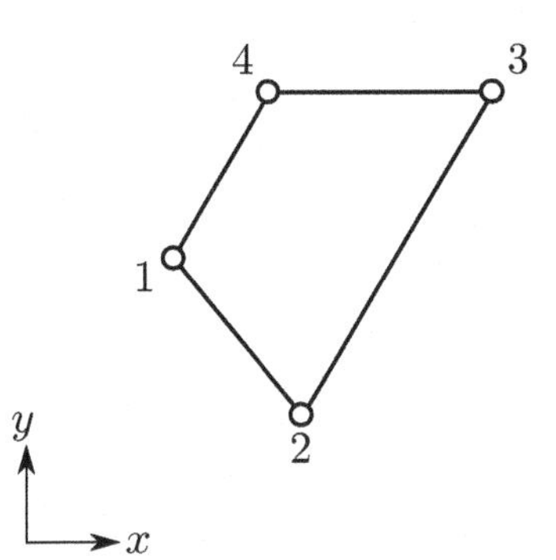

(a) Four-node $\bar{\Omega}^{(e)}$ in xy-space

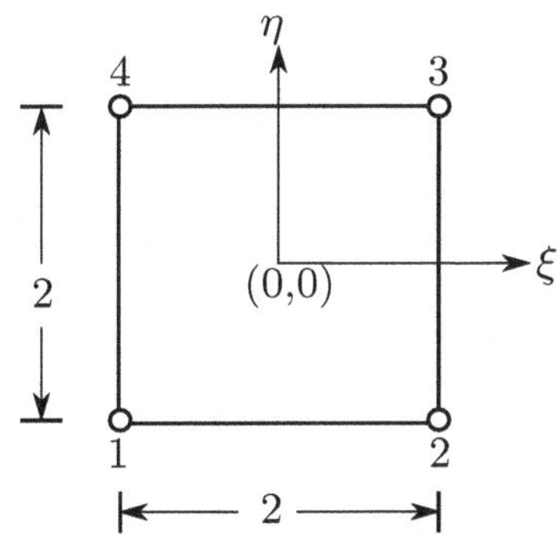

(b) Map of $\bar{\Omega}^{(e)}$ into $\bar{\Omega}^{(\xi,\eta)}$ in a two unit square in $\xi\eta$ coordinate space

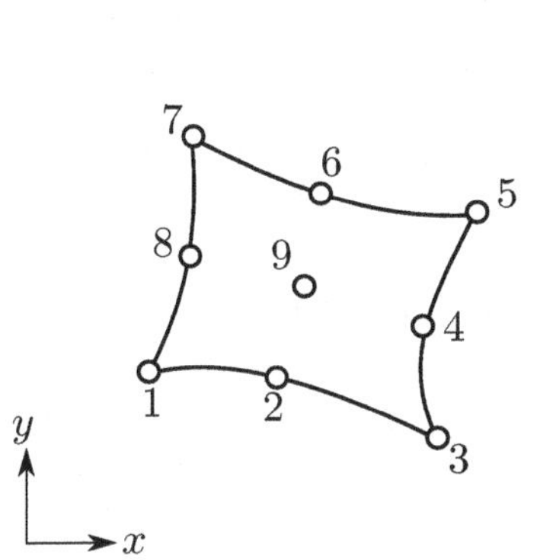

(c) A nine-node $\bar{\Omega}^{(e)}$ in xy-space

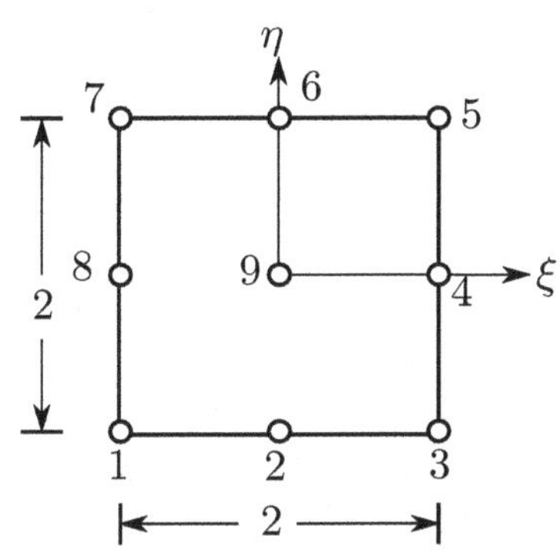

(d) Map of $\bar{\Omega}^{(e)}$ into $\bar{\Omega}^{(\xi,\eta)}$ into a two unit square in $\xi\eta$ coordinate space

Figure 5.9: Maps of $\bar{\Omega}^{(e)}$ in $\bar{\Omega}^{(\xi,\eta)}$

equations resulting from (5.76) after substituting $x(\xi,\eta) = x^*$ and $y(\xi,\eta) = y^*$.

(3) We still need to determine the Lagrange polynomials $L_i(\xi,\eta)$; $i = 1,2,3,4$ that have the desired properties (5.75).

(4) As shown subsequently, this mapping is bilinear due to the fact that $L_i(\xi,\eta)$ in this case are linear in both ξ and η.

(5) In case of mapping of nine-node subdomain (Figures 5.9(c) and 5.9(d)),

we can write:

$$x(\xi,\eta) = \sum_{i=1}^{9} L_i(\xi,\eta) x_i$$
$$y(\xi,\eta) = \sum_{i=1}^{9} L_i(\xi,\eta) y_i \tag{5.77}$$

$L_i(\xi,\eta)$; $i = 1, 2, \dots, 9$ in (5.77) have the same properties as in (5.75) and are completely defined.

(6) Choice of the configurations of nodes (as in Figures 5.9(a) and 5.9(c)) is not arbitrary and is based on the degree of the polynomial desired in ξ and η, and can be determined using Pascal's rectangle.

5.7.3 Pascal's Rectangle: A Polynomial Approach to Determine $L_i(\xi,\eta)$

If we were to express $x(\xi,\eta)$ and $y(\xi,\eta)$ as linear combinations of the monomials in ξ and η, then we could write the following if x and y are linear functions of ξ and η.

$$x(\xi,\eta) = c_0 + c_1\xi + c_2\eta + c_3\xi\eta \tag{5.78}$$

$$y(\xi,\eta) = d_0 + d_1\xi + d_2\eta + d_3\xi\eta \tag{5.79}$$

In this case the choice of monomials 1, ξ, η, and $\xi\eta$ was not too difficult. However in case of nine-node configuration of Figure 5.9(d) the choice of the monomials in ξ and η is not too obvious. Pascal's rectangle facilitates (i) the selection of monomials in ξ and η for complete polynomials of a chosen degree in ξ and η, and (ii) determination of the number of nodes and their location in the two unit square in $\xi\eta$-space.

Consider increasing powers of ξ and η in the horizontal and vertical directions (see Figure 5.10). This arrangement is called Pascal's rectangle.

We can choose up to the desired degree in ξ and η using Figure 5.10. The terms located at the intersections of ξ and η lines are the desired monomial terms. The locations of the monomial terms are the locations of the nodes in the $\xi\eta$ configuration.

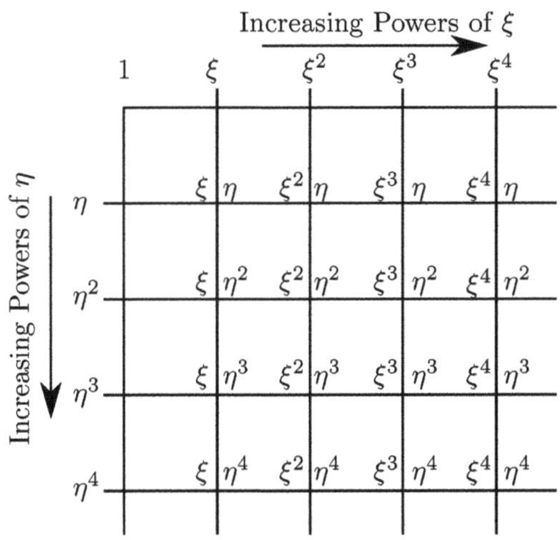

Figure 5.10: Pascal's rectangle in $\mathbb{R}^2$ in $\xi\eta$-coordinates

Example 5.9 (Using Pascal's Rectangle).

(a) From Pascal's rectangle (Figure 5.10), a linear approximation in ξ and η would require four nodes and the terms 1, ξ, η, and $\xi\eta$. We can write:

$$\begin{aligned} x(\xi,\eta) &= c_0 + c_1\xi + c_2\eta + c_3\xi\eta \\ y(\xi,\eta) &= d_0 + d_1\xi + d_2\eta + d_3\xi\eta \end{aligned} \tag{5.80}$$

Using (ξ_i,η_i) ; $i = 1,2,\dots,4$ and the corresponding (x_i,y_i) ; $i = 1,2,\dots,4$ in (5.80) we can determine $x(\xi,\eta)$ and $y(\xi,\eta)$.

$$\begin{aligned} x(\xi,\eta) &= \sum_{i=1}^{4} L_i(\xi,\eta)x_i \\ y(\xi,\eta) &= \sum_{i=1}^{4} L_i(\xi,\eta)y_i \end{aligned} \tag{5.81}$$

$L_i(\xi,\eta)$; $i = 1,2,\dots,4$ have the properties in (5.75).

(b) From Pascal's rectangle (Figure 5.10), a biquadratic approximation in ξ and η would require nine nodes and the monomial terms: 1, ξ, η, $\xi\eta$, ξ^2, $\xi\eta$, η^2, $\xi^2\eta$, $\xi\eta^2$, $\xi^2\eta^2$. In this case we can write:

$$\begin{aligned} x(\xi,\eta) &= c_0 + c_1\xi + c_2\eta + c_3\xi\eta + c_4\xi^2 + c_5\eta^2 \\ &\qquad + c_6\xi^2\eta + c_7\xi\eta^2 + c_8\xi^2\eta^2 \\ y(\xi,\eta) &= d_0 + d_1\xi + d_2\eta + d_3\xi\eta + d_4\xi^2 + d_5\eta^2 \\ &\qquad + d_6\xi^2\eta + d_7\xi\eta^2 + d_8\xi^2\eta^2 \end{aligned} \tag{5.82}$$

Using (5.82) with (ξ_i, η_i) ; $i = 1, 2, \dots, 9$ and the corresponding (x_i, y_i) ; $i = 1, 2, \dots, 9$, we can determine $x(\xi, \eta)$ and $y(\xi, \eta)$

$$
\begin{aligned}
x(\xi, \eta) &= \sum_{i=1}^{9} L_i(\xi, \eta) x_i \\
y(\xi, \eta) &= \sum_{i=1}^{9} L_i(\xi, \eta) y_i
\end{aligned}
\tag{5.83}
$$

$L_i(\xi, \eta)$; $i = 1, 2, \dots, 9$ in (5.83) are completely determined. $L_i(\xi, \eta)$ have the same properties as shown in (5.75).

Remarks.

(1) Pascal's rectangle provides a mechanism to determine functions $L_i(\xi, \eta)$ that allow us to establish mapping between $\bar{\Omega}^{(e)}$ and $\bar{\Omega}^{(\xi\eta)}$ of any desired degree in ξ and η.

(2) This process involves the inverse of a coefficient matrix, an undesirable feature which shall be corrected in the next section using the tensor product of 1D Lagrange polynomials in ξ and η.

(3) Pascal's rectangle is still extremely useful as it can tell us the nodal configurations and the monomials for complete polynomials of desired degrees in ξ and η.

(4) *Complete* implies all terms up to a chosen degree of the polynomial in ξ and η have been considered.

5.7.4 Tensor Product to Generate $L_i(\xi, \eta)$; $i = 1, 2, \dots$

The tensor product is an approach to determine 2D Lagrange polynomials $L_i(\xi, \eta)$ in $\xi\eta$-space corresponding to the desired degrees in ξ and η using 1D Lagrange polynomials in ξ and η.

5.7.4.1 Bilinear $L_i(\xi, \eta)$ in ξ and η

Based on Pascal's rectangle, bilinear $L_i(\xi, \eta)$ in ξ and η would require a four-node configuration (shown in Figure 5.11(a) below).

Consider two-node 1D configurations in ξ and η shown in Figure 5.11(b). Consider 1D Lagrange polynomials in ξ and η corresponding to two-node configurations of Figure 5.11(b).

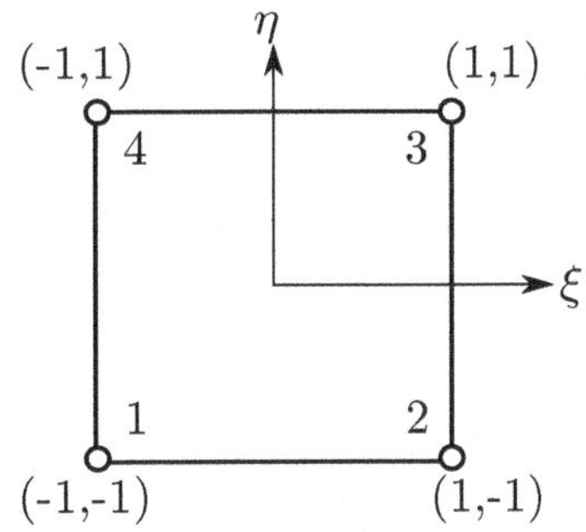

(a) Four-node configuration

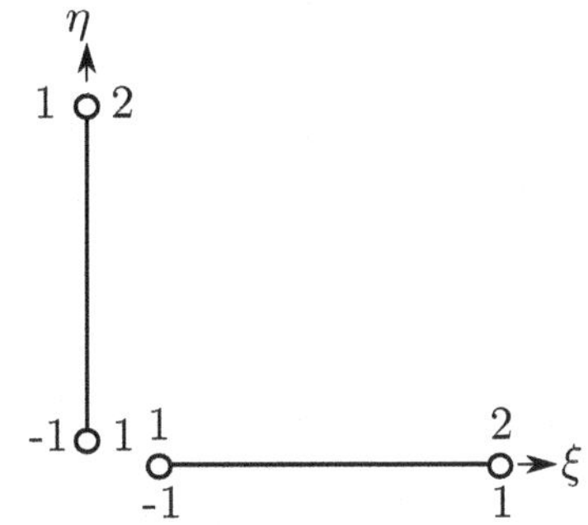

(b) 1D two-node configurations in ξ and η

Figure 5.11: Use of tensor product for four-node subdomains in $\mathbb{R}^2$

In the ξ-direction:

$$L_1^\xi(\xi) = \left(\frac{1-\xi}{2}\right) \quad , \quad L_2^\xi(\xi) = \left(\frac{1+\xi}{2}\right) \tag{5.84}$$

In the η-direction:

$$L_1^\eta(\eta) = \left(\frac{1-\eta}{2}\right) \quad , \quad L_2^\eta(\eta) = \left(\frac{1+\eta}{2}\right) \tag{5.85}$$

Arrange $L_1^\xi(\xi)$ and $L_2^\xi(\xi)$ as a vector along with their ξ coordinates of -1 and $+1$. Note that ± 1 are not elements of the vector, they have been included to indicate the location of the node corresponding to each L_k. In this case, this arrangement gives a 2×1 vector of L_1^ξ and L_2^ξ.

$$\begin{Bmatrix} L_1^\xi(\xi) \\ (-1) \\ L_2^\xi(\xi) \\ (+1) \end{Bmatrix} = \begin{Bmatrix} \left(\frac{1-\xi}{2}\right) \\ (-1) \\ \left(\frac{1+\xi}{2}\right) \\ (+1) \end{Bmatrix} \tag{5.86}$$

Arrange $L_1^\eta(\eta)$ and $L_2^\eta(\eta)$ as a row matrix along with their η coordinates of -1 and $+1$.

$$\begin{bmatrix} L_1^\eta(\eta) & L_2^\eta(\eta) \\ (-1) & (+1) \end{bmatrix} = \begin{bmatrix} \left(\frac{1-\eta}{2}\right) & \left(\frac{1+\eta}{2}\right) \\ (-1) & (+1) \end{bmatrix} \tag{5.87}$$

Take the product of $L_i^\xi(\xi)$ in (5.86) with $L_j^\eta(\eta)$ in (5.87), keeping their ξ, η coordinates together with the product terms. This is called the tensor product.

$$
\begin{Bmatrix} L_1^\xi(\xi) \\ (-1) \\ L_2^\xi(\xi) \\ (+1) \end{Bmatrix} \begin{bmatrix} L_1^\eta(\eta) & L_2^\eta(\eta) \\ (-1) & (+1) \end{bmatrix} = \begin{bmatrix} L_1^\xi(\xi)L_1^\eta(\eta) & L_1^\xi(\xi)L_2^\eta(\eta) \\ (-1,-1) & (-1,+1) \\ L_2^\xi(\xi)L_1^\eta(\eta) & L_2^\xi(\xi)L_2^\eta(\eta) \\ (+1,-1) & (+1,+1) \end{bmatrix} \tag{5.88}
$$

Substituting for $L_1^\xi(\xi)$, $L_2^\xi(\xi)$, $L_1^\eta(\eta)$, and $L_2^\eta(\eta)$ in (5.88):

$$
\begin{bmatrix} \left(\frac{1-\xi}{2}\right)\left(\frac{1-\eta}{2}\right) & \left(\frac{1-\xi}{2}\right)\left(\frac{1+\eta}{2}\right) \\ (-1,-1) & (-1,+1) \\ \left(\frac{1+\xi}{2}\right)\left(\frac{1-\eta}{2}\right) & \left(\frac{1+\xi}{2}\right)\left(\frac{1+\eta}{2}\right) \\ (+1,-1) & (+1,+1) \end{bmatrix} = \begin{bmatrix} L_1(\xi,\eta) & L_4(\xi,\eta) \\ (-1,-1) & (-1,+1) \\ L_2(\xi,\eta) & L_3(\xi,\eta) \\ (+1,-1) & (+1,+1) \end{bmatrix} \tag{5.89}
$$

The coordinates ξ, η associated with the terms and their comparisons with the ξ, η coordinates of the four nodes in Figure 5.11(a) identifies $L_i(\xi, \eta)$; $i = 1, 2, \dots, 4$ for the four-node configuration of Figure 5.11(a). We could view this process as the two-node configuration in η direction (Figure 5.11(b)) traversing along the ξ-direction. As it encounters a node in the ξ-direction, we obtain a trace of the nodes. Each node of the trace contains products of 1D functions in ξ and η as the two 2D functions in ξ and η. Thus, we have for the four-node configuration of Figure 5.11(a):

$$
\begin{aligned}
L_1(\xi,\eta) &= \left(\frac{1-\xi}{2}\right)\left(\frac{1-\eta}{2}\right) \\
L_2(\xi,\eta) &= \left(\frac{1-\xi}{2}\right)\left(\frac{1+\eta}{2}\right) \\
L_3(\xi,\eta) &= \left(\frac{1+\xi}{2}\right)\left(\frac{1-\eta}{2}\right) \\
L_4(\xi,\eta) &= \left(\frac{1+\xi}{2}\right)\left(\frac{1+\eta}{2}\right)
\end{aligned} \tag{5.90}
$$

The functions $L_i(\xi, \eta)$ in (5.90) satisfy the properties (5.75).

5.7.4.2 Biquadratic $L_i(\xi, \eta)$ in ξ and η

From Pascal's rectangle we have the nine-node configuration in ξ and η.

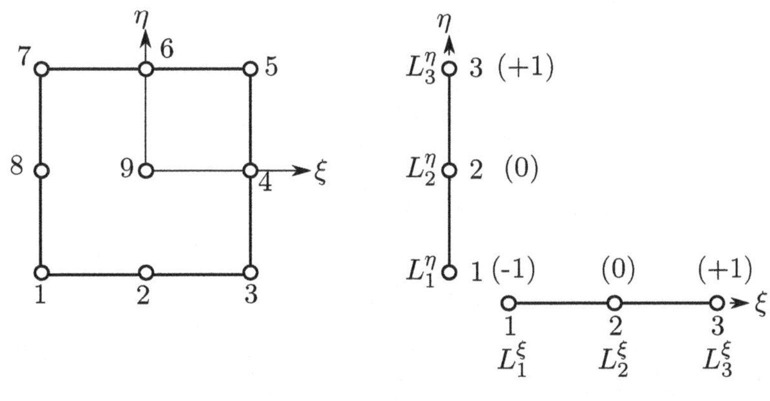

(a) Nine-node configuration

(b) 1D three-node configurations in ξ and η

Figure 5.12: Use of tensor product for nine-node subdomains in $\mathbb{R}^2$

The tensor product of 1D functions in ξ and η gives the following.

$$\begin{Bmatrix} L_1^\xi \; ; (-1) \\ L_2^\xi \; ; \; (0) \\ L_3^\xi \; ; (+1) \end{Bmatrix} \begin{bmatrix} L_1^\eta & L_2^\eta & L_3^\eta \\ (-1) & (0) & (+1) \end{bmatrix} = \begin{bmatrix} L_1^\xi L_1^\eta & L_1^\xi L_2^\eta & L_1^\xi L_2^\eta \\ (-1,-1) & (-1,0) & (-1,+1) \\ L_2^\xi L_1^\eta & L_2^\xi L_2^\eta & L_2^\xi L_3^\eta \\ (0,-1) & (0,-1) & (0,+1) \\ L_3^\xi L_1^\eta & L_3^\xi L_2^\eta & L_3^\xi L_3^\eta \\ (+1,-1) & (+1,0) & (+1,+1) \end{bmatrix} \tag{5.91}$$

$$= \begin{bmatrix} L_1(\xi,\eta) & L_8(\xi,\eta) & L_7(\xi,\eta) \\ L_2(\xi,\eta) & L_9(\xi,\eta) & L_6(\xi,\eta) \\ L_3(\xi,\eta) & L_4(\xi,\eta) & L_5(\xi,\eta) \end{bmatrix}$$

Thus $L_i(\xi,\eta)$; $i = 1, 2, \dots, 9$ are completely determined.

Recall

$$\begin{aligned} L_1^\xi(\xi) &= \frac{\xi(\xi-1)}{2} \quad , \quad L_2^\xi(\xi) = (1-\xi^2) \quad , \quad L_3^\xi(\xi) = \frac{\xi(\xi+1)}{2} \\ L_1^\eta(\eta) &= \frac{\eta(\eta-1)}{2} \quad , \quad L_2^\eta(\eta) = (1-\eta^2) \quad , \quad L_3^\eta(\eta) = \frac{\eta(\eta+1)}{2} \end{aligned} \tag{5.92}$$

Remarks.

(1) Using this procedure it is possible to determine the complete polynomials $L_i(\xi,\eta)$ for any desired degree in ξ and η.

(2) Hence, for mapping of geometry $\bar{\Omega}^{(e)}$ to $\bar{\Omega}^{(\xi,\eta)}$ we can write in general:

$$
\begin{aligned}
x(\xi,\eta) &= \sum_{i=1}^{\widetilde{n}} \widetilde{L}_i(\xi,\eta)x_i \\
y(\xi,\eta) &= \sum_{i=1}^{\widetilde{n}} \widetilde{L}_i(\xi,\eta)y_i
\end{aligned}
\tag{5.93}
$$

Choice of $\widetilde{n}$ and $\widetilde{L}_i(\xi,\eta)$ depend upon the degrees of the polynomial in ξ and η and are defined by Pascal's rectangle. We have intentionally used $\widetilde{n}$ and $\widetilde{L}_i(\xi,\eta)$ for mapping of geometry.

5.7.5 Interpolation of Function Values f_i Over $\bar{\Omega}^{(e)}$ Using $\bar{\Omega}^{(\xi,\eta)}$

The mapping of $\bar{\Omega}^{(e)}$ to $\bar{\Omega}^{(\xi,\eta)}$ is given by:

$$
\begin{aligned}
x(\xi,\eta) &= \sum_{i=1}^{\widetilde{n}} \widetilde{L}_i(\xi,\eta)x_i \\
y(\xi,\eta) &= \sum_{i=1}^{\widetilde{n}} \widetilde{L}_i(\xi,\eta)y_i
\end{aligned}
\tag{5.94}
$$

$\widetilde{n}$ and $\widetilde{L}_i(\xi,\eta)$ are suitably chosen number of nodes and the Lagrange interpolation polynomials for mapping of $\bar{\Omega}^{(e)}$ to $\bar{\Omega}^{(\xi,\eta)}$. The function values f_i at the nodes of $\bar{\Omega}^{(e)}$ or $\bar{\Omega}^{(\xi,\eta)}$ can be interpolated using:

$$
f^{(e)}(\xi,\eta) = \sum_{i=1}^{n} L_i(\xi,\eta)f_i \tag{5.95}
$$

in which

$n = 4$	for bilinear interpolation in ξ,η space (four-node configuration)
$n = 9$	for biquadratic interpolation in ξ,η space (nine-node configuration)
$n = 16$	for bicubic interpolation in ξ,η space (sixteen-node configuration)

and so on. $L_i(\xi,\eta)$ are determined using tensor product of suitable 1D functions in ξ and η.

5.7.6 Mapping of Length, Areas and Derivatives of $f(\xi,\eta)$ with Respect to x, y and ξ, η

In the following we drop the superscript (e) on $f(\cdot)$ (for convenience). Using (5.94):

$$\begin{aligned} dx &= \frac{\partial x}{\partial \xi} d\xi + \frac{\partial x}{\partial \eta} d\eta \\ dy &= \frac{\partial y}{\partial \xi} d\xi + \frac{\partial y}{\partial \eta} d\eta \end{aligned} \tag{5.96}$$

Hence

$$\begin{Bmatrix} dx \\ dy \end{Bmatrix} = \begin{bmatrix} \frac{\partial x}{\partial \xi} & \frac{\partial x}{\partial \eta} \\ \frac{\partial y}{\partial \eta} & \frac{\partial y}{\partial \eta} \end{bmatrix} \begin{Bmatrix} d\xi \\ d\eta \end{Bmatrix} = [J] \begin{Bmatrix} d\xi \\ d\eta \end{Bmatrix} \tag{5.97}$$

where

$$[J] = \begin{bmatrix} \frac{\partial x}{\partial \xi} & \frac{\partial x}{\partial \eta} \\ \frac{\partial y}{\partial \eta} & \frac{\partial y}{\partial \eta} \end{bmatrix} \tag{5.98}$$

The matrix $[J]$ is called the Jacobian of mapping. The matrix $[J]$ provides a relationship between elemental lengths $d\xi$, $d\eta$ and dx, dy in $\xi\eta$- and xy-spaces.

5.7.6.1 Mapping of Areas

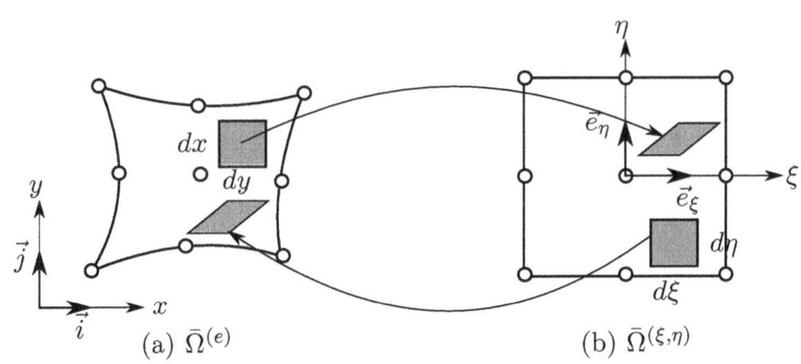

Figure 5.13: $\bar{\Omega}^{(e)}$ and $\bar{\Omega}^{(\xi,\eta)}$ with elemental areas

Consider elemental lengths dx and dy forming elemental area $d\Omega = dxdy$ in xy-space. Likewise, consider lengths $d\xi$, $d\eta$ along ξ- and η-axes forming an area $d\Omega^{(\xi,\eta)} = d\xi d\eta$. In this section we establish a relationship between $d\Omega$ and $d\Omega^{(\xi,\eta)}$.

Let $\vec{i}, \vec{j}$ be unit vectors along x- and y-axes and $\vec{e}_\xi, \vec{e}_\eta$ be the unit vectors along ξ-, η-axes. Then, the cross-product of vectors $dx\vec{i}$ and $dy\vec{j}$ would yield a vector perpendicular to the plane containing the vectors $dx\vec{i}, dy\vec{j}$ and the magnitude of this vector represents the area formed by these two vectors, i.e., $d\Omega$. Thus:

$$dx\vec{i} \times dy\vec{j} = dxdxy\,\vec{i} \times \vec{j} = dxdy\vec{k} \tag{5.99}$$

But

$$dx\vec{i} = \frac{\partial x}{\partial \xi} d\xi \vec{e}_\xi + \frac{\partial x}{\partial \eta} d\eta \vec{e}_\eta \tag{5.100}$$

$$dy\vec{j} = \frac{\partial y}{\partial \xi} d\xi \vec{e}_\xi + \frac{\partial y}{\partial \eta} d\eta \vec{e}_\eta \tag{5.101}$$

$$\therefore \quad dx\vec{i} \times dy\vec{j} = dxdy\vec{k} = \left(\frac{\partial x}{\partial \xi} d\xi \vec{e}_\xi + \frac{\partial x}{\partial \eta} d\eta \vec{e}_\eta \right) \times \left(\frac{\partial y}{\partial \xi} d\xi \vec{e}_\xi + \frac{\partial y}{\partial \eta} d\eta \vec{e}_\eta \right) \tag{5.102}$$

Expanding right side of (5.102):

$$\begin{aligned} dxdy\vec{k} = \frac{\partial x}{\partial \xi} d\xi \frac{\partial y}{\partial \xi} d\xi \, \vec{e}_\xi \times \vec{e}_\xi + \frac{\partial x}{\partial \eta} d\eta \frac{\partial y}{\partial \eta} d\xi \, \vec{e}_\eta \times \vec{e}_\xi \\ + \frac{\partial y}{\partial \xi} d\xi \frac{\partial y}{\partial \eta} d\eta \, \vec{e}_\xi \times \vec{e}_\eta + \frac{\partial y}{\partial \eta} d\eta \frac{\partial y}{\partial \eta} d\eta \, \vec{e}_\eta \times \vec{e}_\eta \end{aligned} \tag{5.103}$$

Noting that:

$$\begin{aligned} \vec{e}_\xi \times \vec{e}_\xi = \vec{e}_\eta \times \vec{e}_\eta = 0 \\ \vec{e}_\xi \times \vec{e}_\eta = \vec{e}_l = \vec{k} \\ \vec{e}_\eta \times \vec{e}_\xi = -\vec{e}_l = -\vec{k} \end{aligned} \tag{5.104}$$

Substituting from (5.104) into (5.102):

$$dxdy\vec{k} = \left(\frac{\partial x}{\partial \xi} \frac{\partial y}{\partial \eta} - \frac{\partial x}{\partial \eta} \frac{\partial y}{\partial \xi} \right) d\xi d\eta \vec{k}$$

$$\therefore \quad dxdy = \left(\frac{\partial x}{\partial \xi} \frac{\partial y}{\partial \eta} - \frac{\partial x}{\partial \eta} \frac{\partial y}{\partial \xi} \right) d\xi d\eta \tag{5.105}$$

But

$$\det[J] = |J| = \left(\frac{\partial x}{\partial \xi} \frac{\partial y}{\partial \eta} - \frac{\partial x}{\partial \eta} \frac{\partial y}{\partial \xi} \right) \tag{5.106}$$

$$\therefore \quad dxdy = |J| d\xi d\eta \tag{5.107}$$

$$\text{or} \quad d\Omega = |J| d\Omega^{(\xi,\eta)} \tag{5.108}$$

5.7.6.2 Obtaining Derivatives of $f(\xi,\eta)$ with Respect to x, y

Since $f = f(\xi,\eta)$ is the interpolation of data over $\bar{\Omega}^{(e)}$, a subdomain of $\bar{\Omega}$, we can write:

$$\frac{df}{dx} = \sum_{i=1}^{n} \frac{\partial L_i(\xi,\eta)}{\partial x} f_i \tag{5.109}$$

$$\frac{df}{dy} = \sum_{i=1}^{n} \frac{\partial L_i(\xi,\eta)}{\partial y} f_i \tag{5.110}$$

Thus, we need to determine $\frac{\partial L_i(\xi,\eta)}{\partial x}$ and $\frac{\partial L_i(\xi,\eta)}{\partial y}$; $i = 1, 2, \dots, n$. We note that:

$$\begin{aligned} \frac{\partial L_i(\xi,\eta)}{\partial \xi} &= \frac{\partial L_i}{\partial x}\frac{\partial x}{\partial \xi} + \frac{\partial L_i}{\partial y}\frac{\partial y}{\partial \xi} \\ \frac{\partial L_i(\xi,\eta)}{\partial \eta} &= \frac{\partial L_i}{\partial x}\frac{\partial x}{\partial \eta} + \frac{\partial L_i}{\partial y}\frac{\partial y}{\partial \eta} \end{aligned} \qquad i = 1, 2, \dots, n \tag{5.111}$$

Arranging (5.111) in matrix and vector form:

$$\begin{Bmatrix} \frac{\partial L_i}{\partial \xi} \\ \frac{\partial L_i}{\partial \eta} \end{Bmatrix} = \begin{bmatrix} \frac{\partial x}{\partial \xi} & \frac{\partial y}{\partial \xi} \\ \frac{\partial x}{\partial \eta} & \frac{\partial y}{\partial \eta} \end{bmatrix} \begin{Bmatrix} \frac{\partial L_i}{\partial x} \\ \frac{\partial L_i}{\partial y} \end{Bmatrix} = [J^T] \begin{Bmatrix} \frac{\partial L_i}{\partial x} \\ \frac{\partial L_i}{\partial y} \end{Bmatrix} \qquad i = 1, 2, \dots, n \tag{5.112}$$

$$\therefore \quad \begin{Bmatrix} \frac{\partial L_i}{\partial x} \\ \frac{\partial L_i}{\partial y} \end{Bmatrix} = [J^T]^{-1} \begin{Bmatrix} \frac{\partial L_i}{\partial \xi} \\ \frac{\partial L_i}{\partial \eta} \end{Bmatrix} \tag{5.113}$$

Hence, $\frac{\partial f}{\partial x}$ and $\frac{\partial f}{\partial y}$ in (5.109) and (5.110) are now explicitly defined, hence can be determined.

Remarks.

(1) Many remarks made in Section 5.6 for mapping and interpolation in $\mathbb{R}^1$ hold here as well.

(2) We recognize that piecewise interpolation over $\bar{\Omega}^{(e)}$ facilitates the process but is not the same as interpolation over the entire $\bar{\Omega}$ due to limited differentiability of $f(x,y) = \bigcup_e f^e(x,y)$ (only C^0 in this case) in the piecewise process.

(3) We could have also used triangular subdomains instead of quadrilateral.

5.8 Serendipity family of C^0 interpolations over square subdomains $\bar{\Omega}^{(\xi\eta)}$

"Serendipity" means discovery by chance. Thus, this family of interpolation has very little theoretical or mathematical basis other than the fact that in generating approximation functions for these we only utilize the two fundamental properties of the approximation functions,

$$N_i(\xi_j, \eta_j) = \begin{cases} 1, & j = i \\ 0, & j \neq i \end{cases} \quad (i = 1, \dots, m) \tag{5.114}$$

and

$$\sum_{i=1}^{m} N_i(\xi, \eta) = 1 \tag{5.115}$$

(a) The main motivation in generating these functions is to possibly eliminate some or many of the internal nodes that appear in generating the interpolations using tensor product for family of higher degree interpolation functions.

(b) For example, in the case of a bi-quadratic local approximation requiring a nine-node element, the corresponding serendipity element will contain eight boundary nodes, as shown in Fig. 5.14.

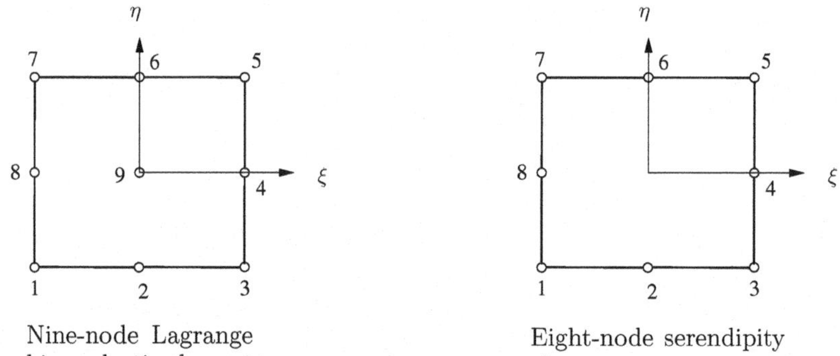

Figure 5.14: Nine-node Lagrange and eight-node serendipity $\bar{\Omega}^{(\xi\eta)}$ domains

(c) In the case of a bi-cubic interpolations requiring 16-nodes with four internal nodes, the corresponding serendipity element will contain 12 boundary nodes (see Fig. 5.15)

(d) While in the case of bi-quadratic and bi-cubic local approximations it was possible to eliminate the internal nodes and thus serendipity elements were possible. This may not always be possible for higher degree local approximations than three.

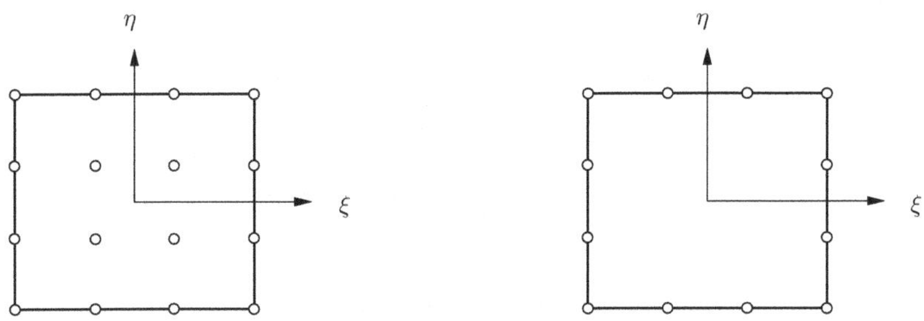

16-node bi-cubic element

12-node cubic serendipity element

Figure 5.15: Sixteen-node Lagrange and twelve-node serendipity $\bar{\Omega}^{(\xi\eta)}$ domains

5.8.1 Method of deriving serendipity interpolation functions

We use the two basic properties that the approximation functions must satisfy (stated by (5.114) and (5.115)). Let us consider a four-node bilinear element. In this case, obviously, non-serendipity and serendipity approximations are identical. Nonetheless, we derive the approximation functions for these using the approach used for serendipity basis functions.

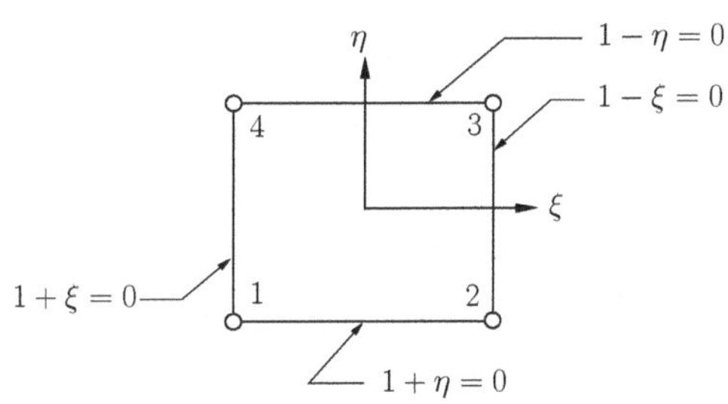

Figure 5.16: Derivation of 2D bilinear serendipity element

(a) First, we note that the four sides of the domain $\bar{\Omega}^{(\xi\eta)}$ are described by the equations of the straight lines as shown in the figure. Consider node 1. $N_1(\xi,\eta)$ is one at node 1 and zero at nodes 2,3 and 4. Hence, equations of the straight lines connecting nodes 2 and 3 and nodes 3 and 4 can be used to derive $N_1(\xi,\eta)$. That is,

$$N_1(\xi,\eta) = c_1(1-\xi)(1-\eta) \tag{5.116}$$

in which c_1 is a constant. But $N_1(-1,-1) = 1$, hence using (5.116) we get

$$N_1(-1,-1) = 1 = c_1(1-(-1))(1-(-1)) \quad \Rightarrow \quad c_1 = \frac{1}{4} \tag{5.117}$$

Thus, we have

$$N_1(\xi,\eta) = \frac{1}{4}(1-\xi)(1-\eta) \tag{5.118}$$

which is the correct approximation function for node 1 of the bilinear approximation functions. Similarly, for nodes 2, 3, and 4 we can write

$$\begin{aligned} N_2(\xi,\eta) &= c_2(1+\xi)(1-\eta) \\ N_3(\xi,\eta) &= c_3(1+\xi)(1+\eta) \\ N_4(\xi,\eta) &= c_4(1-\xi)(1+\eta) \end{aligned} \tag{5.119}$$

But

$$\begin{aligned} N_2(1,-1) = 1 \quad &\Rightarrow \quad c_2 = \frac{1}{4} \\ N_3(1,1) = 1 \quad &\Rightarrow \quad c_3 = \frac{1}{4} \\ N_4(-1,1) = 1 \quad &\Rightarrow \quad c_4 = \frac{1}{4} \end{aligned} \tag{5.120}$$

Thus, from (5.119) and (5.120) we obtain

$$\begin{aligned} N_2(\xi,\eta) &= \frac{1}{4}(1+\xi)(1-\eta) \\ N_3(\xi,\eta) &= \frac{1}{4}(1+\xi)(1+\eta) \\ N_4(\xi,\eta) &= \frac{1}{4}(1-\xi)(1+\eta) \end{aligned} \tag{5.121}$$

(5.118) and (5.121) are the correct approximation functions for the four-node bilinear approximation functions.

(b) In the above derivations we have only utilized the property (5.114), hence we must show that the interpolation functions in (5.118) and (5.121) satisfy (5.115). In this case, obviously they do. However, this may not always be the case.

Eight-node serendipity domain $\bar{\Omega}^{(\xi\eta)}$:

Consider node 1 first. We have $N_1(\xi,\eta)|_{(-1,-1)} = 1$ and zero at all the remaining nodes. Hence, for node 1 we can write

$$N_1(\xi,\eta) = c_1(1-\xi)(1-\eta)(1+\xi+\eta) \tag{5.122}$$

Since

$$N_1(\xi,\eta)|_{(-1,-1)} = 1 \quad \Rightarrow \quad c_1 = -\frac{1}{4} \tag{5.123}$$

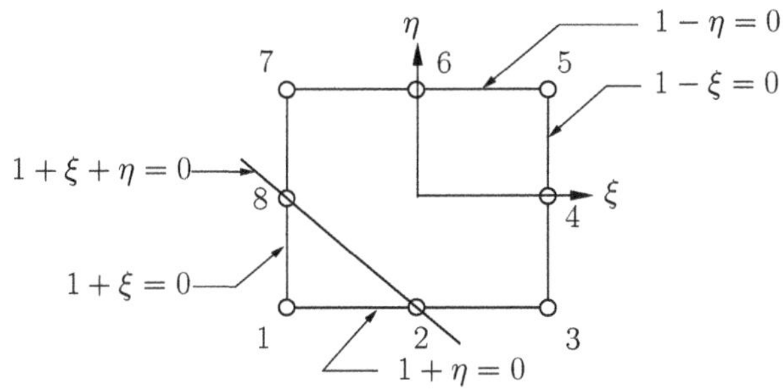

Figure 5.17: Derivation of 2D bilinear serendipity approximation functions: node 1

we obtain

$$N_1(\xi,\eta) = -\frac{1}{4}(1-\xi)(1-\eta)(1+\xi+\eta) \tag{5.124}$$

For nodes 3, 5, and 7 one may use the equations of the lines indicated in Fig. 5.17 and the conditions similar to (5.123) for N_2, N_3, and N_4.

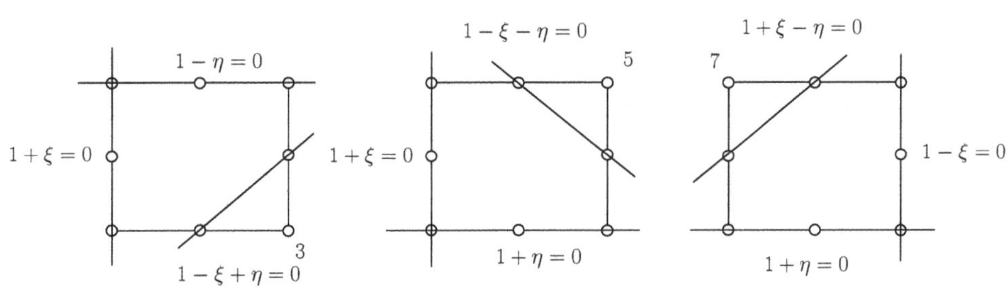

Figure 5.18: Derivation of 2D bi-quadratic serendipity approximation functions: nodes 3, 5, and 7

For the mid-side nodes, the product of the equations of straight lines not containing the mid-side nodes provide the needed expressions and we have

$$
\begin{aligned}
N_1 &= \frac{1}{4}(1-\xi)(1-\eta)(-1-\xi-\eta)\\
N_2 &= \frac{1}{2}(1-\xi^2)(1-\eta)\\
N_3 &= \frac{1}{4}(1+\xi)(1-\eta)(-1+\xi-\eta)\\
N_8 &= \frac{1}{2}(1-\xi)(1-\eta^2)\\
N_4 &= \frac{1}{2}(1+\xi)(1-\eta^2)\\
N_7 &= \frac{1}{4}(1-\xi)(1+\eta)(-1-\xi+\eta)\\
N_6 &= \frac{1}{2}(1-\xi^2)(1+\eta)\\
N_5 &= \frac{1}{4}(1+\xi)(1+\eta)(-1+\xi+\eta)
\end{aligned}
$$

In this case also we must show that $\sum\limits_{i=1}^{8} N_i(\xi,\eta) = 1$, which holds.

Twelve-node serendipity domain $\bar{\Omega}^{(\xi\eta)}$:

Using procedures similar to the four-node bilinear and eight-node bi-quadratic approximations (see Figure 5.19) we can also derive the interpolation functions for the twelve-node serendipity domain $\bar{\Omega}^{(\xi\eta)}$.

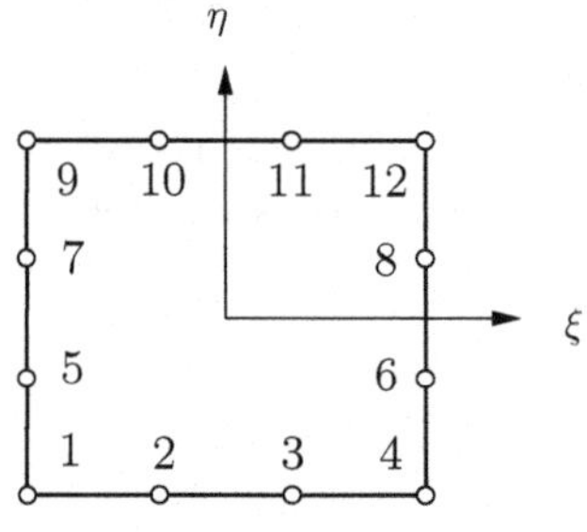

Figure 5.19: Derivation of 2D bi-cubic serendipity domain $\bar{\Omega}^{(\xi\eta)}$

$$
\begin{aligned}
N_1 &= \frac{1}{32}(1-\xi)(1-\eta)[-10+9(\xi^2+\eta^2)] \\
N_2 &= \frac{9}{32}(1-\xi^2)(1-\eta)(1-3\xi) \\
N_3 &= \frac{9}{32}(1-\xi^2)(1-\eta)(1+3\xi) \\
N_4 &= \frac{1}{32}(1+\xi)(1-\eta)[-10+9(\xi^2+\eta^2)] \\
N_5 &= \frac{9}{32}(1-\xi)(1-\eta^2)(1-3\eta) \\
N_6 &= \frac{9}{32}(1+\xi)(1-\eta^2)(1-3\eta) \\
N_7 &= \frac{9}{32}(1-\xi)(1-\eta^2)(1+3\eta) \\
N_8 &= \frac{9}{32}(1+\xi)(1-\eta^2)(1+3\eta) \\
N_9 &= \frac{1}{32}(1-\xi)(1+\eta)[-10+9(\xi^2+\eta^2)] \\
N_{10} &= \frac{9}{32}(1-\xi^2)(1+\eta)(1-3\xi) \\
N_{11} &= \frac{9}{32}(1-\xi^2)(1+\eta)(1+3\xi) \\
N_{12} &= \frac{1}{32}(1+\xi)(1+\eta)[-10+9(\xi^2+\eta^2)]
\end{aligned}
$$

Remarks.

(1) Serendipity interpolations are obviously incomplete polynomials in ξ and η, hence have poorer local approximation compared to the local approximations based on Pascal's rectangle.

(2) There is no particular theoretical basis for deriving them.

(3) In view of p-version hierarchical approximations [49, 50], serendipity approximations are precluded and are of no practical significance.

5.9 Mapping and Interpolation in $\mathbb{R}^3$

Consider $\bar{\Omega} \subset \mathbb{R}^3$ and let $((x_i, y_i, z_i), f_i)$; $i = 1, 2, \dots, n$ be the data points in $\bar{\Omega}$. Our objective is to construct an analytical expression $f(x, y, z)$ that interpolates this data such that $f(x_i, y_i, z_i) = f_i$; $i = 1, 2, \dots, n$.

Following the details presented in Section 5.7 for mapping and interpolation in $\mathbb{R}^2$, here also we choose piecewise interpolations over a subdomain $\bar{\Omega}^{(e)}$ of $\bar{\Omega}$ using its mapping $\bar{\Omega}^{(m)}$ in ξ, η, ζ natural coordinate system. $\bar{\Omega}^{(e)}$ is a suitably chosen volume containing data points. Here we choose $\bar{\Omega}^{(e)}$ to be a hexahedron, shown in Figure 5.20.

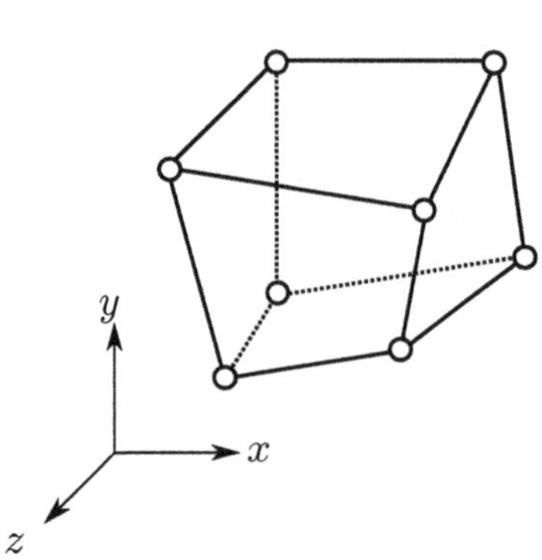

(a) Eight-node irregular hexahedron in xyz-space

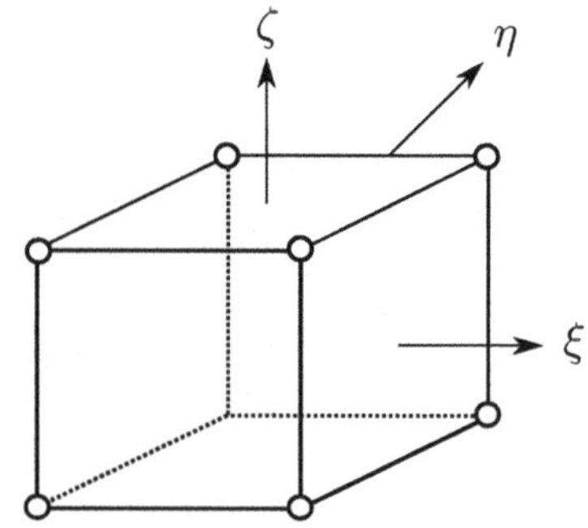

(b) Map of $\bar{\Omega}^{(e)}$ of (a) in a two unit cube in $\xi\eta\zeta$ space

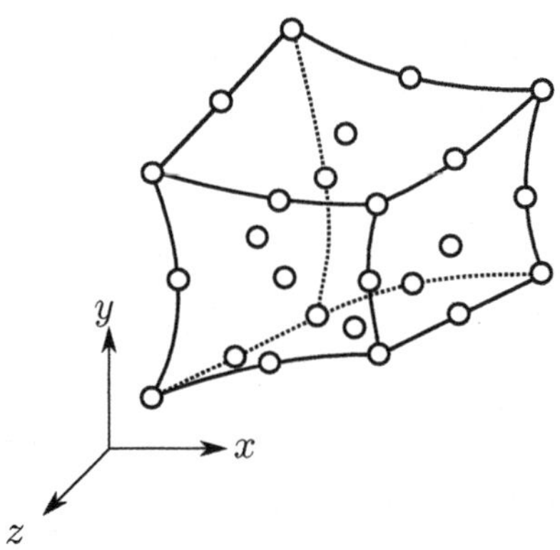

(c) A 27-node distorted hexahedron in xyz-space

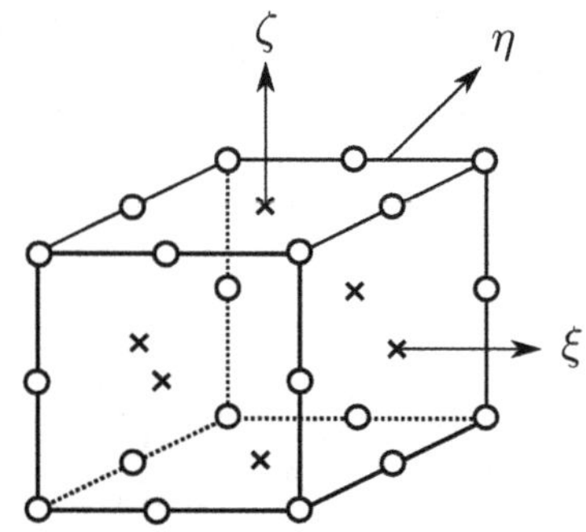

(d) Map of $\bar{\Omega}^{(e)}$ of (c) in $\xi\eta\zeta$ natural coordinate space in a two unit cube

Figure 5.20: $\bar{\Omega}^{(e)}$ in $\mathbb{R}^3$ and their maps $\bar{\Omega}^{(m)}$ in $\xi\eta\zeta$-space

5.9.1 Mapping of $\bar{\Omega}^{(e)}$ into $\bar{\Omega}^{(m)}$ in $\xi\eta\zeta$-Space

Let $((x_i, y_i, z_i), f_i)$; $i = 1, 2, \dots, n$ be the data associated with subdomain $\bar{\Omega}^{(e)}$ in xyz-space. Then following the details of the mapping in $\mathbb{R}^2$, we can write:

$$\begin{aligned}
x(\xi,\eta,\zeta) &= \sum_{i=1}^{\tilde{n}} \widetilde{L}_i(\xi,\eta,\zeta)x_i \\
y(\xi,\eta,\zeta) &= \sum_{i=1}^{\tilde{n}} \widetilde{L}_i(\xi,\eta,\zeta)y_i \\
z(\xi,\eta,\zeta) &= \sum_{i=1}^{\tilde{n}} \widetilde{L}_i(\xi,\eta,\zeta)z_i
\end{aligned} \tag{5.125}$$

n are suitably chosen for mapping depending upon the degrees of polynomials in ξ, η, and ζ and $\widetilde{L}_i(\xi, \eta, \zeta)$ are the Lagrange polynomials associated with n nodes. $\widetilde{L}_i(\xi, \eta, \zeta)$ have the following properties (similar to mapping in $\mathbb{R}^2$).

1. Each $\widetilde{L}_i(\xi, \eta, \zeta)$ is a polynomial of certain degree in ξ, η and ζ
2. $\widetilde{L}_i(\xi_j, \eta_j, \zeta_j) = \begin{cases} 1 & ; \quad j = i \\ 0 & ; \quad j \neq i \end{cases} \quad ; \quad i = 1, 2, \dots, \widetilde{n}$ (5.126)
3. $\displaystyle\sum_{i=1}^{n} \widetilde{L}_i(\xi, \eta, \zeta) = 1$

The importance of these properties have already been discussed for mapping in $\mathbb{R}^2$. The conclusions drawn for $\mathbb{R}^2$ mapping hold here as well. Equations (5.125) map $\bar{\Omega}^{(e)}$ from xyz-space to $\bar{\Omega}^{(m)}$ in a two unit cube in $\xi\eta\zeta$-space. Once we know $\widetilde{L}_i(\xi, \eta, \zeta)$; $i = 1, 2, \dots, \widetilde{n}$, the mapping is completely defined by (5.125).

5.9.1.1 Construction of $\widetilde{L}_i(\xi, \eta, \zeta)$ using Polynomial Approach

We can express x, y, z as a linear combination of monomials in ξ, η and ζ and their products.

$$
\begin{aligned}
x(\xi, \eta, \zeta) &= c_0 + c_1\xi + c_2\eta + c_3\zeta + \dots \\
y(\xi, \eta, \zeta) &= d_0 + d_1\xi + d_2\eta + d_3\zeta + \dots \\
z(\xi, \eta, \zeta) &= b_0 + b_1\xi + b_2\eta + b_3\zeta + \dots
\end{aligned}
\tag{5.127}
$$

Using $x_i = x(\xi_i, \eta_i, \zeta_i)$, $y_i = y(\xi_i, \eta_i, \zeta_i)$, and $z_i = z(\xi_i, \eta_i, \zeta_i)$ for $i = 1, 2, \dots, n$ in (5.127), we obtain n simultaneous algebraic equations in c_i, d_i, b_i; $i = 0, 1, 2, \dots, n-1$ from which we can solve for the coefficients. Substituting these in (5.127) gives us (5.125). The selection of the monomials in ξ, η, and ζ and their products depends upon the degree of the polynomials in ξ, η, and ζ and is facilitated by using Pascal's prism (Figure 5.21).

Figure 5.21 shows progressively increasing degree monomials in ξ, η, and ζ directions (shown orthogonal to each other). We connect these terms by straight lines (only shown for $\xi\eta$ in Figure 5.10) in ξ-, η-, and ζ-directions. We choose degrees of polynomials in ξ, η, and ζ. Based on this choice, using Pascal's prism we have the following information.

(i) The locations of the terms are the locations of the points of the nodes in $\bar{\Omega}^{(m)}$ and $\bar{\Omega}^{(e)}$ configuration.

(ii) The terms or monomials (and their products) are the choice we should use in the linear combination (5.127).

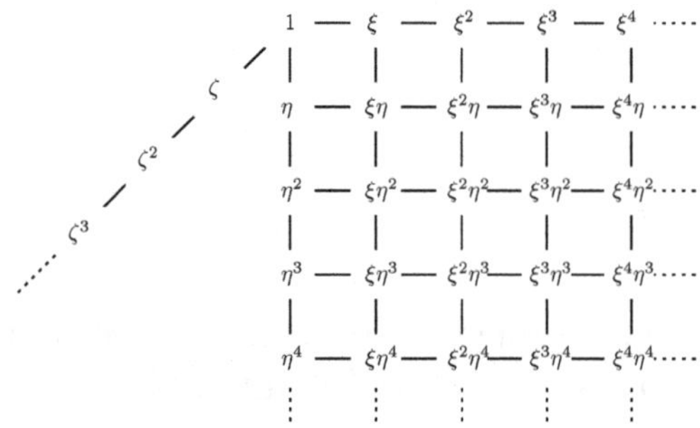

(a) Pascal's prism

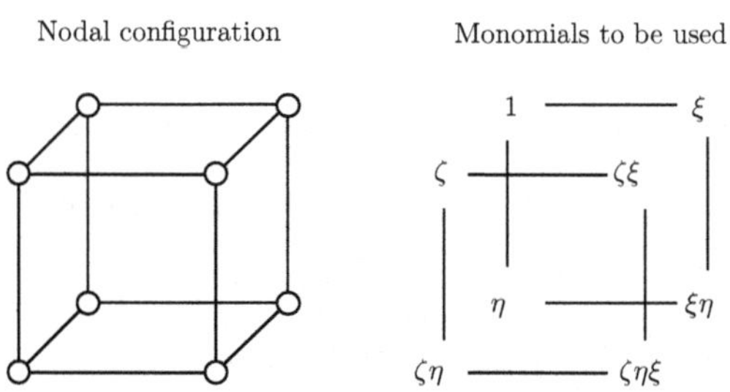

(b) Eight-node configuration

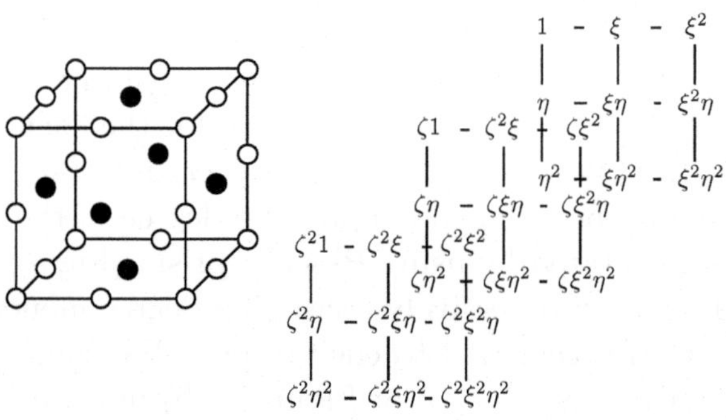

(c) 27-node configuration

Figure 5.21: Pascal's prism and selection of monomials

Thus, for given degrees of the polynomials $\widetilde{L}_i(\xi,\eta,\zeta)$ in ξ, η, ζ, $\widetilde{L}_i(\xi,\eta,\zeta)$ are completely determined from (5.127).

As an example, linear $\widetilde{L}_i(\xi,\eta,\zeta)$ in ξ, η, and ζ would require an eight-node configuration and the monomials shown in Figure 5.21(b). If we require $\widetilde{L}_i(\xi,\eta,\zeta)$ to be quadratic in ξ, η, and ζ, then we need a 27-node configuration and the monomial terms shown in Figure 5.21(c). With this approach quite complex domains $\bar{\Omega}^{(e)}$ can be mapped in $\bar{\Omega}^{(m)}$ by choosing appropriate degrees of $\widetilde{L}_i(\xi,\eta,\zeta)$ in ξ, η, and ζ.

Remarks.

(1) This polynomial approach requires inverse of the coefficient matrix which can be avoided by using tensor product approach similar to $\mathbb{R}^2$.

(2) An important outcome of Pascal's prism is that it tells us the number of nodes based on the choice of the degrees of polynomials $L_i(\xi,\eta,\zeta)$ in ξ, η, and ζ and their locations.

5.9.1.2 Tensor Product to Generate $\widetilde{L}_i(\xi,\eta,\zeta)$

The concept of tensor product used for quadrilateral elements in $\mathbb{R}^2$ can be extended to hexahedron elements in $\mathbb{R}^3$. Consider 1D Lagrange polynomials in ξ, η, and ζ with desired degrees in ξ, η, and ζ. Let n, m and q be the number of points or nodes in ξ-, η-, and ζ-directions that would yield $(n-1)$, $(m-1)$ and $(q-1)$ as degrees of the 1D Lagrange functions $L_i^{\xi}(\xi)$; $i=1,2,\dots,n$; $L_j^{\eta}(\eta)$; $j=1,2,\dots,m$ and $L_k^{\zeta}(\zeta)$; $k=1,2,\dots,q$ associated with n, m, and q nodes in ξ-, η-, and ζ-directions (figure 5.22).

We first take the tensor product of 1D $L_j^{\xi}(\xi)$ and $L_k^{\eta}(\eta)$ in ξ- and η-directions that would yield $(n\times m)$ 2D Lagrange polynomials in $\xi\eta$ with degrees $(n-1)$ and $(m-1)$ in ξ and η (figure 5.23). Tensor product of these functions with 1D Lagrange polynomials in ζ-direction gives

$$\widetilde{L}_i(\xi,\eta,\zeta)\ ;\quad i=1,2,\dots,(n)(m)(q)$$

for a $(n\times m\times q)$ nodal configuration in which $\widetilde{L}(\cdot)$ are polynomials of degrees $(n-1)$, $(m-1)$, and $(q-1)$ in ξ-, η-, and ζ-directions.

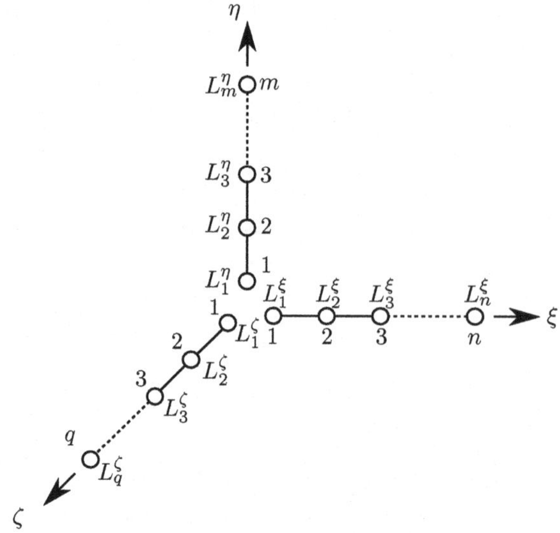

Figure 5.22: 1D Lagrange polynomials in ξ, η, and ζ

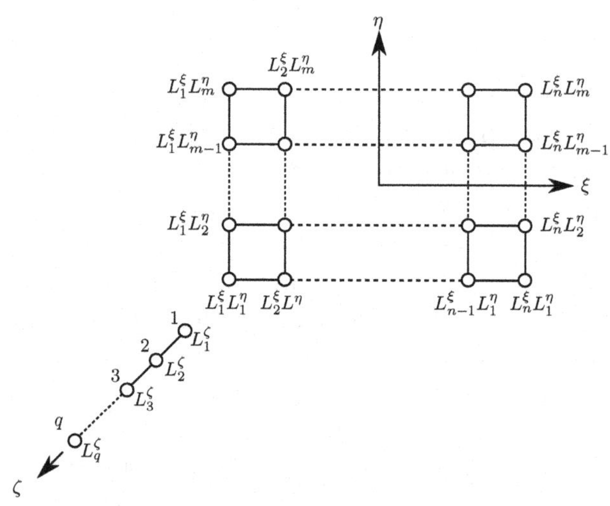

Figure 5.23: Tensor product in ξ, η and 1D functions in ζ

Example 5.10 (3D Lagrange Linear Interpolating Polynomials in $\xi\eta\zeta$-Space). Construct interpolation functions $L_i(\xi, \eta, \zeta)$ that are linear in ξ, η, and ζ. From Pascal's prism $\bar{\Omega}^{(e)}$ is an eight-node domain in xyz-space and $\bar{\Omega}^{(m)}$ is its map (also containing eight vertex nodes) in $\xi\eta\zeta$-space. Figure 5.24 shows details of 1D Lagrange polynomials of degree one in ξ-, η-, and ζ-directions and Figure 5.25 shows their tensor product in $\xi\eta$-directions. The tensor product of $\xi\eta$ functions with the functions in the ζ-direction (shown in Figure 5.26) gives us the final interpolation functions $\tilde{L}_i(\xi, \eta, \zeta)$.

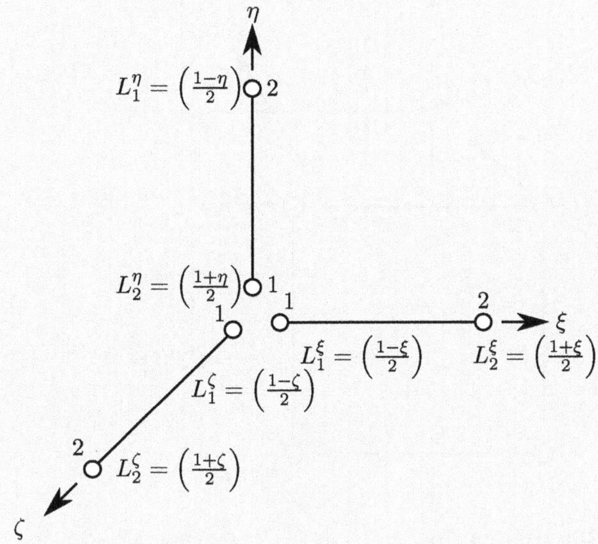

Figure 5.24: 1D Lagrange polynomials of order one

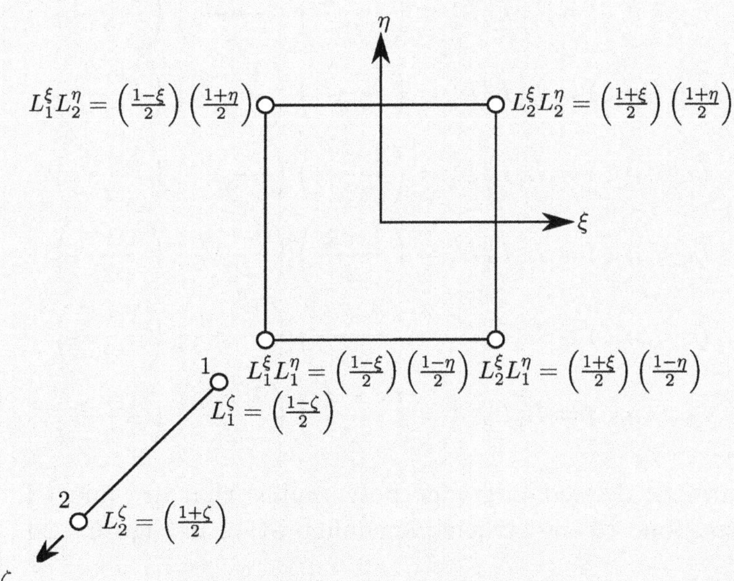

Figure 5.25: Tensor product in $\xi\eta$-space

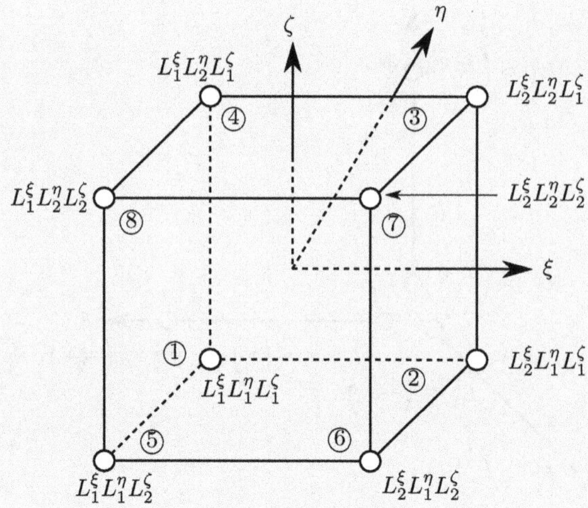

Figure 5.26: Lagrange polynomials $\widetilde{L}_i(\xi, \eta\zeta)$ generalized using tensor product

From Figure 5.26 (using the node numbering shown)

$$
\begin{aligned}
\widetilde{L}_1(\xi,\eta,\zeta) &= L_1^\xi L_1^\eta L_1^\zeta = \left(\frac{1-\xi}{2}\right)\left(\frac{1-\eta}{2}\right)\left(\frac{1-\zeta}{2}\right)\\
\widetilde{L}_2(\xi,\eta,\zeta) &= L_2^\xi L_1^\eta L_1^\zeta = \left(\frac{1+\xi}{2}\right)\left(\frac{1-\eta}{2}\right)\left(\frac{1-\zeta}{2}\right)\\
\widetilde{L}_3(\xi,\eta,\zeta) &= L_2^\xi L_2^\eta L_1^\zeta = \left(\frac{1+\xi}{2}\right)\left(\frac{1+\eta}{2}\right)\left(\frac{1-\zeta}{2}\right)\\
\widetilde{L}_4(\xi,\eta,\zeta) &= L_1^\xi L_2^\eta L_1^\zeta = \left(\frac{1-\xi}{2}\right)\left(\frac{1+\eta}{2}\right)\left(\frac{1-\zeta}{2}\right)\\
\widetilde{L}_5(\xi,\eta,\zeta) &= L_1^\xi L_1^\eta L_2^\zeta = \left(\frac{1-\xi}{2}\right)\left(\frac{1-\eta}{2}\right)\left(\frac{1+\zeta}{2}\right)\\
\widetilde{L}_6(\xi,\eta,\zeta) &= L_2^\xi L_1^\eta L_2^\zeta = \left(\frac{1+\xi}{2}\right)\left(\frac{1-\eta}{2}\right)\left(\frac{1+\zeta}{2}\right)\\
\widetilde{L}_7(\xi,\eta,\zeta) &= L_2^\xi L_2^\eta L_2^\zeta = \left(\frac{1+\xi}{2}\right)\left(\frac{1+\eta}{2}\right)\left(\frac{1+\zeta}{2}\right)\\
\widetilde{L}_8(\xi,\eta,\zeta) &= L_1^\xi L_2^\eta L_2^\zeta = \left(\frac{1-\xi}{2}\right)\left(\frac{1+\eta}{2}\right)\left(\frac{1+\zeta}{2}\right)
\end{aligned}
\tag{5.128}
$$

These are the desired Lagrange polynomials that are linear in ξ, η, and ζ and correspond to the eight-node configuration in Figure 5.26.

5.9.2 Interpolation of Function Values f_i Over $\bar{\Omega}^{(e)}$ Using $\bar{\Omega}^{(m)}$

The mapping of $\bar{\Omega}^{(e)}$ to $\bar{\Omega}^{(m)}$ is given by:

$$
\begin{aligned}
x(\xi,\eta,\zeta) &= \sum_{i=1}^{\tilde{n}} \widetilde{L}_i(\xi,\eta,\zeta)x_i \\
y(\xi,\eta,\zeta) &= \sum_{i=1}^{\tilde{n}} \widetilde{L}_i(\xi,\eta,\zeta)y_i \\
z(\xi,\eta,\zeta) &= \sum_{i=1}^{\tilde{n}} \widetilde{L}_i(\xi,\eta,\zeta)z_i
\end{aligned}
\tag{5.129}
$$

$\tilde{n}$ and $\widetilde{L}_i(\xi,\eta,\zeta)$ are suitably chosen number of nodes and the Lagrange interpolation polynomials for mapping of $\bar{\Omega}^{(e)}$ to $\bar{\Omega}^{(m)}$ in a two unit cube. If f_i are the function values at the nodes of $\bar{\Omega}^{(e)}$ or $\bar{\Omega}^{(m)}$ then these can be interpolated using

$$
f^{(e)}(\xi,\eta,\zeta) = \sum_{i=1}^{n} L_i(\xi,\eta,\zeta)f_i
$$

in which

$$
\begin{aligned}
n &= 8 \quad \text{for} \quad \text{linear } L_i(\xi,\eta,\zeta) \text{ in } \xi,\ \eta, \text{ and } \zeta \\
n &= 27 \quad \text{for} \quad \text{quadratic } L_i(\xi,\eta,\zeta) \text{ in } \xi,\ \eta, \text{ and } \zeta \\
n &= 64 \quad \text{for} \quad \text{cubic } L_i(\xi,\eta,\zeta) \text{ in } \xi,\ \eta, \text{ and } \zeta
\end{aligned}
$$

and so on. $L_i(\xi,\eta,\zeta)$ are determined using the tensor product. The functions $\widetilde{L}_i(\xi,\eta,\zeta)$ and $L_i(\xi,\eta,\zeta)$ are generally not the same but can be the same if so desired. The choice of $\widetilde{L}_i(\xi,\eta,\zeta)$ depends on the geometry mapping considerations, whereas $L_i(\xi,\eta,\zeta)$ are chosen based on data points to be interpolated.

5.9.3 Mapping of Lengths, Volumes and Derivatives of $f(\xi,\eta,\zeta)$ with Respect to x,y,z and ξ,η,ζ in $\mathbb{R}^3$

5.9.3.1 Mapping of Lengths

We establish a relationship between dx, dy and dz and $d\xi$, $d\eta$, $d\zeta$ in the xyz- and $\xi\eta\zeta$-spaces. Since $x = x(\xi,\eta,\zeta)$, $y = y(\xi,\eta,\zeta)$ and $z = z(\xi,\eta,\zeta)$, we can write

$$
\begin{aligned}
dx &= \frac{\partial x}{\partial \xi}d\xi + \frac{\partial x}{\partial \eta}d\eta + \frac{\partial x}{\partial \zeta}d\zeta \\
dy &= \frac{\partial y}{\partial \xi}d\xi + \frac{\partial y}{\partial \eta}d\eta + \frac{\partial y}{\partial \zeta}d\zeta \\
dz &= \frac{\partial z}{\partial \xi}d\xi + \frac{\partial z}{\partial \eta}d\eta + \frac{\partial z}{\partial \zeta}d\zeta
\end{aligned}
\tag{5.130}
$$

or

$$\begin{Bmatrix} dx \\ dy \\ dz \end{Bmatrix} = [J] \begin{Bmatrix} d\xi \\ d\eta \\ d\zeta \end{Bmatrix} \tag{5.131}$$

where

$$[J] = \begin{bmatrix} \frac{\partial x}{\partial \xi} & \frac{\partial x}{\partial \eta} & \frac{\partial x}{\partial \zeta} \\ \frac{\partial y}{\partial \xi} & \frac{\partial y}{\partial \eta} & \frac{\partial y}{\partial \zeta} \\ \frac{\partial z}{\partial \xi} & \frac{\partial z}{\partial \eta} & \frac{\partial z}{\partial \zeta} \end{bmatrix} \tag{5.132}$$

$[J]$ is called the Jacobian of mapping.

5.9.3.2 Mapping of Volumes

In this section we derive a relationship between the elemental volume $dxdydz$ in xyz-space and $d\xi d\eta d\zeta$ in $\xi\eta\zeta$-space. Let $\vec{i}$, $\vec{j}$, and $\vec{k}$ be unit vectors along x-, y-, and z-axes and $\vec{e}_\xi$, $\vec{e}_\eta$, and $\vec{e}_\zeta$ be unit vectors along ξ-, η-, and ζ-axes, then:

$$\begin{aligned} dx\vec{i} &= \frac{\partial x}{\partial \xi} d\xi \vec{e}_\xi + \frac{\partial x}{\partial \eta} d\eta \vec{e}_\eta + \frac{\partial x}{\partial \zeta} d\zeta \vec{e}_\zeta \\ dy\vec{j} &= \frac{\partial y}{\partial \xi} d\xi \vec{e}_\xi + \frac{\partial y}{\partial \eta} d\eta \vec{e}_\eta + \frac{\partial y}{\partial \zeta} d\zeta \vec{e}_\zeta \\ dz\vec{k} &= \frac{\partial z}{\partial \xi} d\xi \vec{e}_\xi + \frac{\partial z}{\partial \eta} d\eta \vec{e}_\eta + \frac{\partial z}{\partial \zeta} d\zeta \vec{e}_\zeta \end{aligned} \tag{5.133}$$

We note that:

$$dx\vec{i} \cdot (d\vec{j} \times dz\vec{k}) = dx\vec{i} \cdot (dydz\vec{i}) = dxdydz\vec{i} \cdot \vec{i} = dxdydz \tag{5.134}$$

Substituting for $dx\vec{i}$, $dy\vec{j}$ and $dz\vec{k}$ from (5.133) into (5.134) and using the properties of the dot product and cross product in $\xi\eta\zeta$- and xyz-spaces, we obtain:

$$dxdydz = \det[J] d\xi d\eta d\zeta \tag{5.135}$$

We note that for (5.135) to hold $\det[J] > 0$ must hold. Thus for the mapping between $\bar{\Omega}^{(e)}$ and $\bar{\Omega}^{(m)}$ to be valid (one-to-one and onto) $\det[J] > 0$ must hold. This is an important conclusion from (5.135).

5.9.3.3 Obtaining Derivatives of $f(\xi, \eta, \zeta)$ with Respect to x, y, z

We note that $f = f(\xi, \eta, \zeta)$ and since

$$f(\xi, \eta, \zeta) = \sum_{i=1}^{n} L_i(\xi, \eta, \zeta) f_i \tag{5.136}$$

we have:

$$
\begin{aligned}
\frac{\partial f}{\partial x} &= \sum_{i=1}^{n} \frac{\partial L_i(\xi,\eta,\zeta)}{\partial x} f_i \\
\frac{\partial f}{\partial y} &= \sum_{i=1}^{n} \frac{\partial L_i(\xi,\eta,\zeta)}{\partial y} f_i \\
\frac{\partial f}{\partial z} &= \sum_{i=1}^{n} \frac{\partial L_i(\xi,\eta,\zeta)}{\partial z} f_i
\end{aligned}
\tag{5.137}
$$

Thus, $\frac{\partial f}{\partial x}$, $\frac{\partial f}{\partial y}$, $\frac{\partial f}{\partial z}$ are deterministic if we know:

$$
\frac{\partial L_i(\xi,\eta,\zeta)}{\partial x} \quad , \quad \frac{\partial L_i(\xi,\eta,\zeta)}{\partial y} \quad \text{and} \quad \frac{\partial L_i(\xi,\eta,\zeta)}{\partial z} \quad ; \quad i = 1,2,\dots,n
$$

Since $L_i = L_i(\xi,\eta,\zeta)$ and $x = x(\xi,\eta,\zeta)$, $y = y(\xi,\eta,\zeta)$, and $z = z(\xi,\eta,\zeta)$, we can use the chain rule of differentiation to obtain:

$$
\left.
\begin{aligned}
\frac{\partial L_i(\xi,\eta,\zeta)}{\partial \xi} &= \frac{\partial L_i}{\partial x}\frac{\partial x}{\partial \xi} + \frac{\partial L_i}{\partial y}\frac{\partial y}{\partial \xi} + \frac{\partial L_i}{\partial z}\frac{\partial z}{\partial \xi} \\
\frac{\partial L_i(\xi,\eta,\zeta)}{\partial \eta} &= \frac{\partial L_i}{\partial x}\frac{\partial x}{\partial \eta} + \frac{\partial L_i}{\partial y}\frac{\partial y}{\partial \eta} + \frac{\partial L_i}{\partial z}\frac{\partial z}{\partial \eta} \\
\frac{\partial L_i(\xi,\eta,\zeta)}{\partial \zeta} &= \frac{\partial L_i}{\partial x}\frac{\partial x}{\partial \zeta} + \frac{\partial L_i}{\partial y}\frac{\partial y}{\partial \zeta} + \frac{\partial L_i}{\partial z}\frac{\partial z}{\partial \zeta}
\end{aligned}
\right. \quad ; \quad i = 1,2,\dots,n
\tag{5.138}
$$

or

$$
\begin{Bmatrix} \frac{\partial L_i(\xi,\eta,\zeta)}{\partial \xi} \\ \frac{\partial L_i(\xi,\eta,\zeta)}{\partial \eta} \\ \frac{\partial L_i(\xi,\eta,\zeta)}{\partial \zeta} \end{Bmatrix} = [J^T] \begin{Bmatrix} \frac{\partial L_i}{\partial x} \\ \frac{\partial L_i}{\partial y} \\ \frac{\partial L_i}{\partial z} \end{Bmatrix} \quad ; \quad i = 1,2,\dots,n
\tag{5.139}
$$

$$
\therefore \quad \begin{Bmatrix} \frac{\partial L_i}{\partial x} \\ \frac{\partial L_i}{\partial y} \\ \frac{\partial L_i}{\partial z} \end{Bmatrix} = [J^T]^{-1} \begin{Bmatrix} \frac{\partial L_i}{\partial \xi} \\ \frac{\partial L_i}{\partial \eta} \\ \frac{\partial L_i}{\partial \zeta} \end{Bmatrix} \quad ; \quad i = 1,2,\dots,n
\tag{5.140}
$$

Using (5.140) and (5.137) $\frac{\partial f}{\partial x}$, $\frac{\partial f}{\partial y}$, and $\frac{\partial f}{\partial z}$ are deterministic. We note that in the computation of $[J]$ we use (5.129) to find components of $[J]$.

Example 5.11 (Mapping and Interpolation in $\mathbb{R}^2$). Consider $\bar{\Omega}^{(e)}$ to be a four-node quadrilateral domain in xy-space shown in Figure 5.27(a) with xy-coordinates given.

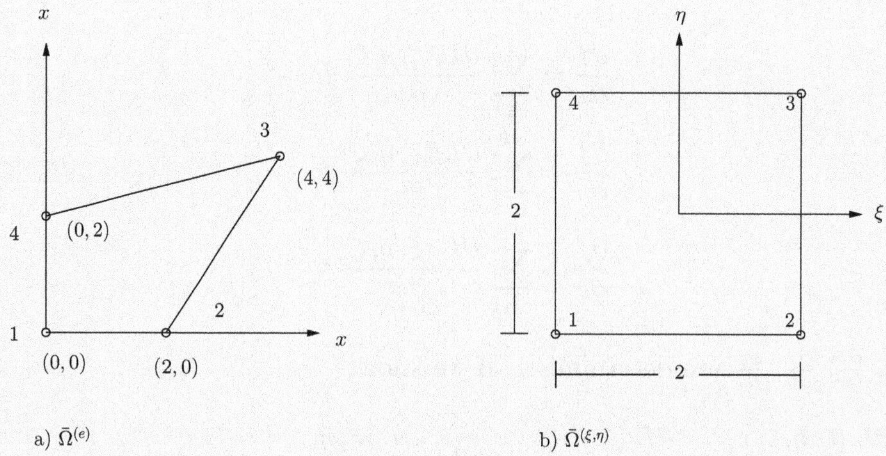

Figure 5.27: $\bar{\Omega}^{(e)}$

Figure 5.27(b) shows a map of $\bar{\Omega}^{(e)}$ in $\bar{\Omega}^{(\xi,\eta)}$ in $\xi\eta$-space in a two unit square.

(a) Determine equations $x = x(\xi, \eta)$, $y = y(\xi, \eta)$ describing the mapping.

(b) Determine the Jacobian $[J]$ of mapping and $\det[J]$.

(c) Determine the derivatives of the Lagrange polynomials $L_i(\xi, \eta)$; $i = 1, 2, \dots, 4$ with respect to x and y.

(d) If $f_1 = 0$, $f_2 = 1$, $f_3 = 2$ and $f_4 = 1$ are the function values at the four nodes of the quadrilateral, then interpolate this data using $\bar{\Omega}^{(\xi,\eta)}$, i.e., determine $f(\xi, \eta)$ that interpolates this data.

(e) Determine derivatives of $f(\xi, \eta)$ with respect to x, y.

Solution

(a) Equations describing the mapping: The Lagrange polynomials $L_i(\xi, \eta)$; $i = 1, 2, \dots, 4$ are

$$L_1 = \left(\frac{1-\xi}{2}\right)\left(\frac{1-\eta}{2}\right) \quad , \quad L_2 = \left(\frac{1+\xi}{2}\right)\left(\frac{1-\eta}{2}\right)$$

$$L_3 = \left(\frac{1+\xi}{2}\right)\left(\frac{1+\eta}{2}\right) \quad , \quad L_4 = \left(\frac{1-\xi}{2}\right)\left(\frac{1+\eta}{2}\right)$$

$$\therefore \quad x(\xi, \eta) = \sum_{i=1}^{4} L_i x_i \quad , \quad y(\xi, \eta) = \sum_{i=1}^{4} L_i y_i$$

Substituting for x_i, y_i, and L_i ; $i = 1, 2, \dots, 4$:

$$x = \left(\frac{1-\xi}{2}\right)\left(\frac{1-\eta}{2}\right)(0) + \left(\frac{1+\xi}{2}\right)\left(\frac{1-\eta}{2}\right)(2) + \left(\frac{1+\xi}{2}\right)\left(\frac{1+\eta}{2}\right)(4) + \left(\frac{1-\xi}{2}\right)\left(\frac{1+\eta}{2}\right)(0)$$

or

$$x = \left(\frac{1+\xi}{2}\right)\left(\frac{1-\eta}{2}\right)(2) + \left(\frac{1+\xi}{2}\right)\left(\frac{1+\eta}{2}\right)(4)$$

or

$$x = \left(\frac{1+\xi}{2}\right)(3+\eta)$$

Similarly

$$y(\xi,\eta) = \left(\frac{1-\xi}{2}\right)\left(\frac{1-\eta}{2}\right)(0) + \left(\frac{1+\xi}{2}\right)\left(\frac{1-\eta}{2}\right)(0) + \left(\frac{1+\xi}{2}\right)\left(\frac{1+\eta}{2}\right)(4) + \left(\frac{1-\xi}{2}\right)\left(\frac{1+\eta}{2}\right)(2)$$

or

$$y(\xi,\eta) = \left(\frac{1+\xi}{2}\right)\left(\frac{1+\eta}{2}\right)(4) + \left(\frac{1-\xi}{2}\right)\left(\frac{1+\eta}{2}\right)(2)$$

or

$$y(\xi,\eta) = \left(\frac{1+\eta}{2}\right)(3+\xi)$$

$$\therefore \quad \left.\begin{aligned} x(\xi,\eta) &= \left(\frac{1+\xi}{2}\right)(3+\eta) \\ y(\xi,\eta) &= \left(\frac{1+\eta}{2}\right)(3+\xi) \end{aligned}\right\} \text{ Equations describing mapping}$$

(b) Jacobian of mapping $[J]$ and its determinant $|J|$:

$$[J] = \begin{bmatrix} \frac{\partial x}{\partial \xi} & \frac{\partial x}{\partial \eta} \\ \frac{\partial y}{\partial \xi} & \frac{\partial y}{\partial \eta} \end{bmatrix}$$

$$\frac{\partial x}{\partial \xi} = \left(\frac{3+\eta}{2}\right), \quad \frac{\partial x}{\partial \eta} = \left(\frac{1+\xi}{2}\right), \quad \frac{\partial y}{\partial \xi} = \left(\frac{1+\eta}{2}\right), \quad \frac{\partial y}{\partial \eta} = \left(\frac{3+\xi}{2}\right)$$

$$\therefore \quad [J] = \begin{bmatrix} \left(\frac{3+\eta}{2}\right) & \left(\frac{1+\xi}{2}\right) \\ \left(\frac{1+\eta}{2}\right) & \left(\frac{3+\xi}{2}\right) \end{bmatrix}$$

$$\det[J] = \left(\frac{3+\eta}{2}\right)\left(\frac{1+\xi}{2}\right) - \left(\frac{1+\eta}{2}\right)\left(\frac{3+\xi}{2}\right) = \frac{1}{4}(8+2\xi+2\eta)$$

(c) Derivatives of L_i with respect to x, y:

$$[J^T] = \begin{bmatrix} \left(\frac{3+\eta}{2}\right) & \left(\frac{1+\eta}{2}\right) \\ \left(\frac{1+\xi}{2}\right) & \left(\frac{3+\xi}{2}\right) \end{bmatrix}$$

$$[J^T]^{-1} = \frac{1}{\det[J]}\begin{bmatrix} \left(\frac{3+\eta}{2}\right) & -\left(\frac{1+\eta}{2}\right) \\ -\left(\frac{1+\xi}{2}\right) & \left(\frac{3+\xi}{2}\right) \end{bmatrix} =$$

$$\frac{1}{\frac{1}{4}(8+2\xi+2\eta)}\begin{bmatrix} \left(\frac{3+\eta}{2}\right) & -\left(\frac{1+\eta}{2}\right) \\ -\left(\frac{1+\xi}{2}\right) & \left(\frac{3+\xi}{2}\right) \end{bmatrix}$$

$$\begin{Bmatrix} \frac{\partial L_i}{\partial x} \\ \frac{\partial L_i}{\partial y} \end{Bmatrix} = [J^T]^{-1}\begin{Bmatrix} \frac{\partial L_i}{\partial \xi} \\ \frac{\partial L_i}{\partial \eta} \end{Bmatrix} \quad ; \quad [J^T]^{-1} \text{ is defined above}$$

and

$$\begin{aligned}
\frac{\partial L_1}{\partial \xi} &= -\frac{1}{2}\left(\frac{1-\eta}{2}\right) & \frac{\partial L_1}{\partial \eta} &= -\frac{1}{2}\left(\frac{1-\xi}{2}\right) \\
\frac{\partial L_2}{\partial \xi} &= \frac{1}{2}\left(\frac{1-\eta}{2}\right) & \frac{\partial L_2}{\partial \eta} &= -\frac{1}{2}\left(\frac{1+\xi}{2}\right) \\
\frac{\partial L_3}{\partial \xi} &= \frac{1}{2}\left(\frac{1+\eta}{2}\right) & \frac{\partial L_3}{\partial \eta} &= \frac{1}{2}\left(\frac{1+\xi}{2}\right) \\
\frac{\partial L_4}{\partial \xi} &= -\frac{1}{2}\left(\frac{1+\eta}{2}\right) & \frac{\partial L_4}{\partial \eta} &= \frac{1}{2}\left(\frac{1-\xi}{2}\right)
\end{aligned}$$

Hence, $\frac{\partial L_i}{\partial x}$, $\frac{\partial L_i}{\partial y}$; $i = 1, 2, 3, 4$ are completely defined.

(d) Determination of $f(\xi, \eta)$:

$$\begin{aligned}
f(\xi,\eta) &= \sum L_i(\xi,\eta) f_i \\
&= \left(\frac{1-\xi}{2}\right)\left(\frac{1-\eta}{2}\right)(0) + \left(\frac{1+\xi}{2}\right)\left(\frac{1-\eta}{2}\right)(1) \\
&\quad + \left(\frac{1+\xi}{2}\right)\left(\frac{1+\eta}{2}\right)(2) + \left(\frac{1-\xi}{2}\right)\left(\frac{1+\eta}{2}\right)(1)
\end{aligned}$$

After simplifying,

$$f(\xi,\eta) = \frac{1}{4}(4+2\xi+2\eta)$$

(e) Derivatives of $f(\xi,\eta)$ with respect to x, y

$$\frac{\partial f}{\partial \xi} = \frac{\partial f}{\partial x}\frac{\partial x}{\partial \xi} + \frac{\partial f}{\partial y}\frac{\partial y}{\partial \xi}$$

$$\frac{\partial f}{\partial \eta} = \frac{\partial f}{\partial x}\frac{\partial x}{\partial \eta} + \frac{\partial f}{\partial y}\frac{\partial y}{\partial \eta}$$

$$\therefore \quad \begin{Bmatrix} \frac{\partial f}{\partial \xi} \\ \frac{\partial f}{\partial \eta} \end{Bmatrix} = \begin{bmatrix} \frac{\partial x}{\partial \xi} & \frac{\partial y}{\partial \xi} \\ \frac{\partial x}{\partial \eta} & \frac{\partial y}{\partial \eta} \end{bmatrix} \begin{Bmatrix} \frac{\partial f}{\partial x} \\ \frac{\partial f}{\partial y} \end{Bmatrix} = [J^T] \begin{Bmatrix} \frac{\partial f}{\partial x} \\ \frac{\partial f}{\partial y} \end{Bmatrix}$$

$$\frac{\partial f}{\partial \xi} = \frac{1}{2} \quad , \quad \frac{\partial f}{\partial \eta} = \frac{1}{2}$$

$$\therefore \quad \begin{Bmatrix} \frac{\partial f}{\partial x} \\ \frac{\partial f}{\partial y} \end{Bmatrix} = \frac{1}{\frac{1}{4}(8+2\xi+2\eta)} \begin{bmatrix} \left(\frac{3+\xi}{2}\right) & -\left(\frac{1+\eta}{2}\right) \\ -\left(\frac{1+\xi}{2}\right) & \left(\frac{3+\eta}{2}\right) \end{bmatrix} \begin{Bmatrix} \frac{1}{2} \\ \frac{1}{2} \end{Bmatrix}$$

$$\begin{Bmatrix} \frac{\partial f}{\partial x} \\ \frac{\partial f}{\partial y} \end{Bmatrix} = \frac{1}{\frac{1}{4}(8+2\xi+2\eta)} \begin{Bmatrix} \frac{(3+\xi)}{4} - \frac{(1+\eta)}{4} \\ -\frac{(1+\xi)}{4} + \frac{(3+\eta)}{4} \end{Bmatrix}$$

5.10 Newton's Interpolating Polynomials in $\mathbb{R}^1$

Let (x_i, f_i) ; $i = 1, 2, \dots, n+1$ be given data points. Newton's interpolating polynomial is another method of determining an nth degree polynomial that passes through these data points. Recall that when we construct an nth degree polynomial, we write:

$$f(n) = C_0 + C_1 x + C_2 x^2 + \dots + C_n x^n \tag{5.141}$$

We can also write (5.141) in an alternate way using the locations of the data points.

$$f(x) = a_0 + a_1(x - x_1) + a_2(x - x_1)(x - x_2) + a_3(x - x_1)(x - x_2)(x - x_3) + \dots . \tag{5.142}$$

We can show that (5.141) and (5.142) are equivalent. Consider $f(x)$ to be quadratic (for simplicity) in x, then from (5.142), we have:

$$f(x) = a_0 + a_1(x - x_1) + a_2(x - x_1)(x - x_2) \tag{5.143}$$

Expanding (5.143):

$$f(x) = a_0 + a_1 x - a_1 x_1 + a_2 x^2 + a_2 x x_1 - a_2 x x_2 + a_2 x_1 x_2 \tag{5.144}$$

Collecting constant terms, coefficients of x and x^2:

$$f(x) = (a_0 + a_1x_1 + a_2x_1x_2) + (a_1 + a_2x_1 - a_2x_2)x + a_2x^2 \tag{5.145}$$

If we define:

$$\begin{aligned} C_0 &= a_0 + a_1x_1 + a_2x_1x_2 \\ C_1 &= a_1 + a_2x_1 - a_2x_2 \\ C_2 &= a_2 \end{aligned} \tag{5.146}$$

then (5.145) becomes:

$$f(x) = C_0 + C_1x + C_2x^2 \tag{5.147}$$

which is exactly the same as (5.141) when $f(x)$ is a quadratic polynomial in x. The same holds true when $f(x)$ is a higher degree polynomial.

Thus, we conclude that (5.141) and (5.142) are exactly equivalent. We consider (5.142) in the following.

5.10.1 Determination of Coefficients in (5.142)

The coefficients a_i ; $i = 0, 1, \dots, n$ must be determined using the data (x_i, f_i) ; $i = 1, 2, \dots, n+1$.

(i) If we let $x = x_1$, then except for a_0, all other terms become zero due to the fact that they all contain $(x - x_1)$. Thus, we obtain:

$$f(x_1) = a_0 = f_1 \tag{5.148}$$

The coefficient a_0 is determined.

(ii) If we let $x = x_2$, then except the first two terms on the right side of (5.142), all others are zero and we obtain (after substituting for a_0 from (5.148)):

$$f(x_2) = b_0 + a_1(x_2 - x_1) = f(x_1) + a_1(x_2 - x_1) \tag{5.149}$$

$$\therefore \quad a_1 = \frac{f(x_2) - f(x_1)}{x_2 - x_1} = \frac{f_2 - f_1}{x_2 - x_1} = f[x_2, x_1] \tag{5.150}$$

$f[x_2, x_1]$ is a convenient notation. It is called *first divided difference* between the points x_1 and x_2. Thus, a_1 is determined.

(iii) If we let $x = x_3$, then except the first three terms on the right side of (5.142), all others are zero and we obtain:

$$f(x_3) = a_0 + a_1(x_3 - x_1) + a_2(x_3 - x_1)(x_3 - x_2) \tag{5.151}$$

Substituting for a_0 and a_1 from (5.148) and (5.150):

$$f(x_3) = f(x_1) + \frac{f(x_2) - f(x_1)}{x_2 - x_1}(x_3 - x_1) + a_2(x_3 - x_1)(x_3 - x_2) \quad (5.152)$$

Solving for a_2, we obtain:

$$a_2 = \frac{\frac{f(x_3)-f(x_2)}{x_3-x_2} - \frac{f(x_2)-f(x_1)}{x_2-x_1}}{x_3 - x_1} \quad (5.153)$$

Introducing the notation for the first divided difference in (5.153):

$$a_2 = \frac{f[x_3, x_2] - f[x_2, x_1]}{x_3 - x_3} = f[x_3, x_2, x_1] \quad (5.154)$$

$f[x_3, x_2, x_1]$ is called second divided difference.

(iv) Following this procedure we obtain:

$$\begin{aligned} a_0 &= f(x_1) \\ a_1 &= f[x_2, x_1] \\ a_2 &= f[x_3, x_2, x_1] \\ &\vdots \\ a_n &= f[x_{n+1}, x_n, \dots, x_1] \end{aligned} \quad (5.155)$$

in which the f values in the square brackets are the divided differences defined by:

$$\begin{aligned} f[x_2, x_1] &= \frac{f(x_2) - f(x_1)}{x_2 - x_1} \ ; \quad \text{first divided difference} \\ f[x_3, x_2, x_1] &= \frac{f[x_3, x_2] - f[x_2, x_1]}{x_3 - x_1} \ ; \quad \text{second divided difference} \\ f[x_4, x_3, x_2, x_1] &= \frac{f[x_4, x_3, x_2] - f[x_3, x_2, x_1]}{x_4 - x_1} \ ; \quad \text{third divided difference} \\ \vdots \quad & \vdots \qquad \vdots \\ f[x_{n+1}, x_n, \dots, x_1] &= \frac{f[x_{n+1}, x_n, \dots, x_2] - f[x_n, x_{n-1}, \dots, x_1]}{x_{n+1} - x_1} \end{aligned} \quad (5.156)$$

The details of calculating $a_0, a_1, \dots, a_n$ described above can be made more systematic by using a tabular presentation (Table 5.1). Consider (x_i, f_i) ; $i = 1, 2, \dots, 4$, i.e., four data points for the purpose of this illustration.

Table 5.1: Divided differences in Newton's interpolating polynomials

i	x_i	$f(x_i) = f_i$	First Divided Difference	Second Divided Difference	Third Divided Difference
1	x_1	$f(x_1)$			
			$f[x_2, x_1]$		
2	x_2	$f(x_2)$		$f[x_3, x_2, x_1]$	
			$f[x_3, x_2]$		$f[x_4, x_3, x_2, x_1]$
3	x_3	$f(x_3)$		$f[x_4, x_3, x_2]$	
			$f[x_4, x_3]$		
4	x_4	$f(x_4)$			

For this data:

$$f(x) = a_0 + a_1(x - x_1) + a_2(x - x_1)(x - x_2) + a_3(x - x_3)(x - x_2)(x - x_3) \tag{5.157}$$

where

$$\begin{aligned} & f(x_i) = f_i \qquad ; \qquad i = 1, 2, \dots, 4 \\ & \text{and} \\ & a_0 = f(x_1) \\ & a_1 = f[x_2, x_1] \\ & a_2 = f[x_3, x_2, x_1] \\ & a_3 = f[x_4, x_3, x_2, x_1] \end{aligned} \tag{5.158}$$

Example 5.12 (Newton's Interpolating Polynomial). Consider the following data set:

i	1	2	3	4
x_i	1	2	3	4
f_i	0	10	0	-5

First Divided Differences:

$$f[x_2, x_1] = \frac{f(x_2) - f(x_1)}{x_2 - x_1} = \frac{10 - 0}{2 - 1} = 10$$

$$f[x_3, x_2] = \frac{f(x_3) - f(x_2)}{x_3 - x_2} = \frac{0 - 10}{3 - 2} = -10$$

$$f[x_4, x_3] = \frac{f(x_4) - f(x_3)}{x_4 - x_3} = \frac{-5 - 0}{4 - 3} = -5$$

Second Divided Differences:

$$f[x_3, x_2, x_1] = \frac{f[x_3, x_2] - f[x_2, x_1]}{x_3 - x_1} = \frac{-10 - 10}{3 - 1} = -10$$

$$f[x_4, x_3, x_2] = \frac{f[x_4, x_3] - f[x_3, x_2]}{x_4 - x_2} = \frac{-5 - (-10)}{4 - 2} = \frac{5}{2}$$

Third Divided Difference:

$$f[x_4, x_3, x_2, x_1] = \frac{f[x_4, x_3, x_2] - f[x_3, x_2, x_1]}{x_4 - x_1} = \frac{\frac{5}{2} - (-10)}{4 - 1} = \frac{25}{6}$$

Table 5.2: Newton's interpolating polynomial example

i	x_i	$f(x_i) = f_i$	First Divided Difference	Second Divided Difference	Third Divided Difference
1	1	(a_0) 0			
			(a_1) 10		
2	2	10		(a_2) -10	
			-10		(a_3) $\frac{25}{6}$
3	3	0		$\frac{5}{2}$	
			-5		
4	4	-5			

$$\therefore \quad f(x) = a_0(x - x_1) + a_2(x - x_1)(x - x_2) + a_3(x - x_1)(x - x_2)(x - x_3)$$

$$f(x) = (0) + 10(x - 1) - 1(x - 1)(x - 2) + \frac{25}{6}(x - 1)(x - 2)(x - 3)$$

or

$$f(x) = 10(x - 1) - 10(x - 1)(x - 3) + \frac{25}{6}(x - 1)(x - 2)(x - 3)$$

is the interpolation of the given data.

5.11 Approximation Errors in Interpolations

By considering Newton's interpolating polynomials it is possible to establish the order of the truncation errors in the interpolation process. Consider Newton's interpolating polynomial.

For equally spaced data, if h is the data spacing in x-space, then:

$$\begin{aligned} x_2 &= x_1 + h \\ x_3 &= x_2 + 2h \\ &\vdots \\ x_n &= x_1 + nh \end{aligned} \tag{5.159}$$

For this case:

$$f[x_2, x_1] = \frac{f(x_2) - f(x_1)}{x_2 - x_1} = \frac{f(x_2) - f(x_2)}{h} \tag{5.160}$$

$$f[x_3, x_2, x_1] = \frac{f[x_3, x_2] - f[x_2, x_1]}{x_3 - x_1} = \frac{\frac{f(x_3)-f(x_2)}{x_3-x_2} - \frac{f(x_2)-f(x_2)}{x_2-x_1}}{(x_3 - x_1)}$$

or

$$f[x_3, x_2, x_1] = \frac{\frac{f(x_3)-f(x_2)}{h} - \frac{f(x_2)-f(x_1)}{h}}{2h} = \frac{f(x_3) - 2f(x_2) + f(x_1)}{2h^2} \tag{5.161}$$

If we use a Taylor series expansion about x_1 in the interval $[x_1, x_2]$ to approximate the derivatives of $f(x)$, we obtain:

$$\begin{aligned} f'(x_1) &= \frac{f(x_2) - f(x_1)}{h} = f[x_2, x_1] \\ f''(x_1) &= \frac{f(x_3) - 2f(x_2) + f(x_1)}{h^2} = \frac{f[x_3, x_2, x_1]}{2!} \end{aligned} \tag{5.162}$$

and so on. Hence,

$$f(x) = f(x_1) + f[x_2, x_1](x - x_1) + f[x_3, x_2, x_1](x - x_1)(x - x_2) + \dots . \tag{5.163}$$

can be written as

$$f(x) = f(x_1) + f'(x_1)(x - x_1) + \frac{f''(x_1)}{2!}(x - x_1)(x - x_2) + \dots \tag{5.164}$$

Equation (5.164) is an important form of the Newton's interpolating polynomial.

If we let

$$\frac{x - x_1}{h} = \alpha \qquad \therefore \quad x - x - 1 = h\alpha$$

then

$$\begin{aligned} x - x_2 &= x - (x_1 - h) = x - x_1 - h = \alpha h - h = h(\alpha - 1) \\ x - x_3 &= x - (x_2 + h) = x - (x_1 + h + h) = x - x_1 - 2h \\ &= \alpha h - 2h = h(\alpha - 2) \end{aligned} \tag{5.165}$$

Hence, (5.164) can be written as:

$$f(x) = f(x_1) + f'(x_1)h\alpha + \frac{f'(x_1)}{2!}h^2\alpha(\alpha-1) + \cdots + \frac{f^n h^n}{n!}\alpha(\alpha-1)\dots(\alpha-(n-1)) + R^n \tag{5.166}$$

$$R^n = \frac{f^{n+1}(\xi)}{(n+1)!}h^{n+1}\alpha(\alpha-1)(\alpha-2)\dots(\alpha-n) \quad ; \quad \text{remainder} \tag{5.167}$$

Remarks.

(1) R^n is the remainder in (5.166) and is a measure of the order of truncation error.

(2) Equation (5.166) for interpolation $f(x)$ suggests that:

   (i) If $f(x)$ is linear ($n = 2$), then the truncation error R^2 is $O(h^2)$.

   (ii) If $f(x)$ is quadratic ($n = 3$), then the truncation error R^3 is $O(h^3)$ and so on.

(3) This conclusion drawn using Newton's interpolating polynomials also holds for other methods of interpolation.

(4) We note that all interpolation processes are methods of approximation. The interpolating polynomials (in $\mathbb{R}^1$, $\mathbb{R}^2$, and $\mathbb{R}^3$) only approximate the real behavior of the data.

5.12 Concluding Remarks

Interpolation theory and mapping of lengths, areas, and volumes from physical coordinate spaces x, xy and xyz to natural coordinate spaces ξ, $\xi\eta$ and $\xi\eta\zeta$ are presented in this chapter. Mapping of irregular domains in $\mathbb{R}^2$ and $\mathbb{R}^3$ in the physical spaces xy and xyz to the natural coordinate spaces $\xi\eta$ and $\xi\eta\zeta$ spaces facilitates interpolation theory. Lagrange interpolation in $\mathbb{R}^1$, $\mathbb{R}^2$ and $\mathbb{R}^3$ is considered in ξ, $\xi\eta$ and $\xi\eta\zeta$ spaces. The tensor product is highly meritorious for generating interpolation functions in the natural coordinate space in $\mathbb{R}^2$ and $\mathbb{R}^3$.

Problems

5.1 Consider the following table of data

i	1	2	3
x_i	0	3	6
f_i	f_1	f_2	f_3

We can express

$$f(x) = \sum_{k=1}^{3} L_k(x) f_k \tag{1}$$

Where $L_k(x)$ are Lagrange interpolation functions.

(a) State properties of Lagrange interpolation functions $L_k(x)$. Why are these properties necessary.

(b) Construct an analytical expression for $f(x)$ using Lagrange interpolating polynomial (1). If $f_1 = 1$, $f_2 = 13$ and $f_3 = 43$, then determine numerical value of $f(4)$.

5.2 Consider the following table of data

i	1	2	3	4	5
x_i	0	1	2	3	4
f_i	1	3	7	13	21

Use Newton's interpolating polynomial to calculate $f(2.5)$ and $f'(2.5)$ i.e. $\left.\frac{df}{dx}\right|_{x=2.5}$.

5.3 Consider the following table of data

i	1	2	3
x_i	-1.0	-0.5	1.0
f_i	f_1	f_2	f_3

(a) Determine an analytical expression for $f(x)$; $-1 \leq x \leq 1$ using the data in the table and using Lagrange interpolating polynomial.

(b) Tabulate the values of Lagrange interpolation functions $L_k(x)$; $k = 1, 2, 3$ at x_i; $i = 1, 2, 3$. Comment on their behavior. Why is this behavior necessary? Show that $\sum_{k=1}^{3} L_k(x) = 1$. Why is this property necessary?

5.4 Consider the following table of data

i	1	2	3
x_i	0	2	4
f_i	1	5	17

Determine $f(3)$ using Lagrange interpolating polynomial for the data in the table.

5.5 Consider the following table of data

i	1	2	3	4
x_i	3	4	2.5	5
f_i	8	2	7	1

Calculate $f(3.4)$ using Newton's interpolating polynomials of degrees 1, 2 and 3.

5.6 Given the following table

i	1	2	3	4	5
x_i	0	1	2	3	4
f_i	0	2	6	12	20

Use Newton's interpolating polynomials of degrees 1, 2, 3 and 4 to calculate $f(2.5)$.

5.7 Consider a two node configuration $\bar{\Omega}^{(e)}$ in $\mathbb{R}^1$ shown in Figure (a) with coordinates. Figure (b) shows its map $\bar{\Omega}^{(\xi)}$ in the natural coordinate space ξ.

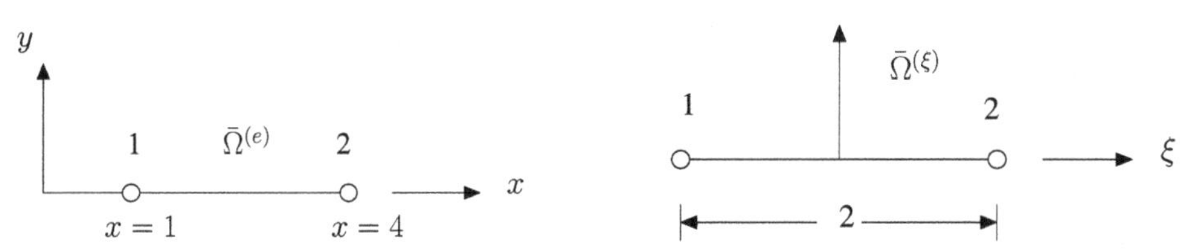

(a) A two node domain $\bar{\Omega}^{(e)}$

(b) Map of $\bar{\Omega}^{(e)}$ in $\bar{\Omega}^{(\xi)}$

(a) Derive equation describing the mapping of points between x and ξ spaces.

(b) Derive equation describing mapping of lengths between x and ξ spaces.

(c) If f_1 and f_2 are function values at nodes 1 and 2 of Figure (a), then establish interpolation $f(\xi)$ in the natural coordinate space ξ i.e. $\bar{\Omega}^{(\xi)}$.

(d) Derive an expression for $\frac{df(\xi)}{dx}$ using the interpolation $f(\xi)$ derived in (c).

(e) Using $f_1 = 10$ and $f_2 = 20$ calculate values of f at $x = 1.75$ and $x = 3.25$ using the interpolation $f(\xi)$ in (c).

(f) Also calculate $\frac{df}{dx}$ at $x = 1.75$ and $x = 3.25$.

5.8 Consider a three node configuration $\bar{\Omega}^{(e)}$ in $\mathbb{R}^1$ shown in Figure (a). Figure (b) shows a map $\bar{\Omega}^{(\xi)}$ of $\bar{\Omega}^{(e)}$ in the natural coordinate space ξ.

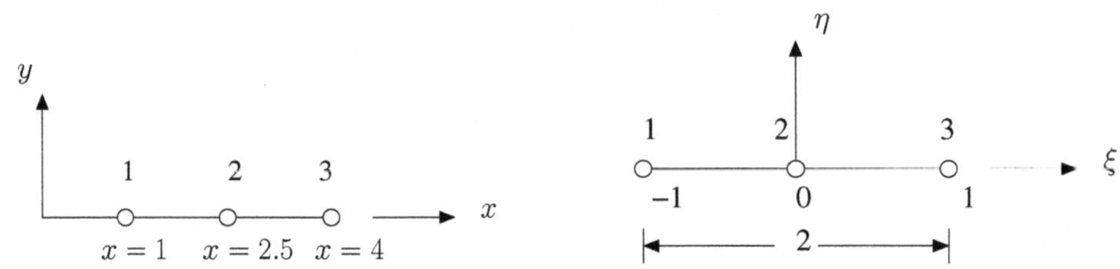

(a) A three node configuration $\bar{\Omega}^{(e)}$ (b) Map $\bar{\Omega}^{(\xi)}$ of $\bar{\Omega}^{(e)}$

(a) Derive equation describing the mapping of points between x and ξ spaces.

(b) Derive equation describing mapping of lengths between x and ξ spaces.

(c) If f_1, f_2 and f_3 are function values at nodes 1, 2 and 3 of Figure (a), then establish interpolation $f(\xi)$ of f in the natural coordinate space ξ i.e. $\bar{\Omega}^{(\xi)}$.

(d) Derive an expression for $\frac{df(\xi)}{dx}$ using the interpolation $f(\xi)$ derived in (c).

(e) Using $f_1 = 2$, $f_2 = 6$ and $f_3 = 1$ calculate values of f and $\frac{df(\xi)}{dx}$ at $x = 1.375$ and $x = 3.625$.

5.9 Consider a three node configuration $\bar{\Omega}^{(e)}$ in $\mathbb{R}^1$ shown in Figure (a). Figure (b) shows a map $\bar{\Omega}^{(e)}$ of $\bar{\Omega}^{(\xi)}$ in the natural coordinate space ξ.

(a) Derive equation describing the mapping of points between x and ξ spaces.

(b) Derive equation describing mapping of lengths between x and ξ spaces.

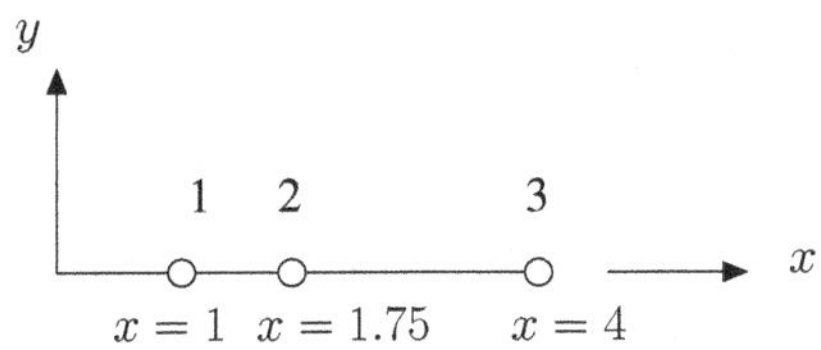

(a) A three node configuration $\bar{\Omega}^{(e)}$

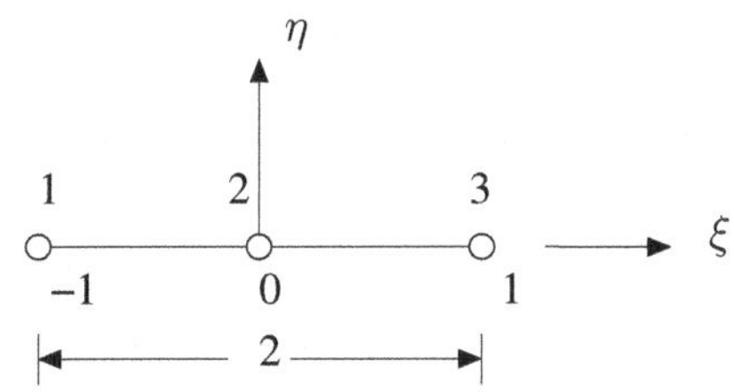

(b) Map $\bar{\Omega}^{(\xi)}$ of $\bar{\Omega}^{(e)}$

(c) If f_1, f_2 and f_3 are function values at nodes 1, 2 and 3 of Figure (a), then establish interpolation $f(\xi)$ of f in the natural coordinate space ξ i.e. $\bar{\Omega}^{(\xi)}$.

(d) Derive an expression for $\frac{df(\xi)}{dx}$ using the interpolation $f(\xi)$ derived in (c).

(e) Using $f_1 = 2$, $f_2 = 6$ and $f_3 = 1$ calculate values of f and $\frac{df(\xi)}{dx}$ at $x = 1.375$ and $x = 2.875$. Also calculate $\frac{df(\xi)}{dx}$ at nodes 1, 2, and 3 of the configuration in Figure (a).

(f) Plot graphs of f versus x and $\frac{df(\xi)}{dx}$ versus x for $x \in [1, 4]$. Take at least twenty points between $x = 1$ and $x = 4$. Do not curve fit the calculated values.

5.10 Consider a three node configuration $\bar{\Omega}^{(e)}$ in $\mathbb{R}^1$ shown in Figure (a). Figure (b) shows a map $\bar{\Omega}^{(e)}$ of $\bar{\Omega}^{(\xi)}$ in the natural coordinate space ξ.

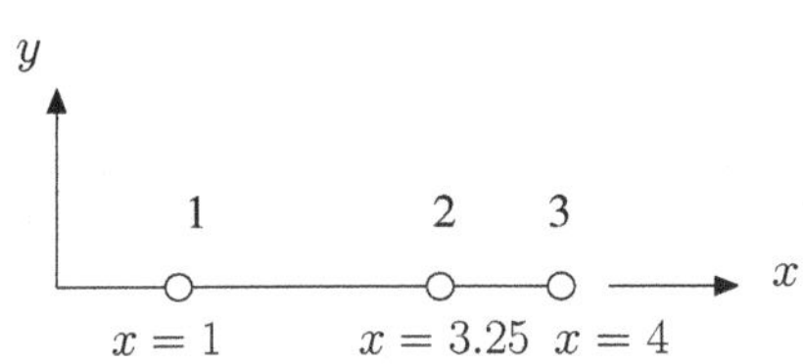

(a) A three node configuration $\bar{\Omega}^{(e)}$

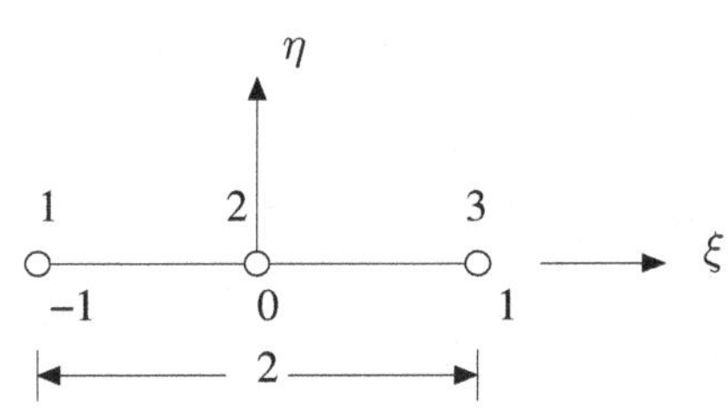

(b) Map $\bar{\Omega}^{(\xi)}$ of $\bar{\Omega}^{(e)}$

(a) Derive equation describing the mapping of points between x and ξ spaces.

(b) Derive equation describing mapping of lengths between x and ξ spaces.

(c) If f_1, f_2 and f_3 are function values at nodes 1, 2 and 3 of Figure (a), then establish interpolation $f(\xi)$ of f in the natural coordinate space ξ i.e. $\bar{\Omega}^{(\xi)}$.

(d) Derive an expression for $\frac{df(\xi)}{dx}$ using the interpolation $f(\xi)$ derived in (c).

(e) Using $f_1 = 2$, $f_2 = 6$ and $f_3 = 1$ calculate values of f and $\frac{df(\xi)}{dx}$ at $x = 2.125$ and $x = 3.625$. Also calculate $\frac{df(\xi)}{dx}$ at nodes of the configuration in Figure (a).

(f) Plot graphs of f versus x and $\frac{df(\xi)}{dx}$ versus x for $x \in [1, 4]$. Take at least twenty points between $x = 1$ and $x = 4$. Do not curve fit the calculated values.

5.11 Consider a three node configuration $\bar{\Omega}^{(e)}$ in $\mathbb{R}^1$ shown in Figure (a). The coordinates of the nodes are x_1, x_2, x_3. The map of the element $\bar{\Omega}^{(\xi)}$ in the natural coordinate space ξ is shown in Figure (b).

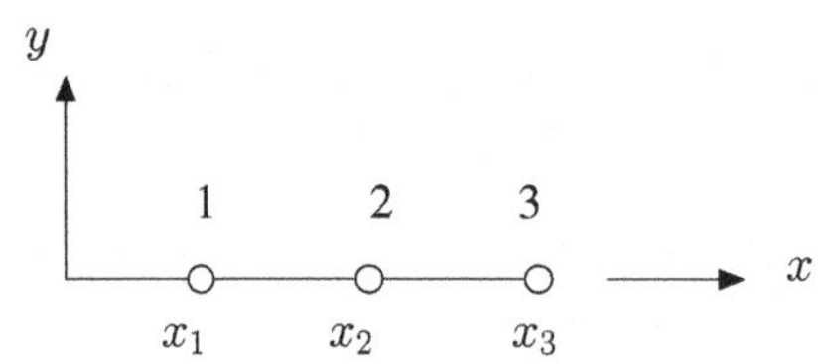

(a) A three node configuration $\bar{\Omega}^{(e)}$

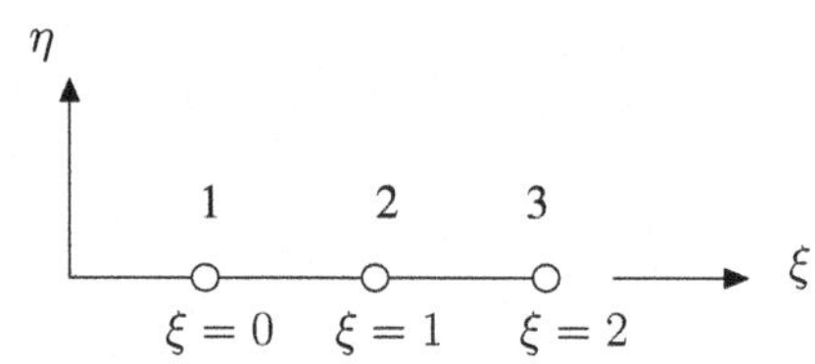

(b) Map $\bar{\Omega}^{(\xi)}$ of $\bar{\Omega}^{(e)}$

(a) Derive expression describing the mapping of points between $\bar{\Omega}^{(e)}$ and $\bar{\Omega}^{(\xi)}$ i.e. derive $x = x(\xi)$.

(b) Derive an expression for mapping of lengths between x and ξ spaces.

(c) Determine the length between nodes 1 and 3 in $\bar{\Omega}^{(e)}$ i.e. Figure (a) using its map in the natural coordinate space.

5.12 Figure (a) shows a four node quadrilateral $\bar{\Omega}^{(e)}$ in $\mathbb{R}^2$. Coordinates of the nodes are given. Figure (b) shows a map $\bar{\Omega}^{(\xi\eta)}$ of $\bar{\Omega}^{(e)}$ in natural coordinate space $\xi\eta$.

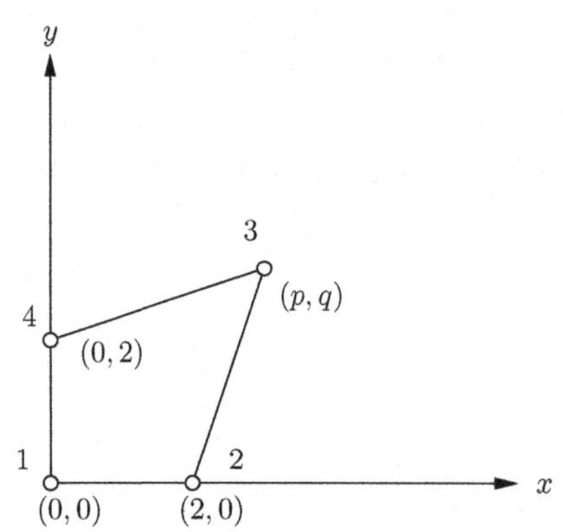

(a) $\bar{\Omega}^{(e)}$ in x, y space

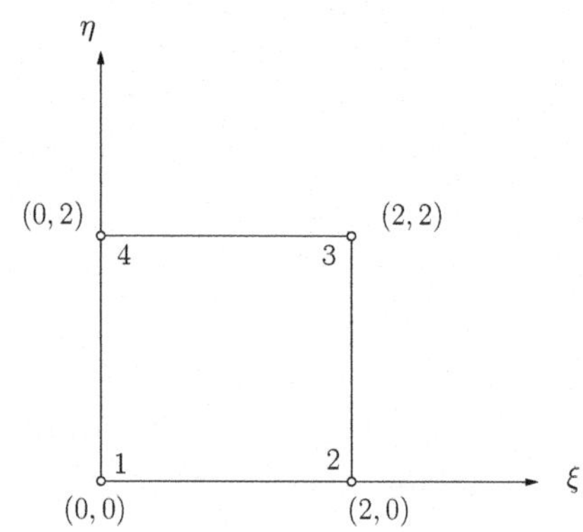

(b) Map $\bar{\Omega}^{(\xi,\eta)}$ of $\bar{\Omega}^{(e)}$

The coordinates of the nodes are also given in the two spaces in Figures (a) and (b).

(a) Determine the equations describing the mapping of points in xy and $\xi\eta$ spaces for $\bar{\Omega}^{(e)}$ and $\bar{\Omega}^{(\xi\eta)}$ i.e. determine $x = x(\xi, \eta)$, $y = y(\xi, \eta)$. Simplify the expressions till no further simplification is possible.

(b) Determine the relationship between p and q (the Cartesian coordinates of node 3) for their admissibility in the geometric description of the geometry $\bar{\Omega}^{(e)}$ in the xy space. Simplify the final expression or equation.

5.13 Consider a four node quadrilateral bilinear geometry $\bar{\Omega}^{(e)}$ in $\mathbb{R}^2$ shown in Figure (a).

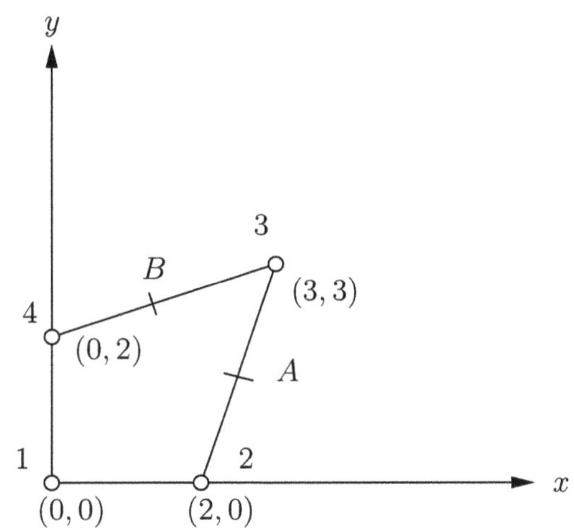

(a) A four node $\bar{\Omega}^{(e)}$ in $\mathbb{R}^2$

Let f_i; $i = 1, 2, \ldots, 4$ be the function values at nodes $1, 2, \ldots, 4$. Locations A and B represent the mid points of sides 2,3 and 4,3.
If $f_1 = 100$, $f_2 = 200$, $f_A = 300$ and $f_B = 275$, then calculate f_3 and f_4 using interpolation theory.

5.14 Consider a six node $\bar{\Omega}^{(e)}$ in $\mathbb{R}^2$ shown in Figure (a). Its map $\bar{\Omega}^{(\xi\eta)}$ in the natural coordinate space $\xi\eta$ is shown in Figure (b).

(a) Construct interpolation functions for the $\bar{\Omega}^{(\xi\eta)}$ in the natural coordinate space $\xi\eta$.

(b) If a function f is interpolated using the functions generated in (a) and using values of f at nodes 1 – 6 of Figure (a), then determine whether $f(\xi, \eta)$ so interpolated is a complete polynomial in ξ and η. Explain, why or why not.

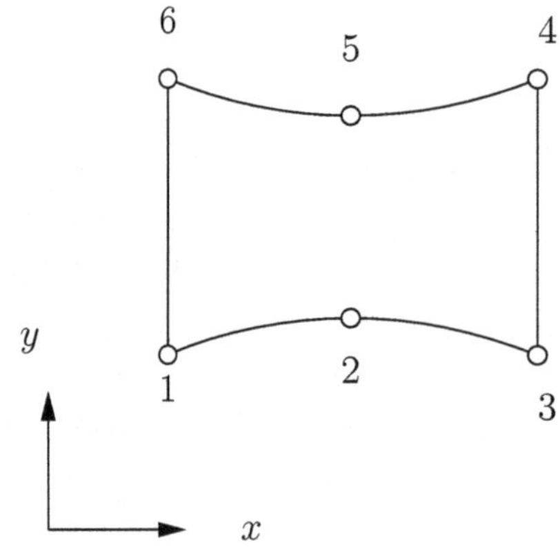

(a) $\bar{\Omega}^{(e)}$ in $\mathbb{R}^2$

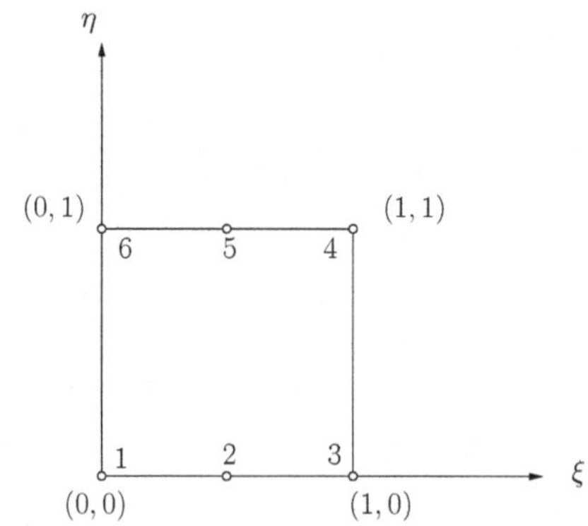

(b) Map $\bar{\Omega}^{(\xi,\eta)}$ of $\bar{\Omega}^{(e)}$

(c) Determine degrees of interpolation of $f(\xi,\eta)$ in ξ and η i.e. p_ξ and p_η. Explain your reasoning.

5.15 Consider a four node bilinear $\bar{\Omega}^{(e)}$ in $\mathbb{R}^2$ shown in Figure (a). Its map $\bar{\Omega}^{(\xi\eta)}$ in $\xi\eta$ space is shown in Figure (b).

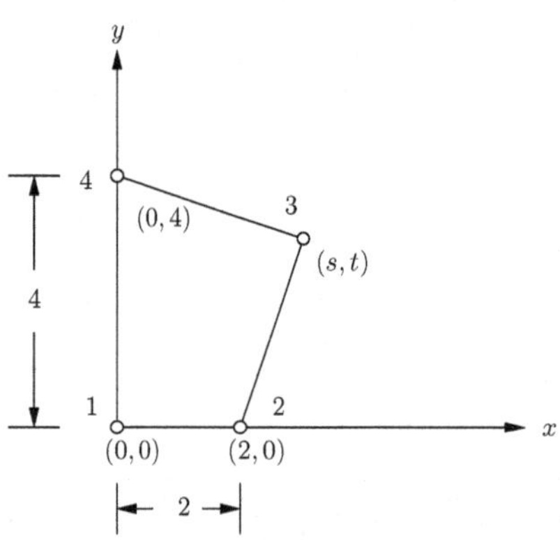

(a) $\bar{\Omega}^{(e)}$ in $\mathbb{R}^2$

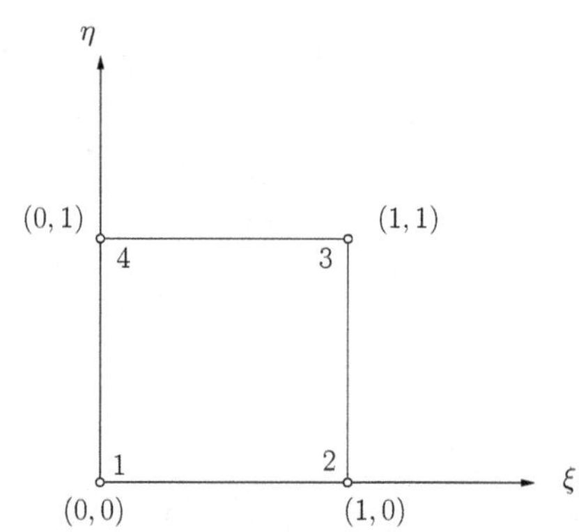

(b) Map $\bar{\Omega}^{(\xi,\eta)}$ of $\bar{\Omega}^{(e)}$

(a) Determine equations describing mapping of points between $\bar{\Omega}^{(\xi\eta)}$ and $\bar{\Omega}^{(e)}$. Simplify the resulting expressions.

(b) Determine Jacobian of mapping $[J]$.

(c) Can the location of node 3 in xy space be arbitrary (i.e. can the values of s and t be arbitrary) or are their restrictions on them? If there are, then determine them.

5.16 Consider a six node para-linear $\bar{\Omega}^{(e)}$ in $\mathbb{R}^2$ shown in Figure (a). Its map $\bar{\Omega}^{(\xi\eta)}$ in $\xi\eta$ space is shown in Figure (b).

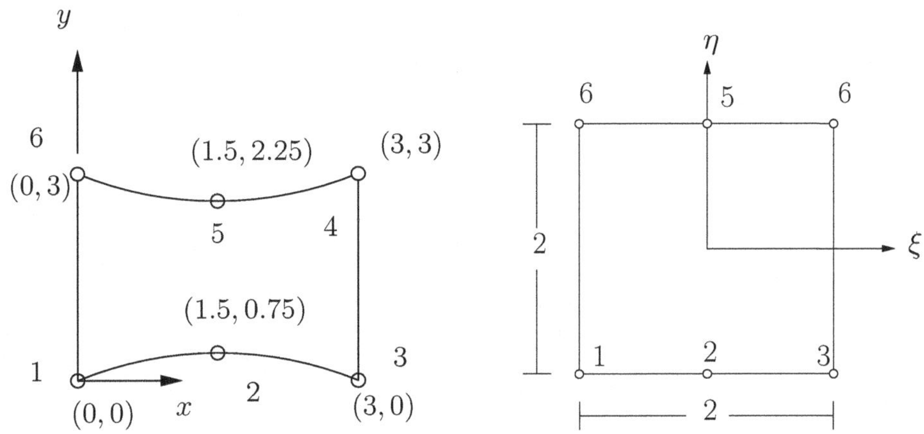

(a) $\bar{\Omega}^{(e)}$ in $\mathbb{R}^2$ (b) Map $\bar{\Omega}^{(\xi,\eta)}$ of $\bar{\Omega}^{(e)}$

Calculate length of the face of $\bar{\Omega}^{(e)}$ containing nodes 1, 2, 3 in the Cartesian coordinate space by utilizing its map $\bar{\Omega}^{(\xi\eta)}$ in the natural coordinate space $\xi\eta$.

5.17 Consider a four node bilinear $\bar{\Omega}^{(e)}$ in $\mathbb{R}^2$ shown in Figure (a).

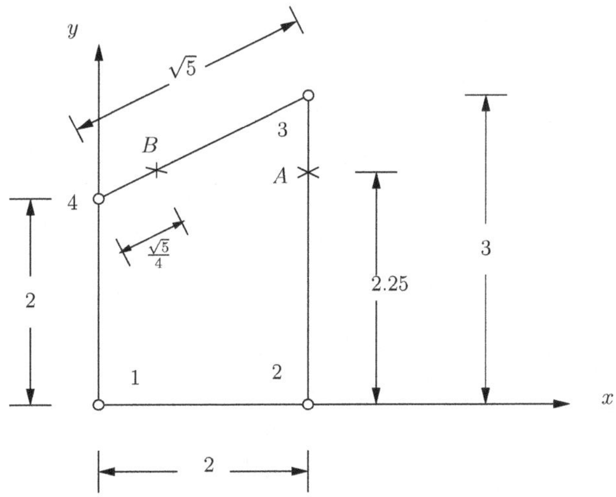

(a) A four node $\bar{\Omega}^{(e)}$ in $\mathbb{R}^2$

Let f_1, f_2, f_3 and f_4 be the function values at nodes 1, 2, 3 and 4. Locations of points A and B on sides $2-3$ and $4-3$ are shown in Figure (a). If $f_1 = 10$, $f_2 = 20$, $f_A = 30$ and $f_B = 27.5$, then calculate f_3 and f_4 using interpolation theory.

5.18 Consider two-dimensional $\bar{\Omega}^{(e)}$ in $\mathbb{R}^2$ shown in Figure (a), (b), and (c).

The Cartesian coordinates of the nodes are given. The domains $\bar{\Omega}^{(e)}$ are mapped into $\xi\eta$-space into a two-unit square.

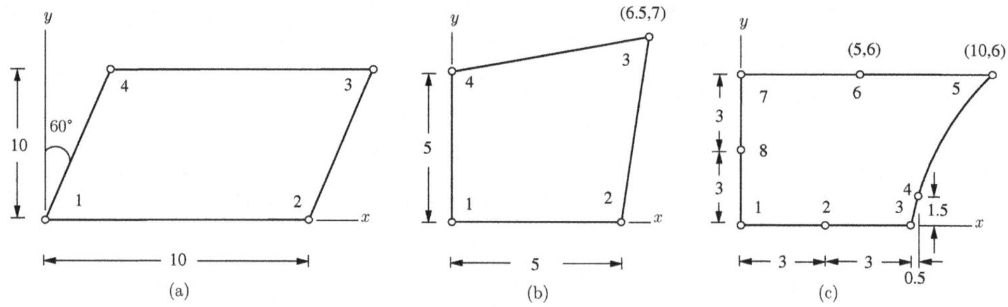

Figure 1: $\bar{\Omega}^{(e)}$ in $\mathbb{R}^2$

(a) Determine the Jacobian matrix of transformation and its determinant for each $\bar{\Omega}^{(e)}$. Calculate and tabulate the value of the determinant of the Jacobian at the nodes of each $\bar{\Omega}^{(e)}$.

(b) Calculate the derivatives of the approximation function with respect to x and y for node 3 (i.e. $\frac{\partial N_3(\xi,\eta)}{\partial x}$ and $\frac{\partial N_3(\xi,\eta)}{\partial y}$) for each of the three $\bar{\Omega}^{(e)}$ shown in Figures (a) – (c).

5.19 Consider a two-dimensional eight-node $\bar{\Omega}^{(e)}$ shown in Figure (a). The Cartesian coordinates of the nodes are given in Figure (a). The domain $\bar{\Omega}^{(e)}$ is mapped into natural coordinate space $\xi\eta$ into a two-unit square $\bar{\Omega}^{(\xi\eta)}$ with the origin of the $\xi\eta$ coordinate system at the center of $\bar{\Omega}^{(\xi\eta)}$.

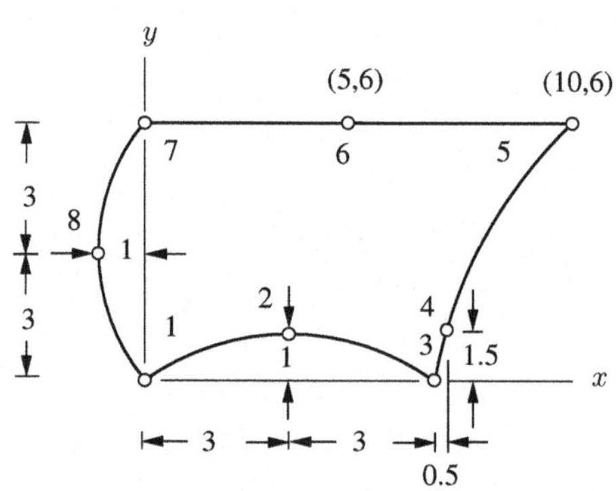

Figure 1: $\bar{\Omega}^{(e)}$ in $\mathbb{R}^2$

(a) Write a computer program (or calculate otherwise) to determine the Cartesian coordinates of the points midway between the nodes. Tabulate the xy coordinates of these points. Plot the sides of $\bar{\Omega}^{(e)}$ in xy-space by taking more intermediate points.

(b) Determine the area of $\bar{\Omega}^{(e)}$ using Gauss quadrature. Select and use the minimum number of quadrature points in ξ and η directions to calculate the area exactly. Show that increasing the order of the quadrature does not affect the area.

(c) Determine the locations of the quadrature points (used in (b)) in the Cartesian space. Provide a table of these points and their locations in xy-space. Also mark their locations on the plot generated in part (a).

Provide program listing, results, tables, and plots along with a write-up on the equations used as part of the report. Also provide a discussion of your results.

6

Numerical Integration or Quadrature

6.1 Introduction

In many situations, due to the complexity of integrands and irregularity of the domain in definite integrals it becomes necessary to approximate the value of the integral. The numerical integration methods or quadrature methods are methods of obtaining approximate values of the definite integrals. Many simple numerical integration methods are derived using the simple fact that if we wish to calculate the integral of $f(x)$ between the limits $x = A$ to $x = B$, i.e.,

$$I = \int_A^B f(x)\, dx \tag{6.1}$$

then the value of the integral of $f(x)$ between $x = A$ to $x = B$ is the area under the curve $f(x)$ versus x (Figure 6.1). Thus, the numerical integration

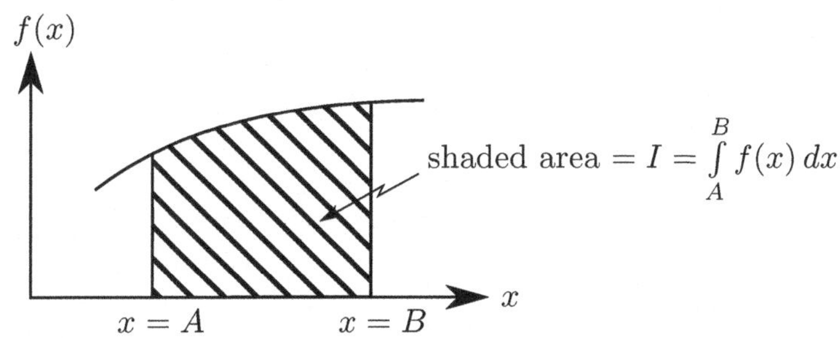

Figure 6.1: Plot of $f(x)$ versus x

methods are based on approximating the actual area under the curve $f(x)$ versus x between $x = A$ and $x = B$.

Numerical integration methods such as trapezoid rule, Simpson's $1/3$ and $3/8$ rules, Newton-Cotes integration, Richardson's extrapolation, and Romberg method presented in the following sections are all methods of approximation. These methods are only effective in $\mathbb{R}^1$. Gauss quadrature is equally effective in $\mathbb{R}^1$, $\mathbb{R}^2$, and $\mathbb{R}^3$. Gauss quadrature a numerical method without approximation when the integrand is an algebraic polynomial. When the integrand is not an algebraic polynomial, Gauss quadrature is also a method of approximation.

In this chapter we consider numerical integration methods in $\mathbb{R}^1$, $\mathbb{R}^2$, as well as $\mathbb{R}^3$. First we consider numerical integration in $\mathbb{R}^1$.

6.1.1 Numerical Integration in $\mathbb{R}^1$

We consider two classes of methods.

(1) In the first category of methods the integration interval $[A, B]$ is divided into subintervals (may be considered of equal width for convenience). The integration methods are developed for calculating the approximate value of the integral for a subinterval. The sum of the approximated integral values for each subinterval then yields the approximate value of the integral over the entire interval of integration $[A, B]$. We consider two methods:

   (a) In a typical subinterval $[a, b]$, $f(x)$ in (6.1) is approximated by a polynomial of degree one, two, etc. and then integrated explicitly to obtain the approximate value of the integral for this subinterval $[a, b]$. The trapezoid rule, Simpson's $\frac{1}{3}$ method, Simpson's $\frac{3}{8}$ method, and Newton-Cotes integration techniques fall into this category.

   (b) The second class of methods using subintervals includes Richardson's extrapolation and Romberg method. In these methods the initially calculated values of the integral using a subinterval size are improved based on the order of truncation errors and their elimination and the integral estimate based on reduced subinterval size.

(2) In the second category of methods, the integration interval $[A, B]$ is not subdivided into subintervals. A numerical integration method is designed such that if $f(x)$ is an algebraic polynomial in x, then it is integrated exactly. This method is called Gauss quadrature. Gauss quadrature can also be used to integrate $f(x)$ even if it is not an algebraic polynomial, but in this case the calculated integral value is approximate (however, it can be improved to desired accuracy).

6.1.2 Numerical Integration in $\mathbb{R}^2$ and $\mathbb{R}^3$:

Gauss quadrature in $\mathbb{R}^1$ can be easily extended to numerical integration in $\mathbb{R}^2$ and $\mathbb{R}^3$ by using mapping from physical domain (x, y or x, y, z) to natural coordinate space. Gauss quadrature in $\mathbb{R}^2$ and $\mathbb{R}^3$ also can be used to integrate algebraic polynomial integrands exactly.

6.2 Numerical Integration in $\mathbb{R}^1$: Methods Based on Approximating $f(x)$ by a Polynomial

Consider the integrand $f(x)$ between $x = A$ and $x = B$ in equation (6.1). We subdivide the interval $[A, B]$ into n subintervals. Let $[a_i, b_i]$ be the integration limits for a subinterval i, then (6.1) can be written as:

$$I = \int_A^B f(x)\,dx = \sum_{i=1}^{n} \left(\int_{a_i}^{b_i} f(x)\,dx \right) = \sum_{i=1}^{n} I_i \tag{6.2}$$

in which

$$I_i = \int_{a_i}^{b_i} f(x)\,dx \tag{6.3}$$

is the integral of $f(x)$ for a subinterval $[a_i, b_i]$. We consider methods of approximating the integral I_i for each subinterval i ($i = 1, 2, \dots, n$) and thereby approximating I in (6.2).

For a subinterval we consider (6.3). We approximate $f(x)$ $\forall x \in [a_i, b_i]$ by linear, quadratic, cubic, etc. polynomials. This leads to various methods of approximating the integral I_i for $[a_i, b_i]$. We consider details of various methods resulting from these approximations of $f(x)$ in the following.

6.2.1 Trapezoid Rule

Consider the integral (6.3). Calculate $f(a_i)$ and $f(b_i)$ using $f(x)$ (given integrand). Using $(a_i, f(a_i))$ and $(b_i, f(b_i))$, we approximate $f(x)$ $\forall x \in [a_i, b_i]$ by a linear polynomial in x, i.e., a straight line.

$$f(x) \approx \widetilde{f}(x) = f(a_i) + \frac{f(b_i) - f(a_i)}{b_i - a_i}(x - a_i) \tag{6.4}$$

$$I_i \approx \widetilde{I}_i = \int_{a_i}^{b_i} \widetilde{f}(x)\,dx \tag{6.5}$$

Substituting for $\widetilde{f}(x)$ from (6.4) into (6.5).

$$I_i \approx \int_{a_i}^{b_i} \left(f(a_i) + \frac{f(b_i) - f(a_i)}{(b_i - a_i)}(x - a_i) \right) dx \tag{6.6}$$

or

$$I_i \approx (b_i - a_i) \left(\frac{f(a_i) + f(b_i)}{2} \right) \tag{6.7}$$

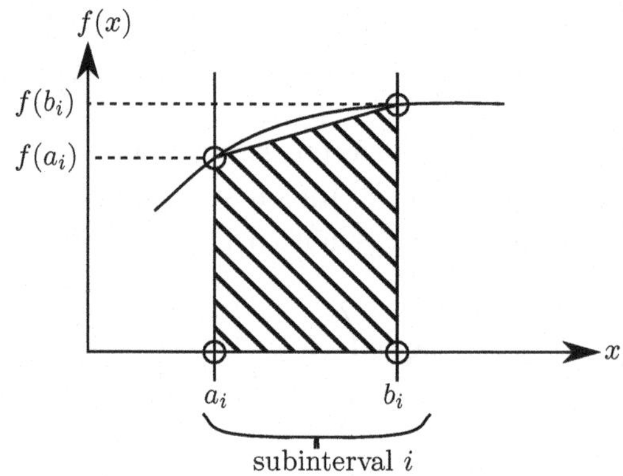

Figure 6.2: Trapezoidal Rule for subinterval $[a_i, b_i]$

This is called the trapezoid rule (Figure 6.2).

$\widetilde{I}_i$ is the area of the trapezoid between $[a_i, b_i]$, shown in Figure 6.2. We calculate $\widetilde{I}_i$ for each subinterval $[a_i, b_i]$ using (6.7) and then use $\widetilde{I} = \sum_{i=1}^{n} \widetilde{I}_i$ to obtain the approximate value $\widetilde{I}$ of the integral (6.1)

Remarks.

(1) The accuracy of the method is dependent on the size of the subinterval $[a_i, b_i]$. The smaller the subinterval, the better the accuracy of the approximated value of the integral I.

(2) It can be shown that in the trapezoid rule, the truncation error in calculating I using (6.7) for a subinterval $(b_i - a_i) = h_i$ is on the order of $O(h_i^2)$, where $h_i = b_i - a_i$.

(3) This definition of h_i is different than used in Sections 6.2.2 and 6.2.3.

6.2.2 Simpson's $\frac{1}{3}$ Rule

Consider I_i for a subinterval $[a_i, b_i]$.

$$I_i = \int_{a_i}^{b_i} f(x)\,dx \tag{6.8}$$

We calculate $f(a_i)$, $f(\frac{a_i+b_i}{2})$, and $f(b_i)$ using $f(x)$, the integrand in (6.8). For convenience of notation we let:

$$\begin{aligned} x_1 &= a_i &;&\quad f(a_i) = f(x_1) \\ x_2 &= \frac{a_i + b_2}{2} &;&\quad f\left(\frac{a_i + b_i}{2}\right) = f(x_2) \\ x_3 &= b_i &;&\quad f(b_i) = f(x_3) \end{aligned} \tag{6.9}$$

Using $(x_1, f(x_1))$, $(x_2, f(x_2))$, and $(x_3, f(x_3))$, we establish a quadratic interpolating polynomial $\widetilde{f}(x)$ (say, using Lagrange polynomials) that is considered to approximate $f(x)$ $\forall x \in [a_i, b_i]$.

$$\widetilde{f}(x) = \frac{(x-x_2)(x-x_3)}{(x_1-x_2)(x_1-x_3)} f(x_1) + \frac{(x-x_1)(x-x_3)}{(x_2-x_1)(x_2-x_3)} f(x_2) + \frac{(x-x_1)(x-x_2)}{(x_3-x_1)(x_3-x_2)} f(x_3) \tag{6.10}$$

We approximate $f(x)$ in (6.8) by $\widetilde{f}(x)$ in (6.10).

$$I_i \approx \widetilde{I}_i = \int_{a_i}^{b_i} \widetilde{f}(x)\, dx \tag{6.11}$$

Substituting $\widetilde{f}(x)$ from (6.10) into (6.11) and integrating:

$$I_i \approx \widetilde{I}_i = \frac{h_i}{3}\left(f(x_1) + 4f(x_2) + f(x_3)\right) \qquad h_i = \left(\frac{b_i - a_i}{2}\right) \tag{6.12}$$

This is called Simpson's $\frac{1}{3}$ Rule. Figure 6.3 shows $\widetilde{f}(x)$ and the true $f(x)$ for the subinterval $[a_i, b_i]$. We note that (6.12) can also be written as:

$$\widetilde{I}_i = \underbrace{(b_i - a_i)}_{\text{width}} \underbrace{\left(\frac{f(x_1) + 4f(x_2) + f(x_3)}{6}\right)}_{\text{average height}} = (b_i - a_i) H_i \tag{6.13}$$

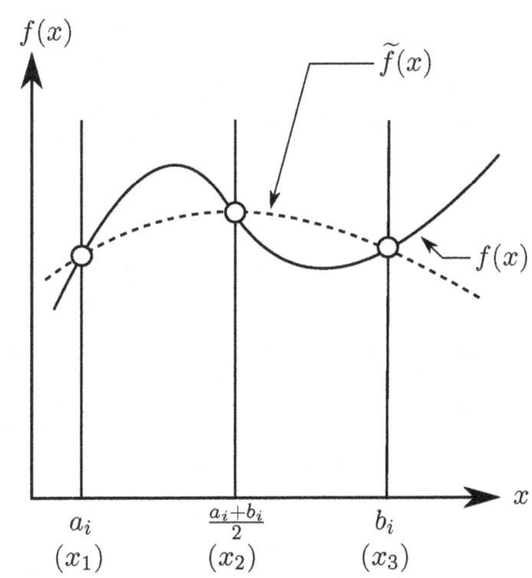

Figure 6.3: $f(x)$ and $\widetilde{f}(x)$, quadratic approximation of $f(x)$ for the subinterval $[a_i, b_i]$

From (6.13), we note that Simpson's $\frac{1}{3}$ can be interpreted as the area of a rectangle with base $(b_i - a_i)$ and height H_i (given in equation (6.13)). $\widetilde{I} = \sum_{i=1}^{n} \widetilde{I}_i$ is used to obtain an approximate value of the integral (6.1). As shown in Figure 6.3, the approximation $\widetilde{f}(x)$ may be quite different than the true $f(x)$.

Remarks.

(1) The accuracy of the method is dependent on the size of the subinterval $[a_i, b_i]$. The smaller the subinterval, the better the accuracy of the approximated value of the integral I.

(2) The truncation error in calculating I_i using (6.13) for a subinterval $(b_i - a_i) = h_i$ is of the order $O(h_i^4)$ (proof omitted).

(3) This definition of h_i is different than used in Sections 6.2.1 and 6.2.3.

6.2.3 Simpson's $\frac{3}{8}$ Rule

Consider I_i for a subinterval $[a_i, b_i]$. We divide the subinterval in three equal parts and define the coordinates as:

$$x_1 = a_i \quad ; \quad x_2 = a_i + \left(\frac{b_i - a_i}{3}\right) \quad ; \quad x_3 = a_i + \frac{2(b_i - a_i)}{3} \quad ; \quad x_4 = b_4 \tag{6.14}$$

We calculate $f(x_1)$, $f(x_2)$, $f(x_3)$, and $f(x_4)$ using $f(x)$ in the integrand of I_i.

$$I_i = \int_{a_i}^{b_i} f(x)\, dx \tag{6.15}$$

Using $(x_i, f(x_i))$; $i = 1, 2, \dots, 4$, we construct a cubic interpolating polynomial $\widetilde{f}(x)$ (say, using Lagrange polynomials) that is assumed to approximate $f(x)$ $\forall x \in [a_i, b_i]$.

$$\begin{aligned}\widetilde{f}(x) =& \frac{(x - x_2)(x - x_3)(x - x_4)}{(x_1 - x_2)(x_1 - x_3)(x_1 - x_4)} f(x_1) + \frac{(x - x_1)(x - x_3)(x - x_4)}{(x_2 - x_1)(x_2 - x_3)(x_2 - x_4)} f(x_2) + \\ & \frac{(x - x_1)(x - x_2)(x - x_4)}{(x_3 - x_1)(x_3 - x_2)(x_3 - x_4)} f(x_3) + \frac{(x - x_1)(x - x_2)(x - x_3)}{(x_4 - x_1)(x_4 - x_2)(x_4 - x_3)} f(x_4)\end{aligned} \tag{6.16}$$

We approximate $f(x)$ in (6.15) by $\widetilde{f}(x)$ in (6.16).

$$I_i \approx \widetilde{I}_i = \int_{a_i}^{b_i} \widetilde{f}(x)\, dx \tag{6.17}$$

Substituting for $\widetilde{f}(x)$ in (6.17) and integrating yields:

$$I_i \approx \widetilde{I}_i = \frac{3}{8}h_i(f(x_1)+3f(x_2)+3f(x_3)+f(x_4)) \quad ; \quad h_i = \frac{b_i - a_i}{3} \tag{6.18}$$

This method of approximating I_i by $\widetilde{I}_i$ is called Simpson's $\frac{3}{8}$ rule. We can also write (6.18) as:

$$I_i \approx \widetilde{I}_i = \underbrace{(b_i - a_i)}_{\text{width}} \underbrace{\left(\frac{f(x_1) + 3f(x_2) + 3f(x_3) + f(x_4)}{8} \right)}_{\text{average height}} = (b_i - a_i)H_i \tag{6.19}$$

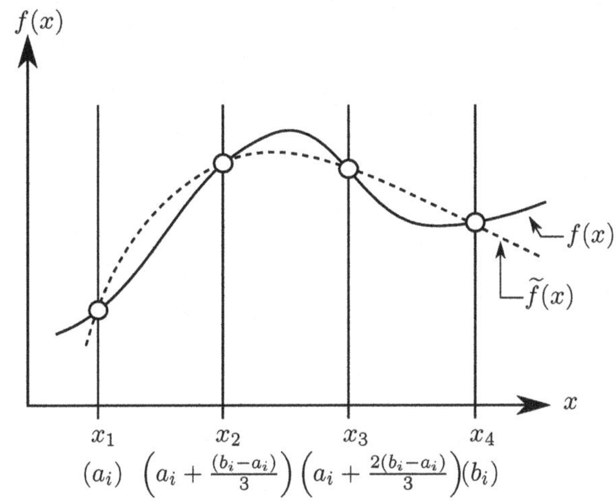

Figure 6.4: $f(x)$ and $\widetilde{f}(x)$, cubic approximation of $f(x)$ for the interval $[a_i, b_i]$

We note that $\widetilde{f}(x)$ can be quite different than $f(x)$. From (6.19), we can interpret Simpson's $\frac{3}{8}$ rule as the area of a rectangle with base $(b_i - a_i)$ and height H_i (given by (6.19)). We calculate $\widetilde{I}_i$ for a subinterval and use $\widetilde{I} = \sum_{i=1}^{n} \widetilde{I}_i$ to obtain approximate value of the integral (6.1).

Remarks.

(1) As in the other methods discussed, here also the accuracy of the method is dependent on the size of the subinterval. The smaller the subinterval, the better the accuracy of the approximated value of the integral I.

(2) It can be shown that the truncation error in calculating I_i using (6.19) is of the order of $O(h^6)$ (proof omitted).

(3) This definition of h_i is different than used in Sections 6.2.1 and 6.2.2.

6.2.4 Newton-Cotes Iteration

In trapezoid rule, Simpson's $\frac{1}{3}$ rule, and Simpson's $\frac{3}{8}$ rule we approximate the integrand $f(x)$ by $\widetilde{f}(x)$, a linear, quadratic, and cubic polynomial (respectively) over a subinterval $[a_i, b_i]$. Based on these approaches, it is possible to construct $\widetilde{f}(x) \forall x \in [a_i, b_i]$ as a higher degree polynomial than three and then proceed with the approximation $\widetilde{I}_i$ of I_i for subinterval $[a_i, b_i]$. These methods or schemes are called Newton-Cotes integration schemes. Details are straightforward and follow what has already been presented.

6.2.4.1 Numerical Examples

In this section we consider a numerical example using Trapezoid rule, Simpson's $\frac{1}{3}$ rule, and Simpson's $\frac{3}{8}$ rule.

Consider the integral

$$I = \int_0^2 (\sin(x))^2 \, e^x \, dx \tag{6.20}$$

In all these three methods we divide the interval $[0, 2]$ into subintervals. In this example we consider uniform subdivision of $[0, 2]$ interval into one, two, four, eight, and sixteen subintervals of widths 2, 1 , 0.5, 0.25, and 0.125. We represent the integral I as sum of the integrals over the subintervals.

$$I = \sum_{i=1}^{n} I_i = \sum_{i=1}^{n} \int_{a_i}^{b_i} (\sin(x))^2 \, e^x \, dx = \sum_{i=1}^{n} \int_{a_i}^{b_i} f(x) \, dx$$

In each of the three methods, we calculate I for each subinterval $[a_i, b_i]$ and sum them to obtain I.

Example 6.1 (Trapezoid Rule).

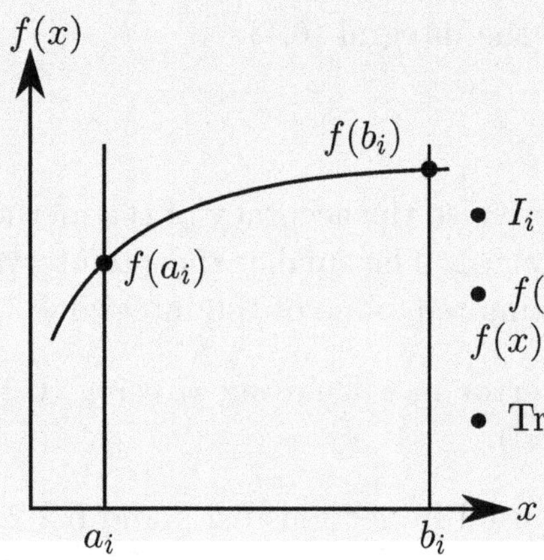

- $I_i \approx \frac{b_i - a_i}{2} \left(f(a_i) + f(b_i) \right)$
- $f(a_i)$ and $f(b_i)$ are calculated using $f(x) = (\sin x)^2 e^x$
- Truncation error $O(h^2)$

Table 6.1: Results of trapezoid rule for (6.20) using one subinterval

subintervals = 1
$b_i - a_i$ = 0.200000E+01

i	a_i	b_i	I_i
1	0.000000E+00	0.200000E+01	0.610943E+01
TOTAL			0.610943E+01

Table 6.2: Results of trapezoid rule for (6.20) using two subintervals

subintervals = 2
$b_i - a_i$ = 0.100000+01

i	a_i	b_i	I_i
1	0.000000E+00	0.100000E+01	0.962371E+00
2	0.100000E+00	0.200000E+01	0.401709E+01
TOTAL			0.497946E+01

Table 6.3: Results of trapezoid rule for (6.20) using four subintervals

subintervals = 4
$b_i - a_i$ = 0.500000+00

i	a_i	b_i	I_i
1	0.000000E+00	0.500000E+00	0.947392E−01
2	0.500000E+00	0.100000E+01	0.575925E+00
3	0.100000E+01	0.150000E+01	0.159600E+01
4	0.150000E+01	0.200000E+01	0.264217E+01
TOTAL			0.490884E+01

Table 6.4: Results of trapezoid rule for (6.20) using eight subintervals

subintervals = 8
$b_i - a_i =$ 0.250000+00

i	a_i	b_i	I_i
1	0.000000E+00	0.250000E+00	0.982419E−02
2	0.250000E+00	0.500000E+00	0.571938E−01
3	0.500000E+00	0.750000E+00	0.170323E+00
4	0.750000E+00	0.100000E+01	0.363546E+00
5	0.100000E+01	0.125000E+01	0.633506E+00
6	0.125000E+01	0.150000E+01	0.950321E+00
7	0.150000E+01	0.175000E+01	0.125388E+01
8	0.175000E+01	0.200000E+01	0.146015E+01
TOTAL			0.489874E+01

Table 6.5: Results of trapezoid rule for (6.20) using 16 subintervals

subintervals = 16
$b_i - a_i =$ 0.250000+00

i	a_i	b_i	I_i
1	0.000000E+00	0.125000E+00	0.110084E−02
2	0.125000E+00	0.250000E+00	0.601293E−02
3	0.250000E+00	0.375000E+00	0.171118E−01
4	0.375000E+00	0.500000E+00	0.358845E−01
5	0.500000E+00	0.625000E+00	0.636581E−01
6	0.625000E+00	0.750000E+00	0.101450E+00
7	0.750000E+00	0.875000E+00	0.149804E+00
8	0.875000E+00	0.100000E+01	0.208623E+00
9	0.100000E+01	0.112500E+01	0.277019E+00
10	0.112500E+01	0.125000E+01	0.353179E+00
11	0.125000E+01	0.137500E+01	0.434293E+00
12	0.137500E+01	0.150000E+01	0.516540E+00
13	0.150000E+01	0.162500E+01	0.595174E+00
14	0.162500E+01	0.175000E+01	0.664705E+00
15	0.175000E+01	0.187500E+01	0.719221E+00
16	0.187500E+01	0.200000E+01	0.752825E+00
TOTAL			0.489660E+01

Example 6.2 (Simpson's $\frac{1}{3}$ Rule).

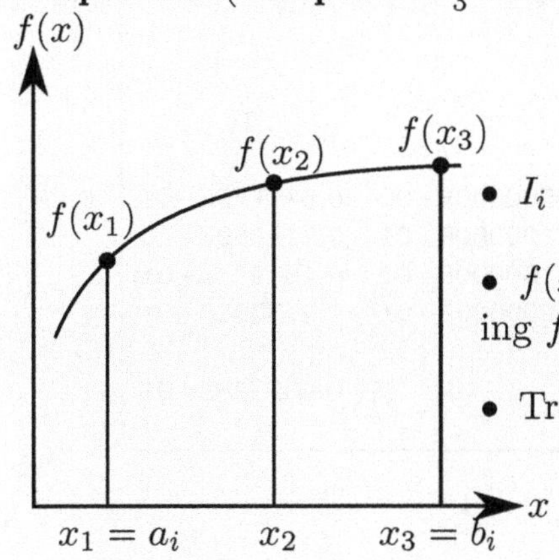

- $I_i \approx \left(\frac{(b_i - a_i)}{6}\right)(f(x_1) + 4f(x_2) + f(x_3))$
- $f(x_1)$, $f(x_2)$, and $f(x_3)$ are calculated using $f(x) = (\sin x)^2 e^x$
- Truncation error $O(h^4)$

Table 6.6: Results of Simpson's $\frac{1}{3}$ rule for (6.20) using one subinterval

subintervals =	1		
$b_i - a_i$ =	0.200000E+01		
i	a_i	b_i	I_i
1	0.000000E+00	0.200000E+01	0.460280E+01
TOTAL			0.460280E+01

Table 6.7: Results of Simpson's $\frac{1}{3}$ rule for (6.20) using two subintervals

subintervals =	2		
$b_i - a_i$ =	0.100000+01		
i	a_i	b_i	I_i
1	0.000000E+00	0.100000E+01	0.573428E+00
2	0.100000E+00	0.200000E+01	0.431187E+01
TOTAL			0.488530E+01

Table 6.8: Results of Simpson's $\frac{1}{3}$ rule for (6.20) using four subintervals

subintervals = 4
$b_i - a_i =$ 0.500000+00

i	a_i	b_i	I_i
1	0.000000E+00	0.500000E+00	0.577776E−01
2	0.500000E+00	0.100000E+01	0.519850E+00
3	0.100000E+01	0.150000E+01	0.157977E+01
4	0.150000E+01	0.200000E+01	0.273798E+01
TOTAL			0.489538E+01

Table 6.9: Results of Simpson's $\frac{1}{3}$ rule for (6.20) using eight subintervals

subintervals = 8
$b_i - a_i =$ 0.250000+00

i	a_i	b_i	I_i
1	0.000000E+00	0.250000E+00	0.621030E−02
2	0.250000E+00	0.500000E+00	0.515971E−01
3	0.500000E+00	0.750000E+00	0.163370E+00
4	0.750000E+00	0.100000E+01	0.356720E+00
5	0.100000E+01	0.125000E+01	0.629096E+00
6	0.125000E+01	0.150000E+01	0.951004E+00
7	0.150000E+01	0.175000E+01	0.126188E+01
8	0.175000E+01	0.200000E+01	0.147601E+01
TOTAL			0.489589E+01

Table 6.10: Results of Simpson's $\frac{1}{3}$ rule for (6.20) using 16 subintervals

subintervals = 16

$b_i - a_i$ = 0.250000+00

i	a_i	b_i	I_i
1	0.000000E+00	0.125000E+00	0.713010E−03
2	0.125000E+00	0.250000E+00	0.549697E−02
3	0.250000E+00	0.375000E+00	0.164699E−01
4	0.375000E+00	0.500000E+00	0.351296E−01
5	0.500000E+00	0.625000E+00	0.628159E−01
6	0.625000E+00	0.750000E+00	0.100560E+00
7	0.750000E+00	0.875000E+00	0.148919E+00
8	0.875000E+00	0.100000E+01	0.207811E+00
9	0.100000E+01	0.112500E+01	0.276356E+00
10	0.112500E+01	0.125000E+01	0.352750E+00
11	0.125000E+01	0.137500E+01	0.434184E+00
12	0.137500E+01	0.150000E+01	0.516830E+00
13	0.150000E+01	0.162500E+01	0.595925E+00
14	0.162500E+01	0.175000E+01	0.665956E+00
15	0.175000E+01	0.187500E+01	0.720973E+00
16	0.187500E+01	0.200000E+01	0.755030E+00
TOTAL			0.489591E+01

Example 6.3 (Simpson's $\frac{3}{8}$ Rule).

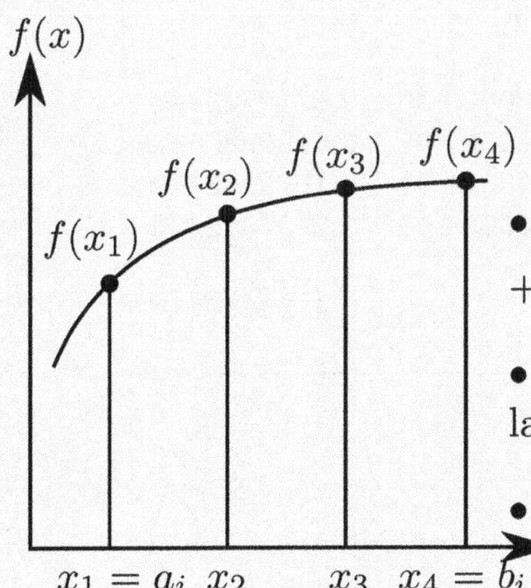

- $I_i \approx \left(\frac{(b_i-a_i)}{8}\right)(f(x_1) + 3f(x_2) + 3f(x_3) + f(x_4))$
- $f(x_1)$, $f(x_2)$, $f(x_3)$, and $f(x_4)$ are calculated using $f(x) = (\sin x)^2 e^x$
- Truncation error $O(h^6)$

$$x_2 = a_i + \left(\frac{b_i - a_i}{3}\right)$$

$$x_3 = a_i + \left(\frac{2(b_i - a_i)}{3}\right)$$

Table 6.11: Results of Simpson's $\frac{3}{8}$ rule for (6.20) using one subinterval

subintervals =	1		
$b_i - a_i$ =	0.200000E+01		
i	a_i	b_i	I_i
1	0.000000E+00	0.200000E+01	0.477375E+01
TOTAL			0.477375E+01

Table 6.12: Results of Simpson's $\frac{3}{8}$ rule for (6.20) using two subintervals

subintervals =	2		
$b_i - a_i$ =	0.100000+01		
i	a_i	b_i	I_i
1	0.000000E+00	0.100000E+01	0.575912E+00
2	0.100000E+00	0.200000E+01	0.431542E+01
TOTAL			0.488530E+01

Table 6.13: Results of Simpson's $\frac{3}{8}$ rule for (6.20) using four subintervals

subintervals =	4		
$b_i - a_i$ =	0.500000+00		
i	a_i	b_i	I_i
1	0.000000E+00	0.500000E+00	0.577952E−01
2	0.500000E+00	0.100000E+01	0.519993E+00
3	0.100000E+01	0.150000E+01	0.157996E+01
4	0.150000E+01	0.200000E+01	0.273793E+01
TOTAL			0.489568E+01

Table 6.14: Results of Simpson's $\frac{3}{8}$ rule for (6.20) using eight subintervals

subintervals = 8
$b_i - a_i$ = 0.250000+00

i	a_i	b_i	I_i
1	0.000000E+00	0.250000E+00	0.621011E−02
2	0.250000E+00	0.500000E+00	0.515985E−01
3	0.500000E+00	0.750000E+00	0.163373E+00
4	0.750000E+00	0.100000E+01	0.356726E+00
5	0.100000E+01	0.125000E+01	0.629102E+00
6	0.125000E+01	0.150000E+01	0.951009E+00
7	0.150000E+01	0.175000E+01	0.126188E+01
8	0.175000E+01	0.200000E+01	0.147601E+01
TOTAL			0.489591E+01

Table 6.15: Results of Simpson's $\frac{3}{8}$ rule for (6.20) using 16 subintervals

subintervals = 16
$b_i - a_i$ = 0.250000+00

i	a_i	b_i	I_i
1	0.000000E+00	0.125000E+00	0.712995E−03
2	0.125000E+00	0.250000E+00	0.549698E−02
3	0.250000E+00	0.375000E+00	0.164699E−01
4	0.375000E+00	0.500000E+00	0.351297E−01
5	0.500000E+00	0.625000E+00	0.628160E−01
6	0.625000E+00	0.750000E+00	0.100560E+00
7	0.750000E+00	0.875000E+00	0.148919E+00
8	0.875000E+00	0.100000E+01	0.207811E+00
9	0.100000E+01	0.112500E+01	0.276357E+00
10	0.112500E+01	0.125000E+01	0.352751E+00
11	0.125000E+01	0.137500E+01	0.434184E+00
12	0.137500E+01	0.150000E+01	0.516830E+00
13	0.150000E+01	0.162500E+01	0.595925E+00
14	0.162500E+01	0.175000E+01	0.665956E+00
15	0.175000E+01	0.187500E+01	0.720973E+00
16	0.187500E+01	0.200000E+01	0.755029E+00
TOTAL			0.489592E+01

Numerical values of the integral I obtained using Trapezoid rule (Example 6.1), Simpson's $\frac{1}{3}$ rule (Example 6.2), and Simpson's $\frac{3}{8}$ rule (Example

6.3) are summarized in Tables 6.1 – 6.15. In these studies all subintervals are of uniform widths, however use of non-uniform width subintervals presents no problem. In this case one needs to be careful to establish a_i and b_i based on subinterval widths for evaluating I_i for the subinterval $[a_i, b_i]$.

In each method, as the number of subintervals are increased, accuracy of the value of the integral improves. For the same number of subintervals Simpson's $\frac{1}{3}$ method produces integral values with better accuracy (error $O(h^4)$) compared to trapezoid rule (error $O(h^2)$), and Simpson's $\frac{3}{8}$ rule (error $O(h^6)$) is more accurate than Simpson's $\frac{1}{3}$ rule. In Simpson's $\frac{3}{8}$ rule the integral values for 8 and 16 subintervals are accurate up to four decimal places.

6.2.5 Richardson's Extrapolation

When we approximate $f(x)$ by $\widetilde{f}(x)$, an algebraic polynomial in a subinterval $[a_i, b_i]$, then the truncation error is $O(h^N)$, where N depends upon the polynomial approximation made for the actual $f(x)$, the integrand.

Thus, if we consider a uniform subdivision of the integration interval $[A, B]$, then for two subinterval sizes h_1 and h_2, where $h_2 < h_1$, we can write:

$$I \approx I_{h_1} + Ch_1^N \tag{6.21}$$

$$I \approx I_{h_2} + Ch_2^N \tag{6.22}$$

where I_{h_1} is the value of the integral I using subinterval size h_1, I_{h_2} is the value of the integral I using subinterval size h_2. N depends upon the polynomial approximation used in approximating actual $f(x)$. Ch_1^N is the error in I_{h_1} and Ch_2^N is the error in I_{h_2}. The expressions (6.21) and (6.22) are based on several assumptions:

(i) The constant C is not the same in (6.21) and (6.22), but we assume it to be.

(ii) Since I_{h_2} is based on $h_2 < h_1$, I_{h_2} is more accurate than I_{h_1} and hence we expect I in (6.22) to have better accuracy than I in (6.21).

First, assuming I to be the same in (6.21) and (6.22), we can solve for C.

$$C \approx \frac{I_{h_1} - I_{h_2}}{h_2^N - h_1^N} \tag{6.23}$$

We substitute C from (6.23) in I in equation (6.22) (as it is based on $h_2 < h_1$, hence more accurate).

$$\therefore \quad I \approx I_{h_2} + \left(\frac{I_{h_1} - I_{h_2}}{h_2^N - h_1^N}\right) h_2^N \tag{6.24}$$

or

$$I \approx I_{h_2} + \frac{I_{h_2} - I_{h_1}}{\left(\frac{h_1}{h_2}\right)^N - 1} = \frac{\left(\frac{h_1}{h_2}\right)^N I_{h_2} - I_{h_1}}{\left(\frac{h_1}{h_2}\right)^N - 1} \tag{6.25}$$

Value of N:

1. In trapezoid rule ; $N = 2$
2. In Simpson's $\frac{1}{3}$ rule ; $N = 4$
3. In Simpson's $\frac{3}{8}$ rule ; $N = 6$ and so on

Remarks.

(1) Use of (6.25) requires I_{h_1} for subinterval width h_1 and I_{h_2} for subinterval width $h_2 < h_1$. The integral value I in (6.25) is an improved approximation of the integral.

(a) When $N = 2$, we have eliminated errors $O(h^2)$ in (6.25)

(b) When $N = 4$, we have eliminated errors $O(h^4)$ in (6.25)

(c) When $N = 6$, we have eliminated errors $O(h^6)$ in (6.25)

(2) Thus, we can view the truncation error in the numerical integration process when the integrand $f(x)$ is approximated by a polynomial in a subinterval to be of the following form, a series in h:

$$E_t = C_1 h^2 + C_2 h^4 + C_3 h^6 + \dots \tag{6.26}$$

where h is the width of the subinterval.

(3) In Richardson's extrapolation if I_{h_1} and I_{h_2} are obtained using trapezoid rule, then $N = 2$ and by using (6.25), we eliminate errors $O(h^2)$.

(4) On the other hand if I_{h_1} and I_{h_2} are obtained using Simpson's $\frac{1}{3}$ rule, then $N = 4$ and by using (6.25), we eliminate errors of the order of $O(h^4)$ and so on.

6.2.6 Romberg Method

Romberg method is based on successive application of Richardson's extrapolation to eliminate errors of various orders of h. Consider the integral

$$I = \int_A^B f(x)\,dx \tag{6.27}$$

Let us consider trapezoid rule and let us calculate numerical values of the integral I using one, two, three, etc. uniform subintervals. Then all these

Table 6.16: Romberg method

Subintervals	Integral Value Trapezoid Rule Error $O(h^2)$	Integral Value Error $O(h^4)$	Integral Value Error $O(h^6)$	Integral Value Error $O(h^8)$
1	I_1			
		I^1_{12}		
2	I_2		$I^2_{12,24}$	
		I^1_{24}		$I^3_{(12,24),(24,48)}$
4	I_4		$I^2_{24,48}$	
		I^1_{48}		
8	I_8			

integral values have truncation error of the order of $O(h^2)$, shown in Table 6.16 below.

We use values of the integral in column two that contain errors $O(h^2)$ in Richardson's extrapolation to eliminate errors $O(h^2)$. In doing so we use

$$I \approx I_{h_2} + \frac{I_{h_2} - I_{h_1}}{\left(\frac{h_1}{h_2}\right)^N - 1} \tag{6.28}$$

in which N is the order ot the leading truncation error (2, 4, 6, 8, ...etc).

We use values in column one with error $O(h^2)$ and (6.28) with $N = 2$ and $h_1/h_2 = 2$ to obtain column three of table 6.16 in which errors $O(h^2)$ are eliminated. Thus, the leading order of the error in the integral values in column three is $O(h^4)$. We use integral values in column three, and (6.28) with $N = 4$, $h_1/h_2 = 2$ to obtain integral values in column four in which leading orders of the error is $O(h^6)$. We use integral values in column four and (6.28) with $N = 6$, $h_1/h_2 = 2$ to obtain the final integral value in column five that contains leading error of the order of $O(h^8)$. We will consider a numerical example to illustrate the details.

Example 6.4 (Romberg Method with Richardson's Extrapolation). In this example we consider repeated use of Richardson's extrapolation in Romberg method to obtain progressively improved numerical values of the integral. Consider

$$I = \int_A^B f(x)\,dx$$

in which

$$A = 0$$
$$B = 0.8$$

and

$$f(x) = 0.2 + 25x - 200x^2 + 675x^3 - 900x^4 + 400x^5$$

Consider trapezoid rule with 1, 2, 4, and 8 subintervals of uniform width for calculating the numerical value of the integrals. These numerical values of the integral contain truncation errors $0(h^2)$. The numerical values of this integral I are listed in Table 6.17 in column two.

Table 6.17: Romberg method example

Subintervals	Integral Value Trapezoid Rule Error $O(h^2)$	Integral Value Error $O(h^4)$	Integral Value Error $O(h^6)$	Integral Value Error $O(h^8)$
1	0.172800			
		1.367467		
2	1.068800		1.640533	
		1.623467		1.640533
4	1.484800		1.640533	
		1.639467		
8	1.600800			

Using the integral values in column two containing truncation errors $O(h^2)$, in Richardson's extrapolation we can eliminate truncation errors of $O(h^2)$ to obtain integral values that contain leading truncation error $O(h^4)$. We use

$$I \approx \frac{\left(\frac{h_1}{h_2}\right)^N I_{h_2} - I_{h_1}}{\left(\frac{h_1}{h_2}\right)^N - 1}$$

with $N = 2$, I_{h_1} and I_{h_2} corresponding to $h_2 < h_1$. Since the subinterval is uniformly reduced to half the previous width we have:

$$\frac{h_1}{h_2} = 2$$

Hence, we can write:

$$I \approx \frac{(2)^2 I_{h_2} - I_{h_1}}{(2)^2 - 1} = \frac{4I_{h_2} - I_{h_1}}{4 - 1}$$

We use this with values of the integral in column 2 to calculate the integral values in column three of Table 6.17, which contain leading truncation error $O(h^4)$. With the integral values in column three we use:

$$I \approx \frac{(2)^4 I_{h_2} - I_{h_1}}{(2)^4 - 1} = \frac{16I_{h_2} - I_{h_1}}{16 - 1}$$

to obtain integral values in column four of table 6.17, which contain leading truncation error $O(h^6)$. Using integral values in column four and

$$I \approx \frac{(2)^6 I_{h_2} - I_{h_1}}{(2)^6 - 1} = \frac{64 I_{h_2} - I_{h_1}}{64 - 1}$$

we obtain the integral value in column five that contains leading truncation error $O(h^8)$. This is the final most accurate value of the integral I based on Romberg method employing Richardson's extrapolation.

Remarks.

(1) Numerical integration methods such as the Newton-Cotes methods discussed in $\mathbb{R}^1$ are difficult to extend to integrals in $\mathbb{R}^2$ and $\mathbb{R}^3$.

(2) Richardson's extrapolation and Romberg method are specifically designed to improve the accuracy of the numerically calculated values of the integrals from trapezoid rule, Simpson's $\frac{1}{3}$ method, Simpson's $\frac{3}{8}$ method, and in general Newton-Cotes methods. Thus, their extensions to $\mathbb{R}^2$ and $\mathbb{R}^3$ are not possible either.

(3) In the next section we present *Gauss quadrature* that overcomes these shortcomings.

6.3 Numerical Integration in $\mathbb{R}^1$ using Gauss Quadrature for $[-1, 1]$

The Gauss quadrature is designed to integrate algebraic polynomials exactly. This method can also be used to integrate functions that are not algebraic polynomials, but in such cases the calculated value of the integral may not be exact. The method is based on indeterministic coefficients. To understand the basic principles of the method, we recall that in trapezoid rule we use:

$$I = \int_a^b f(x)\,dx \approx \left(\frac{a-b}{2}\right)(f(a) + f(b)) \tag{6.29}$$

We can rewrite (6.29) as:

$$I \approx \left(\frac{b-a}{2}\right) f(a) + \left(\frac{b-a}{2}\right) f(b) \tag{6.30}$$

In (6.30) the integrand $f(x)$ is calculated at $x = a$ and $x = b$. The calculated values at $x = a$ and $x = b$ are multiplied with $\left(\frac{b-a}{2}\right)$ and $\left(\frac{b-a}{2}\right)$ and then added to obtain the approximate value of the integral I. We can write a more general form of (6.30).

$$I \approx w_1 f(x_1) + w_2 f(x_2) \tag{6.31}$$

If we choose $w_1 = \left(\frac{b-a}{2}\right)$, $w_2 = \left(\frac{b-a}{2}\right)$, $x_1 = a$, and $x_2 = b$ in (6.31), then we recover (6.30) for trapezoid rule.

Consider (6.31) as integral of $f(x)$ between the limits $[a,b]$ in which we will treat w_1, w_2, x_1, and x_2 as unknowns. Obviously to determine w_1, w_2, x_1, and x_2 we need four conditions. We present the details in the following. To make the derivation of determining w_1, w_2, x_1, and x_2 general so that these would be applicable to any arbitrary integration limits $[a,b]$, we consider the following integral with integration limits $[-1,1]$.

$$I = \int_{-1}^{1} f(\xi)\, d\xi \tag{6.32}$$

6.3.1 Two-Point Gauss Quadrature

Let

$$I \approx w_1 f(\xi_1) + w_2 f(\xi_2) \tag{6.33}$$

in which w_1, w_2, ξ_1, and ξ_2 are yet to be determined. Determination of w_1, w_2, ξ_1, and ξ_2 requires four conditions. We assume that (6.33) integrates a constant, a linear, a quadratic, and a cubic function *exactly*, i.e., when $f(\xi) = 1$, $f(\xi) = \xi$, $f(\xi) = \xi^2$, and $f(\xi) = \xi^3$, (6.33) gives exact values of their integrals in the interval $[-1,1]$.

$$\begin{aligned}
&\text{when } f(\xi) = 1 \quad ; \quad w_1 + w_2 &&= \int_{-1}^{1} 1\, d\xi &&= 2\\
&\text{when } f(\xi) = \xi \quad ; \quad w_1\xi_1 + w_2\xi_2 &&= \int_{-1}^{1} \xi\, d\xi &&= 0\\
&\text{when } f(\xi) = \xi^2 \quad ; \quad w_1\xi_1^2 + w_2\xi_2^2 &&= \int_{-1}^{1} \xi^2\, d\xi &&= \tfrac{2}{3}\\
&\text{when } f(\xi) = \xi^3 \quad ; \quad w_1\xi_1^3 + w_2\xi_2^3 &&= \int_{-1}^{1} \xi^3\, d\xi &&= 0
\end{aligned} \tag{6.34}$$

Thus we have four equations in four unknowns: w_1, w_2, ξ_1, and ξ_2.

$$\begin{aligned}
w_1 + w_2 &= 2\\
w_1\xi_1 + w_2\xi_2 &= 0\\
w_1\xi_1^2 + w_2\xi_2^2 &= \frac{2}{3}\\
w_1\xi_1^3 + w_2\xi_2^3 &= 0
\end{aligned} \tag{6.35}$$

Equations (6.35) are a system of nonlinear equations in w_1, w_2, ξ_1, and ξ_2. Their solution gives:

$$
\begin{aligned}
w_1 &= 1 \quad ; \quad \xi_1 = -\frac{1}{\sqrt{3}} \\
w_2 &= 1 \quad ; \quad \xi_2 = \frac{1}{\sqrt{3}}
\end{aligned}
\tag{6.36}
$$

w_1 and w_2 are called weight factors and ξ_1, ξ_2 are called sampling points or quadrature points.

Thus for integrating $f(\xi)$ in (6.32) using (6.33) with (6.36), we have a two-point integration scheme referred to as two-point Gauss quadrature that integrates an algebraic polynomial of up to degree three exactly. We note two-point Gauss quadrature is the minimum quadrature rule for algebraic polynomials of up to degree three.

Thus, in summary, to integrate

$$
I = \int_{-1}^{1} f(\xi)\, d\xi \tag{6.37}
$$

using a two-point Gauss quadrature we write

$$
I = \sum_{i=1}^{2} w_i f(\xi_i) \tag{6.38}
$$

in which w_i ; $i = 1, 2$ are weight factors and ξ_i ; $i = 1, 2$ are the sampling or quadrature points given by (6.37).

Remarks.

(1) w_i ; $i = 1, 2$ and ξ_i ; $i = 1, 2$ are derived using the fact that 1, ξ, ξ^2, and ξ^3 are integrated exactly. Therefore, given a cubic algebraic polynomial in ξ:

$$
f(\xi) = C_0 + C_1\xi + C_2\xi^2 + C_3\xi^3 \quad ; \quad C_0, C_1, C_2, C_3 : \text{Constants} \tag{6.39}
$$

the two-point quadrature rule (6.38) would integrate $f(\xi)$ in (6.39) exactly. That is, a two-point Gauss quadrature integrates up to a cubic algebraic polynomial exactly and is the minimum quadrature rule.

6.3.2 Three-Point Gauss Quadrature

Consider:

$$
I = \int_{-1}^{1} f(\xi)\, d\xi \tag{6.40}
$$

In three-point Gauss quadrature we have three weight factors w_1, w_2, w_3 and three sampling or quadrature points ξ_1, ξ_2, ξ_3, and we write:

$$I = w_1 f(\xi_1) + w_2 f(\xi_2) + w_3 f(\xi_3) = \sum_{i=1}^{3} w_i f(\xi_i) \tag{6.41}$$

In this case (6.41) requires determination of w_1, w_2, w_3, ξ_1, ξ_2, and ξ_3, hence we need six conditions.

Let (6.41) integrate $f(\xi) = 1$, $f(\xi) = \xi^i$; $i = 1, 2, \dots, 5$ exactly, then using (6.41) and (6.40) we can write:

$$\begin{aligned}
&\text{when } f(\xi) = 1 \quad &; \quad & w_1 + w_2 + w_3 &= \int_{-1}^{1} 1 \, d\xi &= 2 \\
&\text{when } f(\xi) = \xi \quad &; \quad & w_1\xi_1 + w_2\xi_2 + w_3\xi_3 &= \int_{-1}^{1} \xi \, d\xi &= 0 \\
&\text{when } f(\xi) = \xi^2 \quad &; \quad & w_1\xi_1^2 + w_2\xi_2^2 + w_3\xi_3^2 &= \int_{-1}^{1} \xi^2 \, d\xi &= \tfrac{2}{3} \\
&\text{when } f(\xi) = \xi^3 \quad &; \quad & w_1\xi_1^3 + w_2\xi_2^3 + w_3\xi_3^3 &= \int_{-1}^{1} \xi^3 \, d\xi &= 0 \\
&\text{when } f(\xi) = \xi^4 \quad &; \quad & w_1\xi_1^4 + w_2\xi_2^4 + w_3\xi_3^4 &= \int_{-1}^{1} \xi^4 \, d\xi &= \tfrac{2}{5} \\
&\text{when } f(\xi) = \xi^5 \quad &; \quad & w_1\xi_1^5 + w_2\xi_2^5 + w_3\xi_3^5 &= \int_{-1}^{1} \xi^5 \, d\xi &= 0
\end{aligned} \tag{6.42}$$

Equations (6.42) are six simultaneous nonlinear algebraic equations in six unknowns: w_1, w_2, w_3, ξ_1, ξ_2, ξ_3. The solution of these six equations gives weight factors w_i; $i = 1, 2, 3$ and the locations of the quadrature points ξ_i; $i = 1, 2, 3$.

$$\begin{aligned}
w_1 &= 0.5555556 \quad ; \quad & \xi_1 &= -0.774596669 \\
w_2 &= 0.8888889 \quad ; \quad & \xi_2 &= 0.0 \\
w_3 &= 0.5555556 \quad ; \quad & \xi_3 &= 0.774596669
\end{aligned} \tag{6.43}$$

This is a three-point Gauss quadrature.

In summary, to evaluate

$$I = \int_{-1}^{1} f(\xi) \, d\xi \tag{6.44}$$

using a three-point Gauss quadrature, we write

$$I = \sum_{i=1}^{3} w_i f(\xi_i) \tag{6.45}$$

in which (w_i, ξ_i) ; $i = 1, 2, 3$ are given by (6.43).

Remarks.

(1) w_i, ξ_i ; $i = 1, 2, 3$ are derived using the fact that 1, ξ, ξ^2, ξ^3, ξ^4, and ξ^5 are integrated exactly. Thus, given:

$$f(\xi) = C_0 + C_1\xi + C_2\xi^2 + C_3\xi^3 + C_4\xi^4 + C_5\xi^5 \tag{6.46}$$

where C_i; $i = 1, 2, \dots, 5$ are constants, a fifth degree polynomial in ξ, then the three-point quadrature rule will integrate it exactly. Three-point Gauss quadrature is the minimum for algebraic polynomials up to fifth degree.

(2) From remark (1) it is clear that if we had used three-point rule for a polynomial ξ of degree one or two or three or four (less than five), the three-point Gauss quadrature will integrate them exactly also.

6.3.3 n-Point Gauss Quadrature

Consider the following integral.

$$I = \int_{-1}^{1} f(\xi)\, d\xi \tag{6.47}$$

Let N, the degree of the polynomial $f(\xi)$, be such that the n-point quadrature rule integrates it exactly. Then, we can write:

$$I = \sum_{i=1}^{n} w_i f(\xi_i) \tag{6.48}$$

where w_i; $i = 1, 2, \dots, n$ are weight factors and ξ_i ; $i = 1, 2, \dots, n$ are the sampling or quadrature points. Clearly N and n are related for the integral value to be exact.

Remarks.

(1) Since two- and three-point Gauss quadratures integrate polynomials of up to degree 3 and 5 exactly, we conclude that n-point quadrature must integrate an algebraic polynomial of up to degree $(2n-1)$ exactly (proof omitted). Thus, we have the following rule for determining minimum number of quadrature points n for an algebraic polynomial of maximum degree N.

$$\begin{aligned} N &= 2n - 1 \\ \text{or } n &= \frac{N+1}{2}\text{, round up to the next integer} \end{aligned} \tag{6.49}$$

Knowing the highest degree of the polynomial $f(\xi)$ in ξ, i.e., knowing N, we can determine n, the minimum number of quadrature points needed integrate it exactly.

(2) Values of w_i and ξ_i for various values of n are generally tabulated (see Table 6.18). Since the locations of the quadrature points in the interval $[-1,1]$ are symmetric about $\xi = 0$, the values of ξ_i in the table are listed only in the itnerval $[0,1]$. For example, for $n = 3$ we have $-\xi_1$, ξ_2 $(=0)$, and ξ_1, thus only ξ_1 and ξ_2 need to be listed as given in Table 6.18. For $n = 4$, we have $-\xi_1$, $-\xi_2$, ξ_2, and ξ_1, thus only ξ_1 and ξ_2 ned to be listed in Table 6.18. The weight factors for $\pm\xi_i$ are the same, i.e., w_i applies to $+\xi_i$ as well as $-\xi_i$. The values of w_i and ξ_i are listed up to fifteen decimal places.

6.3.4 Using Gauss Quadrature in $\mathbb{R}^1$ with $[-1,1]$ Limits for Integrating Algebraic Polynomials and Other Functions

Consider:

$$I = \int_{-1}^{1} f(\xi)\, d\xi \tag{6.50}$$

(a) When $f(\xi)$ is an algebraic polynomial in ξ:

(i) Determine the highest degree N of the polynomial $f(\xi)$.

(ii) Use $n = \frac{N+1}{2}$ (round to the next highest integer) to find out the minimum number of quadrature points n.

(iii) Use Table 6.18 to determine the weight factors and the locations of the quadrature points.

$$(w_i, \xi_i) \quad ; \quad i = 1, 2, \dots, n$$

(iv) Then

$$I = \sum_{i=1}^{n} w_i f(\xi_i) \tag{6.51}$$

is the exact value of the integral (6.50).

(v) If we use a higher number of quadrature points than n, then obviously the accuracy of I does not deteriorate or improve.

(vi) Obviously the choice of n lower than $n = \frac{N+1}{2}$ will not integrate $f(\xi)$ exactly.

(b) When $f(\xi)$ is not an algebraic polynomial in ξ:

Table 6.18: Sampling points and weight factors for Gauss quadrature for integration limits $[-1,1]$

$$I = \int_{-1}^{+1} f(x)\,dx = \sum_{i=1}^{n} W_i\, f(x_i)$$

$\pm x_i$			W_i		
		$n=1$			
0			2.00000	00000	00000
		$n=2$			
0.57735	02691	89626	1.00000	00000	00000
		$n=3$			
0.77459	66692	41483	0.55555	55555	55556
0.00000	00000	00000	0.88888	88888	88889
		$n=4$			
0.86113	63115	94053	0.34785	48451	37454
0.33998	10435	84856	0.65214	51548	62546
		$n=5$			
0.90617	98459	38664	0.23692	68850	56189
0.53846	93101	05683	0.47862	86704	99366
0.00000	00000	00000	0.56888	88888	88889
		$n=6$			
0.93246	95142	03152	0.17132	44923	79170
0.66120	93864	66265	0.36076	15730	48139
0.23861	91860	83197	0.46791	39345	72691
		$n=7$			
0.94910	79123	42759	0.12948	49661	68870
0.74153	11855	99394	0.27970	53914	89277
0.40584	51513	77397	0.38183	00505	05119
0.00000	00000	00000	0.41795	91836	73469
		$n=8$			
0.96028	98564	97536	0.10122	85362	90376
0.79666	64774	13627	0.22238	10344	53374
0.52553	24099	16329	0.31370	66458	77887
0.18343	46424	95650	0.36268	37833	78362
		$n=9$			
0.96816	02395	07626	0.08127	43883	61574
0.83603	11073	26636	0.18064	81606	94857
0.61337	14327	00590	0.26061	06964	02935
0.32425	34234	03809	0.31234	70770	40003
0.00000	00000	00000	0.33023	93550	01260
		$n=10$			
0.97390	65285	17172	0.06667	13443	08688
0.86506	33666	88985	0.14945	13491	50581
0.67940	95682	99024	0.21908	63625	15982
0.43339	53941	29247	0.26926	67193	09996
0.14887	43389	81631	0.29552	42247	14753

(i) If $f(\xi)$ is not an algebraic polynomial in ξ, we can still use Gauss quadrature to integrate it.

(ii) In this case determination of minimum required n is not possible. We can begin with lowest possible value of n and progressively increase it by one.

(iii) The integral values calculated for progressively increasing n are progressively better approximations of the integral of $f(\xi)$. When a desired decimal place accuracy is achieved we can stop the integration process.

(iv) Thus, when $f(\xi)$ is not an algebraic polynomial, Gauss quadrature can not integrate $f(\xi)$ exactly, but we do have a mechanism of obtaining an integral value of $f(\xi)$ within any desired accuracy by progressively increasing n.

6.3.5 Gauss Quadrature in $\mathbb{R}^1$ for Arbitrary Integration Limits

Consider the integral:

$$I = \int_a^b f(x)\,dx \tag{6.52}$$

When we compare (6.52) with:

$$\underset{\sim}{I} = \int_{-1}^{1} f(\xi)\,d\xi \tag{6.53}$$

we find that integration variables are x and ξ in (6.52) and (6.53), but that makes no difference. Secondly, the limits of integration in (6.52) (the integral we want to evaluate) are $[a,b]$, whereas in (6.53) they are $[-1,1]$. By performing a change of variable from ξ to x in (6.53), we obtain (6.52). We proceed as follows

(i) Determine the highest degree of the polynomial $f(x)$ in (6.52), say N, then the minimum number of quadrature points n are determined using $n = \frac{N+1}{2}$ (round to the next highest integer).

(ii) From Table 6.18 determine w_i ; $i = 1,2,\dots,n$ and ξ_i ; $i = 1,2,\dots,n$ for the integration interval $[-1,1]$.

(iii) Transform (w_i,ξ_i) ; $i = 1,2,\dots,n$ for the integration interval $[a,b]$ in (6.52) using:

$$x_i = \frac{a+b}{2} + \left(\frac{b-a}{2}\right)\xi_i \; ; \quad w_i^x = \left(\frac{b-a}{2}\right) w_i \; ; \qquad i = 1,2,\dots,n \tag{6.54}$$

(iv) Now using the weight factors w_i^x ; $i = 1, 2, \dots, n$ and quadrature points x_i ; $i = 1, 2, \dots, n$ for the integration interval $[a, b]$ we can integrate $f(x)$ in (6.52).

$$I = \sum_{i=1}^{n} w_i^x f(x_i) \tag{6.55}$$

6.4 Gauss Quadrature in $\mathbb{R}^2$

6.4.1 Gauss Quadrature in $\mathbb{R}^2$ over $\bar{\Omega} = [-1, 1] \times [-1, 1]$

The basic principles of Gauss quadrature in $\mathbb{R}^1$ can be extended for numerical quadrature in $\mathbb{R}^2$. First, let us consider integration of $f(\xi, \eta)$, a polynomial in ξ and η, over a square domain $\Omega = [-1, 1] \times [-1, 1]$.

$$I = \int_{-1}^{1}\int_{-1}^{1} f(\xi, \eta)\, d\xi\, d\eta \tag{6.56}$$

We can rewrite (6.56) as:

$$I = \int_{-1}^{1}\left(\int_{-1}^{1} f(\xi, \eta)\, d\xi\right) d\eta \tag{6.57}$$

If N^ξ and N^η are the highest degrees of the polynomial $f(\xi, \eta)$ in ξ and η, then the minimum number of quadrature points n^ξ and n^η in ξ and η can be determined using:

$$n^\xi = \frac{N^\xi + 1}{2} \; ; \quad n^\eta = \frac{N^\eta + 1}{2} \quad \text{round to the next higher integers} \tag{6.58}$$

Using Table 6.18, determine (w_i^ξ, ξ_i); $i = 1, 2, \dots, n^\xi$ and (w_j^η, n_j); $j = 1, 2, \dots, n^\eta$ in the ξ- and η-directions. Using (6.57), first integrate with respect to ξ using Gauss quadrature, holding η constant. This gives:

$$I = \int_{-1}^{1}\left(\sum_{i=1}^{n^\xi} w_i^\xi f(\xi_i, \eta)\right) d\eta \tag{6.59}$$

Now integrate with respect to η using (6.59).

$$I = \sum_{j=1}^{n^\eta} w_j^\eta \left(\sum_{i=1}^{n^\xi} w_i^\xi f(\xi_i, \eta_j)\right) \tag{6.60}$$

This is the exact numerical value of the integral (6.56) using Gauss quadrature.

6.4.2 Gauss Quadrature in $\mathbb{R}^2$ Over Arbitrary Rectangular Domains $\bar{\Omega} = [a,b] \times [c,d]$

The Gauss quadrature in $\mathbb{R}^1$ over arbitrary $\Omega = [a,b]$ can be easily extended for $\mathbb{R}^2$ over arbitrary rectangular domain $\Omega = [a,b] \times [c,d]$. Consider:

$$I = \int_c^d \int_a^b f(x,y)\,dx\,dy = \int_c^d \left(\int_a^b f(x,y)\,dx \right) dy \tag{6.61}$$

Determine highest degrees of the polynomial $f(x,y)$ in x and y, say N^x and N^y, then the minimum number of quadrature points in x and y are determined using:

$$n^x = \frac{N^x+1}{2}\ ; \quad n^y = \frac{N^y+1}{2} \quad \text{round to the next higher integers} \tag{6.62}$$

Determine (w_i^ξ, ξ_i); $i = 1,2,\dots,n^x$ and (w_j^η, η_j); $j = 1,2,\dots,n^y$ using Table 6.18 for the interval $[-1,1]$ in ξ and η. Transform (w_i^ξ, ξ_i); $i = 1,2,\dots,n^x$ to (w_i^x, x_i); $i = 1,2,\dots,n^x$ and (w_j^η, η_j); $j = 1,2,\dots,n^y$ to (w_j^y, y_j) ; $j = 1,2,\dots,n^y$ for the integration intervals $[a,b]$ and $[c,d]$ in x and y using:

$$\begin{aligned} x_i &= \frac{a+b}{2} + \left(\frac{b-a}{2}\right)\xi_i\ ; \quad w_i^x = \left(\frac{b-a}{2}\right) w_i^\xi\ ; \qquad i = 1,2,\dots,n^x \\ y_j &= \frac{c+d}{2} + \left(\frac{d-c}{2}\right)\eta_j\ ; \quad w_j^y = \left(\frac{d-c}{2}\right) w_j^\eta\ ; \qquad j = 1,2,\dots,n^y \end{aligned} \tag{6.63}$$

Now using (6.63) in (6.61), first we integrate with respect to x using Gauss quadrature holding y constant.

$$I = \int_c^d \left(\sum_{i=1}^{n^x} w_i^x f(x_i, y) \right) dy \tag{6.64}$$

Now we integrate (6.64) with respect to y using Gauss quadrature.

$$I = \sum_{j=1}^{n^y} w_j^y \left(\sum_{i=1}^{n^x} w_i^x f(x_i, y_j) \right) \tag{6.65}$$

This is the exact numerical value of the integral (6.61) obtained using Gauss quadrature.

6.5 Gauss Quadrature in $\mathbb{R}^3$

6.5.1 Gauss Quadrature in $\mathbb{R}^3$ over $\bar{\Omega} = [-1,1] \times [-1,1] \times [-1,1]$

Consider:

$$I = \int_{-1}^{1}\int_{-1}^{1}\int_{-1}^{1} f(\xi,\eta,\zeta)\,d\xi\,d\eta\,d\zeta \tag{6.66}$$

or

$$I = \int_{-1}^{1}\left(\int_{-1}^{1}\left(\int_{-1}^{1} f(\xi,\eta,\zeta)\,d\xi\right)d\eta\right)d\zeta \tag{6.67}$$

If N^ξ, N^η, and N^ζ are the highest degrees of the polynomial $f(\xi,\eta,\zeta)$ in ξ, η, and ζ, then n^ξ, n^η, and n^ζ, the minimum number of quadrature points in ξ, η, and ζ, are determined using:

$$n^\xi = \frac{N^\xi+1}{2}, \quad n^\eta = \frac{N^\eta+1}{2}, \quad n^\zeta = \frac{N^\zeta+1}{2} \tag{6.68}$$
$$\text{round to the next higher integers}$$

Determine (w_i^ξ,ξ_i); $i = 1,2,\dots,n^\xi$, (w_j^η,η_j); $j = 1,2,\dots,n^\eta$, and (w_k^ζ,ζ_k); $k = 1,2,\dots,n^\zeta$ using Table 6.18.

Using (6.67), first integrate with respect to ξ using Gauss quadrature, holding η and ζ constant.

$$I = \int_{-1}^{1}\left(\int_{-1}^{1}\left(\sum_{i=1}^{n^\xi} w_i^\xi f(\xi_i,\eta,\zeta)\right)d\eta\right)d\zeta \tag{6.69}$$

Now integrate with respect to η using (6.69) holding ζ constant.

$$I = \int_{-1}^{1}\left(\sum_{j=1}^{n^\eta} w_j^\eta\left(\sum_{i=1}^{n^\xi} f(\xi_i,\eta_j,\zeta)\right)\right)d\zeta \tag{6.70}$$

Lastly, integrate with respect to ζ using (6.70).

$$I = \sum_{k=1}^{n^\zeta} w_k^\zeta\left(\sum_{j=1}^{n^\eta} w_j^\eta\left(\sum_{i=1}^{n^\xi} w_i^\xi f(\xi_i,\eta_j,\zeta_k)\right)\right) \tag{6.71}$$

This is the exact numerical value of the integral using Gauss quadrature.

6.5.2 Gauss Quadrature in $\mathbb{R}^3$ Over Arbitrary Prismatic Domains $\Omega = [a,b] \times [c,d] \times [e,f]$

Consider the following integral.

$$I = \int_e^f \int_c^d \int_a^b f(x,y,z)\,dx\,dy\,dz \tag{6.72}$$

or

$$I = \int_e^f \left(\int_c^d \left(\int_a^b f(x,y,z)\,dx \right) dy \right) dz \tag{6.73}$$

Let N^x, N^y, and N^z be the highest degrees of the polynomial $f(x,y,z)$ in x, y, and z, then n^x, n^y, and n^z, the minimum number of quadrature points in x, y, and z, are determined using:

$$n^x = \frac{N^x+1}{2}, \quad n^y = \frac{N^y+1}{2}, \quad n^z = \frac{N^z+1}{2} \quad \text{round up to the next higher integer} \tag{6.74}$$

Determine (w_i^ξ, ξ_i); $i = 1, 2, \dots, n^x$, (w_j^η, η_j); $j = 1, 2, \dots, n^y$, and (w_k^ζ, ζ_k); $k = 1, 2, \dots, n^z$ for $[-1, 1]$ interval in ξ, η, and ζ using Table 6.18. Transform (w_i^ξ, ξ_i), (w_j^η, η_j), and (w_k^ζ, ζ_k) to (w_i^x, x_i), (w_j^y, y_j), and (w_k^z, z_k) using the following.

$$\begin{aligned}
x_i &= \frac{a+b}{2} + \left(\frac{b-a}{2}\right)\xi_i\,; \quad w_i^x = \left(\frac{b-a}{2}\right) w_i^\xi\,; \quad && i = 1, 2, \dots, n^x \\
y_j &= \frac{c+d}{2} + \left(\frac{d-c}{2}\right)\eta_j\,; \quad w_j^y = \left(\frac{d-d}{2}\right) w_j^\eta\,; \quad && j = 1, 2, \dots, n^y \\
z_k &= \frac{e+f}{2} + \left(\frac{f-e}{2}\right)\zeta_k\,; \quad w_k^z = \left(\frac{f-e}{2}\right) w_k^\zeta\,; \quad && k = 1, 2, \dots, n^z
\end{aligned} \tag{6.75}$$

Now using (6.73) and (6.75) we can integrate $f(x,y,z)$ with respect to x, y, and z using Gauss quadrature.

$$I = \sum_{k=1}^{n^z} w_k^z \left(\sum_{j=1}^{n^y} w_j^y \left(\sum_{i=1}^{n^x} w_i^x f(x_i, y_j, z_k) \right) \right) \tag{6.76}$$

This is the exact value of the integral using Gauss quadrature.

Remarks.

(1) When $f(x)$, $f(x, y)$, or $f(x, y, z)$ are algebraic polynomials in x ; x, y ; or x, y, z we can determine the minimum number of quadrature points required to integrate them exactly.

(2) If the integrand is not an algebraic polynomial in any one or more of the variables, then we must proceed with the minimum number of quadrature points in those variables and progressively increase the number of quadrature points until the desired accuracy is achieved.

(3) The Gauss quadratures discussed in $\mathbb{R}^2$ and $\mathbb{R}^3$ only hold for rectangular and prismatic domains but of arbitrary size.

6.5.3 Numerical Examples

In this section we consider numerical examples for Gauss quadrature in $\mathbb{R}^1$ and $\mathbb{R}^2$ for integration intervals $[-1, 1]$, $[-1, 1]\times[-1, 1]$ as well as arbitrary integration intervals $[a, b]$, $[a, b] \times [c, d]$.

Example 6.5 (Gauss Quadrature in $\mathbb{R}^1$: Integration Interval $[-1, 1]$). Consider:

$$I = \int_{-1}^{1} (1 - 0.1x^2 + x^3)\, dx = \int_{-1}^{1} f(x)\, dx$$

The highest degree of the polynomial in the integrand is three ($N = 3$) hence the minimum number of quadrature points n are given by:

$$n = \frac{N+1}{2} = \frac{3+1}{2} = 2$$

From Table 6.18:

$$\begin{aligned} x_1 &= -0.5773502590 \quad ; \quad w_1 = 1.0 \\ x_2 &= 0.5773502590 \quad ; \quad w_2 = 1.0 \end{aligned}$$

$$\therefore \quad I = \sum_{i=1}^{2} w_i f(x_i) = w_1 f(x_1) + w_2 f(x_2)$$

or

$$\begin{aligned} I = (1)(1.0 - 0.1(-0.5773502590)^2 + (-0.5773502590)^3) \\ + (1)(1.0 - 0.1(0.5773502590)^2 + (0.5773502590)^3) \end{aligned}$$

or

$$I = 1.9333334000$$

This value agrees with the theoretical value of I up to six decimal places.

We could check that if we use $n = 3$ (one order higher than minimum quadrature rule) the value of the integral remains unaffected up to six decimal places. Details are given in the following.

$$I = \sum_{i=1}^{3} w_i f(x_i)$$

From Table 6.18, for $n = 3$, we have:

$$\begin{aligned} x_1 &= -0.7745966910 \quad ; \quad w_1 = 0.5555555820 \\ x_2 &= 0.0 \quad ; \quad w_2 = 0.8888888960 \\ x_3 &= 0.7745966910 \quad ; \quad w_3 = 0.5555555820 \end{aligned}$$

Using these values of w_i, x_i; $i = 1, 2, 3$ and $I = \sum\limits_{i=1}^{3} w_i f(x_i)$, we obtain:

$$I = 1.9333334000$$

which agrees with the integral value calculated using $n = 2$ up to all computed decimal places. Thus, using $n = 3$, the integral value neither improved nor deteriorated.

Example 6.6 (Gauss Quadrature in $\mathbb{R}^1$: Arbitrary Integration Interval). Consider the integral:

$$I = \int_{1.5}^{3.7} (1 - 0.1x^2 + x^3)\, dx$$

The integrand in this case is the same as in Example 6.5, but the limits of integration are [1.5,3.7] as opposed to $[-1, 1]$ in Example 6.5. Thus, in this case also $N = 3$ and $n = 2$ (minimum), and we have the following from Table 6.18 for the integration interval $[-1, 1]$.

$$\begin{aligned} \xi_1 &= -0.5773502590 \quad ; \quad w_1^\xi = 1.0 \\ \xi_2 &= 0.5773502590 \quad ; \quad w_2^\xi = 1.0 \end{aligned}$$

We transform (w_i^ξ, ξ_i); $i = 1, 2$ for the interval $[1.5, 3.7]$ using:

$$\left.\begin{aligned} x_i &= \left(\frac{1.5 + 3.7}{2}\right) + \left(\frac{3.7 + 1.5}{2}\right)\xi_i \\ w_i^x &= \left(\frac{3.7 - 1.5}{2}\right) \end{aligned}\right\} \quad ; \quad i = 1, 2$$

This gives us:

$$x_1 = 1.964914680 \quad ; \quad w_1^x = 1.100000020$$
$$x_2 = 3.235085250 \quad ; \quad w_2^x = 1.100000020$$

$$\therefore \quad I = \sum_{i=1}^{2} w_i^x f(x_i)$$

or

$$\begin{aligned} I &= (1.100000020)\big(1 - 0.1(1.964914680)^2 + (1.964914680)^3\big) \\ &\quad + (1.00000020)\big(1 - 0.1(3.235085250)^2 + (3.235085250)^3\big) \\ \text{or} \quad I &= 46.212463400 \end{aligned}$$

This value is accurate up to six decimal places when compared with the theoretical value of I.

As in Example 6.5, here also if we use $n = 3$ instead of $n = 2$ (minimum) and repeat the integration process, we find that I remains unaffected.

Example 6.7 (Gauss Quadrature in $\mathbb{R}^2$: Integration Interval $[-1,1] \times [-1,1]$).

Consider the following integral.

$$I = \int_{-1}^{1}\int_{-1}^{1} (xy + x^2y^2)\, dx\, dy = \int_{-1}^{1}\int_{-1}^{1} f(x,y)\, dx\, dy \tag{6.77}$$

We can rewrite this as:

$$I = \int_{-1}^{1} \left(\int_{-1}^{1} (xy + x^2y^2)\, dx \right) dy \tag{6.78}$$

The highest degrees of the polynomial in the integrand in x and y are $N^x = 2$, $N^y = 2$, hence the minimum number of quadrature points n^x and n^y in x and y are:

$$\begin{aligned} n^x &= \frac{N^x + 1}{2} = \frac{2+1}{2} = \frac{3}{2}\,; \qquad n^x = 2 \\ n^y &= \frac{N^y + 1}{2} = \frac{2+1}{2} = \frac{3}{2}\,; \qquad n^y = 2 \end{aligned} \tag{6.79}$$

Determine the quadrature point location and weight factors in x and y using Table 6.18 for the interval $[-1, 1]$.

$$\begin{aligned} x_1 &= -0.5773502590 \quad ; \quad w_1^x = 1.0 \\ x_2 &= 0.5773502590 \quad ; \quad w_2^x = 1.0 \end{aligned} \tag{6.80}$$

$$\begin{aligned} y_1 &= -0.5773502590 \quad ; \quad w_1^y = 1.0 \\ y_2 &= 0.5773502590 \quad ; \quad w_2^y = 1.0 \end{aligned} \tag{6.81}$$

Using (6.78) and (6.80) we integrate with respect to x holding y constant.

$$I = \int_{-1}^{1} \left(w_1^x (x_1 y + x_1^2 y^2) + w_2^x (x_2 y + x_2^2 y^2) \right) dy \tag{6.82}$$

Now using (6.82) and (6.81) integrate with respect to y.

$$\begin{aligned} I = & w_1^y \left(w_1^x (x_1 y_1 + x_1^2 y_1^2) + w_2^x (x_2 y_1 + x_2^2 y_1^2) \right) + \\ & w_2^y \left(w_1^x (x_1 y_2 + x_1^2 y_2^2) + w_2^x (x_2 y_2 + x_2^2 y_2^2) \right) \end{aligned} \tag{6.83}$$

Substituting numerical values of w_i^x, x_i; $i = 1, 2$ and w_j^y, y_j; $j = 1, 2$ from (6.80) and (6.81) in (6.83), we obtain:

$$I = 0.4444443880$$

This value agrees with theoretical values up to at least seven decimal places.

It can be verified that using Gauss quadrature higher than $(n^x \times n^y) = (2 \times 2)$, the value I of the integral remains unaffected.

Example 6.8 (Gauss Quadrature in $\mathbb{R}^2$: Arbitrary Rectangular Domain). Consider the following integral:

$$I = \int_{0.31}^{1.13} \int_{0.11}^{1.05} (xy + x^2 y^2) \, dx \, dy \tag{6.84}$$

We rewrite (6.84) as:

$$I = \int_{0.31}^{1.13} \left(\int_{0.11}^{1.05} (xy + x^2 y^2) \, dx \right) dy \tag{6.85}$$

In this example the integrand is the same as in Example 6.7, but the limits of integration are not $[-1,1]$ in x and y as in Example 6.7. In this case also $N^x = 2$, $N^y = 2$, hence:

$$n^x = \frac{N^x+1}{2} = \frac{2+1}{2} = \frac{3}{2}\,; \qquad n^x = 2$$
$$n^y = \frac{N^y+1}{2} = \frac{2+1}{2} = \frac{3}{2}\,; \qquad n^y = 2$$

Determine the quadrature points and the weight function factors in ξ and η for the integration interval $[-1,1]$ using Table 6.18.

$$\begin{aligned} \xi_1 &= -0.5773502590\ ;\ w_1^\xi = 1.0 \\ \xi_1 &= 0.5773502590\ \ ;\ w_2^\xi = 1.0 \end{aligned} \tag{6.86}$$

$$\begin{aligned} \eta_1 &= -0.5773502590\ ;\ w_1^\eta = 1.0 \\ \eta_1 &= 0.5773502590\ \ ;\ w_2^\eta = 1.0 \end{aligned} \tag{6.87}$$

Transform (ξ_i, w_i^ξ); $i = 1,2$ to (x_i, w_i^x); $i = 1,2$ using:

$$\left.\begin{aligned} x_i &= \left(\frac{1.05+0.11}{2}\right) + \left(\frac{1.05-0.11}{2}\right)\xi_i \\ w_i^x &= \left(\frac{1.05-0.11}{2}\right) w_i^\xi \end{aligned}\right\}\ ; \qquad i = 1,2 \tag{6.88}$$

Also transform (η_j, w_j^η); $j = 1,2$ to (y_j, w_j^y); $j = 1,2$ using:

$$\left.\begin{aligned} y_j &= \left(\frac{1.12+0.31}{2}\right) + \left(\frac{1.13-0.31}{2}\right)\eta_j \\ w_j^y &= \left(\frac{1.13-0.31}{2}\right) w_j^\eta \end{aligned}\right\}\ ; \qquad j = 1,2 \tag{6.89}$$

We obtain the following:

$$\begin{aligned} x_1 &= 0.3086453680\ ;\ w_1^x = 0.4699999690 \\ x_2 &= 0.8513545990\ ;\ w_2^x = 0.4699999690 \end{aligned} \tag{6.90}$$

$$\begin{aligned} y_1 &= 0.4832863810\ ;\ w_1^y = 0.4099999960 \\ y_2 &= 0.9567136170\ ;\ w_2^y = 0.4099999960 \end{aligned} \tag{6.91}$$

Using (6.85) and (6.90) we integrate with respect to x.

$$I = \int_{0.31}^{1.13} \left(w_1^x(x_1 y + x_1^2 y^2) + w_2^x(x_2 y + x_2^2 y^2) \right) dy \tag{6.92}$$

Now using (6.92) and (6.91), we integrate with respect to y.

$$\begin{aligned} I =& w_1^y\left(w_1^x(x_1y_1 + x_1^2y_1^2) + w_2^x(x_2y_1 + x_2^2y_1^2)\right) + \\ & w_2^y\left(w_1^x(x_1y_2 + x_1^2y_2^2) + w_2^x(x_2y_2 + x_2^2y_2^2)\right) \end{aligned} \tag{6.93}$$

Substituting numerical values of w_i^x, x_i; $i = 1,2$ and w_j^y, y_j; $j = 1,2$ from (6.90) and (6.91) in (6.93), we obtain the value of the integral.

$$I = 0.5034378170$$

This value agrees with the theoretical value up to at least seven decimal places.

Example 6.9 (Integrating Functions that are not Polynomials Using Gauss Quadrature). Consider the following integral.

$$I = \int_0^1 \left(1 - \left(\frac{e^x + e^{(1-x)}}{1+e}\right)\right)^2 dx$$

In this case the integrand is not an algebraic polynomial, hence it is not possible to determine N^x or n^x. In such cases we begin with the lowest order Gauss quadrature and progressively increase the order of the Gauss quadrature until desired accuracy of the computed integral is obtained. We begin with $n = 2$ and progressively increase it by one. The number of quadrature points and the integral values are listed in Table 6.19.

Table 6.19: Results of Gauss quadrature for Example 6.9

Quadrature Points	I_n
2	0.577189866000E−02
3	0.686041173000E−02
4	0.687233079000E−02
5	0.687239133000E−02
6	0.687239319000E−02
7	0.687239412000E−02

From the integral values for $n = 6$ and $n = 7$, we observe at least six decimal place accuracy.

6.5.4 Concluding Remarks

In this chapter we have presented various methods of obtaining numerical values of definite integrals. All numerical integration methods presented in this chapter are methods of approximation except Gauss quadrature for integrals in which integrands that are algebraic polynomials. Gauss quadrature integrates algebraic polynomials in $\mathbb{R}^1$, $\mathbb{R}^2$, and $\mathbb{R}^3$ exactly, hence is a numerical method without approximation. However, when the integrand is not an algebraic polynomial, Gauss quadrature is also a method of approximation. Gauss quadrature is most meritorious out of all the other methods, even when the integrand is not an algebraic polynomial. When integration functions that are not algebraic polynomials using Gauss quadrature, we progressively increase the number of quadrature points until the desired decimal place accuracy is achieved in the value of the integral.

Problems

6.1 Calculate the value of the integral

$$I = \int_1^2 \left(x + \frac{1}{x}\right)^2 dx$$

numerically.

(a) Using trapezoidal rule with 1, 2 and 4 strips. Tabulate results.

(b) Using the integral values calculated in (a), apply Romberg method to improve the accuracy the integral value.

6.2 Use Romberg method to evaluate the following integral with accuracy of the order of $O(h^8)$

$$I = \int_0^3 xe^{2x}\, dx$$

Hint: Use trapezoidal rule with 1, 2, 4 and 8 steps then apply Romberg method.

6.3 Use lowest order Gauss quadrature to obtain exact value of the following integral.

$$I = \int_{-1}^{1} \left(10 + 5x^2 + 2.5x^3 + 1.25x^4 + 0.62x^5\right) dx$$

6.4 Use lowest order Gauss quadrature to obtain exact value of the following integral.

$$I = \int_1^2 \left(4x^2 + 2x^4 + x^6\right) dx$$

Provide details of the sampling point locations and the weight functions.

6.5 Use two, three and four point Gauss quadrature to evaluate the following integral.

$$I = \int_0^{1/2} \sin(\pi x)\, dx$$

Will the value of the integral improve with 5, 6 and higher order quadrature and why? Can you determine the order of the Gauss quadrature that will yield exact value of the integral I, explain.

6.6 Write a computer program to calculate the following integral numerically using Gauss quadrature.

$$I = \int_{c}^{d} \left(\int_{a}^{b} f(x,y)\,dx \right) dy$$

$f(x,y)$ and the limits $[a,b]$ and $[c,d]$ are given in the following. Use lowest order Gauss quadrature in each case.

(a)

$$\begin{aligned} f(x,y) &= 1 + 4x^2y^2 + 2x^4y^4 \\ [a,b] &= [-1,1] \; ; \; [c,d] = [-1,1] \end{aligned}$$

(b)

$$\begin{aligned} f(x,y) &= 1 + x + y + xy + x^2 + y^2 + x^3 + x^2y + xy^2 + y^3 \\ [a,b] &= [-1,1] \; ; \; [c,d] = [1,2] \end{aligned}$$

(c)

$$\begin{aligned} f(x,y) &= x^2y^2e^{xy} \\ [a,b] &= [1,2] \; ; \; [c,d] = [1.2, 2.1] \end{aligned}$$

Use 1×1, 2×2, 3×3, 4×4, 5×5 and 6×6 Gauss quadrature. Tabulate your results and comment on the accuracy of the integral. Can it be improved further using higher order quadrature? Explain.

Provide a listing of the computer program and a writeup documenting your work.

6.7 Write a computer program to calculate the following integral numerically .

$$I = \int_{a}^{b} f(x)\,dx \tag{1}$$

using:

(a) Trapezoid rule

(b) Simpson's 1/3 rule

(c) Simpson's 3/8 rule

Calculate numerical values of the integral I in (1) using (a), (b) and (c) with 1, 2, 4, 8, 16 and 32 number of strips for the following $f(x)$ and $[a,b]$.

I. $f(x) = \left(x + \frac{1}{x}\right)^2$; $[a,b] = [1,2]$

II. $f(x) = xe^{2x}$; $[a,b] = [0,3]$

III. $f(x) = 10 + 5x^2 + 2.5x^3 + 1.25x^4 + 0.625x^5$; $[a,b] = [-1,1]$

IV. $f(x) = 4x^2 + 2x^4 + x^6$; $[a,b] = [1,2]$

V. $f(x) = \sin(\pi x)$; $[a,b] = [1, 1/2]$

For each $f(x)$, tabulate your results in the following form. Provide a listing of your program and document your work.

Number of steps	Integral Value		
	Trapezoid rule	Simpson's 1/3 rule	Simpson's 3/8 rule
1			
2			
⋮			
32			

For each of the three methods ((a), (b) and (c)) apply Romberg method to obtain the most improved values of the integrals. When using Romberg method, provide a separate table for each of the three methods for each $f(x)$.

6.8 Consider the following integral

$$I = \int_2^3 \int_1^2 \left(x^2 + y^2\right)^{1/3} dx\, dy$$

obtain numerical values of the integral I using Gauss quadrature: 1×1, 2×2, ..., $n \times n$ to obtain the integral value with seven decimal place accuracy. Tabulate your calculations.

6.9 Consider the following integral

$$I = \int_0^{\pi/2} \cos(x)\, dx$$

Calculate numerical value of I using progressively increasing Gauss quadrature till the I value is accurate up to seven decimal places.

6.10 Given

$$I = \int_{-1}^{1.5} \int_{1.5}^{2.5} \int_{1}^{2} x^3(x^2+1)^2(y^2-1)(y+4)z\, dx\, dy\, dz$$

Use the lowest order of Gauss quadrature in x, y and z to obtain exact numerical value of I.

6.11 (a) Why is it that Gauss quadrature can be used to integrate algebraic polynomials exactly?

(b) Describe relationship between the highest degree of the polynomial and the minimum number of quadrature points needed to integrate it exactly. Justify your answers based on (a).

(c) Let

$$I = \int_{-1}^{1}\int_{-1}^{1} \left(\frac{x^5(x^2+4x-12)^2(y^3+2y-6)y^4}{(x-2)^2} \right) dx\, dy$$

Can this be integrated exactly using Gauss quadrature? If yes, then find the minimum number of quadrature points in x and y. Clearly illustrate and justify your answers.

6.12 Consider the following table of data.

i	1	2	3	4	5
x_i	0	0.25	0.5	0.75	1.0
$f_i = f(x_i)$	0.3989	0.3867	0.3521	0.3011	0.2410

(a) Compute $I = \int_0^1 f(x)\, dx$ with strip widths of $h = 0.25$, $h = 0.5$ and $h = 1.0$ using Trapezoid rule. Using these computed values employ Romberg method to compute the most accurate value of the integral I. Tabulate your calculations. Show the orders of the errors being eliminated.

(b) given

$$I = \int_{0.5}^{1}\int_{2.5}^{3.9}\int_{1.6}^{2.7} x^{4.2}(1+x^{1.2})^2 y^3 e^{-y} z^{3.6}\left(1+\frac{1}{z^{2.6}}\right) dx\, dy\, dz$$

What is the lowest order of Gauss quadrature in x, y and z to calculate exact value of I. Explain your reasoning.

6.13 Consider the integral

$$I = \int_1^3 \left(\int_1^2 x^y\, dx \right) dy$$

Calculate numerical values of I using progressively increasing equal order (in x and y) Gauss quadrature that is accurate up to seven decimal places.

7
Curve Fitting

7.1 Introduction

In the interpolation theory, we construct an analytical expression, say $f(x)$, for the data points (x_i, f_i); $i = 1, 2, \ldots, n$. The function $f(x)$ is such that it passes through the data points, i.e., $f(x_i) = f_i$; $i = 1, 2, \ldots, n$. This polynomial representation $f(x)$ of the data points may some times be a poor representation of the functional relationship described by the data (see Figure 7.1).

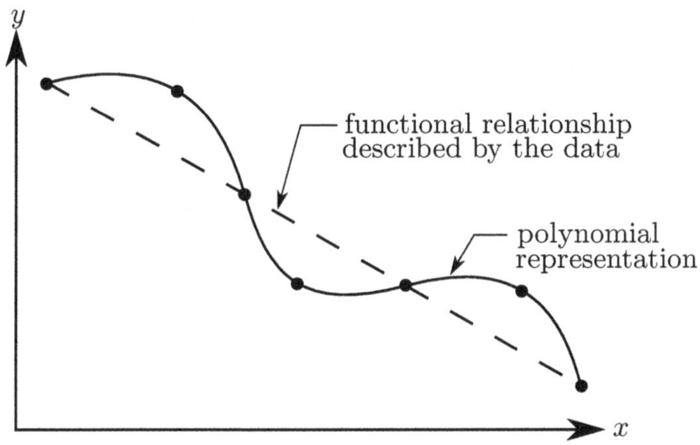

Figure 7.1: Polynomial representation of data and comparison with true functional relationship

Thus, a polynomial representation of data points may not be the best analytical form describing the behavior represented by the data points. In such cases we need to find an analytical expression that represents the best fit to the data points. In doing so, we assume that we know the analytical form of the function $g(x)$ that best describes the data. Generally $g(x)$ is represented by a linear or non-linear combination of the suitably chosen functions using unknown constants or coefficients. The unknown constants or coefficients in $g(x)$ are determined to ensure that $g(x)$ is the best fit to the data. The method of least squares fit is one such method.

When $g(x)$ is a linear function of the unknown constants or coefficients, we have a linear least squares fit to the data. When $g(x)$ is a non-linear function of the unknown constants or coefficients, we obviously have a non-linear least squares fit. In this chapter we consider linear as well as non-linear least squares fits. In case of linear least squares fit we also consider weighted least squares fit in which the more accurate data points can be assigned larger weight factors to ensure that the resulting fit is biased towards these data points. The non-linear least squares fit is first presented for a special class of $g(x)$ in which taking log or ln of both sides of $g(x)$ yields a form that is suitable for linear least squares fit with appropriate correction so that true residual is minimized. This is followed by a general non-linear least squares fit process that is applicable to any form of $g(x)$ in which $g(x)$ is a desired non-linear function of the unknown constants or coefficients. A weighted non-linear least squares formulation of this non-linear least squares fit is also presented. It is shown that this non-linear least squares fit formulation naturally degenerates to linear and weighted linear least squares fit when $g(x)$ is a linear function of the unknown constants or coefficients.

7.2 Linear Least Squares Fit (LLSF)

Let (x_i, f_i); $i = 1, 2, \dots, n$ be the given data points. Let $g_1(x)$, $g_2(x), \dots,$ $g_m(x)$ be known functions to be used in the least squares fit $g(x)$ of the data points (x_i, f_i); $i = 1, 2, \dots, n$ such that $g(x)$ is a linear combination of g_k; $k = 1, 2, \dots, m$, hence $g(x)$ is a linear function of c_i.

$$g(x) = \sum_{k=1}^{m} c_k g_k(x) \tag{7.1}$$

Since $g(x)$ does not necessarily pass through the data points, if x_i; $i = 1, 2, \dots, n$ are substituted in $g(x)$ to obtain $g(x_i)$; $i = 1, 2, \dots, n$, these may not agree with f_i; $i = 1, 2, \dots, n$. Let r_1, r_2, $\dots$, r_n be the differences between $g(x_1)$, $g(x_2)$, $\dots$, $g(x_n)$ and f_1, f_2, $\dots$, f_n, called residuals at each of the location x_i; $i = 1, 2, \dots, n$.

$$\sum_{k=1}^{m} c_k g_k(x_i) - f_i = r_i \quad ; \quad i = 1, 2, \dots, n \tag{7.2}$$

or

$$[G]_{n \times m} \{c\}_{m \times 1} - \{f\}_{n \times 1} = \{r\}_{n \times 1} \tag{7.3}$$

where

$$[G] = \begin{bmatrix} g_1(x_1) & g_2(x_1) & \dots & g_m(x_1) \\ g_1(x_2) & g_2(x_2) & \dots & g_m(x_2) \\ & & \vdots & \\ g_1(x_n) & g_2(x_n) & \dots & g_m(x_n) \end{bmatrix} \tag{7.4}$$

$$\{c\} = \begin{Bmatrix} c_1 \\ c_2 \\ \vdots \\ c_m \end{Bmatrix}_{m\times 1} \quad ; \quad \{f\} = \begin{Bmatrix} f_1 \\ f_2 \\ \vdots \\ f_n \end{Bmatrix}_{n\times 1} \quad ; \quad \{r\} = \begin{Bmatrix} r_1 \\ r_2 \\ \vdots \\ r_n \end{Bmatrix}_{n\times 1} \tag{7.5}$$

The vector $\{r\}$ is called the residual vector. It represents the difference between the assumed fit $g(x)$ and the actual function values f_i. In the least squares fit we minimize the sum of squares of the residuals, i.e., we consider minimization of the sum of the squares of the residuals R.

$$(R)_{\text{minimize}} = \left(\sum_{i=1}^{n}(r_i)^2\right)_{\text{minimize}} \tag{7.6}$$

From (7.3) we note that r_i are functions of c_k ; $k = 1, 2, \dots, m$. Hence, minimization in (7.6) implies the following.

$$\frac{\partial R}{\partial c_k} = \frac{\partial}{\partial c_k}\left(\sum_{i=1}^{n}(r_i)^2\right) = 0 \quad ; \quad k = 1, 2, \dots, m \tag{7.7}$$

or

$$\sum_{i=1}^{n} 2r_i \frac{\partial r_i}{\partial c_k} = 0 \quad ; \quad k = 1, 2, \dots, m$$

or

$$\sum_{i=1}^{n} r_i \frac{\partial r_i}{\partial c_k} = 0 \quad ; \quad k = 1, 2, \dots, m \tag{7.8}$$

But

$$\frac{\partial r_i}{\partial c_k} = g_k(x_i) \quad ; \quad \text{from (7.2)} \tag{7.9}$$

Hence, (7.8) can be written as:

$$\sum_{i=1}^{n} r_i g_k(x_i) = 0 \quad ; \quad k = 1, 2, \dots, m \tag{7.10}$$

or

$$\{r\}^T[G] = [0, 0, \dots, 0] \tag{7.11}$$

or

$$[G]^T\{r\} = \{0\} \tag{7.12}$$

Substituting for $\{r\}$ from (7.3) into (7.12):

$$[G]^T[G]\{c\} - [G]^T\{f\} = \{0\} \tag{7.13}$$

or

$$[G]^T[G]\{c\} = [G]^T\{f\} \tag{7.14}$$

Using (7.14), the unknowns $\{c\}$ can be calculated. Once $\{c\}$ are known, the desired least squares fit is given by $g(x)$ in (7.1).

Example 7.1. Linear least squares fit
Consider the following data:

i	1	2	3	4
x_i	0	1	2	3
f_i	2.4	3.4	13.8	39.5

Determine the constants c_1 and c_2 for $g(x)$ given below to be a least squares fit to the data in the table.

$$g(x) = c_1 + c_2 x^3$$

$$\text{Here } g_1(x) = 1 \quad , \quad g_2(x) = x^3$$

$$[G] = \begin{bmatrix} g_1(x_1) & g_2(x_2) \\ g_1(x_2) & g_2(x_2) \\ g_1(x_3) & g_2(x_3) \\ g_1(x_4) & g_2(x_4) \end{bmatrix} = \begin{bmatrix} 1 & 0 \\ 1 & 1 \\ 1 & 8 \\ 1 & 27 \end{bmatrix}$$

$$\{f\}^T = [2.4 \;\; 3.4 \;\; 13.8 \;\; 39.5]$$

$$[G]^T[G] = \begin{bmatrix} 4.0 & 36.0 \\ 36.0 & 794.0 \end{bmatrix}$$

$$[G]^T\{f\} = \begin{Bmatrix} 59.1 \\ 1180.0 \end{Bmatrix}$$

$$\therefore \quad [G]^T[G]\begin{Bmatrix} c_1 \\ c_2 \end{Bmatrix} = [G]^T\{f\}$$

or

$$\begin{bmatrix} 4.0 & 36.0 \\ 36.0 & 794 \end{bmatrix}\begin{Bmatrix} c_1 \\ c_2 \end{Bmatrix} = \begin{Bmatrix} 59.1 \\ 1180.0 \end{Bmatrix}$$

$$\therefore \quad c_1 = 2.35883$$
$$c_2 = 1.37957$$

Hence,

$$g(x) = 2.35883 + 1.37957x^3$$

is the least squares fit to the data in the table.
For this least squares fit we have

$$R = \sum_{i=1}^{3} r_i^2 = 0.291415$$

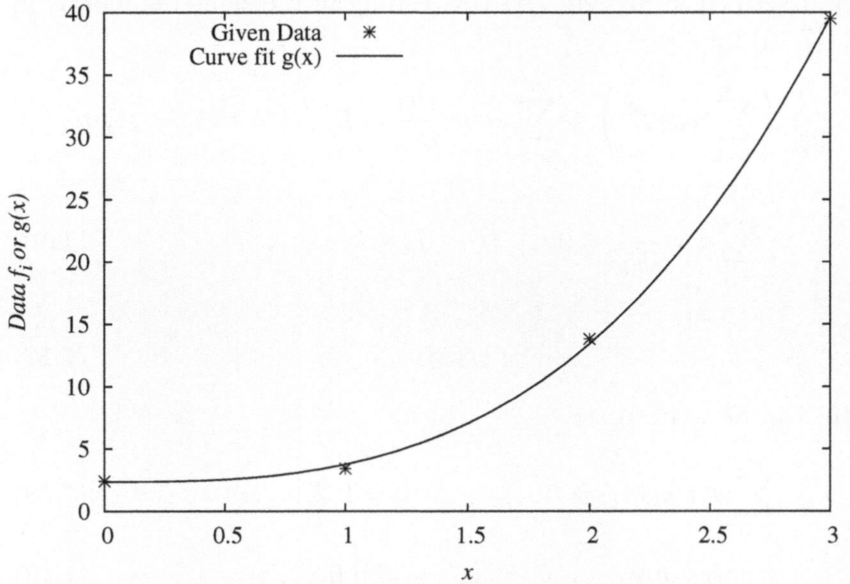

Figure 7.2: f_i versus x_i or $g(x)$ versus x: example 7.1

Figure 7.2 shows plots of data points (x_i, f_i) and $g(x)$ versus x. In this case $g(x)$ provides good approximation to the data (x_i, f_i).

7.3 Weighted Linear Least Squares Fit (WLLSF)

In the weighted least squares method or fit, the residual r_i for a data point i may be assigned a weight factor w_i and then we minimize the weighted sum of squares of the residuals. Thus in this case we consider

$$(R)_{\text{minimize}} = \left(\sum_{i=1}^{n} w_i (r_i)^2\right)_{\text{minimize}} \tag{7.15}$$

w_1, w_2, ..., w_n are the weight factors assigned to data points (x_i, f_i); $i = 1, 2, \dots, n$. These weight factors are positive real numbers. This process allows us to assign extra weight by assigning a weight factor > 1 to some individual data points that may be more accurate or relevant than others. This procedure allows us to bias the curve fitting processes towards data points that we feel are more important or more accurate. When $w_i = 1$; $i =$

$1, 2, \ldots, n$, then the weighted least squares curve fitting reduces to standard curve fitting described in Section 7.2. Proceeding in the same manner as in Section 7.2, (7.15) implies:

$$\frac{\partial R}{\partial c_k} = \frac{\partial}{\partial c_k}\left(\sum_{i=1}^{n} w_i(r_i)^2\right) = \sum_{i=1}^{n} 2w_i r_i \frac{\partial r_i}{\partial c_k} = 0 \quad ; \quad k = 1, 2, \ldots, m$$

or

$$\sum_{i=1}^{n} w_i r_i \frac{\partial r_i}{\partial c_k} = 0 \quad ; \quad k = 1, 2, \ldots, m \tag{7.16}$$

But

$$\frac{\partial r_i}{\partial c_k} = g_k(x_i) \tag{7.17}$$

Hence, (7.16) can be written as:

$$\sum_{i=1}^{n} w_i r_i g_k(x_i) = 0 \quad ; \quad k = 1, 2, \ldots, m \tag{7.18}$$

$$[w_1 r_1 \; w_2 r_2 \; \ldots \; w_n r_n][G] = [0 \; 0 \; 0 \; \ldots \; 0] \tag{7.19}$$

or

$$[G]^T [W]\{r\} = 0 \tag{7.20}$$

where $[W]$ is a diagonal matrix of the weight factors. Substituting for $\{r\}$ from (7.3) in (7.20), we obtain:

$$[G]^T [W]\left([G]\{c\} - \{f\}\right) = \{0\} \tag{7.21}$$

or

$$[G]^T [W][G]\{c\} = [G]^T [W]\{f\} \tag{7.22}$$

This is weighted least squares fit. We use (7.22) to calculate $\{c\}$. When $[W] = [I]$, (7.22) reduces to standard least squares fit given by (7.14).

Example 7.2. Weighted linear least squares fit
Consider the following data:

i	1	2	3
x_i	1	2	3
f_i	4.5	9.5	19.5

Here we demonstrate the use of weight factors (considered unity in this case). Determine the constants c_1 and c_2 for $g(x)$ given below to be a least squares fit to the data in the table. Use weight factors of 1.0 for each data point.

$$g(x) = c_1 + c_2 x^2$$

$$g_1(x) = 1 \quad , \quad g_2(x) = x^2$$

$$[G] = \begin{bmatrix} g_1(x_1) & g_2(x_2) \\ g_1(x_2) & g_2(x_2) \\ g_1(x_3) & g_2(x_3) \end{bmatrix} = \begin{bmatrix} 1 & 1 \\ 1 & 4 \\ 1 & 9 \end{bmatrix}$$

$$[W] = \begin{bmatrix} 1 & 0 & 0 \\ 0 & 1 & 0 \\ 0 & 0 & 1 \end{bmatrix} \quad ; \quad \{f\} = \begin{Bmatrix} 4.5 \\ 9.5 \\ 19.5 \end{Bmatrix}$$

$$\therefore \quad [G]^T[W][G]\begin{Bmatrix} c_1 \\ c_2 \end{Bmatrix} = [G]^T[W]\{f\}$$

Where

$$[G]^T[W][G] = \begin{bmatrix} 3.0 & 14.0 \\ 14.0 & 98.0 \end{bmatrix}$$

$$[G]^T[W]\{f\} = \begin{Bmatrix} 33.5 \\ 218.0 \end{Bmatrix}$$

$$\therefore \quad \begin{bmatrix} 3.0 & 14.0 \\ 14.0 & 98.0 \end{bmatrix}\begin{Bmatrix} c_1 \\ c_2 \end{Bmatrix} = \begin{Bmatrix} 33.5 \\ 218.0 \end{Bmatrix}$$

$$\therefore \quad c_1 = 2.35714$$
$$c_2 = 1.88775$$

Hence,

$$g(x) = 2.35714 + 1.88775x^2$$

is a least squares fit to the data in the table with weight factors of 1.0 assigned to each data point. We note that the least squares with or without the weight factors will yield the same results due to the fact that the weight factors are unity in this example.

In this case we have

$$R = \sum_{i=1}^{3} r_i^2 = 0.255102$$

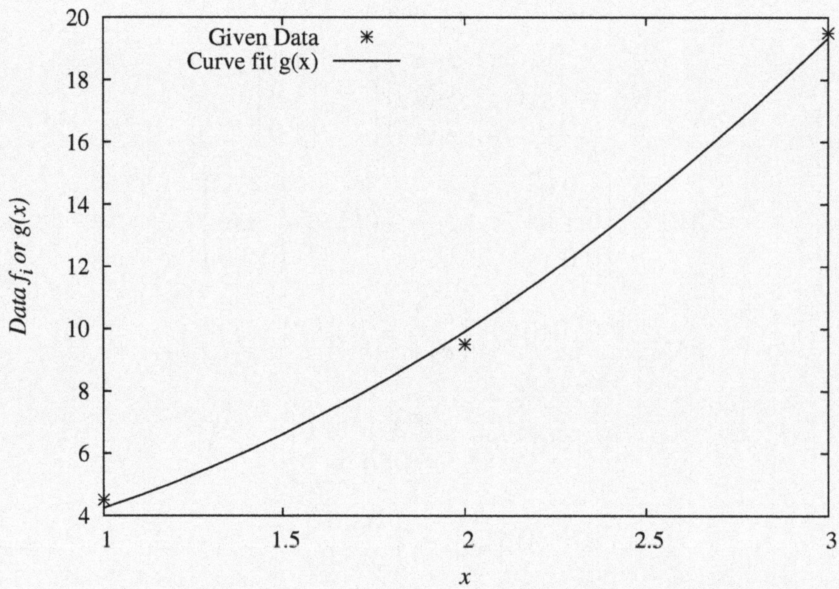

Figure 7.3: f_i versus x_i or $g(x)$ versus x: example 7.2

Figure 7.3 shows plots of data points (x_i, f_i) and $g(x)$ versus x. We note that $g(x)$ is a good fit to the data (x_i, f_i).

Example 7.3. Weighted linear least squares fit
Consider the same problem as in example 7.2 except that we use weight factor of 2 for the second data point.

i	1	2	3
x_i	1	2	3
f_i	4.5	9.5	19.5

Consider $g(x) = c_1 + c_2x^2$ as a least squares fit to the data. Determine c_1 and c_2 when weight factors 1, 2, and 1 are assigned to the three data points. Thus, here we create a bias to data point two as it has a weight factor $w_2 = 2$ compared to weight factors of 1 for the other two data points.

$$g_1(x) = 1 \qquad ; g_2(x) = x^2$$

$$[G] = \begin{bmatrix} g_1(x_1) \; g_2(x_2) \\ g_1(x_2) \; g_2(x_2) \\ g_1(x_3) \; g_2(x_3) \end{bmatrix} = \begin{bmatrix} 1 \; 1 \\ 1 \; 4 \\ 1 \; 9 \end{bmatrix}$$

$$[W] = \begin{bmatrix} 1 & 0 & 0 \\ 0 & 2 & 0 \\ 0 & 0 & 1 \end{bmatrix} \quad ; \quad \{f\} = \begin{Bmatrix} 4.5 \\ 9.5 \\ 19.5 \end{Bmatrix}$$

$$[G]^T[W][G] = \begin{bmatrix} 4.0 & 18.0 \\ 18.0 & 114.0 \end{bmatrix}$$

$$[G]^T[W]\{f\} = \begin{Bmatrix} 43.0 \\ 256.0 \end{Bmatrix}$$

$$\therefore \quad \begin{bmatrix} 4.0 & 18.0 \\ 18.0 & 114.0 \end{bmatrix} \begin{Bmatrix} c_1 \\ c_2 \end{Bmatrix} = \begin{Bmatrix} 43.0 \\ 256.0 \end{Bmatrix}$$

$$\therefore \quad c_1 = 2.227273$$

$$c_2 = 1.89394$$

$$\therefore \quad g(x) = 2.227273 + 1.89394x^2$$

This is a least squares fit to the data in the table with weight factors of 1, 2, and 1 for the three data points. Due to $w_2 = 2$, c_1 and c_2 have changed slightly compared to Example 7.2. In this case we have

$$R = \sum_{i=1}^{3} w_i r_i^2 = 0.378788$$

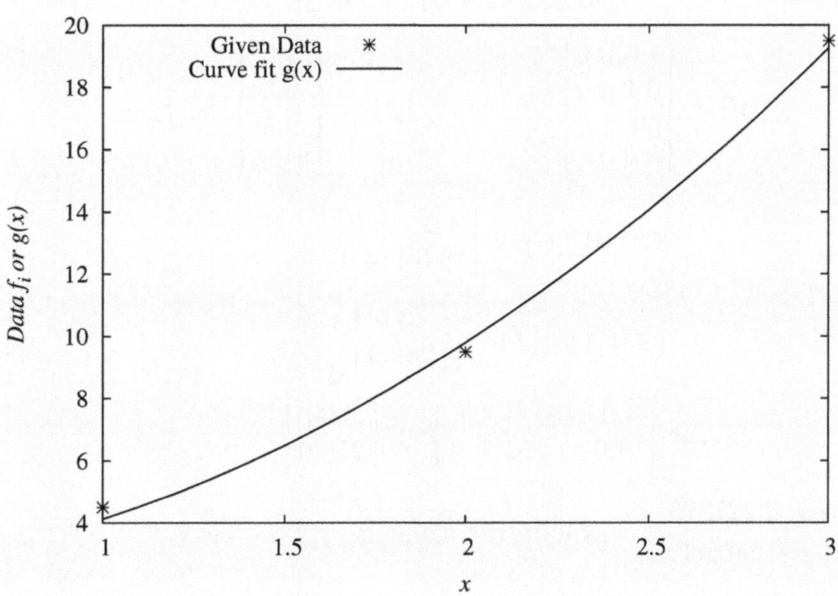

Figure 7.4: f_i versus x_i or $g(x)$ versus x: example 7.3

Figure 7.4 shows plots of (x_i, f_i) and $g(x)$ versus x. Weight factors of 2 for data point two does not appreciatively alter $g(x)$.

Example 7.4. Weighted linear least squares fit
We consider the same problem as in Example 7.3, but rather than assigning a weight factor of 2.0 to the second data point, we repeat this data point and assign weight factors of 1 to all data points. Thus we have:

i	1	2	3	4
x_i	1	2	2	3
f_i	4.5	9.5	9.5	19.5

$$g(x) = c_1 + c_2x^2$$

$$g_1(x) = 1 \quad , \quad g_2(x) = x^2$$

$$[G] = \begin{bmatrix} g_1(x_1) \; g_2(x_2) \\ g_1(x_2) \; g_2(x_2) \\ g_1(x_3) \; g_2(x_3) \\ g_1(x_4) \; g_2(x_4) \end{bmatrix} = \begin{bmatrix} 1 \; 1 \\ 1 \; 4 \\ 1 \; 4 \\ 1 \; 9 \end{bmatrix}$$

$$[W] = \begin{bmatrix} 1\,0\,0\,0 \\ 0\,1\,0\,0 \\ 0\,0\,1\,0 \\ 0\,0\,0\,1 \end{bmatrix} \quad ; \quad \{f\} = \begin{Bmatrix} 4.5 \\ 9.5 \\ 9.5 \\ 19.5 \end{Bmatrix}$$

$$[G]^T[W][G] = \begin{bmatrix} 4.0 & 18.0 \\ 18.0 & 114.0 \end{bmatrix}$$

$$[G]^T[W]\{f\} = \begin{Bmatrix} 43.0 \\ 256.0 \end{Bmatrix}$$

$$\therefore \quad \begin{bmatrix} 4.0 & 18.0 \\ 18.0 & 114.0 \end{bmatrix} \begin{Bmatrix} c_1 \\ c_2 \end{Bmatrix} = \begin{Bmatrix} 43.0 \\ 256.0 \end{Bmatrix}$$

$$\therefore \quad \begin{aligned} c_1 &= 2.227273 \\ c_2 &= 1.89394 \end{aligned} \qquad \text{exactly the same as in Example 7.3}$$

$$\therefore \quad g(x) = 2.227273 + 1.89394x^2$$

This is a least squares fit to the data in the table in which data points two and three are identically the same. Thus, assigning a weight factor k (an integer) to the data point is the same as repeating this data point k times with a weight factor of one. In this case we have

$$R = \sum_{i=1}^{3} w_i r_i^2 = 0.378788$$

This value of R is same as in example 7.3 (as expected).

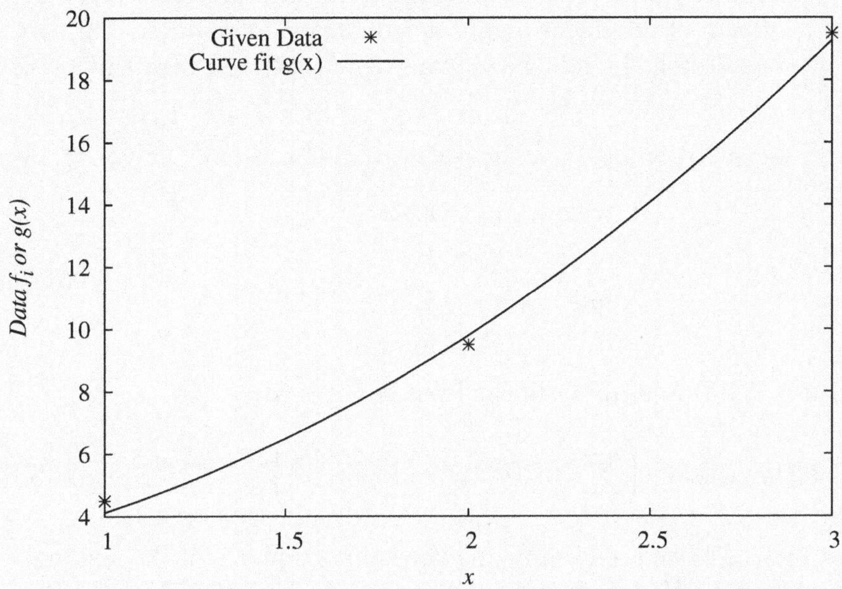

Figure 7.5: f_i versus x_i or $g(x)$ versus x: example 7.4

Figure 7.5 shows plots of (x_i, f_i) and $g(x)$ versus x.

7.4 Non-linear Least Squares Fit: A Special Case (NLSF)

In the least squares fit considered in Sections 7.2 and 7.3, $g(x)$ was a linear function of c_i ; $i = 1, 2, \ldots, m$. In some applications this may not be the case. Let (x_i, f_i) ; $i = 1, 2, \ldots, n$ be the given data points. In this section we consider a special case in which by taking log (or ln) of both sides the least squares fit process will be linear in the log term. This is then followed by correction to account for logs (or ln). Let us assume that

$$g(x) = cx^k \quad ; \quad c \text{ and } k \text{ to be determined} \tag{7.23}$$

describes the fit to the data (x_i, f_i) ; $i = 1, 2, \dots, n$. If we minimize

$$R = \sum_{i=1}^{n} (g(x_i) - f_i)^2 \tag{7.24}$$

then due to the specific form of $g(x)$ in (7.23), the minimization of (7.24) will result in a system of nonlinear algebraic equations in c and k. This can be avoided by considering the following: consider (7.23) and take the log of both sides.

$$\log(g(x)) = \log c + k \log x = c_1 g_1(x) + c_2 g_2(x) \tag{7.25}$$

$$\begin{aligned} \text{where} \quad c_1 &= \log c \\ c_2 &= k \\ \text{and} \quad g_1(x) &= 1 \\ g_2(x) &= \log x \end{aligned} \tag{7.26}$$

Now we can use (7.25) and apply linear least squares fit.

$$(\underset{\sim}{R})_{\text{minimize}} = \left(\sum_{i=1}^{n} (\log(g(x_i)) - \log(f_i))^2 \right)_{\text{minimize}} \tag{7.27}$$

We note that in (7.27), we are minimizing the sum of squares of the residuals of logs of $g(x_i)$ and f_i. However, if we still insist on using (7.27), then some adjustments or corrections must be made so that (7.27) indeed would result in what we want.

Let Δf_i be the error in f_i, then the corresponding error in $\log(g(x_i))$, i.e., $\Delta(\log(g(x_i)))$, can be approximated.

$$\Delta(\log(g(x_i))) = d(\log(g(x_i))) \simeq d(\log(f_i)) = \left(\frac{d(\log(f_i))}{df_i} \right) df_i = \frac{1}{f_i} df_i = \frac{\Delta f_i}{f_i} \tag{7.28}$$

$$\therefore \quad \Delta f_i = f_i \Delta(\log(g(x_i))) \tag{7.29}$$

$$\therefore \quad (\Delta f_i)^2 = (f_i)^2 (\Delta(\log(g(x_i))))^2 \tag{7.30}$$

From (7.30), we note that minimization of the square of the error f_i requires minimization of the error in $\log(g(x_i))$ squared multiplied with f_i^2, i.e., f_i^2 behaves like a weight factor. Thus instead of considering minimization (7.27), if we consider:

$$(R)_{\text{minimize}} = \left(\sum_{i=1}^{n} w_i (\log(g(x_i)) - \log(f_i))^2 \right)_{\text{minimize}} \tag{7.31}$$

in which $w_i = f_i^2$. We minimize the sum of the squares of the residuals between $g(x_i)$ and f_i, which is what we need to do. Equation (7.31) will give:

$$[G]^T[W][G]\{c\} = [G]^T[W]\{\hat{f}\}$$

Where for this particular example we have:

$$[G] = \begin{bmatrix} g_1(x_1) & g_2(x_1) \\ g_1(x_2) & g_2(x_2) \\ \vdots & \vdots \\ g_1(x_n) & g_2(x_n) \end{bmatrix} = \begin{bmatrix} 1 & \log x_1 \\ 1 & \log x_2 \\ \vdots & \vdots \\ 1 & \log x_n \end{bmatrix} \tag{7.32}$$

$$[W] = \begin{bmatrix} f_1^2 & 0 & 0 & \dots & 0 \\ 0 & f_2^2 & 0 & \dots & 0 \\ \vdots & \vdots & \ddots & \ddots & \\ 0 & 0 & 0 & \dots & f_n^2 \end{bmatrix} \tag{7.33}$$

$$\{\hat{f}\} = \begin{Bmatrix} \log(f_1) \\ \log(f_2) \\ \vdots \\ \log(f_n) \end{Bmatrix} \tag{7.34}$$

This procedure described above has approximation due to (7.28), but helps in avoiding nonlinear algebraic equations resulting from the least squares fit.

Example 7.5. Non-linear least squares fit: special case
Consider the following data points:

i	1	2	3
x_i	1	2	3
f_i	1.2	3.63772	6.959455

Let $g(x) = cx^k$ be a least squares fit to the data in the table. Determine c and k using non-linear least squares fit procedure described in section 7.4.

Take $\log_{10}$ of both sides of $g(x) = cx^k$.

$$\log_{10} g(x) = \log_{10} c + k \log_{10} x = c_1 + c_2 \log_{10} x \quad ; \quad c_1 = \log_{10} c \quad , \quad c_2 = k$$

$$\therefore \quad g_1(x) = 1 \quad , \quad g_2(x) = \log_{10} x$$

$$[G] = \begin{bmatrix} g_1(x_1) & g_2(x_2) \\ g_1(x_2) & g_2(x_2) \\ g_1(x_3) & g_2(x_3) \end{bmatrix} = \begin{bmatrix} 1 & \log_{10} 1 \\ 1 & \log_{10} 2 \\ 1 & \log_{10} 3 \end{bmatrix} = \begin{bmatrix} 1 & 0 \\ 1 & 0.30103 \\ 1 & 0.47712 \end{bmatrix}$$

$$[W] = \begin{bmatrix} f_1^2 & 0 & 0 \\ 0 & f_2^2 & 0 \\ 0 & 0 & f_3^2 \end{bmatrix} = \begin{bmatrix} 1.44 & 0 & 0 \\ 0 & 13.233 & 0 \\ 0 & 0 & 48.434 \end{bmatrix}$$

$$\{\hat{f}\} = \begin{Bmatrix} \log_{10} f_1 \\ \log_{10} f_2 \\ \log_{10} f_3 \end{Bmatrix} = \begin{Bmatrix} \log_{10} 1.2 \\ \log_{10} 3.63772 \\ \log_{10} 6.959455 \end{Bmatrix}$$

$$\therefore \quad [G]^T[W][G] = \begin{bmatrix} 63.1070 & 27.0924 \\ 27.0924 & 12.2249 \end{bmatrix}$$

$$[G]^T[W]\{\hat{f}\} = \begin{Bmatrix} 48.3448 \\ 21.7051 \end{Bmatrix}$$

$$\therefore \quad [G]^T[W][G] \begin{Bmatrix} c_1 \\ c_2 \end{Bmatrix} = [G]^T[W]\{\hat{f}\} \quad \text{gives}$$

$$\begin{bmatrix} 63.1070 & 27.0924 \\ 27.0924 & 12.2249 \end{bmatrix} \begin{Bmatrix} c_1 \\ c_2 \end{Bmatrix} = \begin{Bmatrix} 48.3448 \\ 21.7051 \end{Bmatrix}$$

$$c_1 = 0.0791822 = \log_{10} c \qquad \therefore \quad c = 1.2$$

$$c_2 = 1.6 = k$$

Hence,

$$g(x) = 1.2x^{1.6}$$

is the least squares fit to the data. For this least squares fit we have

$$R = \sum_{i=1}^{3} r_i^2 = \sum_{i=1}^{3} (f_i - g(x_i))^2 = 1.16858 \times 10^{-11}$$

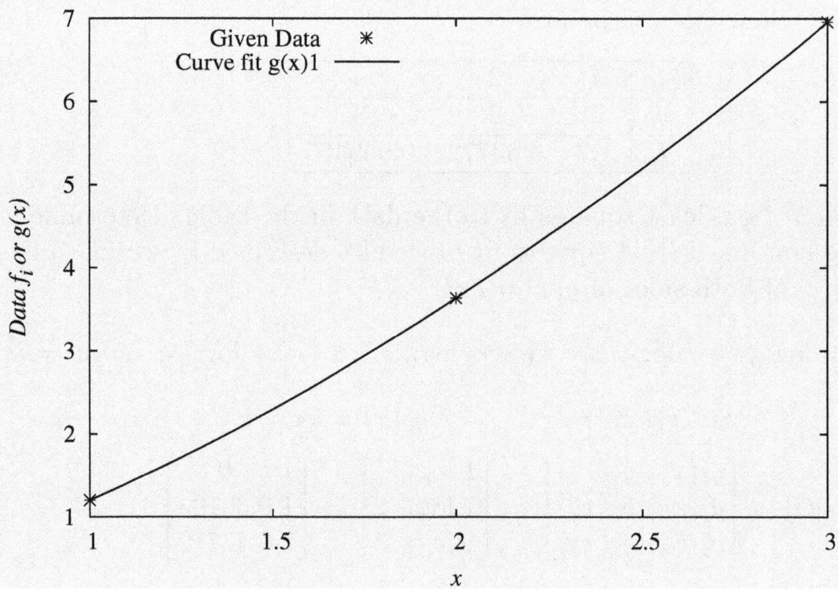

Figure 7.6: f_i versus x_i or $g(x)$ versus x: example 7.5

Figure 7.6 shows plots of (x_i, f_i) and $g(x)$ versus x. The fit by $g(x)$ is almost exact. This is also obvious from such low value of R.

Example 7.6. Non-linear least squares fit: special case
We consider the same data as in Example 7.5 and the same function $g(x) = cx^k$, but obtain a solution for c and k using nonlinear least squares fit employing natural logarithm instead of $\log_{10}$.

Take natural log, ln, of both sides of $g(x) = cx^k$.

$$\ln(g(x)) = \ln(c) + k\ln(x) = c_1 + c_2\ln(x) \quad ; \quad c_1 = \ln(c) \quad , \quad c_2 = k$$

$$g_1(x) = 1 \quad , \quad g_2(x) = \ln(x)$$

$$[W] = \begin{bmatrix} f_1^2 & 0 & 0 \\ 0 & f_2^2 & 0 \\ 0 & 0 & f_3^2 \end{bmatrix} = \begin{bmatrix} 1.44 & 0 & 0 \\ 0 & 13.233 & 0 \\ 0 & 0 & 48.434 \end{bmatrix}$$

$$\{\hat{f}\} = \begin{Bmatrix} \ln(f_1) \\ \ln(f_2) \\ \ln(f_3) \end{Bmatrix} = \begin{Bmatrix} \ln(1.2) \\ \ln(3.63772) \\ \ln(6.959455) \end{Bmatrix}$$

$$\therefore \quad [G]^T[W][G]\begin{Bmatrix} c_1 \\ c_2 \end{Bmatrix} = [G]^T[W]\{\hat{f}\}$$

where

$$[G]^T[W][G] = \begin{bmatrix} 63.107 & 62.3826 \\ 62.3826 & 64.8152 \end{bmatrix}$$

$$[G]^T[W]\{\hat{f}\} = \begin{Bmatrix} 111.1318 \\ 115.078 \end{Bmatrix}$$

$$\therefore \quad \begin{bmatrix} 63.107 & 62.3826 \\ 62.3826 & 64.8152 \end{bmatrix}\begin{Bmatrix} c_1 \\ c_2 \end{Bmatrix} = \begin{Bmatrix} 111.1318 \\ 115.078 \end{Bmatrix}$$

$$\therefore \quad c_1 = 0.182322 \implies \ln(c) = 0.182322 \quad \therefore \quad c = 1.2$$
$$c_2 = 1.6$$

Hence,

$$g(x) = 1.2x^{1.6}$$

is the nonlinear least squares fit to the data. We see that whether we use $\log_{10}$ or ln, it makes no difference. Using calculated $c = 1.2$ and $k = 1.6$ we obtain

$$R = \sum_{i=1}^{3}(f_i - g(x_i))^2 = 2.37556 \times 10^{-12}$$

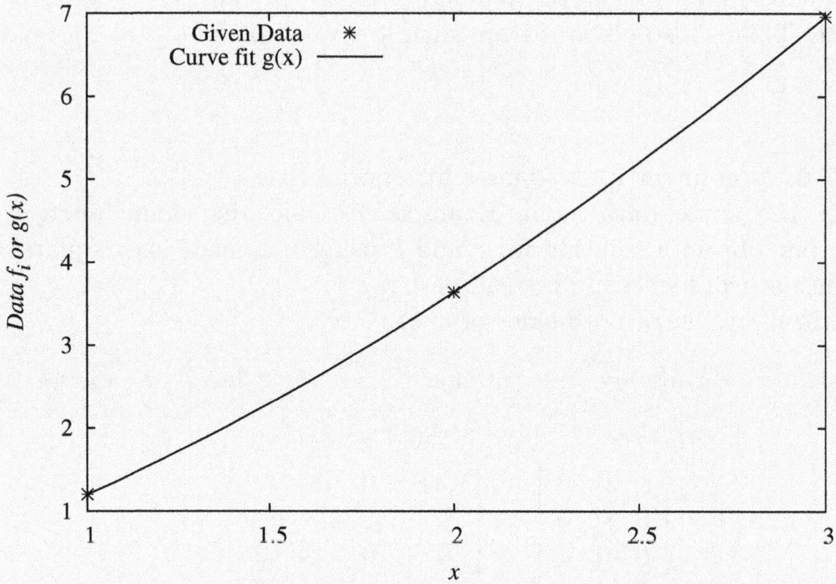

Figure 7.7: f_i versus x_i or $g(x)$ versus x: example 7.6

Figure 7.7 shows plots of (x_i, f_i) and $g(x)$ versus x. $g(x)$ is almost exact fit to the data. This is also obvious from very low R.

Example 7.7. Non-linear least squares fit: special case
Consider the following data:

i	1	2	3
x_i	0	1	2
f_i	1.2	2.67065	5.94364

Let $g(x) = ce^{kx}$ be the least squares fit to the data in the table. Determine c and k using nonlinear least squares fit. In this case it is advantageous to take ln of both sides of $g(x) = ce^{kx}$.

$$\ln(g(x)) = \ln(c) + kx = c_1 + c_2 x \quad ; \quad c_1 = \ln(c) \quad , \quad c_2 = k$$

$$g_1(x) = 1 \quad , \quad g_2(x) = x$$

$$[G] = \begin{bmatrix} g_1(x_1) & g_2(x_2) \\ g_1(x_2) & g_2(x_2) \\ g_1(x_3) & g_2(x_3) \end{bmatrix} = \begin{bmatrix} 1 & 0 \\ 1 & 1 \\ 1 & 2 \end{bmatrix}$$

$$[W] = \begin{bmatrix} f_1^2 & 0 & 0 \\ 0 & f_2^2 & 0 \\ 0 & 0 & f_3^2 \end{bmatrix} = \begin{bmatrix} 1.44 & 0 & 0 \\ 0 & 7.1324 & 0 \\ 0 & 0 & 35.327 \end{bmatrix}$$

$$\{\hat{f}\} = \begin{Bmatrix} \ln(f_1) \\ \ln(f_2) \\ \ln(f_3) \end{Bmatrix} = \begin{Bmatrix} \ln(1.2) \\ \ln(2.67065) \\ \ln(5.94364) \end{Bmatrix}$$

Therefore we have

$$[G]^T[W][G]\begin{Bmatrix} c_1 \\ c_2 \end{Bmatrix} = [G]^T[W]\{\hat{f}\}$$

where

$$[G]^T[W][G] = \begin{bmatrix} 43.8992 & 77.7861 \\ 77.7861 & 148.440 \end{bmatrix}$$

$$[G]^T[W]\{\hat{f}\} = \begin{Bmatrix} 70.2327 \\ 132.93 \end{Bmatrix}$$

$$\therefore \quad \begin{bmatrix} 43.8992 & 77.7861 \\ 77.7861 & 148.440 \end{bmatrix}\begin{Bmatrix} c_1 \\ c_2 \end{Bmatrix} = \begin{Bmatrix} 70.2327 \\ 132.93 \end{Bmatrix}$$

$$\therefore \quad c_1 = 0.1823226 = \ln(c) \implies c = 1.2$$

$$c_2 = 0.8 = k$$

Hence

$$g(x) = 1.2e^{0.8x}$$

is the least squares fit to the data given in the table. For this least squares fit we have

$$R = \sum_{i=1}^{3} r_i^2 = \sum_{i=1}^{3}(f_i - g(x_i))^2 = 9.59730 \times 10^{-13}$$

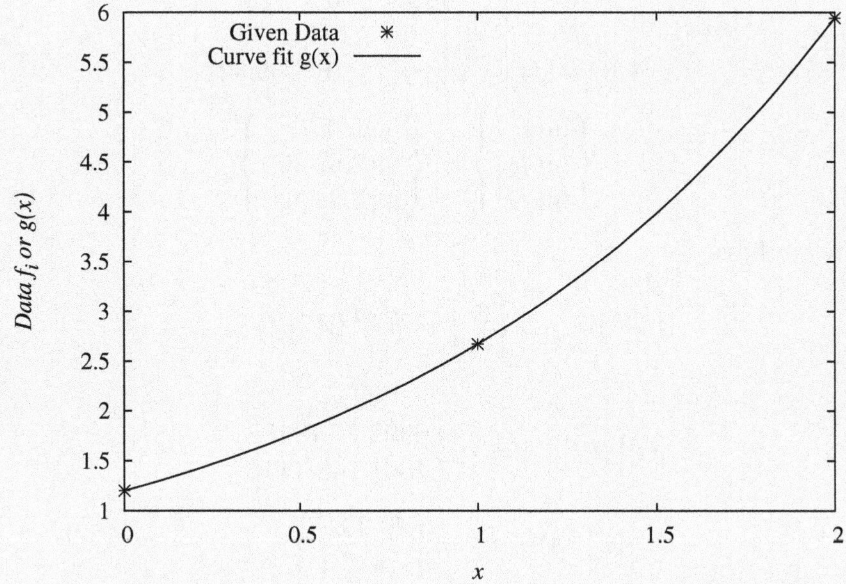

Figure 7.8: f_i versus x_i or $g(x)$ versus x: example 7.7

Figure 7.8 shows graphs of (x_i, f_i) and $g(x)$ versus x. In this case also $g(x)$ is almost exact fit to the data

7.5 General formulation for non-linear least squares fit (GNLSF)

We note that in general for nonlinear least squares fit the forms required may not always be such that the technique of taking log will always be beneficial as shown in section 7.4. In this section we consider a general nonlinear least square fit formulation that is applicable to all non-linear least squares fit regardless of the forms.

Let (x_i, f_i); $i = 1, 2, \ldots, n$ be the data points and $g(x, c_1, c_2, \ldots, c_m)$ be the desired least squares fit to the data in which $g(\cdot)$ is a non-linear function of $c_1, c_2, \ldots, c_m$ and x. Thus, we need to minimize the sum of squares of the residuals between the given data and the non-linear equation $g(\cdot)$ the residuals r_i are given by

$$g(x_i, c_1, c_2, \ldots, c_m) - f_i = r_i \ ; \ i = 1, 2, \ldots, n \tag{7.35}$$

Since $g(\cdot)$ is a non-linear function of $c_1, c_2, \ldots, c_m$, amongst all other techniques of solving for $c_1, c_2, \ldots, c_m$, Gauss-Newton method is the simplest and straight forward. The main concept in this approach is to obtain an approximate linear form of (7.35) by using Taylor series. It is obvious that since $g(x, c_1, c_2, \dots, c_m)$ is a non-linear function of c_i; $i = 1, 2, \dots, m$ we will have to determine c_i; $i = 1, 2, \dots, m$ iteratively.

Let k and $k+1$ be two successive iterations, then we can write (7.35) as

$$g(x_i, c_1, c_2, \dots, c_m)_{k+1} - f_i = (r_i)_k \; ; \; i = 1, 2, \dots, n \tag{7.36}$$

At the beginning of the iterative procedure k refers to initial guess $\{c\}_k$ of $c_1, c_2, \dots, c_m$ and $\{c\}_{k+1}$ the improved values of $c_1, c_2, \dots, c_m$.

We expand $g(x_i, c_1, c_2, \dots, c_m)_{k+1}$ in Taylor series in $\{c\}$ about $\{c\}_k$

$$\begin{aligned} g(x_i, c_1, c_2, \dots, c_m)_{k+1} &= g(x_i, c_1, c_2, \dots, c_m)_k + \\ \frac{\partial g(x_i, c_1, c_2, \dots, c_m)_k}{\partial c_1}(\Delta c_1)_k &+ \frac{\partial g(x_i, c_1, c_2, \dots, c_m)_k}{\partial c_2}(\Delta c_2)_k + \dots \end{aligned} \tag{7.37}$$

in which

$$\begin{aligned} (\Delta c_1)_k &= (c_1)_{k+1} - (c_1)_k \\ (\Delta c_2)_k &= (c_2)_{k+1} - (c_2)_k \;\; \dots \;\; \text{etc.} \end{aligned} \tag{7.38}$$

Substituting (7.37) in (7.36) we obtain

$$\begin{aligned} &\frac{\partial g(x_i, c_1, c_2, \dots, c_m)_k}{\partial c_1}(\Delta c_1)_k + \frac{\partial g(x_i, c_1, c_2, \dots, c_m)_k}{\partial c_2}(\Delta c_2)_k + \dots + \\ &\qquad g(x_i, c_1, c_2, \dots, c_m)_k - f_i = r_i \; ; \; i = 1, 2, \dots, n \end{aligned} \tag{7.39}$$

Equation (7.39) can be written in the matrix and vector form.

$$[G]_k \{\Delta c\}_k - \{d\}_k = \{r\}_k \tag{7.40}$$

in which

$$[G]_k = \begin{bmatrix} \frac{\partial g(x_1,c_1,c_2,\dots,c_m)_k}{\partial c_1} & \frac{\partial g(x_1,c_1,c_2,\dots,c_m)_k}{\partial c_2} & \cdots & \frac{\partial g(x_1,c_1,c_2,\dots,c_m)_k}{\partial c_m} \\ \frac{\partial g(x_2,c_1,c_2,\dots,c_m)_k}{\partial c_1} & \frac{\partial g(x_2,c_1,c_2,\dots,c_m)_k}{\partial c_2} & \cdots & \frac{\partial g(x_2,c_1,c_2,\dots,c_m)_k}{\partial c_m} \\ \vdots & \vdots & & \vdots \\ \frac{\partial g(x_n,c_1,c_2,\dots,c_m)_k}{\partial c_1} & \frac{\partial g(x_n,c_1,c_2,\dots,c_m)_k}{\partial c_2} & \cdots & \frac{\partial g(x_n,c_1,c_2,\dots,c_m)_k}{\partial c_m} \end{bmatrix}_{n \times m} \tag{7.41}$$

$$\{\Delta c\}_k^T = [(\Delta c_1)_k \; (\Delta c_2)_k \; \dots \; (\Delta c_m)_k] \tag{7.42}$$

$$\begin{aligned} \{d\}_k^T = [f_1 - g(x_1, c_1, c_2, \dots, c_m)_k \, , \; f_2 - g(x_2, c_1, c_2, \dots, c_m)_k \, , \; \dots \, , \\ f_n - g(x_n, c_1, c_2, \dots, c_m)_k] \end{aligned} \tag{7.43}$$

$$(R_k)_{\text{minimization}} = \left(\sum_{i=1}^{n} (r_i)_k^2 \right)_{\text{minimization}} \tag{7.44}$$

We note that (7.40) is similar to (7.3) when $\{d\}_k$ in (7.40) takes the place of $\{f\}$ in (7.3), hence the least squares fit becomes

$$[G]_k^T [G]_k \{\Delta c\}_k = [G]_k^T \{d\}_k \tag{7.45}$$

We solve for $\{\Delta c\}_k$ using (7.45). Improved values of $\{c\}$ i.e. $\{c\}_{k+1}$ are given by

$$\{c\}_{k+1} = \{c\}_k + \{\Delta c\}_k \tag{7.46}$$

Convergence check for the iterative process (7.45) and (7.46) is given by

$$\left| \frac{(c_i)_{k+1} - (c_i)_k}{(c_i)_{k+1}} \right| 100 \leq \Delta \; ; \; i = 1, 2, \dots, m \tag{7.47}$$

or simply $|(c_i)_{k+1} - (c_i)_k| \leq \Delta_1$.
We note that the method requires initial or starting values of c_i; $i = 1, 2, \dots, m$ i.e. $\{c\}_k$ so that coefficients of $[G]_k$ and $g(x, c_1, c_2, \dots, c_m)_k$ in $\{d\}_k$ in (7.5) can be calculated.
When the convergence criteria in (7.47) is satisfied we have the solution $\{c\}_{k+1}$ for c_i; $i = 1, 2, \dots, m$ in $g(x, c_1, c_2, \dots, c_m)$ otherwise we increment k by one and repeat (7.45) and (7.46) till (7.47) is satisfied.

7.5.1 Weighted general non-linear least squares fit (WGNLSF)

In this case we consider

$$(R)_{\text{minimizing}} = \left(\sum_{i=1}^{n} w_i (r_i)_k^2 \right)_{\text{minimizing}} \tag{7.48}$$

in which w_i are weight factors and r_i; $i = 1, 2, \dots, n$ are given by (7.40). Following the derivations in section 7.3 and using (7.40) and (7.48), we obtain the following instead of (7.45).

$$[G]_k^T [W][G]_k \{\Delta c\}_k = [G]_k^T [W] \{d\}_k \tag{7.49}$$

in which $[W]$ is the diagonal matrix of weight factors. Rest of the details remains same as in section 7.3.

7.5.1.1 Using general non-linear least squares fit for linear least squares fit

In case of linear least squares fit we have

$$g(x, c_1, c_2, \dots, c_m) = \sum_{i=1}^{m} c_i g_i(x) \tag{7.50}$$

Hence

$$\frac{\partial g}{\partial c_k} = g_k(x) \tag{7.51}$$

Hence, $[G]_k$ in (7.45) reduces to $[G]$ ((7.4)) in linear least squares fit and we have (omitting subscript k)

$$[G]^T[G]\{\Delta c\} = [G]^T\{d\} \tag{7.52}$$

in which $d_i = f_i - g(x_i)$; $i = 1, 2, \dots, n$.

(1) With the initial choice of $c_i = 0$; $i = 1, 2, \dots, m$, $\{d\}$ becomes $\{f\}$, hence (7.52) reduces to

$$[G]^T[G]\{\Delta c\} = \{f\} \tag{7.53}$$

we note that (7.53) is same as (7.4) in linear least squares fit. Clearly $\{\Delta c\} = \{c\}$.

(2) With any non-zero choices of $\{c\}$, the iterative process converges in two iterations as $[G]$ is not a function of $\{c\}$.

Example 7.8. We consider the same problem as example 7.5 but apply the general formulation for non-linear least squares fit presented in section 7.5.

$$g(x, c, k) = cx^k$$

$$\frac{\partial g}{\partial c} = x^k \ , \ \frac{\partial g}{\partial k} = ckx^{k-1}$$

we consider matrix $[G]$ and vector $\{d\}$ using (7.41) and (7.43)

$$[G] = \begin{bmatrix} \frac{\partial g(x_1, c_k)}{\partial c} & \frac{\partial g(x_1, c_k)}{\partial k} \\ \frac{\partial g(x_2, c_k)}{\partial c} & \frac{\partial g(x_2, c_k)}{\partial k} \\ \vdots & \vdots \\ \frac{\partial g(x_n, c_k)}{\partial c} & \frac{\partial g(x_n, c_k)}{\partial k} \end{bmatrix}$$

$$\{d\}^T = [f_1 - g(x_1, c_k), f_2 - g(x_2, c_k), \dots, f_n - g(x_n, c_k)]$$

From example 7.5 we know that the correct values of c and k are 1.2 and 1.6.

We begin with (7.45) for $k = 1$ and choose $c_1 = 1.2$ and $k_1 = 1.6$ as initial values of c and k at $k = 1$. We consider tolerance $\Delta_1 = 10^{-6}$ for convergence. Details of the computations using (7.45) are given below.

$$[G]_1 = \begin{bmatrix} 1.0 & 1.92 \\ 3.03143 & 2.91018 \\ 5.79955 & 3.71171 \end{bmatrix}$$

$$[G]_1^T[G]_1 = \begin{bmatrix} 43.8243 & 32.2682 \\ 32.2682 & 25.9323 \end{bmatrix}$$

$$\{[G]_1^T\{d\}_1\}^T = [-3.65106 \times 10^{-6} \; -2.10685 \times 10^{-6}]$$

$$\{c\}_2^T = [1.1999998 \; 1.6000003]$$

$$[G]_1^T[G]_1\{\Delta c\}_1 = [G]_1^T\{d\}_1$$

gives

$$\{\Delta c\}_1^T = [-2.08034 \times 10^{-7} \; 2.6759 \times 10^{-7}]$$

are

$$\{c\}_2^T = \{c\}_1^T + \{\Delta c\}_1^T = [1.1999998 \; 1.600003] = [c, k]$$

$\{c\}_2$ is converged solution based on tolerance $\Delta_1 = 10^{-6}$.
Since $\{c\}_1$ (initial values of c and k) are the correct values, the non-linear iterations solution procedure converges in only one iteration and we have

$$g(x) = 1.2x^{1.6} \quad \text{with} \quad R = 1.38914 \times 10^{-12}$$

the desired least squares fit.

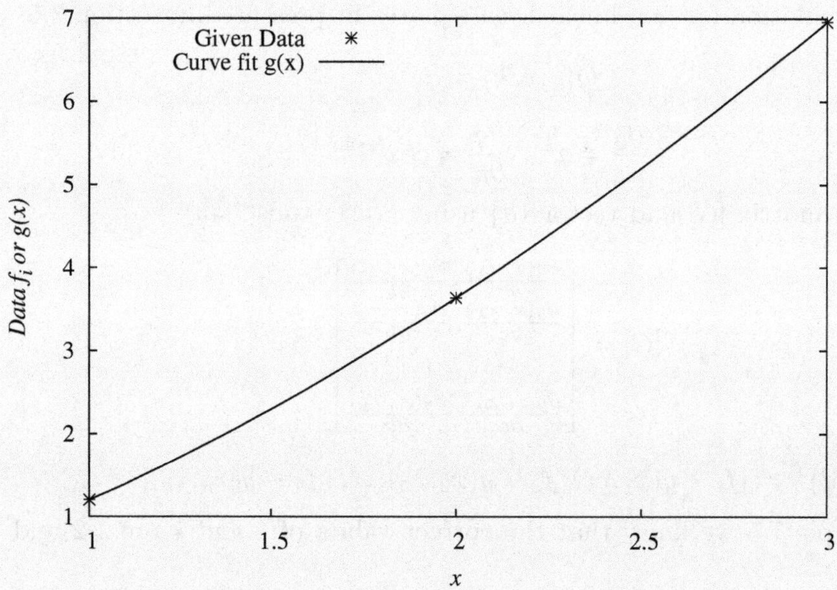

Figure 7.9: f_i versus x_i or $g(x)$ versus x: example 7.8

Figure 7.9 shows plots of (x_i, f_i) and $g(x)$ versus x.

Example 7.9. Here we consider the same problem as example 7.7 but apply the general formulation for non-linear least squares fit.

$$g(x, c, k) = ce^{kx}$$

$$\frac{\partial g}{\partial c} = e^{kx} \, , \quad \frac{\partial g}{\partial k} = cke^{kx}$$

Matrix $[G]$ and vector $\{d\}$ are constructed using using (7.41) and (7.43)

$$[G] = \begin{bmatrix} \frac{\partial g(x_1, c_k)}{\partial c} & \frac{\partial g(x_1, c_k)}{\partial k} \\ \frac{\partial g(x_2, c_k)}{\partial c} & \frac{\partial g(x_2, c_k)}{\partial k} \\ \vdots & \vdots \\ \frac{\partial g(x_n, c_k)}{\partial c} & \frac{\partial g(x_n, c_k)}{\partial k} \end{bmatrix}$$

$$\{d\}^T = [f_1 - g(x_1, c_k), f_2 - g(x_2, c_k), \dots, f_n - g(x_n, c_k)]$$

From example 7.7, the correct values of c and k are $c = 1.2$ and $k = 0.8$. We choose $c_1 = 1.1$ and $k_1 = 0.7$, i.e. $\{c\}$ in (7.41) as $\{c\}_1^T = [1.1\,, 0.7]$ and $\Delta_1 = 10^{-6}$ as convergence tolerance for $\{\Delta c\}$. Details are given in the following.

For $k = 1$ (iteration one)

$$[G]_1 = \begin{bmatrix} 1.0 & 0.0 \\ 2.01375 & 2.21513 \\ 4.05520 & 8.92144 \end{bmatrix}$$

$$[G]_1^T[G]_1 = \begin{bmatrix} 21.4998 & 40.6389 \\ 40.6389 & 84.4989 \end{bmatrix}$$

$$\{[G]_1^T\{d\}_1\}^T = [7.63085 \ 14.2388]$$

$$[G]_1^T[G]_1\{\Delta c\}_1 = [G]_1^T\{d\}_1$$

gives

$$\{\Delta c\}_1^T = [0.093517 \ 0.12353]$$

Hence,

$$\{c\}_2^T = \{c\}_1^T + \{\Delta c\}_1^T = [1.1935166 \ 0.823533]$$

and $R = 6.63185 \times 10^{-2}$

For $k = 2$ (iteration two)

$$[G]_2 = \begin{bmatrix} 1.0 & 0.0 \\ 2.27854 & 2.71947 \\ 5.19173 & 12.3928 \end{bmatrix}$$

$$[G]_2^T[G]_2 = \begin{bmatrix} 33.1457 & 70.5365 \\ 70.5365 & 160.978 \end{bmatrix}$$

$$\{[G]_2^T\{d\}_2\}^T = [-1.41707 \quad -3.26531]$$

$$[G]_2^T[G]_2\{\Delta c\}_2 = [G]_2^T\{d\}_2$$

gives

$$\{\Delta c\}_2^T = [6.1246 \times 10^{-3} \quad -2.2968 \times 10^{-2}]$$

Hence,

$$\{c\}_3^T = \{c\}_2^T + \{\Delta c\}_2^T = [1.19964123 \; 0.800565183]$$

with $R = 2.50561 \times 10^{-5}$

For $k = 3$ (iteration three)

$$[G]_3 = \begin{bmatrix} 1.0 & 0.0 \\ 2.22680 & 2.67136 \\ 4.95863 & 11.8972 \end{bmatrix}$$

$$[G]_3^T[G]_3 = \begin{bmatrix} 30.5467 & 64.9423 \\ 64.9423 & 148.679 \end{bmatrix}$$

$$\{[G]_3^T\{d\}_3\}^T = [-2.57275 \times 10^{-2} \quad -6.06922 \times 10^{-2}]$$

$$[G]_3^T[G]_3\{\Delta c\}_3 = [G]_3^T\{d\}_3$$

gives

$$\{\Delta c\}_3^T = [3.5895 \times 10^{-4} \quad -5.6500 \times 10^{-4}]$$

Hence,

$$\{c\}_4^T = \{c\}_3^T + \{\Delta c\}_3^T = [1.20000017 \; 0.80000019]$$

and $R = 3.53244 \times 10^{-12}$

For $k = 4$ (iteration four)

$$[G]_4 = \begin{bmatrix} 1.0 & 0.0 \\ 2.22554 & 2.67065 \\ 4.95303 & 11.8873 \end{bmatrix}$$

$$[G]_4^T[G]_4 = \begin{bmatrix} 30.4856 & 64.8218 \\ 64.8218 & 148.44 \end{bmatrix}$$

$$\{[G]_4^T\{d\}_4\}^T = [-9.38711 \times 10^{-6} \quad -2.22698 \times 10^{-5}]$$

$$[G]_4^T[G]_4\{\Delta c\}_4 = [G]_4^T\{d\}_4$$

gives

$$\{\Delta c\}_4^T = [1.5505 \times 10^{-7} \quad -2.1773 \times 10^{-7}]$$

Hence,

$$\{c\}_5^T = \{c\}_4^T + \{\Delta c\}_4^T = [1.200000290 \; 0.799999952]$$

with $R = 0.313247 \times 10^{-13}$

Absolute value of each components of $\{c\}_4 - \{c\}_3$ is less than or equal to $\Delta_1 = 10^{-6}$, hence

$$\{c\}_2^T = [c, k] = [1.2 \; 0.8]$$

Thus we have

$$g(x) = 1.2e^{0.8x}$$

is the desired least squares fit. This is same as in example 7.7.

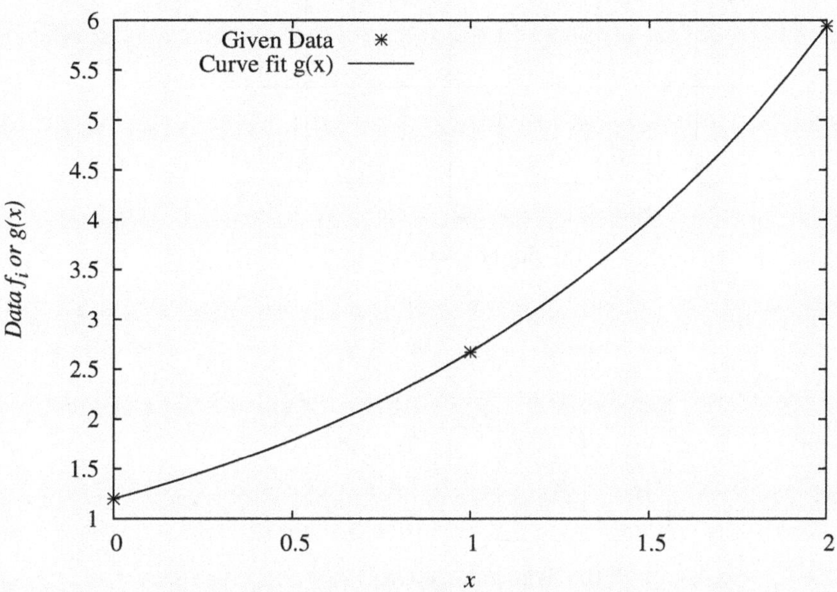

Figure 7.10: f_i versus x_i or $g(x)$ versus x: example 7.9

Figure 7.10 shows plots of (x_i, f_i) and $g(x)$ versus x.

Remarks.

(1) Examples 7.1 - 7.4, linear least squares fit have also been solved using the general non-linear least squares fit (section 7.5), the results are identical to those obtained in examples 7.1 - 7.4, hence are not repeated here.

(2) In examples 7.1 - 7.4 when using formulations of section 7.5 the initial

or starting values of the unknown coefficients is irrelevant. The process always converges in two iterations as the problem is linear.

7.6 Least squares fit using sinusoidal functions (LS-FSF)

Use of trigonometric functions in the least squares fit is useful for least squares fit of periodic data. Square waves, sawtooth waves, triangular waves, etc. are examples of periodic functions encountered in many applications.

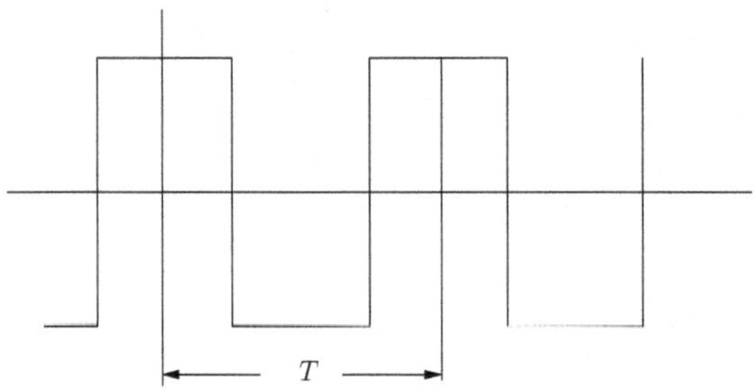

(a) Square wave

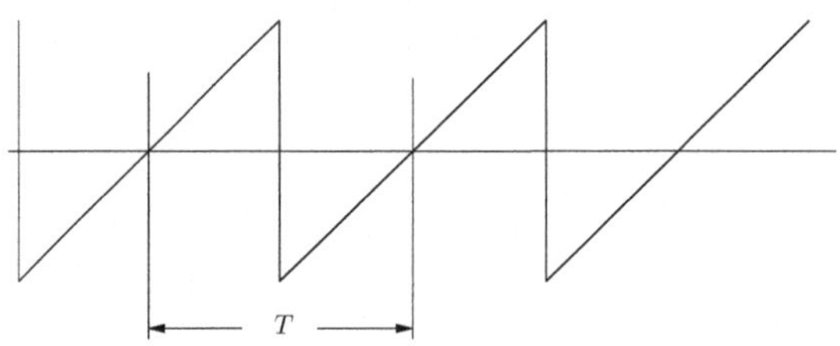

(b) Triangular wave

Figure 7.11: Periodic functions

For a periodic function $f(t)$ we have

$$f(t) = f(t+T) \tag{7.54}$$

in which T is called period. T is the smallest value of time for which (7.54) holds i.e. $f(\cdot)$ repeats after every value of time as a multiple of T.

In least squares fit we can generally use functions in time of the forms

$$\begin{aligned} g(t) &= \tilde{c}_1 + \tilde{c}_2 \cos(\omega t + \phi) \\ \text{or} \quad g(t) &= \underset{\sim}{c}_1 + \underset{\sim}{c}_2 \sin(\omega t + \phi) \end{aligned} \tag{7.55}$$

$\tilde{c}_1$ or $\underset{\sim}{c}_1$ is mean value, $\tilde{c}_2$ or $\underset{\sim}{c}_2$ is the peak value of the oscillating function $\cos(\omega t + \phi)$ or $\sin(\omega t + \phi)$, ω is the frequency i.e. how often the cycle repeats and ϕ is called phase shift that defines how the function is shifted horizontally. Negative ϕ implies lag whereas positive ϕ results in lead.

An alternative to (7.55) is to use

$$g(t) = c_1 + c_2 \cos \omega t + c_3 \sin \omega t \tag{7.56}$$

We note that (7.56) can be obtained from (7.55) by expanding $\cos(\omega t + \phi)$ or $\sin(\omega t + \phi)$ and by defining new coefficients. For example

$$g(t) = \tilde{c}_1 + \tilde{c}_2 \cos(\omega t + \phi) \tag{7.57}$$

or

$$\begin{aligned} g(t) &= \tilde{c}_1 + \tilde{c}_2(\cos \omega t \cos \phi - \sin \omega t \sin \phi) \\ &= \tilde{c}_1 + (\tilde{c}_2 \cos \phi) \cos \omega t + (-\tilde{c}_2 \sin \phi) \sin \omega t \end{aligned} \tag{7.58}$$

Let

$$c_1 = \tilde{c}_1 \ , \ c_2 = \tilde{c}_2 \cos \phi \ , \ c_3 = -\tilde{c}_2 \sin \phi \tag{7.59}$$

Then, we can write (7.57) as

$$g(t) = c_1 + c_2 \cos \omega t + c_3 \sin \omega t \tag{7.60}$$

on the other hand if we consider

$$g(t) = \underset{\sim}{c}_1 + \underset{\sim}{c}_2 \sin(\omega t + \phi) \tag{7.61}$$

or

$$g(t) = \underset{\sim}{c}_1 + \underset{\sim}{c}_2(\sin \omega t \cos \phi + \cos \omega t \sin \phi) \tag{7.62}$$

Let

$$c_1 = \underset{\sim}{c}_1 \ , \ c_2 = \underset{\sim}{c}_2 \sin \phi \ , \ c_3 = \underset{\sim}{c}_2 \cos \phi \tag{7.63}$$

Then, we can write (7.61) as

$$g(t) = c_1 + c_2 \cos(\omega t) + c_3 \sin(\omega t) \tag{7.64}$$

Thus, instead of using (7.55) we consider (7.60)(or (7.64)).

Let $(t_i, f(t_i))$ or (t_i, f_i); $i = 1, 2, \dots, n$ be given data with time period T, hence with $\omega = \frac{2\pi}{T}$ (radians/sec), the angular frequency.
Let

$$g(t) = c_1 + c_2 \cos \omega t + c_3 \sin \omega t \tag{7.65}$$

be the desired least squares fit to the data (t_i, f_i); $i = 1, 2, \dots, n$. We can rewrite (7.65) in standard form (7.1) by letting

$$g_1(t) = 1 \ , \ g_2(t) = \cos \omega t \ \text{ and } \ g_3(t) = \sin \omega t \tag{7.66}$$

Then, (7.65) can be written as

$$g(t) = \sum_{k=1}^{3} c_k g_k(t) \tag{7.67}$$

The residuals r_i; $i = 1, 2, \dots, n$ are given by

$$g(t_i) - f_i = \sum_{k=1}^{3} c_k g_k(t_i) - f_i = r_i \ ; \ i = 1, 2, \dots, n \tag{7.68}$$

In matrix form we can write (7.66) as

$$\begin{bmatrix} g_1(t_1) & g_2(t_1) & g_3(t_1) \\ g_1(t_2) & g_2(t_2) & g_3(t_2) \\ \vdots & \vdots & \vdots \\ g_1(t_n) & g_2(t_n) & g_3(t_n) \end{bmatrix} \begin{Bmatrix} c_1 \\ c_2 \\ c_3 \end{Bmatrix} - \begin{Bmatrix} f_1 \\ f_2 \\ \vdots \\ f_n \end{Bmatrix} = \begin{Bmatrix} r_1 \\ r_2 \\ \vdots \\ r_n \end{Bmatrix} \tag{7.69}$$

or

$$[G]\{c\} - \{f\} = \{r\} \tag{7.70}$$

In weighted least squares curve fit we consider

$$(R)_{\text{minimizing}} = (\sum_{i=1}^{n} w_i r_i^2)_{\text{minimizing}} \tag{7.71}$$

Following section 7.3 we obtain the following

$$[G]^T[W][G]\{c\} = [G]^T[W]\{f\} \tag{7.72}$$

which is same as (7.22) in section 7.3.
A more compact form of (7.72) can be derived by substituting from (7.66) in $[G]$ and then carrying out the matrix multiplication in (7.72) and we obtain

$$\begin{bmatrix} \sum_{i=1}^{n} w_i & \sum_{i=1}^{n} w_i \cos\omega t_i & \sum_{i=1}^{n} w_i \sin\omega t_i \\ \sum_{i=1}^{n} w_i \cos\omega t_i & \sum_{i=1}^{n} w_i (\cos\omega t_i)^2 & \sum_{i=1}^{n} w_i \sin\omega t_i \cos\omega t_i \\ \sum_{i=1}^{n} w_i \sin\omega t_i & \sum_{i=1}^{n} w_i \cos\omega t_i \sin\omega t_i & \sum_{i=1}^{n} w_i (\sin\omega t_i)^2 \end{bmatrix} \begin{Bmatrix} c_1 \\ c_2 \\ c_3 \end{Bmatrix} = \begin{Bmatrix} \sum_{i=1}^{n} w_i f_i \\ \sum_{i=1}^{n} w_i f_i \cos\omega t_i \\ \sum_{i=1}^{n} w_i f_i \sin\omega t_i \end{Bmatrix} \tag{7.73}$$

We compute c_1, c_2 and c_3 using (7.73). Once we know c_1, c_2 and c_3, the least squares fit $g(t)$ to the data (t_i, f_i) or $(t_i, f(t_i))$ is defined.

Remarks. Equations (7.73) can be further simplified if weight factors $w_i = 1$; $i = 1, 2, \dots, n$ and if the points $t = t_i$; $i = 1, 2, \dots, n$ are equispaced with time interval Δt i.e. for the time period T we have $T = (n-1)\Delta t$.
Then

$$\begin{aligned} &\sum_{i=1}^{n} \sin \omega t_i = 0\,, \quad \sum_{i=1}^{n} \cos \omega t_i = 0 \\ &\sum_{i=1}^{n} \sin^2 \omega t_i = \frac{n}{2}\,, \quad \sum_{i=1}^{n} \cos^2 \omega t_i = \frac{n}{2} \\ &\sum_{i=1}^{n} \cos \omega t_i \sin \omega t_i = 0 \end{aligned} \tag{7.74}$$

using (7.74) in (7.73) we obtain

$$\begin{bmatrix} n & 0 & 0 \\ 0 & \frac{n}{2} & 0 \\ 0 & 0 & \frac{n}{2} \end{bmatrix} \begin{Bmatrix} c_1 \\ c_2 \\ c_3 \end{Bmatrix} = \begin{Bmatrix} \sum\limits_{i=1}^{n} f_i \\ \sum\limits_{i=1}^{n} f_i \cos \omega t_i \\ \sum\limits_{i=1}^{n} f_i \sin \omega t_i \end{Bmatrix} \tag{7.75}$$

Hence

$$\begin{aligned} c_1 &= \frac{1}{n}\Big(\sum_{i=1}^{n} f_i\Big) \\ c_2 &= \frac{2}{n}\Big(\sum_{i=1}^{n} f_i \cos \omega t_i\Big) \\ c_3 &= \frac{2}{n}\Big(\sum_{i=1}^{n} f_i \sin \omega t_i\Big) \end{aligned} \tag{7.76}$$

Example 7.10. In this example we consider least squares fit using sinusoidal functions.
Consider

$$f(t) = 1.5 + 0.5 \cos 4t + 0.25 \sin 4t$$

for $t \in [0, 1.5]$. We generate t_i and $f(t_i)$ or f_i; $i = 1, 2, \dots, 16$ in equal increment $\Delta t = 0.1$. (t_i, f_i); $i = 1, 2, \dots, 16$ are given in the following ($n = 16$).

i	t_i	f_i
1	0.00000E+00	0.20000E+01
2	0.10000E+00	0.20579E+01
3	0.20000E+00	0.20277E+01
4	0.30000E+00	0.19142E+01
5	0.40000E+00	0.17353E+01
6	0.50000E+00	0.15193E+01
7	0.60000E+00	0.13002E+01
8	0.70000E+00	0.11126E+01
9	0.80000E+00	0.98626E+00
10	0.90000E+00	0.94099E+00
11	0.10000E+01	0.98398E+01
12	0.11000E+01	0.11084E+01
13	0.12000E+01	0.12947E+01
14	0.13000E+01	0.15134E+01
15	0.14000E+01	0.17300E+01
16	0.15000E+01	0.19102E+01

We consider these values of (t_i, f_i); $i = 1, 2, \dots, n$ to obtain a least squares fit $g(t)$ to this data using

$$g(t) = c_1 + c_2 \cos 4t + c_3 \sin 4t = \sum_{k=1}^{3} c_k g_k(t)$$

in which

$$g_1(t) = 1 \ , \ g_2(t) = \cos 4t \ \text{ and } \ g_3(t) = \sin 4t$$

$[G]$ matrix is given by

$$[G] = \begin{bmatrix} g_1(t_1) & g_2(t_1) & g_3(t_1) \\ g_1(t_2) & g_2(t_2) & g_3(t_2) \\ \vdots & \vdots & \vdots \\ g_1(t_n) & g_2(t_n) & g_3(t_n) \end{bmatrix}$$

or

$$[G] = \begin{bmatrix} 1 & \cos 4t_1 & \sin 4t_1 \\ 1 & \cos 4t_2 & \sin 4t_2 \\ \vdots & \vdots & \vdots \\ 1 & \cos 4t_n & \sin 4t_n \end{bmatrix}$$

using t_i; $i = 1, 2, \dots, 16$ in the data (t_i, f_i), we have

$$[G] = \begin{bmatrix} 0.100000E+01 & 0.100000E+01 & 0.000000E+00 \\ 0.100000E+01 & 0.921061E+00 & 0.389418E+00 \\ 0.100000E+01 & 0.696707E+00 & 0.717356E+00 \\ 0.100000E+01 & 0.362358E+00 & 0.932039E+00 \\ 0.100000E+01 & -0.291995E-01 & 0.999574E+00 \\ 0.100000E+01 & -0.416147E+00 & 0.909297E+00 \\ 0.100000E+01 & -0.737394E+00 & 0.675463E+00 \\ 0.100000E+01 & -0.942222E+00 & 0.334988E+00 \\ 0.100000E+01 & -0.998295E+00 & -0.583742E-01 \\ 0.100000E+01 & -0.896758E+00 & -0.442520E+00 \\ 0.100000E+01 & -0.653644E+00 & -0.756802E+00 \\ 0.100000E+01 & -0.307333E+00 & -0.951602E+00 \\ 0.100000E+01 & 0.874992E-01 & -0.996165E+00 \\ 0.100000E+01 & 0.468516E+00 & -0.883455E+00 \\ 0.100000E+01 & 0.775566E+00 & -0.631267E+00 \\ 0.100000E+01 & 0.960170E+00 & -0.279415E+00 \end{bmatrix}$$

$$[G]^T[G] = \begin{bmatrix} 0.160000E+02 & 0.290885E+00 & -0.414649E-01 \\ 0.290885E+00 & 0.814368E+01 & -0.418135E-01 \\ -0.414649E-01 & -0.418135E-01 & 0.785632E+01 \end{bmatrix}$$

$$\{[G]^T\{f\}\}^T = [0.241351E+02 \;\; 0.449774E+01 \;\; 0.188108E+01]$$

Using $[G]^T[G]\{c\} = [G]^T\{f\}$
we obtain

$$\{c\} = \begin{Bmatrix} c_1 \\ c_2 \\ c_3 \end{Bmatrix} = \begin{Bmatrix} 0.1500003340E+01 \\ 0.5000023250E+00 \\ 0.2500131130E+00 \end{Bmatrix}$$

with $R = 6.86627 \times 10^{-9}$
Hence, $g(t) = 1.5 + 0.5\cos\omega t + 0.25\sin\omega t$ is the desired least squares fit. Since the generated data are exact for $c_1 = 1.5$, $c_2 = 0.5$ and $c_3 = 0.25$ the least squares fit using this data produces precisely the same values of the coefficients.
A plot of (x_i, f_i) and $g(x)$ versus x is shown in 7.12.

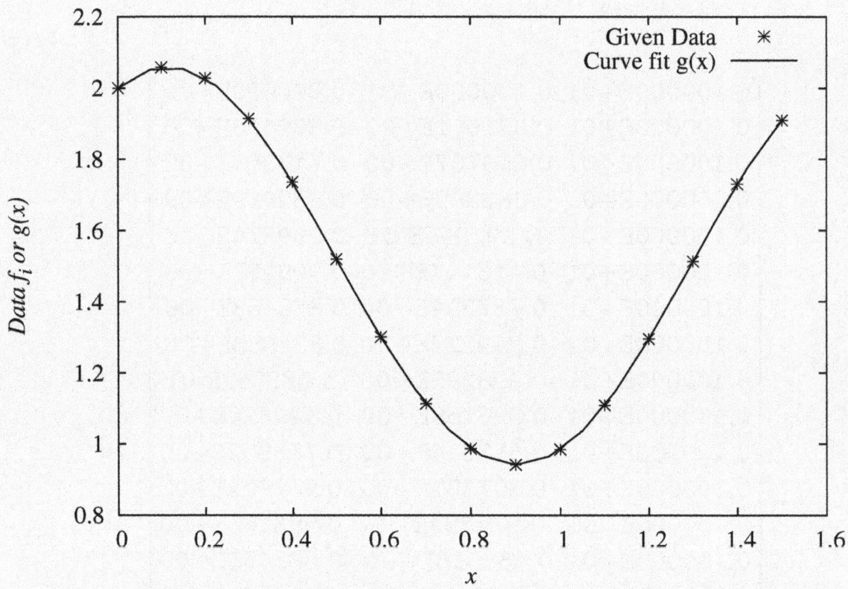

Figure 7.12: f_i versus x_i or $g(x)$ versus x: example 7.10

7.6.1 Concluding remarks

When the data points (x_i, f_i) are close together and when there is large variation in f_i values, the interpolation technique may produce wildly oscillating behavior that may not be a reasonable mathematical representation of this data set. In such cases linear least squares fit is meritorious. As we have seen the least squares fit requires that we know what function and their combinations are a reasonable mathematical description of the data. Weighted linear least squares fit provides means to assign weight factors greater than one to data points that are more accurate so that the least squares fit becomes biased towards these data points. The non-linear least squares fit for special forms suitable for taking log or natural log described in section 7.4 is a convenient way to treat special classes of non-linearities in the coefficients by modifying the linear process provided it is possible to take log natural log of both sides and obtain linear least squares fit in log or natural log. The general non-linear least squares fit presented in section 7.5 is the most general linearized approach to non-linear least squares fit with or without weight factors that is applicable to any non-linear least squares fit. This formulation automatically degenerates to linear least squares fit. Thus, the least squares fit formulation in section 7.5 is meritorious for linear as well as non-linear least squares fit, both weighted as well as without weight factors.

Problems

7.1 Consider the following data

i	1	2	3	4
x_i	−2	−1	0	1
$f_i = f(x_i)$	6	4	3	3

Consider $g(x) = c_1 + c_2 x$ to be least squares fit to this data. Find coefficients c_1 and c_2. Plot graphs of data points and $g(x)$ versus x as well as tabulate.

7.2 Find the constants c_1 and c_2 so that

$$g(x) = c_1 \sin(x) + c_2 \cos(x)$$

is least squares fit to the data in the following table.

i	1	2	3
x_i	0	$\pi/4$	$\pi/2$
$f_i = f(x_i)$	0	1	0

7.3 Find constants c_1 and c_2 so that

$$g(x) = c_1 + c_2 e^x$$

is least squares fit to the data in the following table.

i	1	2	3
x_i	0	1	2
$f_i = f(x_i)$	1	2	2

7.4 Consider the following data

i	1	2	3
x_i	0	1	2
$f_i = f(x_i)$	2	6.40	22.046

Calculate coefficients a and b in $g(x) = ae^{bx}$ for $g(x)$ to be a least squares fit to the data in the table. Calculate values of a and b using formulations in section 7.4 as well as section 7.5. Compare the computed values of a and b from the two formulations and provide some discussion.

7.5 Consider the following data

i	1	2	3
x_i	1	2	4
$f_i = f(x_i)$	1.083	3.394	9.6

Obtain constants a and b in $g(x) = ax^b$ so that $g(x)$ is a least squares fit to the data in the table. Use formulations in section 7.4 and 7.5. Compare and discuss the results obtained from the two formulations.

7.6 Construct a least squares fit to the data

i	1	2	3
x_i	0	1	2
$f_i = f(x_i)$	10	3	1

Using a form of the type $g(x) = k_1 e^{-k_2 x}$. Calculate k_1 and k_2 using the formulations in section 7.4 as well as 7.5. Compare and discuss the values of k_1 and k_2 obtained using the two formulations.

7.7 Consider the data in the following table.

i	1	2	3
x_i	0	1	2
$f_i = f(x_i)$	10	12	18

Determine the constants c_1 and c_2 in $g(x) = c_1 + c_2 x^2$ for $g(x)$ to be a least squares fit to the data in the table. Also calculate c_1 and c_2 using the non-linear formulation in section 7.5.

7.8 Consider data in the following table.

i	1	2	3
x_i	0.1	0.2	0.3
$f_i = f(x_i)$	12.161	13.457	20.332

Determine the constants c_1 and c_2 in $g(x) = c_1 + c_2 3^x$ for $g(x)$ to be a least squares fit to the data in the table. Use formulations in section 7.4 as well as 7.5. Compare and discuss the values of c_1 and c_2 obtained using the two formulations.

8

Numerical Differentiation

8.1 Introduction

In many situations, given the discrete data set (x_i, f_i); $i = 1, 2, \ldots, n$, we are faced with the problem of determining the derivative of a function f with respect to x. The discrete data set (x_i, f_i); $i = 1, 2, \ldots, n$ may be from an experiment in which we have only determined values f_i at discrete locations x_i. In such a situation the value of a function $f \quad \forall x \neq x_i \ ; \ i = 1, 2, \ldots, n$ is not known. Secondly, we only have discrete data points. A function $f(x)$ describing this data set is not known yet.

For the data set (x_i, f_i); $i = 1, 2, \ldots, n$ we wish to determine approximate value of the derivative of f with respect to x. We consider the following two approaches in this chapter.

8.1.1 Determination of Approximate Value of $\frac{d^k f}{dx^k}$; $k = 1, 2, \ldots$ using Interpolation Theory

In this approach we consider the data set (x_i, f_i); $i = 1, 2, \ldots, n$ and establish the interpolating polynomial $f(x)$ using (see Chapter 5):

(a) Polynomial approach

(b) Lagrange interpolating polynomial

(c) Newton's interpolating polynomial

The end result is that we have an analytical expression $f(x)$, a polynomial in x such that:

$$f(x_i) = f_i \qquad ; \qquad i = 1, 2, \ldots, n$$

Now, we can differentiate $f(x)$ and obtain $\frac{d^k f}{dx^k} \quad \forall x \in [x_1, x_n]$ for desired k. The strength of this approach is that once we establish the polynomial $f(x)$, $\frac{d^k f}{dx^k}$ are defined for all values of x between x_1 and x_n. This approach is straightforward and needs no further considerations. Details of determining interpolating polynomials have already been presented in Chapter 5.

8.1.2 Determination of Approximate Values of the Derivatives of f with Respect to x Only at x_i ; $i = 1, 2, \dots, n$

In many applications it is sufficient to know approximate values $\frac{d^k f}{dx^k}$; $k = 1, 2, \dots$ for discrete locations x_i; $i = 1, 2, \dots, n$ for which f_i are given. In such cases we can utilize Taylor series expansions. We consider details of this approach in this chapter. We refer to this approach as numerical differentiation using Taylor series expansions. A major limitation of this approach is that $\frac{d^k f}{dx^k}$; $k = 1, 2, \dots$ are only obtainable at discrete x_i; $i = 1, 2, \dots, n$ values.

8.2 Numerical Differentiation using Taylor Series Expansions

Consider a discrete data set:

$$(x_i, f_i) \quad ; \quad i = 1, 2, \dots, n \tag{8.1}$$

For simplicity, consider x_i ; $i = 1, 2, \dots, n$ to be equally spaced.

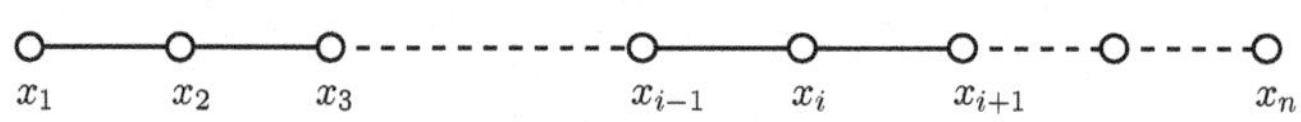

Figure 8.1: Discrete data points (x_i, f_i)

$$x_{i+1} = x_i + h \quad ; \quad i = 1, 2, \dots, n-1 \tag{8.2}$$

The scalar h is the spacing between the two successive data points.

If we pose the problem of determining $\frac{d^k f}{dx^k}$; $k = 1, 2, \dots$ at $x = x_i$, then by letting $i = 1, 2, \dots$ we can determine $\frac{d^k f}{dx^k}$; $k = 1, 2, \dots$ at $x = x_i$; $i = 1, 2, \dots, n$. Consider $x = x_i$ and f_i and two sets (for example) of data points immediately preceding $x = x_i$ as well as immediately following $x = x_i$ (see Figure 8.2). Since f_i is the value of f at x_i, we can define:

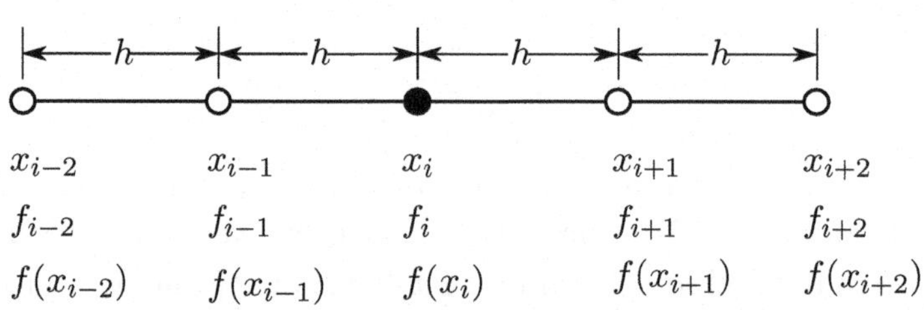

Figure 8.2: Subset of data centered on x_i

$$f_i = f(x_i) \quad ; \quad i = 1, 2, \dots, n \tag{8.3}$$

Our objective is to determine $\frac{d^k f}{dx^k}$ at $x = x_i$; $k = 1, 2, \dots$; $i = 1, 2, \dots, n$.

8.2.1 First Derivative of $\frac{df}{dx}$ at $x = x_i$

(a) *Forward difference method or first forward difference*
Consider Taylor expansion of $f(x_{i+1})$ about $x = x_i$.

$$f(x_{i+1}) = f(x_i) + f'(x_i)h + f''(x_i)\frac{h^2}{2!} + f'''(x_i)\frac{h^3}{3!} + \dots \tag{8.4}$$

or

$$f(x_{i+1}) - f(x_i) = f'(x_i)h + f''x_i\frac{h^2}{2!} + f'''(x_i)\frac{h^3}{3!} \tag{8.5}$$

or

$$f(x_{i+1}) - f(x_i) = f'(x_i)h + O(h^2) \tag{8.6}$$

or

$$\frac{f(x_{i+1}) - f(x_i)}{h} = f'(x_i) + O(h) \tag{8.7}$$

$$\therefore \quad f'(x_i) \simeq \frac{f(x_{i+1}) - f(x_i)}{h} \tag{8.8}$$

The approximate value of the derivative of f with respect to x at $x = x_i$ given by (8.8) has truncation error of the order of h $O(h)$. This is called forward difference approximation of $\frac{df}{dx}$ at $x = x_i$. By letting $i = 1, 2, \dots$ in (8.8), we can obtain $\frac{df}{dx}$ at $x = x_i$; $i = 1, 2, \dots, n-1$.

(b) *Backward difference method or first backward difference*
Consider Taylor series expansion of $f(x_{i-1})$ about $x = x_i$.

$$f(x_{i-1}) = f(x_i) - f'(x_i)h + f''(x_i)\frac{h^2}{2!} - f'''(x_i)\frac{h^3}{3!} \tag{8.9}$$

or

$$f(x_{i-1}) - f(x_i) = -f'(x_i)h + f''(x_i)\frac{h^2}{2!} - f'''(x_i)\frac{h^3}{3!} \tag{8.10}$$

or

$$f(x_{i-1}) - f(x_i) = -f'(x_i)h + O(h^2) \tag{8.11}$$

or

$$\frac{f(x_{i-1}) - f(x_i)}{h} = -f'(x_i) + O(h) \tag{8.12}$$

$$\therefore \quad f'(x_i) \simeq \frac{f(x_i) - f(x_{i-1})}{h} \tag{8.13}$$

The approximate value of the derivative of f with respect to x at $x = x_i$ given by (8.13) has truncation error of the order of h $O(h)$. This is called backward difference approximation of $\frac{df}{dx}$ at $x = x_i$. By letting $i = 1, 2, \ldots$ in (8.13), we can obtain $\frac{df}{dx}$ at $x = x_i$; $i = 2, 3, \ldots$,n.

(c) *First central difference method*
Consider Taylor series expansion (8.5) and (8.10).

$$f(x_{i+1}) - f(x_i) = f'(x_i)h + f''(x_i)\frac{h^2}{2!} + f'''(x_i)\frac{h^3}{3!} + \ldots \tag{8.14}$$

$$f(x_{i-1}) - f(x_i) = -f'(x_i)h + f''(x_i)\frac{h^2}{2!} - f'''(x_i)\frac{h^3}{3!} + \ldots \tag{8.15}$$

Subtracting (8.15) from (8.14):

$$f(x_{i+1}) - f(x_{i-1}) = 2hf'(x_i) + 2f'''(x_i)\frac{h^3}{3!} \tag{8.16}$$

or

$$f(x_{i+1}) - f(x_{i-1}) = 2hf'(x_i) + O(h^3) \tag{8.17}$$

or

$$\frac{f(x_{i+1}) - f(x_{i-1})}{2h} = f'(x_i) + O(h^2) \tag{8.18}$$

$$\therefore \quad f'(x_i) \simeq \frac{f(x_{i+1}) - f(x_{i-1})}{2h} \tag{8.19}$$

The approximate value of $\frac{df}{dx}$ at $x = x_i$ given by (8.19) has truncation error of the order of $O(h^2)$. This is known as central difference approximation of $\frac{df}{dx}$ at $x = x_i$ for $u = 2, 3, \ldots, n-1$.

Remarks.

(1) Forward difference and backward difference approximation have the same order of truncation error $O(h)$, hence we expect similar accuracy in either of these two approaches.

(2) The central difference method has truncation error of the order of $O(h^2)$, hence this method is superior to forward or backward difference method and will yield better accuracy. Thus, this is higher order approximation by one order than (a) and (b).

8.2.2 Second Derivative $\frac{d^2f}{dx^2}$ at $x = x_i$: Central Difference Method

Consider Taylor series expansions (8.5) and (8.10).

$$f(x_{i+1}) - f(x_i) = f'(x_i) + f''(x_i)\frac{h^2}{2!} + f'''(x_i)\frac{h^3}{3!} + \ldots \tag{8.20}$$

$$f(x_{i-1}) - f(x_i) = -f'(x_i) + f''(x)i)\frac{h^2}{2!} - f'''(x_i)\frac{h^3}{3!} + \ldots \tag{8.21}$$

Adding (8.20) and (8.21):

$$f(x_{i+1}) - 2f(x_i) + f(x_{i-1}) = f''(x_i)h^2 + O(h^4) \tag{8.22}$$

or

$$\frac{f(x_{i+1}) - 2f(x_i) + f(x_{i-1})}{h^2} = f''(x_i) + O(h^2) \tag{8.23}$$

$$\therefore \quad f''(x_i) \simeq \frac{f(x_{i+1}) - 2f(x_i) + f(x_{i-1})}{h^2} \tag{8.24}$$

The approximation of $\frac{d^2 f}{dx^2}$ at $x = x_i$; $i = 2, 3, \ldots, n-1$ given by (8.24) has truncation error of the order of $O(h^2)$.

8.2.3 Third Derivative $\frac{d^3 f}{dx^3}$ at $x = x_i$

Recall (8.5) and (8.10) based on Taylor series expansions of $f(x_{i+1})$ and $f(x_{i-1})$ about $x = x_i$.

$$f(x_{i+1}) - f(x_i) = f'(x_i)h + f''(x_i)\frac{h^2}{2!} + f'''(x_i)\frac{h^3}{3!} + \ldots \tag{8.25}$$

$$f(x_{i-1}) - f(x_i) = -f'(x_i)h + f''(x_i)\frac{h^2}{2!} - f'''(x_i)\frac{h^3}{3!} + \ldots \tag{8.26}$$

Also consider Taylor series expansions of $f(x_{i+2})$ and $f(x_{i-2})$ about $x = x_i$.

$$f(x_{i+2}) = f(x_i) + f'(x_i)(2h) + f''(x_i)\frac{(2h)^2}{2!} + f'''(x_i)\frac{(2h)^3}{3!} + \ldots \tag{8.27}$$

$$f(x_{i-2}) = f(x_i) - f'(x_i)(2h) + f''(x_i)\frac{(2h)^2}{2!} - f'''(x_i)\frac{(2h)^3}{3!} + \ldots \tag{8.28}$$

Subtracting (8.26) from (8.25):

$$f(x_{i+1}) - f(x_{i-1}) = 2f'(x_i)h + \frac{1}{3}f'''(x_i)h^3 + O(h^5) \tag{8.29}$$

Subtracting (8.28) from (8.27):

$$f(x_{i+2}) - f(x_{i-2}) = 4f'(x_i)h + \frac{8}{3}f'''(x_i)h^3 + O(h^5) \tag{8.30}$$

Multiply (8.29) by 2 and subtract it from (8.30).

$$f(x_{i+2}) - f(x_{i-2}) - 2f(x_{i+1}) + 2f(x_{i-1}) = 2f'''(x_i)h^3 + O(h^5) \tag{8.31}$$

$$\frac{f(x_{i+2}) - f(x_{i-2}) - 2f(x_{i+1}) + 2f(x_{i-1})}{2h^3} = f'''(x_i) + O(h^2) \tag{8.32}$$

$$\therefore \quad f'''(x_i) \simeq \frac{f(x_{i+2}) - f(x_{i-2}) - 2f(x_{i+1}) + 2f(x_{i-1})}{2h^3} \tag{8.33}$$

The approximation of $\frac{d^3f}{dx^3}$ at $x = x_i$; $i = 3, 4, \ldots, n-2$ using (8.33) has truncation error of $O(h^2)$. Since in this derivation we have considered two data points immediately before and after $x = x_i$, (8.33) can be labeled as central difference approximation of $\frac{d^3f}{dx^3}$ at $x = x_i$; $3, 4, \ldots, n-2$.

Remarks.

(1) It is also possible to derive approximations of the derivatives of f with respect to x of various orders at $x = x_i$ using purely backward differencing approach as well as purely forward differencing approach with truncation errors $O(h)$ or $O(h^2)$. A summary is given in the following:

(a) $\frac{d^kf}{dx^k}$; $k = 1, 2, 3, 4$ with $O(h)$ using Forward difference

$$\begin{aligned}
f'(x_i) &= \frac{f(x_{i+1}) - f(x_i)}{h} \\
f''(x_i) &= \frac{f(x_{i+2}) - 2f(x_{i+1}) + f(x_i)}{h^2} \\
f'''(x_i) &= \frac{f(x_{i+3}) - 3f(x_{i+2}) + 3f(x_{i+1}) - f(x_i)}{h^3} \\
f^{iv}(x_i) &= \frac{f(x_{i+4}) - 4f(x_{i+3}) + 6f(x_{i+2}) - 4f(x_{i+1}) + f(x_i)}{h^4}
\end{aligned} \tag{8.34}$$

Forward difference expressions with truncation error $O(h^2)$ can also be derived.

(b) $\frac{d^kf}{dx^k}$; $k = 1, 2, 3, 4$ with $O(h)$ using backward difference

$$\begin{aligned}
f'(x_i) &= \frac{f(x_i) - f(x_{i-1})}{h} \\
f''(x_i) &= \frac{f(x_i) - 2f(x_{i-1}) + f(x_{i-2})}{h^2} \\
f'''(x_i) &= \frac{f(x_i) - 3f(x_{i-1}) + 3f(x_{i-2}) - f(x_{i-3})}{h^3} \\
f^{iv}(x_i) &= \frac{f(x_i) - 4f(x_{i-1}) + 6f(x_{i-2}) - 4f(x_{i-3}) + f(x_{i-4})}{h^4}
\end{aligned} \tag{8.35}$$

(2) Approximating derivatives using Taylor series expansion works well and is easier to use when the data points are equally spaced or have uniform spacing.

(3) The various differencing expressions are often called *finite difference* approximations of the derivatives of the function f defined by the discrete

data set. The order of approximation n is indicated by $O(h^n)$. It indicates the order of the truncation error in Taylor series.

(4) The finite difference expressions for the derivatives of f with respect to x can also be derived using non-uniform spacing between the data points. However, in such cases it is more advantageous to establish interpolating polynomial $f(x)$ that passes through the data points and then calculate the derivatives of the function by differentiating $f(x)$.

Example 8.1. Consider the following data.

i	1	2	3	4
x_i	0	1	2	3
f_i	0	4	0	-2

Determine:

(a) $\frac{df}{dx}$ at x_i ; $i = 1, 2, \dots, 4$ using numerical differentiation. Use central difference wherever possible.

(b) Determine the Lagrange interpolating polynomial $f(x)$ that passes through the data points such that $f(x) = f_i$; $i = 1, 2, \dots, 4$. Using $f(x)$, determine derivatives of $f(x)$ at $x = 0, 1, 2, 3$ and compare these with those calculated in (a).

Solution:

(a) Using central difference approximation:

$$\left.\frac{df}{dx}\right|_{x_i} = \frac{f_{i+1} - f_{i-1}}{2h}$$

In this case $h = 1$ (spacing between the data points).

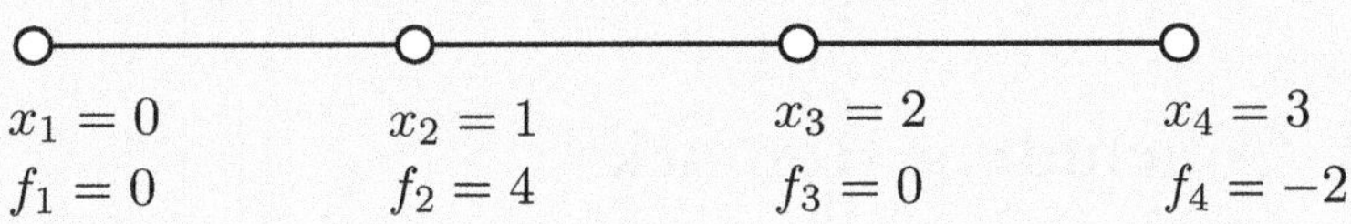

Thus we can determine $\frac{df}{dx}$ using central difference at $x = 1$ and $x = 2$.

$$\left.\frac{df}{dx}\right|_{x=1} = \frac{f_3 - f_1}{2(1)} = \frac{0 - 0}{2(1)} = 0$$

$$\left.\frac{df}{dx}\right|_{x=2} = \frac{f_1 - f_2}{2(1)} = \frac{-2 - (4)}{2} = -3$$

At $x = 0$, we do not have a choice but to use forward difference.

$$\left.\frac{df}{dx}\right|_{x=0} = \frac{f_2 - f_1}{(1)} = \frac{4-0}{1} = 4$$

At $x = 3$, we must use backward difference.

$$\left.\frac{df}{dx}\right|_{x=3} = \frac{f_4 - f_3}{(1)} = \frac{-2-0}{1} = -2$$

i	1	2	3	4
x_i	0	1	2	3
$\left.\frac{df}{dx}\right\|_{x=x_i}$	4	0	-3	-2

(b) Using Lagrange interpolating polynomial:

$$f(x) = \frac{x(x-2)(5x-17)}{3}$$

$$\frac{df}{dx} = \frac{2(5x-17)(x-1)}{3} + \frac{5x(x-2)}{3}$$

i	1	2	3	4
x_i	0	1	2	3
$\left.\frac{df}{dx}\right\|_{x_i}$	$\frac{34}{3}$	$-\frac{5}{3}$	$-\frac{14}{3}$	$\frac{11}{3}$

We note that $\frac{df}{dx}$ values in the two tables are quite different. This is generally the case when only few data points are available and the spacing between them is relatively large as is the case in this example.

8.3 Concluding Remarks

In this chapter two methods have been considered for obtaining derivatives of the function for which only discrete data is given. In the first approach an interpolating polynomial is established using the given data followed by its differentiation to obtain the desired derivative(s). This approach permits calculation of the derivatives of desired orders for any value of x in the range. In the second approach using Taylor series expansion the derivatives can be calculated only at the data point locations x_i.

Problems

8.1 Consider the following table of x_i, $f(x_i)$.

i	1	2	3	4	5	6	7
x_i	0	1/16	1/8	3/16	1/4	3/8	1/2
$f_i = f(x_i)$	0	0.19509	0.38268	0.5556	0.70711	0.9238	1.0

Compute $\frac{df}{dx}$ at $x = 1/8$ and $x = 1/4$ using forward difference and backward difference approximation with truncation error of the order $O(h)$ ($h = 1/16$ in this case). Also compute $\frac{d^2f}{dx^2}$ at $x = 1/8$ and $x = 1/4$ using central difference approximation with truncation error of the order $O(h^2)$.
Using $f(x) = \sin(\pi x)$ as the actual function describing the data in the table, calculate percentage error in the estimates of the first and the second derivatives.

8.2 Consider the following table of x_i, $f(x_i)$.

i	1	2	3	4	5
x_i	0	5	10	15	20
$f_i = f(x_i)$	0	1.60944	2.30259	2.70805	2.99573

Compute $\frac{df}{dx}$ at x =5, 10 and 15 using forward difference and backward difference approximation with truncation error of the order $O(h)$ ($h = 5$ in this case). Also compute $\frac{d^2f}{dx^2}$ at $x = 5$, 10 and 15 using central difference approximation with truncation error of the order $O(h^2)$.
Using $f(x) = \ln(x)$ as the actual function describing the data in the table, calculate percentage error in the estimates of the first and the second derivatives.

8.3 Consider the following table of x_i, $f(x_i)$.

i	1	2	3	4	5
x_i	1	2	3	4	5
$f_i = f(x_i)$	2.71828	7.3896	20.08554	54.59815	148.41316

Compute $\frac{df}{dx}$ at $x = 2$, 3 and 4 using forward difference and backward difference approximation with truncation error of the order $O(h)$ ($h = 1$ in this case). Also compute $\frac{d^2f}{dx^2}$ at $x = 2$, 3 and 4 using central difference approximation with truncation error of the order $O(h^2)$.
Using $f(x) = e^x$ as the actual function describing the data in the table, calculate percentage error in the estimates of the first and the second derivatives.

8.4 Consider the following table of x_i, $f(x_i)$.

i	1	2	3	4	5
x_i	1	2	3	4	5
$f_i = f(x_i)$	0.36788	0.13536	0.04979	0.01832	0.006738

Compute $\frac{df}{dx}$ at $x = 2$, 3 and 4 using forward difference and backward difference approximation with truncation error of the order $O(h)$ ($h = 1$ in this case). Also compute $\frac{d^2f}{dx^2}$ at $x = 2$, 3 and 4 using central difference approximation with truncation error of the order $O(h^2)$.
Using $f(x) = e^{-x}$ as the actual function describing the data in the table, calculate percentage error in the estimates of the first and the second derivatives.

8.5 Consider x_i, $f(x_i)$ given in the table below.

i	1	2	3	4	5
x_i	0	1	2	3	4
$f_i = f(x_i)$	0	0.2	1.6	5.4	32

Compute $\frac{df}{dx}$ at $x = 2$, 3 and 4 using forward difference and backward difference approximation with truncation error of the order $O(h)$ ($h = 1$ in this case). Also compute $\frac{d^2f}{dx^2}$ at $x = 2$, 3 and 4 using central difference approximation with truncation error of the order $O(h^2)$.
Using $f(x) = 0.2x^3$ as the actual function describing the data in the table, calculate percentage error in the estimates of the first and the second derivatives.

8.6 Consider the table of data (x_i, f_i); $i = 1, 2, \dots, 7$ given in problem 8.1. Using the data points (x_i, f_i); $i = 1, 2, \dots, 6$ construct a Lagrange interpolating polynomial $p(x)$ passing through the data points.
Compute $\frac{dp(x)}{dx}$ at $x = 1/8$ and $x = 1/4$, and $\frac{d^2p(x)}{dx^2}$ at $x = 1/8$ and $x = 1/4$, and compare these with $\frac{df(x)}{dx}$ and $\frac{d^2f(x)}{dx^2}$ estimated in problem 8.1 using finite difference approximation as well as with those calculated using $f(x) = \sin(\pi x)$. Also calculate percentage error in $\frac{df(x)}{dx}$ and $\frac{d^2f(x)}{dx^2}$ values using $f(x) = \sin(\pi x)$ as the true behavior of data in the table.

8.7 Consider the table of data (x_i, f_i); $i = 1, 2, \dots, 5$ given in problem 8.2. Using these data points construct a Lagrange interpolating polynomial $p(x)$ passing through the data points.
Compute $\frac{dp(x)}{dx}$ and $\frac{d^2p(x)}{dx^2}$ at $x = 5$, 10 and 15 and compare these with $\frac{df(x)}{dx}$ and $\frac{d^2f(x)}{dx^2}$ estimated in problem 8.2 using finite difference approximation as

well as with those calculated using $f(x) = \ln(x)$. Also calculate percentage error in $\frac{df(x)}{dx}$ and $\frac{d^2f(x)}{dx^2}$ values using $f(x) = \ln(x)$ as the true behavior of data in the table.

8.8 Consider the table of data (x_i, f_i); $i = 1, 2, \dots, 5$ given in problem 8.3. Using these data points construct a Lagrange interpolating polynomial $p(x)$ passing through the data points.
Compute $\frac{dp(x)}{dx}$ and $\frac{d^2p(x)}{dx^2}$ at $x = 2$, 3 and 4 and compare these with $\frac{df(x)}{dx}$ and $\frac{d^2f(x)}{dx^2}$ estimated in problem 8.3 using finite difference approximation as well as with those calculated using $f(x) = e^x$. Also calculate percentage error in $\frac{df(x)}{dx}$ and $\frac{d^2f(x)}{dx^2}$ values using $f(x) = e^x$ as the true behavior of data in the table.

8.9 Consider the table of data (x_i, f_i); $i = 1, 2, \dots, 5$ given in problem 8.4. Using these data points construct a Lagrange interpolating polynomial $p(x)$ passing through the data points.
Compute $\frac{dp(x)}{dx}$ and $\frac{d^2p(x)}{dx^2}$ at $x = 2$, 3 and 4 and compare these with $\frac{df(x)}{dx}$ and $\frac{d^2f(x)}{dx^2}$ estimated in problem 8.4 using finite difference approximation as well as with those calculated using $f(x) = e^{-x}$. Also calculate percentage error in $\frac{df(x)}{dx}$ and $\frac{d^2f(x)}{dx^2}$ values using $f(x) = e^{-x}$ as the true behavior of data in the table.

8.10 Consider the table of data (x_i, f_i); $i = 1, 2, \dots, 5$ given in problem 8.5. Using these data points construct a Lagrange interpolating polynomial $p(x)$ passing through the data points.
Compute $\frac{dp(x)}{dx}$ and $\frac{d^2p(x)}{dx^2}$ at $x = 2$, 3 and 4 and compare these with $\frac{df(x)}{dx}$ and $\frac{d^2f(x)}{dx^2}$ estimated in problem 8.5 using finite difference approximation as well as with those calculated using $f(x) = 0.2x^3$. Also calculate percentage error in $\frac{df(x)}{dx}$ and $\frac{d^2f(x)}{dx^2}$ values using $f(x) = 0.2x^3$ as the true behavior of data in the table.

9

Numerical Solutions of Boundary Value Problems

9.1 Introduction

A boundary value problem (BVP) describes a stationary process in which the state of the process does not change over time, hence the values of the dependent variables remain the same or fixed for all values of time. The mathematical description of BVP result in ordinary or partial differential equations in dependent variables and spatial coordinates x, y, and z but not time t. The BVPs also have boundary conditions that may consist of, specified values of dependent variables and/or their derivatives on the boundaries of the domain of definition of the BVP.

There are many methods currently employed for obtaining approximate numerical solutions of the BVPs:

(a) Finite difference methods

(b) Finite volume methods

(c) Finite element method

(d) Boundary element method

(e) Others

The fundamental question at this stage is 'what is mathematically the correct approach of obtaining solutions (approximate or otherwise) of differential and partial differential equations?' The answer of course is obvious if we realize that differentiation and integration go hand in hand. For example, if ϕ is given, we can obtain $\frac{d\phi}{dx}$ by differentiating ϕ. On the other hand, if $\frac{d\phi}{dx}$ is given then we can recover ϕ by integrating $\frac{d\phi}{dx}$. This fact is crucial in understanding the mathematically correct approach for obtaining solutions of ordinary differential (ODE) and partial differential equations (PDE) describing boundary value problems.

As an example, consider the simple first order ordinary differential equation (ODE):

$$\frac{d\phi}{dx} = x^2 \; ; \quad 0 < x < 2 = \Omega \qquad (9.1)$$
$$\phi(0) = 0$$

In (9.1), Ω is often called the *domain of definition* of the ODE. Ω consists of values of x for which (9.1) holds. In the ODE (9.1) we have $d\phi/dx$, hence to recover ϕ from it (which is the solution to the differential equation (9.1)) we integrate it with respect to x.

$$\int \frac{d\phi}{dx}\, dx = \int x^2\, dx + C \qquad (9.2)$$

or

$$\phi = x^3/3 + C \qquad (9.3)$$

The boundary condition $\phi(0) = 0$ gives $C = 0$, hence the solution ϕ of the ODE (9.1) is:

$$\phi = x^3/3 \qquad (9.4)$$

It is obvious that if we differentiate (9.3) with respect to x, we recover the original ODE (9.1).

Remarks.

(1) We see that integration of the ODE yields its solution and the differentiation of the solution gives back the ODE.

(2) At this stage even though we do not know the specific details of the various methods mentioned above (but regardless of the details), one thing is clear, the methods of solution of ODEs and PDEs must consider their integration in some form or the other over their domain of definition as this is the only mathematically justifiable approach for obtaining their solutions.

(3) We generally represent ODEs and PDEs using differential operators and dependent variable(s). The differential operator contains operations of differentiation (including differentiation of order zero). When the differential operator acts on the dependent variable it produces the original differential or partial differential equations. For example in case of (9.1), we can write

$$A\phi = x^2 \quad \forall x \in \Omega \qquad (9.5)$$

in which the differential operator is $A = d/dx$.

If

$$\frac{d\phi}{dx} - \frac{1}{Pe}\frac{d^2\phi}{dx^2} = f(x) \quad \forall x \in (a,b) = \Omega \qquad (9.6)$$

is the BVP, then we can write (9.6) as

$$\begin{aligned} A\phi &= f(x) \quad \forall x \in \Omega \\ A &= \frac{d}{dx} - \frac{1}{Pe}\frac{d^2}{dx^2} \end{aligned} \tag{9.7}$$

If

$$\phi\frac{d\phi}{dx} - \frac{1}{Re}\frac{d^2\phi}{dx^2} = f(x) \quad \forall x \in (a,b) = \Omega \tag{9.8}$$

is the BVP, then we can write (9.8) as

$$\begin{aligned} A\phi &= f(x) \quad \forall x \in (a,b) = \Omega \\ A &= \phi\frac{d}{dx} - \frac{1}{Re}\frac{d^2}{dx^2} \end{aligned} \tag{9.9}$$

In (9.9) the differential operator is a function of the dependent variable ϕ.

If

$$\frac{d^2\phi}{dx^2} + \phi = f(x) \quad \forall x \in (a,b) = \Omega \tag{9.10}$$

is the BVP, then we can write (9.10) as

$$\begin{aligned} A\phi &= f(x) \quad \forall x \in \Omega \\ A &= \frac{d^2}{dx^2} + 1 \end{aligned} \tag{9.11}$$

(4) If we consider methods of approximation for obtaining approximate solution of a BVP:

$$A\phi - f = 0 \quad \forall x \in \Omega \tag{9.12}$$

in which $\Omega \subset \mathbb{R}^1$ or $\mathbb{R}^2$ or $\mathbb{R}^3$ is the domain over which the BVP is valid, then based on (9.1) we must consider integration of (9.12) in some form over Ω. The approximate solution of (9.12) is then obtained numerically using this integral form. For simplicity consider the differential operator A to be linear.

(5) From (9.4) we see that an approximate solution of a BVP requires an integral form that is constructed using the BVP over the domain Ω. We discuss possible approaches in the following section.

9.2 Integral Form Corresponding to a BVP and Approximate Solution of the BVP

The integral form corresponding to a boundary value problem can be constructed over Ω using [49]: (i) either the fundamental lemma of calculus of variations or (ii) a residual functional and its extremum. We consider both approaches here, first for the entire domain Ω without its discretization. The methods of approximation when considered over the entire domain of definition (undiscretized) are called *classical methods of approximation* based on the fundamental lemma and the residual functional. We consider details in the following.

9.2.1 Integral Form Based on the Fundamental Lemma and the Approximate Solution ϕ_n

Since $A\phi - f = 0$ over Ω, if we choose a function $v(x)$ $\forall x \in \Omega$ such that $v = 0$ where ϕ is given or specified (boundary conditions), then based on the fundamental lemma of the calculus of variations we can write:

$$\int_{\bar{\Omega}} (A\phi - f)v \, dx = 0 \tag{9.13}$$

in which $\bar{\Omega} = \Omega \cup \Gamma$; Γ being the boundary of Ω. The function $v(x)$ is called *test function*. An approximation ϕ_n of ϕ can be obtained using (9.13) by assuming:

$$\phi_n(x) = \psi_0(x) + \sum_{i=1}^{n} C_i \psi_i(x) \tag{9.14}$$

in which $\psi_i(x)$; $i = 0, 1, \dots, n$ are known functions, C_i are unknown coefficients. Since $\phi_n(x)$ is approximation of the solution of the BVP, $\phi_n(x)$ satisfies the boundary conditions of the BVP. The boundary condition requirements on $\phi_n(x)$ and the differentiability and completeness requirements on $\psi_i(x)$; $i = 0, 1, \dots, n$ enable us to choose $\psi_i(x)$; $i = 0, 1, \dots, n$. The requirement that the test function $v(x) = 0$ where ϕ_n is specified implies that $v(x) = \delta\phi_n$, *variation* or change in ϕ_n, is valid as $\delta\phi_n = 0$ on boundaries where ϕ_n is specified and we have:

$$v(x) = \delta\phi_n(C_i) = \frac{\partial \phi_n}{\partial C_i} = \psi_j(x) \; ; \quad j = 1, 2, \dots, n \tag{9.15}$$

First, we rewrite (9.13) using ϕ_n instead of ϕ.

$$\int_{\bar{\Omega}} (A\phi_n - f)v \, dx = 0 \tag{9.16}$$

$$\int_{\bar{\Omega}} (A\phi_n)v \, dx = \int_{\bar{\Omega}} fv \, dx \tag{9.17}$$

Substitute ϕ_n and v from (9.14) and (9.15) into (9.17).

$$\int_{\bar{\Omega}} \left(A(\psi_0(x) + \sum_{i=1}^{n} C_i \psi_i(x)) \right) \psi_j(x) \, dx = \int_{\bar{\Omega}} f \psi_j(x) \, dx \; ; \quad j = 1, 2, \dots, n \tag{9.18}$$

or

$$\int_{\bar{\Omega}} \left(A \sum_{i=1}^{n} C_i \psi_i(x) \right) \psi_j(x) \, dx = \int_{\bar{\Omega}} f \psi_j(x) \, dx - \int_{\bar{\Omega}} A\psi_0(x)\psi_j(x) \, dx \; ; \quad j = 1, 2, \dots, n \tag{9.19}$$

or

$$\int_{\bar{\Omega}} \left(\sum_{i=1}^{n} C_i (A\psi_i(x)) \right) \psi_j(x) \, dx = \int_{\bar{\Omega}} f \psi_j(x) \, dx - \int_{\bar{\Omega}} A\psi_0(x)\psi_j(x) \, dx \; ; \quad j = 1, 2, \dots, n \tag{9.20}$$

We can write (9.20) in the matrix and vector form as:

$$[K]\{C\} = \{F\} \tag{9.21}$$

in which $[K]$ is an $n \times n$ matrix, $\{C\}$ is a vector of n unknowns, and $\{F\}$ is an $n \times 1$ vector of known quantities such that:

$$\left. \begin{aligned} K_{ij} &= \int_{\bar{\Omega}} (A\psi_j(x))\psi_i(x) \, dx \\ F_i &= \int_{\bar{\Omega}} f \psi_i(x) \, dx - \int_{\bar{\Omega}} A\psi_0(x)\psi_i(x) \, dx \end{aligned} \right\} \; ; \quad i, j = 1, 2, \dots, n \tag{9.22}$$

Using (9.21), we calculate $\{C\}$. Then, equation (9.14) defines the approximation $\phi_n(x)$ of $\phi(x)$ over $\bar{\Omega}$.

Remarks.

(1) When $v = \delta\phi_n$, $\int (A\phi_n - f) v \, dx = 0$ is called the *Galerkin method* (GM).

(2) When $v(x) = w(x) = 0$ where ϕ_n is specified but $w(x) \neq \delta\phi_n(x)$, then:

$$\int_{\bar{\Omega}} (A\phi_n(x) - f) v(x) \, dx = \int_{\bar{\Omega}} (A\phi_n(x) - f) w(x) \, dx = 0 \tag{9.23}$$

is called the *Petrov-Galerkin method* (PGM) or the *weighted residual method* (WRM).

(3) If some differentiation is transferred from ϕ_n to v in (9.17) using integration by parts, then the differentiation is lowered on ϕ_n but increased on v. We obtain the following form of (9.17).

$$B(\phi_n, v) - \underset{\sim}{l}(v) = \int_{\bar{\Omega}} fv\,dx \tag{9.24}$$

In $B(\phi_n, v)$ all terms contain both ϕ_n and v are included. The additional expression $\underset{\sim}{l}(v)$ is due to integration by parts and contains those terms that only have v. It is called the *concomitant*. We can combine $\underset{\sim}{l}(v)$ and $\int fv\,dx$ to obtain:

$$\begin{aligned} B(\phi_n, v) &= l(v) \\ l(v) &= \int_{\bar{\Omega}} fv\,dx + \underset{\sim}{l}(v) \end{aligned} \tag{9.25}$$

This method is called the *Galerkin method with weak form* (GM/WF) ($v = \delta\phi_n$) and the integral form (9.25) is called the weak form of (9.17). The reason for transferring differentiation from ϕ_n to v in (9.17) is to ensure that each term of the integrand of $B(\phi_n, v)$ contains equal orders of differentiation of ϕ_n and v. We only perform integration by parts for those terms in $\int_{\bar{\Omega}}(A\phi_n)v\,dx$ that yield this. Thus, integration by parts is performed on those terms that contain even order derivatives of ϕ_n. In such terms, after integration by parts, ϕ_n and v are interchangeable in the integrand in GM/WF in which $v = \delta\phi_n$.

(4) We note that the integrals over $\bar{\Omega}$ are definite integrals, hence produce numbers after the limits are substituted. Such integrals are called *functionals*. Thus, (9.13), (9.17), $B(\phi_n, v)$, $\underset{\sim}{l}(v)$, $l(v)$, and $\int_{\bar{\Omega}} fv\,dx$ are all functionals. In GM, PGM, and WRM also we can write (9.13) as $B(\phi_n, v) = l(v)$ in which:

$$B(\phi_n, v) = \int_{\bar{\Omega}} (A\phi_n)v\,dx \quad \text{and} \quad l(v) = \int_{\bar{\Omega}} fv\,dx \tag{9.26}$$

(5) The domain of definition Ω of the BVP is not discretized, hence GM, PGM, WRM, and GM/WF considered here are often referred to as classical methods of approximation.

(6) These methods, as we have seen here, are rather simple and straightforward in principle. The major difficulty lies in the selection of $\psi_i(x)$;

$i = 0, 1, \dots, n$ such that all boundary conditions of the BVP are satisfied by $\phi_n(x)$. Even in $\mathbb{R}^1$ this may be difficult. In $\mathbb{R}^2$ and $\mathbb{R}^3$ with involved BCs it is virtually impossible to find satisfactory functions $\psi_i(x)$; $i = 0, 1, \dots, n$.

(7) Because of the shortcoming discussed in Remark (6), classical GM, PGM, WRM, and GM/WF are virtually impossible to use in practical applications.

(8) We note that in GM/WF, $[K]$ is symmetric when the differential operator A contains only even order derivatives.

9.2.2 Integral Form Based on the Residual Functional

Let $\phi_n(x)$ given by (9.14) be the approximation of ϕ for the BVP (9.12), then the *residual function* E is defined by:

$$E = A\phi_n - f = A\psi_0 + \sum_{i=1}^{n} C_i A\psi_i - f \neq 0 \tag{9.27}$$

or

$$E = [k]\{c\} - \underset{\sim}{f}\,; \quad E^T = [c]\{k\} - \underset{\sim}{f} \tag{9.28}$$

in which

$$\begin{aligned} k_i &= A\psi_i\,; \quad i = 1, 2, \dots, n \\ \underset{\sim}{f} &= -f + A\psi_0 \end{aligned} \tag{9.29}$$

The *residual functional* I is given by:

$$I = \int_{\bar{\Omega}} E^2\,dx = \int_{\bar{\Omega}} E^T E\,dx = \int_{\bar{\Omega}} \left([c]\{k\} - \underset{\sim}{f}\right)\left([k]\{c\} - \underset{\sim}{f}\right) dx \tag{9.30}$$

or

$$I = \int_{\bar{\Omega}} \left([c]\{k\}[k]\{c\} - \underset{\sim}{f}[k]\{c\} - \underset{\sim}{f}[c]\{k\} + \underset{\sim}{f}^2\right) dx \tag{9.31}$$

Since $[k]\{c\} = [c]\{k\}$, we can write:

$$I = \int_{\bar{\Omega}} \left([c]\{k\}[k]\{c\} - 2\underset{\sim}{f}[c]\{k\} + \underset{\sim}{f}^2\right) dx \tag{9.32}$$

To find an extremum of I we set the first variation of I (i.e. δI) to zero.

$$\delta I = \frac{\partial I}{\partial \{c\}} = 0 \implies 2\int_{\bar{\Omega}} [\{k\}[k]]\,\{c\}dx - 2\int_{\bar{\Omega}} \underset{\sim}{f}\{k\}dx = 0 \tag{9.33}$$

Hence, we have:

$$\left[\int_{\bar{\Omega}} \{k\}[k]dx\right]\{c\} = \int_{\bar{\Omega}} \underset{\sim}{f}dx \tag{9.34}$$

or

$$[K]\{C\} = \{F\} \tag{9.35}$$

in which

$$\begin{aligned} K_{ij} &= \int_{\bar{\Omega}} (A\psi_i \, A\psi_j)dx \\ F_i &= (f - A\psi_0)A\psi_i \end{aligned} \quad ; \qquad i,j = 1,2,\dots,n \tag{9.36}$$

Using (9.35) we can calculate $\{C\}$, hence the approximation ϕ_n is known from (9.14).

Remarks.

(1) $[K]$ is always symmetric, a definite advantage in this method.

(2) Here also we have the same problems associated with the choice of $\psi_i(x)$ as described in Section 9.2.1, hence its usefulness for practical applications is extremely limited.

(3) If we have more than one PDE, then we have a residual function for each PDE, E_i; $i = 1,2,\dots,m$, and the residual functional is defined as:

$$I = \sum_{i=1}^{m} I_i = \sum_{i=1}^{m} \int_{\bar{\Omega}} (E_i)^2 dx \tag{9.37}$$

The details for each I_i follow what has been described for a single residual function.

9.3 Finite Element Method for Solving BVPs

The finite element method borrows its mathematical foundation from the classical methods of approximation based on the integral forms presented in Sections 9.2.1 and 9.2.2. This method eliminates all of the problems associated with the choices of ψ_0 and ψ_i; $i = 1,2,\dots,n$. In the finite element method, we discretize (subdivide) the domain $\bar{\Omega}$ into subdomains of smaller sizes than $\bar{\Omega}$ using subdomain shapes of preference. In $\mathbb{R}^1$, we have line subdomains. In $\mathbb{R}^2$, common choices are triangular or quadrilateral subdomains, and $\mathbb{R}^3$ typically makes use of tetrahedron and hexahedron subdomain shapes. Each subdomain of finite size is called a *finite element.* Figures 9.1 and 9.2 show discretizations in $\mathbb{R}^1$ and $\mathbb{R}^2$.

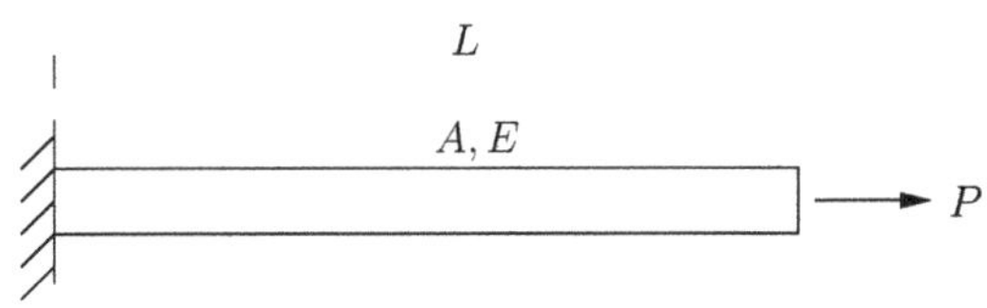

(a) Physical system

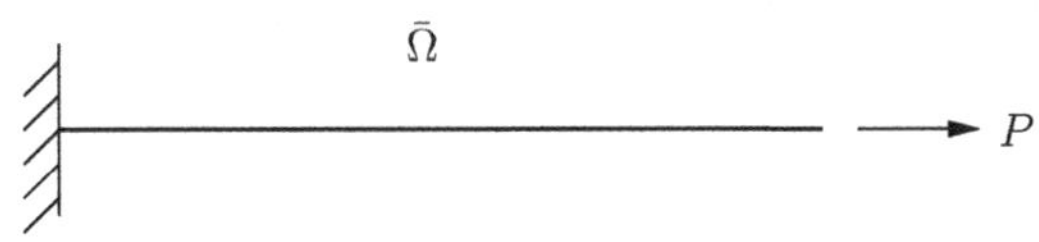

(b) Mathematical idealization

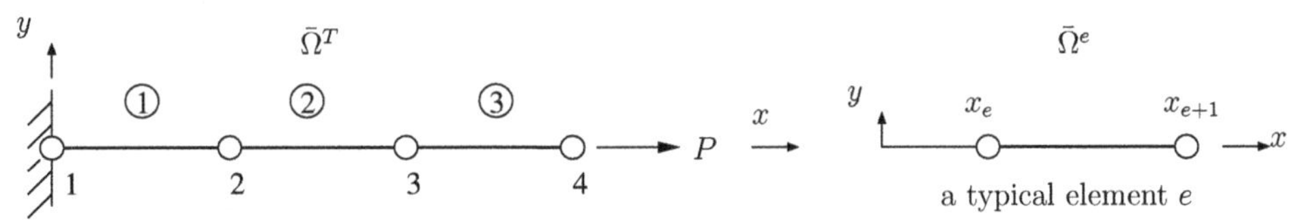

(c) Discretization $\bar{\Omega}^T$ using two-node elements

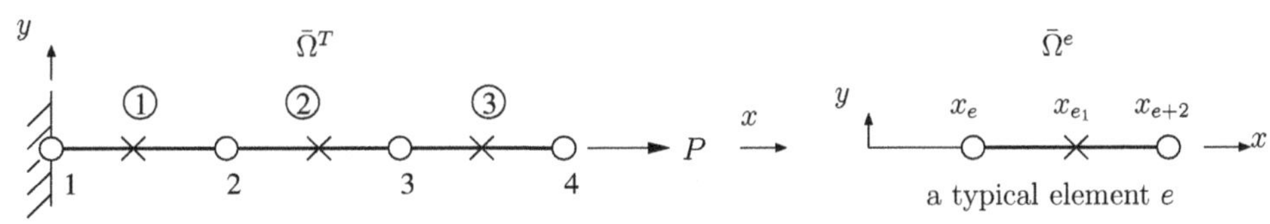

(d) Discretization $\bar{\Omega}^T$ using three-node elements

Figure 9.1: Axial rod with end load P

Each subdomain or finite element contains identifiable and desired points on its boundary and/or interior called node points. A finite element communicates to its neighboring finite elements through the node points and the mating boundaries. Choice of the node points is dictated by geometric considerations as well as considerations for defining the dependent variable ϕ over the element.

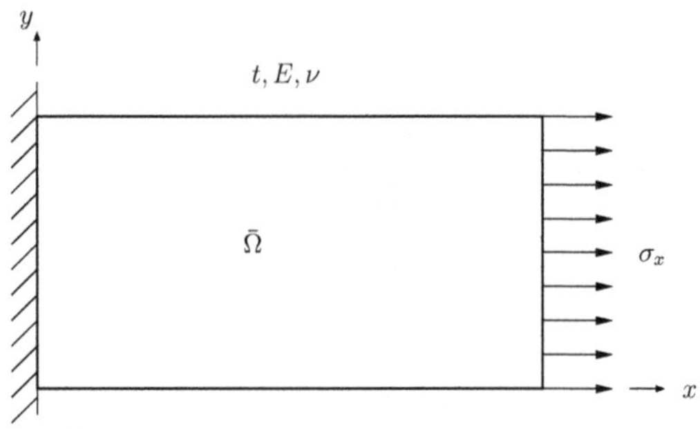

(a) Physical system

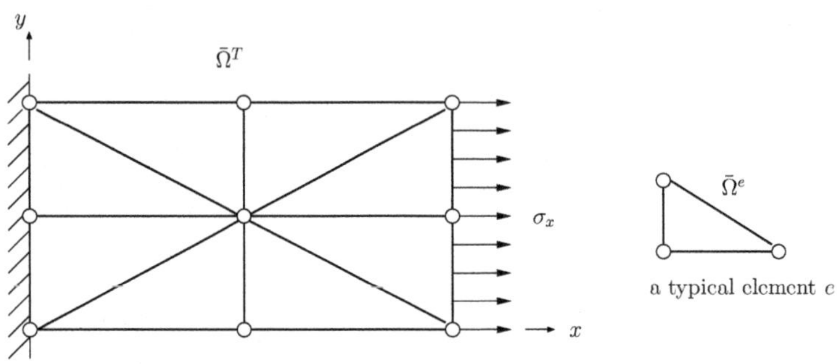

(b) Discretization using 3-node triangular elements

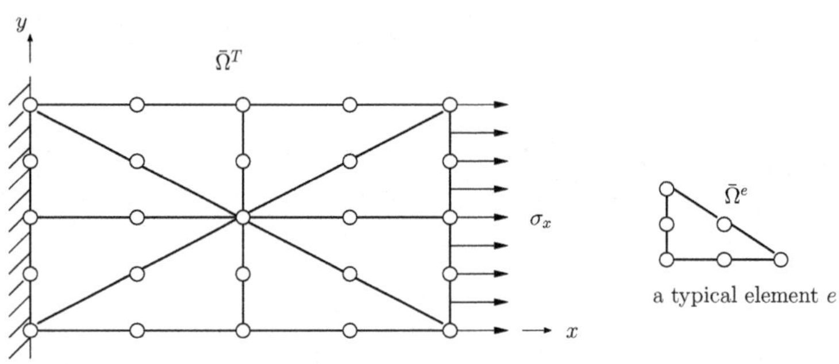

(c) Discretization using 6-node triangular elements

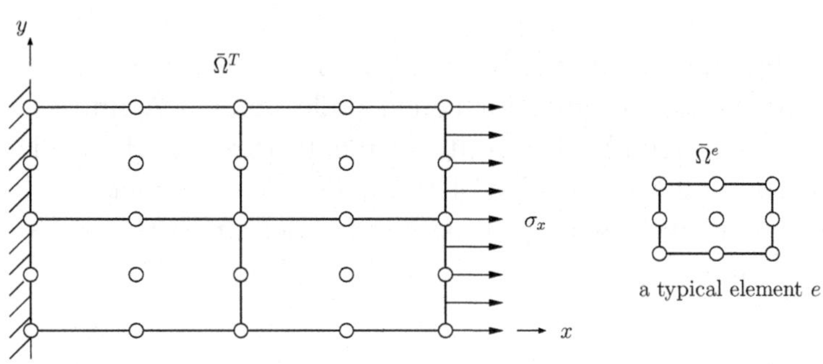

(d) Discretization using 9-node quadrilateral elements

Figure 9.2: Thin plate in tension

Let $\bar{\Omega}^T$ be the discretization of $\bar{\Omega}$, then:

$$\bar{\Omega}^T = \bigcup_e \bar{\Omega}^e \tag{9.38}$$

in which $\bar{\Omega}^e = \Omega^e \cup \Gamma^e$ is the domain of an element with its closed boundary Γ^e (see Figures 9.1 and 9.2). Let $\phi_h(x)$ be the approximation of ϕ over $\bar{\Omega}^T$, then:

$$\phi_h(x) = \bigcup_e \phi_h^e(x) \tag{9.39}$$

in which $\phi_h^e(x)$ is the approximation of $\phi(x)$ over an element e with domain $\bar{\Omega}^e$, called the *local approximation* of ϕ.

9.3.1 Finite Element Processes Based on the Fundamental Lemma

In this section, finite element formulations are constructed using integral methods described in Section 9.2.1, i.e., GM, PGM, WRM, and GM/WF. We begin with (9.13) over $\bar{\Omega}^T$ and use ϕ_h in place of ϕ.

$$\int_{\bar{\Omega}^T} (A\phi_h - f)v\,dx = 0 \tag{9.40}$$

The test function $v = \delta\phi_h$ for GM, GM/WF and $v(x) = w(x) \neq \delta\phi_h$ for in PGM, WRM. Since the definite integral in (9.40) is a functional, we can write this as a sum of the integrals over the elements.

$$\int_{\bar{\Omega}^T} (A\phi_h - f)v\,dx = \sum_e \int_{\bar{\Omega}^e} (A\phi_h^e - f)v\,dx = 0 \tag{9.41}$$

or

$$\sum_e \big(B^e(\phi_h^e, v) - l^e(v)\big) = 0 \tag{9.42}$$

$$\begin{aligned} B^e(\phi_h^e, v) &= \int_{\bar{\Omega}^e} (A\phi_h^e)v\,dx \\ l^e(v) &= \int_{\bar{\Omega}^e} fv\,dx \end{aligned} \tag{9.43}$$

Consider $A\phi - f = 0$ to be an ODE in independent coordinate $x \in (0, L)$. Let us consider local approximation ϕ_h^e of ϕ over $\bar{\Omega}^e$ in which only function values are unknown quantities at the nodes. Figure 9.3(a) and (b) show a two-element discretization in which ϕ_h^e is linear and quadratic, corresponding to the degrees of the polynomials (p-levels) one and two ($p = 1$ and $p = 2$). The nodal values of ϕ are called degrees of freedom. Thus, for element one

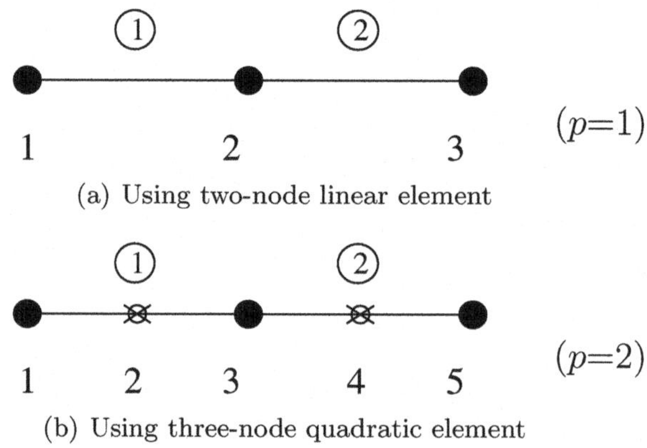

Figure 9.3: Two element uniform discretizations

and two in Figure 9.3(a), the degrees of freedom are $\{\delta^1\}$ and $\{\delta^2\}$. Using Lagrange interpolating polynomials (Chapter 5), we can easily define local approximations ϕ_h^1 and ϕ_h^2 for elements one and two. First, the elements are mapped into ξ-space, i.e., $\bar{\Omega}^e \to \bar{\Omega}^\xi = [-1, 1]$ using (for an element):

$$x(\xi) = \left(\frac{1-\xi}{2}\right) x_e + \left(\frac{1+\xi}{2}\right) x_n \tag{9.44}$$

For elements one and two of Figure 9.3(a), we have $(e, n) = (1, 2)$ and $(2, 3)$, whereas for elements one and two of Figure 9.3(b) we have $(e, n) = (1, 3)$ and $(3, 5)$. The mapping (9.44) is a linear stretch in both cases. The local approximations $\phi_h^1(\xi)$ and $\phi_h^2(\xi)$ can now be established in ξ-space using Lagrange interpolation and $\{\delta^1\}$ and $\{\delta^2\}$.

$$\phi_h^e(\xi) = [N]\{\delta^e\} \; ; \quad e = 1, 2 \tag{9.45}$$

In the case of Figure 9.3(a), the functions $N_1(\xi)$ and $N_2(\xi)$ for both $\phi_h^1(\xi)$ and $\phi_h^2(\xi)$ are:

$$N_1(\xi) = \frac{1-\xi}{2} \; ; \qquad N_2(\xi) = \frac{1+\xi}{2} \tag{9.46}$$

For Figure 9.3(b) we have $N_1(\xi)$, $N_2(\xi)$, and $N_3(\xi)$.

$$N_1(\xi) = \frac{\xi(\xi-1)}{2} \; ; \quad N_2(\xi) = 1 - \xi^2 \; ; \quad N_3(\xi) = \frac{\xi(\xi+1)}{2} \tag{9.47}$$

Thus, we could write (9.45) as:

$$\phi_h^e(\xi) = \sum_{i=1}^{n} N_i(\xi) \delta_i^e \tag{9.48}$$

When using Lagrange interpolation functions, $n = 2, 3, \ldots$ for $p = 1, 2, \ldots$, hence the corresponding elements will contain $p + 1$ nodes (in $\mathbb{R}^1$).

9.3.1.1 Finite Element Processes Based on GM, PGM, WRM

In these methods we consider (9.41) or (9.42) (without integration by parts). Choice of v defines the method. For an element e we consider:

$$\int_{\bar{\Omega}^e} (A\phi_h^e - f)v\,dx = \int_{\bar{\Omega}^e} (A\phi_h^e)v\,dx - \int_{\bar{\Omega}^e} fv\,dx \tag{9.49}$$

in which ϕ_h^e is given by (9.48) and we choose:

$$v = w_j \;;\quad j = 1, 2, \dots, n \tag{9.50}$$

Keep in mind that $w_j = N_j$ in GM but $w_j \neq N_j$ in PGM and WRM. Using (9.50) and (9.42) in (9.49) (and choosing the test function for GM):

$$\int_{\bar{\Omega}^e}(A\phi_h^e - f)v\,dx = \int_{\bar{\Omega}^e} A\Big(\sum_{i=1}^{n} N_i\delta_i^e\Big)N_j\,dx - \int_{\bar{\Omega}^e} fN_j\,dx \;;\; j = 1, 2, \dots, n \tag{9.51}$$

$$= [K^e]\{\delta^e\} - \{f^e\} \tag{9.52}$$

in which

$$\left.\begin{aligned} K_{ij}^e &= \int_{\bar{\Omega}^e} N_i(AN_j)\,dx = \int_{-1}^{1} N_i(AN_j)J\,d\xi \;;\; J = {}^{h_e}\!/_2 \\ f_i^e &= \int_{-1}^{1} fN_iJ\,d\xi \end{aligned}\right\}\quad i, j = 1, 2, \dots, n \tag{9.53}$$

Thus, for elements (1) and (2) of Figure 9.3(a) we have:

$$\int_{\bar{\Omega}^1} (A\phi_h^1)v\,dx = \begin{bmatrix} K_{11}^1 & K_{12}^1 \\ K_{21}^1 & K_{22}^1 \end{bmatrix} \begin{Bmatrix} \phi_1 \\ \phi_2 \end{Bmatrix} - \begin{Bmatrix} f_1^1 \\ f_2^1 \end{Bmatrix} \;;\quad \text{element one} \tag{9.54}$$

$$\int_{\bar{\Omega}^2} (A\phi_h^2)v\,dx = \begin{bmatrix} K_{11}^2 & K_{12}^2 \\ K_{21}^2 & K_{22}^2 \end{bmatrix} \begin{Bmatrix} \phi_2 \\ \phi_3 \end{Bmatrix} - \begin{Bmatrix} f_1^2 \\ f_2^2 \end{Bmatrix} \;;\quad \text{element two} \tag{9.55}$$

These are called the element equations.

For the two-element discretization $\bar{\Omega}^T$ we have:

$$\sum_{e=1}^{2} \int_{\bar{\Omega}^T} (A\phi_h^e - f)v\,dx = 0 \tag{9.56}$$

Equations (9.54) and (9.55) must be substituted into (9.56) to obtain their sum. Since the degrees of freedom for elements are different, the summation process, or *assembly*, of the element equations requires care. From the discretization shown in Figure 9.3(a) and the dofs at the nodes or grid points, we know that (9.56) will yield:

$$[K]\{\delta\} = \{F\} \tag{9.57}$$

in which $[K]$ is a 3×3 matrix, $\{\delta\}^T = [\phi_1 \; \phi_2 \; \phi_3]$, and $\{F\}$ is a 3×1 vector. The contents of $[K]$ and $\{F\}$ are obtained by summing (9.54) and (9.55). The simplest way to do this is to set aside a 3×3 space for $[K]$ and initialize its contents to zero. Label its rows and columns as ϕ_1, ϕ_2, and ϕ_3. Likewise label the rows and columns of $[K^1]$ and $[K^2]$ as ϕ_1, ϕ_2 and ϕ_2, ϕ_3 and the rows of $\{f^1\}, \{f^2\}$ as ϕ_1, ϕ_2 and ϕ_2, ϕ_3. Now add the elements of $[K^1]$, $\{f^1\}$ and $[K^2]$, $\{f^2\}$ to $[K]$ and $\{F\}$ using the row and column identification. The end result is that (9.56) gives (9.57) containing the element contributions.

$$\begin{bmatrix} K_{11}^1 & K_{12}^1 & 0 \\ K_{21}^1 & K_{22}^1 + K_{11}^2 & K_{12}^2 \\ 0 & K_{21}^2 & K_{22}^2 \end{bmatrix} \begin{Bmatrix} \phi_1 \\ \phi_2 \\ \phi_3 \end{Bmatrix} = \begin{Bmatrix} f_1^1 \\ f_2^1 + f_1^2 \\ f_2^2 \end{Bmatrix} \tag{9.58}$$

We impose boundary conditions on one or more of ϕ_1, ϕ_2, and ϕ_3 and solve for the remaining. Thus, now ϕ_1, ϕ_2, and ϕ_3 are known and we have an approximation for the solution ϕ (i.e., ϕ_h) in (9.48), hence ϕ_h^e for each element of the discretization.

Remarks.

(1) The assembly process remains the same for Figure 9.3(b), except that in this case the element matrices and vectors are (3×3) and (3×1) and the assemble $[K]$ and $\{F\}$ are (5×5) and (5×1).

(2) We shall consider a specific example in the following section.

9.3.1.2 Finite Element Processes Based on GM/WF

For an element e we consider $\int_{\bar{\Omega}^e} (A\phi_h^e - f)v \, dx$. For those terms in the integrand that contain even order derivatives of ϕ_h^e, we transfer half of the differentiation to v. By doing so, we can make the order of differentiation on ϕ_h^e and v in these terms the same. This results in a symmetric coefficient matrix for the element corresponding to these terms. The integration by parts results in boundary terms or boundary integrals, called the concomitant. Thus, in this process we have:

$$\int_{\bar{\Omega}^e} (A\phi_h^e - f)v \, dx = \int_{\bar{\Omega}^e} (A\phi_h^e)v \, dx - \int_{\bar{\Omega}^e} fv \, dx \tag{9.59}$$

or

$$\int_{\bar{\Omega}^e} (A\phi_h^e - f)v\,dx = B^e(\phi_h^e, v) - \underset{\sim}{l}^e(v) - \int_{\bar{\Omega}^e} fv\,dx \tag{9.60}$$

We note that (9.60) is due to (9.59) after integration by parts. This is referred to as weak form of (9.59) due to the fact that it contains lower order derivatives of ϕ_h^e compared to the BVP. $B^e(\phi_h^e, v)$ contains only those terms that contain both ϕ_h^e and v. The concomitant $\underset{\sim}{l}^e(v)$ only contains the terms that resulting from integration by parts that have v (and not ϕ_h^e). After substituting for ϕ_h^e and $v = \delta\phi_h^e = N_j$; $j = 1, 2, \dots, n$, we obtain:

$$\int_{\bar{\Omega}^e} (A\phi_h^e - f)v\,dx = [K^e]\{\delta^e\} - \{P^e\} - \{f^e\} \tag{9.61}$$

The vector $\{P^e\}$ is due to the concomitant and is called the vector of *secondary variables.* The assembly process for $[K^e]$ and $\{f^e\}$ follows Section 9.3.1.1. Assembly for $\{P^e\}$, giving $\{P\}$, is the same as that for $\{f^e\}$.

Remarks.

(1) When the differential operator A contains only even order derivatives, $[K^e]$ and $[K]$ are assured to be symmetric. This is not true in GM, PGM, and WRM.

(2) After imposing boundary conditions, $[K]$ is positive-definite, hence a unique solution $\{\delta\}$ is ensured. This may not be the case in GM, PGM, and WRM.

(3) The concomitant resulting due to integration by parts contains valuable and crucial details. For $A\phi = f \quad \forall x \in \Omega$ BVP we designate concomitant by $< A\phi, v >_{\Gamma^e}$ in which $\bar{\Omega}^e = \Omega^e \cup \Gamma^e = [x_e, x_{e+1}]$ or $[x_e, x_n]$ is the domain of the eth element in $\mathbb{R}^1$. Concomitant in $\mathbb{R}^1$ are boundary terms. The precise nature of these depends upon the differential operator A. In general $< A\phi, v >_{\Gamma^e}$ may contain any of the terms $(v(\cdot))|_{x_e}^{x_{e+1}}, ({dv}/{dx}(\cdot))\,|_{x_e}^{x_{e+1}}, \dots,$ *etc.* or all of these depending upon the order of the differentiation of ϕ in $A\phi - f = 0$. For illustration purposes let

$$< A\phi, v >_{\Gamma^e} = v(p)|_{x_e}^{x_{e+1}} + \frac{dv}{dx}(q)|_{x_e}^{x_{e+1}} \tag{9.62}$$

Then,

(1) $\phi, {d\phi}/{dx}$ (due to $v, {dv}/{dx}$) are called primary variables (PV).

(2) p and q are called secondary variables (SV).

(3) $\phi = \phi_0$, ${d\phi}/{dx} = g_0$ (on some boundaries) are called essential boundary conditions (EBC).

(4) $p = p_0$ and $q = q_0$ (on some boundaries) are called natural boundary conditions (NBC).

We shall see that NBC are naturally satisfied or absorbed while EBC need to be specified or imposed on the assembled equations to ensure uniqueness of the solution.

In $\mathbb{R}^2$ the concomitant is a contour integral over closed contour Γ^e of an element e. In $\mathbb{R}^3$ the concomitant is a surface integral. Simplification of concomitant in $\mathbb{R}^2$ and $\mathbb{R}^3$ requires that we split the integral over Γ^e into integral over Γ_1^e, Γ_2^e on which EBC and NBC are specified. For specific details on these see reference [49].

9.3.2 Finite Element Processes Based on the Residual Functional: Least Squares Finite Element Method (LSFEM)

Recall the material presented in Section 9.2.2 related to the classical method based on the residual functional.

$$E = A\phi_h - f \qquad \forall x \in \bar{\Omega}^T \tag{9.63}$$

$$E^e = A\phi_h^e - f \qquad \forall x \in \bar{\Omega}^e \tag{9.64}$$

$$I = \int_{\bar{\Omega}^T} E^2 \, dx = \sum_e \int_{\bar{\Omega}^e} (E^e)^2 \, dx = \sum_e I^e \tag{9.65}$$

An extremum of I requires that:

$$\delta I = 2 \int_{\bar{\Omega}^T} E \delta E \, dx = \sum_e 2 \int_{\bar{\Omega}^e} E^e \delta E^e \, dx = \sum_e \delta I^e = 0 \tag{9.66}$$

Consider δI^e for an element e.

$$E^e = A\phi_h^e - f = \sum_{i=1}^{n} (AN_i)\delta_i^e - f \tag{9.67}$$

$$\delta E^e = \frac{\partial E^e}{\partial \{\delta^e\}} = AN_j \; ; \quad j = 1, 2, \dots, n \tag{9.68}$$

$$\delta I^e = \int_{\bar{\Omega}^e} E^e \delta E^e \, dx = \int_{\bar{\Omega}^e} \Big(\sum_{i=1}^{n} (AN_i)\delta_i^e - f \Big) (AN_j) \, dx \; ; \quad j = 1, 2, \dots, n \tag{9.69}$$

or

$$\delta I^e = [K^e]\{\delta^e\} - \{f^e\} \tag{9.70}$$

in which

$$\left.\begin{aligned} K_{ij}^e &= \int_{\bar{\Omega}^e} (AN_i)(AN_j)\,dx \\ f_i^e &= \int_{\bar{\Omega}^e} f(AN_i)\,dx \end{aligned}\right\} \quad i,j = 1,2,\dots,n \tag{9.71}$$

Assembly of element equations follows the standard procedure, and by substituting (9.70) into (9.66) we obtain:

$$[K]\{\delta\} = \{F\} \tag{9.72}$$

$$[K] = \sum_e [K^e] \; ; \quad \{\delta\} = \bigcup_e \{\delta^e\} \; ; \quad \{F\} = \sum_e \{f^e\} \tag{9.73}$$

Remarks.

(1) When the operator A is linear, $[K^e]$ and $[K]$ are symmetric.

(2) Surana, *et al.* [49] have shown that $[K^e]$ and $[K]$ can also be made symmetric when A is nonlinear.

(3) Numerical examples are presented in the following.

(4) When the differential operator A contains higher order derivatives, we can use auxiliary variables and auxiliary equations to obtain a system of lower order equations. In principle any system of ODEs or PDEs can be reduced to a first order system of equations for which C^0 local approximation can be used.
If

$$\frac{d^2\phi}{dx^2} + \phi = f(x) \quad \forall x \in (a,b) = \Omega \tag{9.74}$$

is the BVP, then we can write

$$\frac{d\alpha}{dx} + \phi = f(x) \quad \forall x \in \Omega \tag{9.75}$$

$$\alpha = \frac{d\phi}{dx} \tag{9.76}$$

in dependent variables ϕ and α. α is called auxiliary variable and equation (9.76) is called auxiliary equation. Using the same approach a higher oder ODE in ϕ can be reduced to a system of first order ODEs.

9.3.3 General Remarks on Various Finite Element Processes

In this section we make some remarks regarding various methods of constructing finite element processes.

(1) Unconditional stability of a computational process is the most fundamental requirement that all computational processes must satisfy. Surana, *et al.* [49] have shown using calculus of variations that a variationally consistent integral form for which a unique extremum principle exists results in unconditional finite element process. This concept can be simply translated with simple guidelines that ensure variationally consistent integral form, hence unconditionally stable finite element processes.

(2) When the differential operator A in $A\phi - f = 0$ only contain even order derivative, then GM/WF yield VC integral form when the functional $B(\phi, v) = B(v, \phi)$ i.e. symmetric. In such cases each term in the integrand of $B(\cdot,\cdot)$ has same orders of derivatives of ϕ and v, hence symmetry of $B(\cdot,\cdot)$. Thus, in linear solid and structural mechanics GM/WF is ideally suited for constructing finite element processes.

(3) When the differential operator A contains odd order derivatives (some or all) or when the BVP is non-linear in which case A is a function of ϕ, then only the least squares method of constructing finite element process yields VC integral form, hence unconditionally stable finite element process.

Example 9.1. Second order non-homogeneous ODE: finite element method

$$\frac{d^2T}{dx^2} + T = f(x) \quad \forall x \in (0,1) = \Omega \subset \mathbb{R}^1 \tag{9.77}$$

with boundary conditions

$$T(0) = 0 \;,\; T(1) = -0.5 \tag{9.78}$$

we can write (9.77) as

$$\begin{aligned} AT &= f(x) \quad \forall x \in \Omega \\ A &= \frac{d^2}{dx^2} + 1 \end{aligned} \tag{9.79}$$

Since A contains even order derivatives, GM/WF is ideally suited for designing finite element process for (9.77). Let $\bar{\Omega}^T = \underset{e}{\cup}\bar{\Omega}^e$ be discretization of $\bar{\Omega} = [0,1]$ in which $\bar{\Omega}^e = [x_e, x_{e+1}]$ is an element e.
Let T_h be approximation of T over $\bar{\Omega}^T$ and T_h^e be approximation of T over $\bar{\Omega}^e$, then

$$T_h = \underset{e}{\cup} T_h^e \tag{9.80}$$

we consider

$$\int_{\bar{\Omega}^T} (AT_h - f(x))v(x)\,dx = 0\,; \quad v = \delta T_h \tag{9.81}$$

or

$$\sum_e \int_{\bar{\Omega}^e} (AT_h^e - f)v(x)\,dx = 0\,; \quad v = \delta T_h^e \tag{9.82}$$

consider

$$\int_{\bar{\Omega}^e} (AT_h^e - f)v(x)\,dx = \int_{\bar{\Omega}^e} \left(\frac{d^2T_h^e}{dx^2} + T_h^e\right) v(x)\,dx - \int_{\bar{\Omega}^e} fv\,dx \tag{9.83}$$

We transfer one order of differentiation from ${d^2T_h^e}/{dx^2}$ to $v(x)$ using integration by parts.

$$\int_{\bar{\Omega}^e} (AT_h^e - f(x))v(x)\,dx = \int_{\bar{\Omega}^e} \left(-\frac{dv}{dx}\frac{dT_h^e}{dx} + T_h^e v\right) dx + \left[v(x)\left(\frac{dT_h^e}{dx}\right)\right]\Bigg|_{x_e}^{x_{e+1}} - \int_{\bar{\Omega}^e} fv\,dx = \int_{\bar{\Omega}^e} \left(-\frac{dv}{dx}\frac{dT_h^e}{dx} + T_h^e v\right) dx + \langle AT_h^e, v\rangle_{\Gamma^e} - \int_{\bar{\Omega}^e} fv\,dx \tag{9.84}$$

In which

$$\langle AT_h^e, v\rangle_{\Gamma^e} = \left[v(x)\left(\frac{dT_h^e}{dx}\right)\right]\Bigg|_{x_e}^{x_{e+1}} \tag{9.85}$$

is the concomitant resulting due to integration by parts. In this case since we have an ODE in $\mathbb{R}^1$, the concomitant consists of boundary terms. From (9.85), we find that

- T is PV and $T = T_0$ (given) on some boundary Γ_1^* is EBC.
- ${dT}/{dx}$ is SV and ${dT}/{dx} = q_0$ on some boundary Γ_2^* is NBC.

We expand the boundary term in (9.85)

$$\langle AT_h^e, v\rangle_{\Gamma^e} = v(x_{e+1}) \left.\frac{dT_h^e}{dx}\right|_{x_{e+1}} - v(x_e) \left.\frac{dT_h^e}{dx}\right|_{x_e} \tag{9.86}$$

Let

$$\left.\frac{dT_h^e}{dx}\right|_{x_{e+1}} = -P_2^e \quad \text{and} \quad \left.\frac{dT_h^e}{dx}\right|_{x_e} = P_1^e \tag{9.87}$$

Using (9.87) in (9.86)

$$\langle AT_h^e, v\rangle_{\Gamma^e} = -v(x_{e+1})P_2^e - v(x_e)P_1^e \tag{9.88}$$

Substituting from (9.88) in (9.84) we obtain

$$(AT_h^e, v)_{\Omega^e} = \int\limits_{\bar{\Omega}^e} \left(-\frac{dv}{dx}\frac{dT_h^e}{dx} + T_h^e v \right) dx - \int\limits_{\bar{\Omega}^e} fv\, dx - v(x_e)P_1^e - v(x_{e+1})P_2^e \tag{9.89}$$

or

$$(AT_h^e, v)_{\Omega^e} = B^e(T_h^e, v) - l^e(v) \tag{9.90}$$

in which

$$B^e(T_h^e, v) = \int\limits_{\bar{\Omega}^e} \left(-\frac{dv}{dx}\frac{dT_h^e}{dx} + T_h^e v \right) dx \tag{9.91}$$

$$l^e(v) = \int\limits_{\bar{\Omega}^e} fv\, dx + v(x_e)P_1^e + v(x_{e+1})P_2^e \tag{9.92}$$

$B^e(T_h^e, v) = B^e(v, T_h^e)$ i.e. interchanging the roles of T_h^e and v does not change $B^e(\cdot,\cdot)$, hence $B^e(\cdot,\cdot)$ is symmetric. (9.90) is the weak form of the integral from (9.83).
Consider a five element uniform discretization using two node linear element

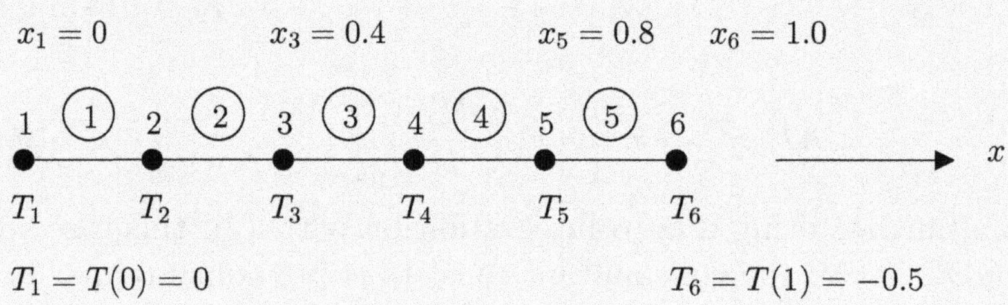

Figure 9.4: A five element uniform discretization using two node elements

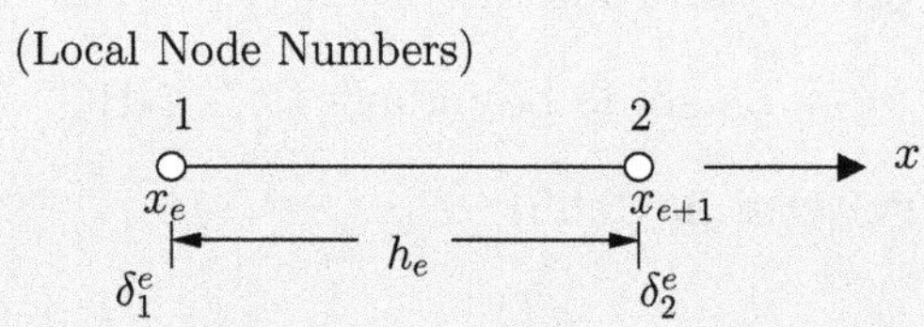

Figure 9.5: A two node linear element $\bar{\Omega}^e$

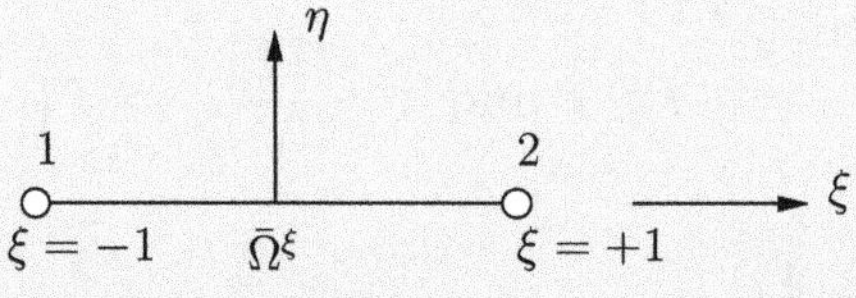

Figure 9.6: Map of $\bar{\Omega}^e$ in $\bar{\Omega}^\xi$

Following section 9.3.1

$$T_h^e(\xi) = \left(\frac{1-\xi}{2}\right)\delta_1^e + \left(\frac{1+\xi}{2}\right)\delta_2^e = N_1(\xi)\delta_1^e + N_2(\xi)\delta_2^e \tag{9.93}$$

in which δ_1^e and δ_2^e are nodal degrees of freedom for nodes 1 and 2 (local node numbers) of element e. Mapping of points is defined by

$$x(\xi) = \left(\frac{1-\xi}{2}\right)x_e + \left(\frac{1+\xi}{2}\right)x_{e+1} \tag{9.94}$$

Hence,

$$dx = \frac{dx}{d\xi}d\xi = Jd\xi \tag{9.95}$$

Where

$$J = \frac{d}{d\xi}\left(\frac{1-\xi}{2}\right)x_e + \frac{d}{d\xi}\left(\frac{1+\xi}{2}\right)x_{e+1} = \left(\frac{x_{e+1}-x_e}{2}\right) = \frac{h_e}{2} \tag{9.96}$$

$$\frac{dN_j}{d\xi} = \frac{dN_j}{dx}\frac{dx}{d\xi} = \frac{dN_j}{dx}J \; ; \quad j = 1,2 \tag{9.97}$$

$$\frac{dN_i}{dx} = \frac{dN_i}{d\xi}\frac{1}{J} = \frac{2}{h_e}\frac{dN_i}{d\xi} \; ; \quad i = 1,2 \tag{9.98}$$

$$v = \delta T_h^e = N_j(\xi) \; ; \quad j = 1,2 \tag{9.99}$$

we now return back to weak form (9.90)

$$\begin{aligned}
B^e(T_h^e, v) &= \int\limits_{\bar{\Omega}^e} \left(-\frac{dv}{dx}\frac{dT_h^e}{dx} + T_h^e v\right) dx \\
&= \int\limits_{\bar{\Omega}^e} \left(-\frac{dN_j}{dx}\left(\sum_{i=1}^{2}\frac{dN_i}{dx}\delta_i^e\right) + \left(\sum_{i=1}^{2}N_i\delta_i^e\right)N_j\right) dx \\
&= \int\limits_{\bar{\Omega}^e} \left(-\frac{1}{J}\frac{dN_j}{d\xi}\left(\sum_{i=1}^{2}\frac{1}{J}\frac{dN_i}{d\xi}\delta_i^e\right) + \left(\sum_{i=1}^{2}N_i\delta_i^e\right)N_j\right) Jd\xi \\
&= \frac{2}{h_e}\int\limits_{i=1}^{2} \left(-\frac{dN_j}{d\xi}\left(\sum_{i=1}^{2}\frac{dN_i}{d\xi}\delta_i^e\right)\right) d\xi + \frac{h_e}{2}\int\limits_{-1}^{+1}\left(\sum_{i=1}^{2}N_i\delta_i^e\right)N_j\, d\xi \\
&= [{}^1K^e]\{\delta^e\} + [{}^2K^e]\{\delta^e\}
\end{aligned} \tag{9.100}$$

in which

$$ {}^1K_{ij}^e = -\frac{2}{h_e}\int_{-1}^{+1} \frac{dN_i}{d\xi}\frac{dN_j}{d\xi}\,d\xi\ ;\quad i,j=1,2 \tag{9.101}$$

$$ {}^2K_{ij}^e = \frac{h_e}{2}\int_{-1}^{+1} N_iN_j\,d\xi\ ;\quad i,j=1,2 \tag{9.102}$$

and $l^e(v)$ is given by (after substituting $v = N_j$)

$$ l^e(v) = \int_{\bar{\Omega}^e} f(x)N_j\,dx + N_j(x_e)P_1^e + N_j(x_{e+1})P_2^e\ ;\quad j=1,2 \tag{9.103}$$

$$ = \int_{-1}^{+1} f(\xi)N_j(\xi)J\,d\xi + N_j(-1)P_1^e + N_j(1)P_2^e\ ;\quad j=1,2 \tag{9.104}$$

for $j = 1$ $\quad (x_e \to \xi = -1,\ x_{e+1} \to \xi = +1)$, we have

$$ l^e(N_1) = \int_{-1}^{+1} f(\xi)N_1J\,d\xi + N_1(-1)P_1^e + N_1(1)P_2^e \tag{9.105}$$

for $j = 2$

$$ l^e(N_2) = \int_{-1}^{+1} f(\xi)N_2J\,d\xi + N_2(-1)P_1^e + N_2(1)P_2^e \tag{9.106}$$

Since

$$ \begin{aligned} N_1(-1) = 1\ ,\ N_1(1) = 0 \\ N_2(-1) = 0\ ,\ N_2(1) = 1 \end{aligned} \tag{9.107}$$

We can write

$$ \begin{aligned} l^e(v) &= \{F^e\} + \{P^e\} \\ F_i^e &= \int_{-1}^{+1} f(\xi)N_iJ\,d\xi\ ;\quad i=1,2 \\ \{P^e\}^T &= [P_1^e \quad P_2^e] \end{aligned} \tag{9.108}$$

$\{F^e\}$ are loads at the element nodes due to $f(x)$ and $\{P^e\}$ is a vector of secondary variables at the element nodes. Secondary variable $\{P^e\}$ at the

element nodes are still unknown.
Using (9.93)

$$\frac{dN_1}{d\xi} = -\frac{1}{2} \; ; \; \frac{dN_2}{d\xi} = \frac{1}{2} \tag{9.109}$$

Hence,

$$[{}^1K^e] = -\frac{2}{h_e}\int_{-1}^{+1}\begin{bmatrix} \frac{dN_1}{d\xi}\frac{dN_1}{d\xi} & \frac{dN_1}{d\xi}\frac{dN_2}{d\xi} \\ \frac{dN_2}{d\xi}\frac{dN_1}{d\xi} & \frac{dN_2}{d\xi}\frac{dN_2}{d\xi} \end{bmatrix} d\xi \tag{9.110}$$

or

$$[{}^1K^e] = -\frac{1}{h_e}\begin{bmatrix} 1 & -1 \\ -1 & 1 \end{bmatrix} \tag{9.111}$$

and

$$[{}^2K^e] = \frac{h_e}{2}\int_{-1}^{+1}\begin{bmatrix} N_1N_1 & N_1N_2 \\ N_2N_1 & N_2N_2 \end{bmatrix} d\xi \tag{9.112}$$

or

$$[{}^2K^e] = \frac{h_e}{6}\begin{bmatrix} 2 & 1 \\ 1 & 2 \end{bmatrix} \tag{9.113}$$

$$\{F^e\}^T = [F_1^e \quad F_2^e] \tag{9.114}$$

Where

$$\begin{aligned} F_1^e &= \frac{1}{h_e}\left(\frac{x_{e+1}}{4}(x_{e+1}^4 - x_e^4) - \frac{1}{5}(x_{e+1}^5 - x_e^5)\right) \\ F_2^e &= \frac{1}{h_e}\left(\frac{1}{5}(x_{e+1}^5 - x_e^5) - \frac{x_e}{4}(x_{e+1}^4 - x_e^4)\right) \end{aligned} \tag{9.115}$$

Thus, we have for element e

$$\begin{aligned} \int_{\bar{\Omega}^e}(AT_h^e - f)v\,dx &= \left[-\frac{1}{h_e}\begin{bmatrix} 1 & -1 \\ -1 & 1 \end{bmatrix} + \frac{h_e}{6}\begin{bmatrix} 2 & 1 \\ 1 & 2 \end{bmatrix}\right]\begin{Bmatrix} \delta_1^e \\ \delta_2^e \end{Bmatrix} - \begin{Bmatrix} F_1^e \\ F_2^e \end{Bmatrix} - \begin{Bmatrix} P_1^e \\ P_2^e \end{Bmatrix} \\ &= [K^e]\begin{Bmatrix} \delta_1^e \\ \delta_2^e \end{Bmatrix} - \begin{Bmatrix} F_1^e \\ F_2^e \end{Bmatrix} - \begin{Bmatrix} P_1^e \\ P_2^e \end{Bmatrix} \end{aligned} \tag{9.116}$$

For elements 1−5 of figure (9.4), we can write (9.116) as

$$\int_{\bar{\Omega}^e}(AT_h^e - f)v\,dx = [K^e]\begin{Bmatrix} T_e \\ T_{e+1} \end{Bmatrix} - \begin{Bmatrix} F_1^e \\ F_2^e \end{Bmatrix} - \begin{Bmatrix} P_1^e \\ P_2^e \end{Bmatrix} \; ; \quad e = 1, 2, \dots, 5 \tag{9.117}$$

in which

$$[K^e] = \begin{bmatrix} -4.93333 & 5.03333 \\ 5.03333 & -4.93333 \end{bmatrix} \; ; \quad e = 1, 2, \dots, 5 \tag{9.118}$$

$$
\begin{Bmatrix} F_1^1 \\ F_2^1 \end{Bmatrix} = \begin{Bmatrix} 0.48889 \times 10^{-4} \\ 0.31111 \times 10^{-4} \end{Bmatrix} , \begin{Bmatrix} F_1^2 \\ F_2^2 \end{Bmatrix} = \begin{Bmatrix} 0.20889 \times 10^{-2} \\ 0.39111 \times 10^{-2} \end{Bmatrix} ,
$$
$$
\begin{Bmatrix} F_1^3 \\ F_2^3 \end{Bmatrix} = \begin{Bmatrix} 0.10489 \times 10^{-1} \\ 0.15511 \times 10^{-1} \end{Bmatrix} , \begin{Bmatrix} F_1^4 \\ F_2^4 \end{Bmatrix} = \begin{Bmatrix} 0.30089 \times 10^{-1} \\ 0.399111 \times 10^{-1} \end{Bmatrix} ,
$$
$$
\begin{Bmatrix} F_1^5 \\ F_2^5 \end{Bmatrix} = \begin{Bmatrix} 0.65689 \times 10^{-1} \\ 0.81911 \times 10^{-1} \end{Bmatrix} \tag{9.119}
$$

Assembly of element equations is given by

$$
\begin{aligned}
\sum_e \int_{\bar{\Omega}^e} (AT_h^e - f) v \, dx &= \left[\sum_e [K^e] \right] \{\delta\} - \sum_e \{F^e\} - \sum_e \{P^e\} = \{0\} \\
&= [K]\{\delta\} - \{F\} - \{P\} = 0 \\
\text{in which } \{\delta\} &= \bigcup_e \{\delta^e\}
\end{aligned} \tag{9.120}
$$

Assembled $[K]$, $\{F\}$ and $\{P\}$ and degrees of freedom $\{\delta\}$ are shown in the following. $T_1, T_2, \dots, T_6$ are arranged in $\{\delta\}$ in such a way that known values of T_1 and T_6 appear as first two elements of $\{\delta\}$ so that the assembled equations remain in partitioned form. We note that rows and columns of assembled $[K]$ are identified as T_1, T_6, T_2, T_3, T_4 and T_5 for ease of assembly and solutions.

$$
[K] = \begin{array}{c}
\begin{array}{cccccc} T_1 & T_6 & T_2 & T_3 & T_4 & T_5 \end{array} \\
\left[\begin{array}{cccccc}
-4.933 & 0 & 5.033 & 0 & 0 & 0 \\
0 & -4.933 & 0 & 0 & 0 & 5.033 \\
5.033 & 0 & (-4.933 - 4.933) & 5.033 & 0 & 0 \\
0 & 0 & 5.033 & (-4.933 - 4.933) & 5.033 & 0 \\
0 & 0 & 0 & 5.033 & (-4.933 - 4.933) & 5.033 \\
0 & 5.033 & 0 & 0 & 5.033 & (-4.933 - 4.933)
\end{array} \right]
\end{array}
\begin{array}{c} \\ T_1 \\ T_6 \\ T_2 \\ T_3 \\ T_4 \\ T_5 \end{array}
\tag{9.121}
$$

$$
\{F\} = \begin{Bmatrix}
0.88889 \times 10^{-4} \\
0.819711 \times 10^{-1} \\
0.31111 \times 10^{-4} + 0.20889 \times 10^{-2} \\
0.39111 \times 10^{-2} + 0.10489 \times 10^{-1} \\
0.15511 \times 10^{-1} + 0.30089 \times 10^{-1} \\
0.39911 \times 10^{-1} + 0.65689 \times 10^{-1}
\end{Bmatrix} \tag{9.122}
$$

$$
\{\delta\}^T = [T_1 \; T_6 \; T_2 \; T_3 \; T_4 \; T_5] \tag{9.123}
$$

Rules regarding secondary variables $\{P\}$ are as follows. The secondary variables in solid mechanics are forces or moment, in heat transfer these are fluxes.

(1) The sum of the secondary variables at a node is zero if there is no externally applied disturbance. Thus,

$$\begin{aligned} P_2^1 + P_1^2 &= P_2 = 0 \\ P_2^2 + P_1^3 &= P_3 = 0 \\ P_2^3 + P_1^4 &= P_4 = 0 \\ P_2^4 + P_1^5 &= P_5 = 0 \end{aligned} \tag{9.124}$$

(2) Where primary variables are specified (or given) at a node i.e. when the essential boundary conditions are given at a node, the sum of the secondary variables is unknown at that node. Thus $P_1^1 = P_1$ and $P_2^5 = P_6$ are unknown.

(3) If there is externally applied disturbance at a node, then the sum of the secondary variables is equal to the externally applied disturbance at that node. This condition does not exist in this example.

Thus in $\{\delta\}$ vector T_1 and T_6 are known (0.0 and −0.5) and $T_2, T_3, \dots, T_5$ are unknown. $\{F\}$ vector is completely known. In the secondary variable vector $\{P\}$, P_1 and P_6 are unknown but $P_2, P_3, \dots, P_5$ are zero (known). Assembled equations (9.120) with $[K]$, $\{F\}$, $\{P\}$, $\{\delta\}$ in (9.121) - (9.123) can be written in partitioned form.

$$\begin{bmatrix} [K_{11}] & [K_{12}] \\ [K_{21}] & [K_{22}] \end{bmatrix} \begin{Bmatrix} \{\delta\}_1 \\ \{\delta\}_2 \end{Bmatrix} = \begin{Bmatrix} \{F\}_1 \\ \{F\}_2 \end{Bmatrix} + \begin{Bmatrix} \{P\}_1 \\ \{P\}_2 \end{Bmatrix} \tag{9.125}$$

in which

$$\begin{aligned} &\{\delta\}_1^T = [T_1 \; T_6] = [0.0 \quad -0.5] \; ; \quad \text{known} \\ &\{\delta\}_2^T = [T_2 \; T_3 \; T_4 \; T_5] \; ; \quad \text{unknown} \\ &\{F\}_1^T = [F_1 \; F_6] \; ; \; \{F\}_2^T = [F_2 \; F_3 \; F_4 \; F_5] \\ &\{P\}_1^T = [P_1 \; P_6] \; ; \quad \text{unknown} \\ &\{P\}_2^T = [P_2 \; P_3 \; P_4 \; P_5] \; ; \quad \text{known} \end{aligned} \tag{9.126}$$

Using (9.125) we can write

$$\begin{aligned} [K_{11}]\{\delta\}_1 + [K_{12}]\{\delta\}_2 &= \{F\}_1 + \{P\}_1 \\ [K_{21}]\{\delta\}_1 + [K_{22}]\{\delta\}_2 &= \{F\}_2 + \{P\}_2 \end{aligned} \tag{9.127}$$

Using second set of equation in (9.127) we can solve for $\{\delta\}_2$

$$[K_{22}]\{\delta\}_2 = \{F\}_2 + \{P\}_2 - [K_{21}]\{\delta\}_1 \tag{9.128}$$

and from the first set of equations in (9.127) we can solve for $\{P\}_1$.

$$\{P\}_1 = [K_{11}]\{\delta\}_1 + [K_{12}]\{\delta\}_2 - \{F\}_1 \tag{9.129}$$

First, we solve for $\{\delta\}_2$ using (9.128) and calculate $\{P\}_1$ using (9.129). in which

$$\begin{aligned}
[K_{11}] &= \begin{bmatrix} -4.93333 & 0 \\ 0 & -4.93333 \end{bmatrix} \\
[K_{12}] &= \begin{bmatrix} 5.03333 & 0 & 0 & 0 \\ 0 & 0 & 0 & 5.03333 \end{bmatrix} \\
[K_{21}] &= [K_{12}]^T
\end{aligned} \tag{9.130}$$

$$[K_{22}] = \begin{bmatrix} -9.86666 & 5.03333 & 0 & 0 \\ 5.03333 & -9.86666 & 5.03333 & 0 \\ 0 & 5.03333 & -9.86666 & 5.03333 \\ 0 & 0 & 5.03333 & -9.86666 \end{bmatrix} \tag{9.131}$$

$$\{F\}_1^T = \begin{bmatrix} 0.88889 \times 10^{-4} & 0.818711 \times 10^{-1} \end{bmatrix} \tag{9.132}$$

$$\{F\}_2^T = \begin{bmatrix} 0.212 \times 10^{-2} \; 0.144 \times 10^{-1} \; 0.456 \times 10^{-1} \; 1.056 \times 10^{-1} \end{bmatrix} \tag{9.133}$$

$$\{[K_{21}]\{\delta\}_1\}^T = \begin{bmatrix} 0.0 \; 0.0 \; 0.0 \; -2.516667 \end{bmatrix} \tag{9.134}$$

$$\{P\}_2^T = \begin{bmatrix} 0 \; 0 \; 0 \; 0 \end{bmatrix} \tag{9.135}$$

Thus, using (9.128), we now have using (9.131), (9.133), (9.134) and (9.135) in (9.128) we can solve for $\{\delta\}_2$.

$$\{\delta\}_2^T = [T_2 \; T_3 \; T_4 \; T_5] = \begin{bmatrix} -0.12945 \; -0.25328 \; -0.36418 \; -0.45155 \end{bmatrix} \tag{9.136}$$

Now using (9.129) we can calculate the unknown secondary variables $\{P\}_1$.

$$\begin{aligned}
\{P\}_1 = \begin{Bmatrix} P_1 \\ P_6 \end{Bmatrix} &= \begin{bmatrix} -4.93333 & 0 \\ 0 & -4.93333 \end{bmatrix} \begin{Bmatrix} 0.0 \\ -0.5 \end{Bmatrix} + \\
& \begin{bmatrix} 5.03333 \; 0 \; 0 & 0 \\ 0 & 0 \; 0 \; 5.03333 \end{bmatrix} \begin{Bmatrix} -0.12945 \\ -0.25328 \\ -0.36418 \\ -0.45155 \end{Bmatrix} - \\
& \begin{Bmatrix} 0.88889 \times 10^{-4} \\ 0.819711 \times 10^{-1} \end{Bmatrix}
\end{aligned} \tag{9.137}$$

or

$$\{P\}_1 = \begin{Bmatrix} P_1 \\ P_6 \end{Bmatrix} = \begin{Bmatrix} -0.651643 \\ 0.111952 \end{Bmatrix} \tag{9.138}$$

Thus, now $\{\delta\}^T = [T_1, T_2, \dots, T_6]$ is known, hence using local approximation for each element we can describe T over each element domain $\bar{\Omega}^\xi$.

$$T(\xi) = \left(\frac{1-\xi}{2}\right) T_e + \left(\frac{1+\xi}{2}\right) T_{e+1} \; ; \quad e = 1, 2, \dots, 5 \tag{9.139}$$

The local approximation (9.139) describes T over each element $\bar{\Omega}^e = [x_e, x_{e+1}] \to \bar{\Omega}^\xi = [-1, 1]$. Using (9.139) we can also calculate derivative of $T(\xi)$ with respect to x for each element.

$$\frac{dT}{dx} = \frac{1}{J}\frac{dT(\xi)}{d\xi} = \frac{1}{h_e}(T_{e+1} - T_e) \; ; \; J = \frac{h_e}{2} \quad e = 1, 2, \dots, 5 \tag{9.140}$$

Table 9.1: Nodal values of the solution (example 9.1)

node	x	T
1	0.0	0.0
2	0.2	-0.12945
3	0.46	-0.25328
4	0.6	-0.36418
5	0.8	-0.45155
6	1.0	-0.5

Table 9.2: dT_h^e/dx versus x (example 9.1)

Element	nodes	x-coordinate	dT_h^e/dx
1	1	0.0	-0.64725
	2	0.2	-0.64725
2	2	0.2	-0.6195
	3	0.4	-0.6195
3	3	0.4	-0.5545
	4	0.5	-0.5545
4	4	0.6	-0.43685
	5	0.8	-0.43685
5	5	0.8	-0.24225
	6	1.0	-0.24225

Table 9.1 gives values of temperature at the nodes of the discretization.

Table 9.2 gives dT_h^e/dx values calculated at each of the five elements of the discretization.

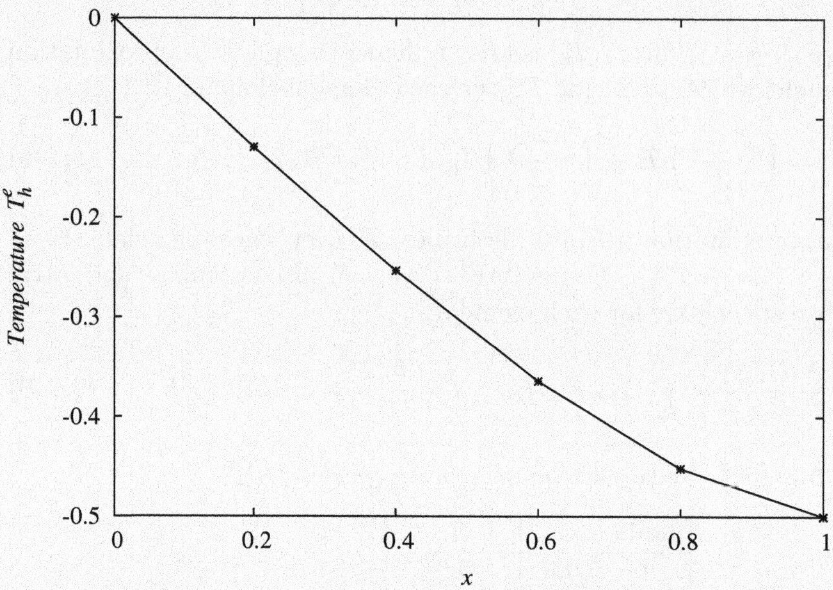

Figure 9.7: T versus x (example 9.1)

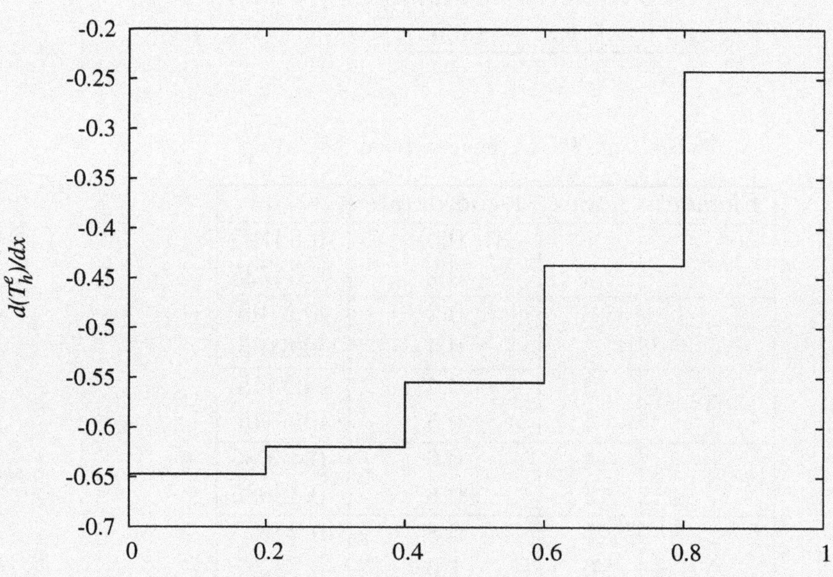

Figure 9.8: dT/dx versus x (example 9.1)

Figures 9.7 and 9.8 show plots of T versus x and ${dT}/{dx}$ versus x. Since T_h^e is of class $C^0(\bar{\Omega}^e)$, we observe inter element discontinuity of ${dT_h^e}/{dx}$ at the inter element boundaries. Upon mesh refinement the jumps in ${dT_h^e}/{dx}$ diminishes at the inter element boundaries.

Example 9.2. Second order non-homogeneous ODE in $\mathbb{R}^1$: finite element method
We consider same ODE as in example 9.1 but with different boundary conditions.

$$\frac{d^2T}{dx^2} + T = f(x) \quad \forall x \in (0,1) = \Omega \subset \mathbb{R}^1 \tag{9.141}$$

$$T(0) = 0\,, \quad \left.\frac{dT}{dx}\right|_{x=1} = 20 \tag{9.142}$$

The weak form using GM/WF in this case is same as in example 9.1. The integral form $\int\limits_{\bar{\Omega}^e} (AT_h^e - f)v\,dx$ yields (9.116) or (9.117). For a uniform discretization consisting of five two node linear elements (Figure 9.4), the element matrices and $\{F^e\}$ vectors given by (9.118) and (9.119). Assembly of the element equations in symbolic form are shown in (9.120). For this example problem the BC $T(0) = 0$ is Essential Boundary Condition (EBC) whereas $\left.\frac{dT}{dx}\right|_{x=1} = 20$ is the Natural Boundary Condition (NBC). Thus, for the five element discretization $T_1 = 0.0$, but T_6 is not known and secondary variable at node 6 ($x = 1$) is -20 (as the SV at $\xi = +1$ is defined as $-{dT}/{dx}$). Thus, when assembling element equations for this example we can choose the following order for $\{\delta\}$, the degrees of freedom at the nodes.

$$\{\delta\}^T = \{T_1\, T_2 \dots T_6\} \tag{9.143}$$

For the assembly equations, we have (rows and columns are identified as $T_1, T_2, \dots, T_6$)

$$[K] = \begin{array}{c} \begin{array}{cccccc} T_1 & T_2 & T_3 & T_4 & T_5 & T_6 \end{array} \\ \begin{bmatrix} -4.933 & 5.033 & 0 & 0 & 0 & 0 \\ 5.033 & (-4.933 - 4.933) & 5.033 & 0 & 0 & 0 \\ 0 & 5.033 & (-4.933 - 4.933) & 5.033 & 0 & 0 \\ 0 & 0 & 5.033 & (-4.933 - 4.933) & 5.033 & 0 \\ 0 & 0 & 0 & 5.033 & (-4.933 - 4.933) & 5.033 \\ 0 & 0 & 0 & 0 & 5.033 & -4.933 \end{bmatrix} \end{array} \begin{array}{c} T_1 \\ T_2 \\ T_3 \\ T_4 \\ T_5 \\ T_6 \end{array} \tag{9.144}$$

$$\{F\} = \begin{Bmatrix} 0.88889 \times 10^{-4} \\ 0.31111 \times 10^{-4} + 0.20889 \times 10^{-2} \\ 0.39111 \times 10^{-4} + 0.10489 \times 10^{-1} \\ 0.15511 \times 10^{-1} + 0.30089 \times 10^{-1} \\ 0.39911 \times 10^{-1} + 0.65689 \times 10^{-1} \\ 0.819711 \times 10^{-1} \end{Bmatrix} \tag{9.145}$$

Using the rules for defining the sum of the secondary variables in example (9.1), we have

$$P_2 = 0\,,\ P_3 = 0\,,\ P_4 = 0\,,\ P_5 = 0\,,\ P_6 = -20 \tag{9.146}$$

and P_1 is unknown.
Due to EBC, we have

$$T(0) = 0\,,\ T_1 = 0.0 \ \text{and}\ T_2, T_3, \dots, T_6 \tag{9.147}$$

are unknown.
Assembled equations (9.120) with $[K]$, $\{F\}$ and $\{P\}$ defined by (9.144) – (9.147) can be written in partitioned form

$$\begin{bmatrix} [K_{11}] & [K_{12}] \\ [K_{21}] & [K_{22}] \end{bmatrix} \begin{Bmatrix} \{\delta\}_1 \\ \{\delta\}_2 \end{Bmatrix} = \begin{Bmatrix} \{F\}_1 \\ \{F\}_2 \end{Bmatrix} + \begin{Bmatrix} \{P\}_1 \\ \{P\}_2 \end{Bmatrix} \tag{9.148}$$

in which

$$\begin{aligned} &\{\delta\}_1^T = \{0.0\}\ ; \quad \text{known} \\ &\{\delta\}_2^T = \begin{bmatrix} T_2\ T_3\ T_4\ T_5\ T_6 \end{bmatrix}\ ; \quad \text{unknown} \\ &\{F\}_1^T = \{F_1\}\ ;\ \{F\}_2^T = \begin{bmatrix} F_2\ F_3\ F_4\ F_5\ F_6 \end{bmatrix} \\ &\{P\}_1^T = [P_1]\ ; \quad \text{unknown} \\ &\{P\}_2^T = \begin{bmatrix} 0.0\ 0.0\ 0.0\ 0.0\ {-20} \end{bmatrix}\ ; \quad \text{known} \end{aligned} \tag{9.149}$$

Using (9.148) we can write

$$\begin{aligned} [K_{11}]\{\delta\}_1 + [K_{12}]\{\delta\}_2 &= \{F\}_1 + \{P\}_1 \\ [K_{21}]\{\delta\}_1 + [K_{22}]\{\delta\}_2 &= \{F\}_2 + \{P\}_2 \end{aligned} \tag{9.150}$$

Using second set of equations in (9.150) we can solve for $\{\delta\}_2$

$$[K_{22}]\{\delta\}_2 = \{F\}_2 + \{P\}_2 - [K_{21}]\{\delta\}_1 \tag{9.151}$$

From the first set of equations in (9.150) we can solve for $\{P\}_1$.

$$\{P\}_1 = [K_{11}]\{\delta\}_1 + [K_{12}]\{\delta\}_2 - \{F\}_1 \tag{9.152}$$

First, we solve for $\{\delta\}_2$ using (9.151) and then calculate $\{P\}_1$ using (9.152). in which

$$\begin{aligned}[K_{11}] &= [-4.93333] \\ [K_{12}] &= \big[5.03333\ 0.0\ 0.0\ 0.0\ 0.0\big] \\ [K_{21}] &= [K_{12}]^T\end{aligned} \tag{9.153}$$

$$[K_{22}] = \begin{bmatrix} -9.86666 & 5.03333 & 0 & 0 & 0 \\ 5.03333 & -9.86666 & 5.03333 & 0 & 0 \\ 0 & 5.03333 & -9.86666 & 5.03333 & 0 \\ 0 & 0 & 5.03333 & -9.86666 & 5.03333 \\ 0 & 0 & 0 & 5.03333 & -4.93333 \end{bmatrix} \tag{9.154}$$

$$\{F\}_1^T = \{0.88889 \times 10^{-4}\} \tag{9.155}$$

$$\{F\}_2^T = [0.212 \times 10^{-2} \quad 0.144 \times 10^{-1} \quad 0.456 \times 10^{-1} \quad 1.056 \times 10^{-1} \quad 0.818711 \times 10^{-1}] \tag{9.156}$$

$$\{[K_{21}]\{\delta\}_1\}^T = \big[0.0\ 0.0\ 0.0\ 0.0\ 0.0\big] \tag{9.157}$$

$$\{P\}_2^T = \big[0.0\ 0.0\ 0.0\ 0.0\ -20.0\big] \tag{9.158}$$

Thus, using (9.151), we now have using (9.154), (9.156), (9.157) and (9.158) we can solve for $\{\delta\}_2$.

$$\{\delta\}_2^T = \big[T_2\ T_3\ T_4\ T_5\ T_6\big] = \big[7.2469\ 14.206\ 20.604\ 26.192\ 30.761\big] \tag{9.159}$$

and using (9.152) we can calculate the unknown secondary variable $\{P\}_1$.

$$\{P\}_1 = P_1 = [-4.93333]\{0.0\} + \big[5.03333\ 0.0\ 0.0\ 0.0\ 0.0\big] \begin{Bmatrix} 7.2469 \\ 14.206 \\ 20.604 \\ 26.192 \\ 30.761 \end{Bmatrix} - \{0.88889 \times 10^{-4}\} \tag{9.160}$$

or

$$\{P\}_1 = 36.47599 \tag{9.161}$$

Since $\{\delta\}^T = [T_1, T_2, \dots, T_6]$ is known, hence using local approximations for each element we can describe T over each element domain $\bar{\Omega}^e \to \bar{\Omega}^\xi$.

$$T(\xi) = \left(\frac{1-\xi}{2}\right) T_e + \left(\frac{1+\xi}{2}\right) T_{e+1}\ ; \quad e = 1, 2, \dots, 5 \tag{9.162}$$

The local approximation (9.162) describes T over each element $\bar{\Omega}^e = [x_e, x_{e+1}] \to \bar{\Omega}^\xi = [-1, 1]$. Using (9.162) we can also calculate derivative of $T(\xi)$ with respect to x for each element.

$$\frac{dT}{dx} = \frac{1}{J}\frac{dT(\xi)}{d\xi} = \frac{1}{h_e}(T_{e+1} - T_e) \; ; \; J = \frac{h_e}{2} \quad e = 1, 2, \ldots, 5 \tag{9.163}$$

Table 9.3: Nodal values of the solution (example 9.2)

node	x	T
1	0.0	0.0
2	0.2	7.2469
3	0.46	14.206
4	0.6	20.604
5	0.8	26.192
6	1.0	30.761

Table 9.4: dT_h^e/dx versus x (example 9.2)

Element	nodes	x-coordinate	dT_h^e/dx
1	1	0.0	36.2345
	2	0.2	36.2345
2	2	0.2	34.7955
	3	0.4	34.7955
3	3	0.4	31.99
	4	0.5	31.99
4	4	0.6	27.94
	5	0.8	27.94
5	5	0.8	22.845
	6	1.0	22.845

Table 9.3 gives values of temperature at the nodes of the discretization. Table 9.4 gives dT_h^e/dx values calculated at each of the five elements of the discretization.

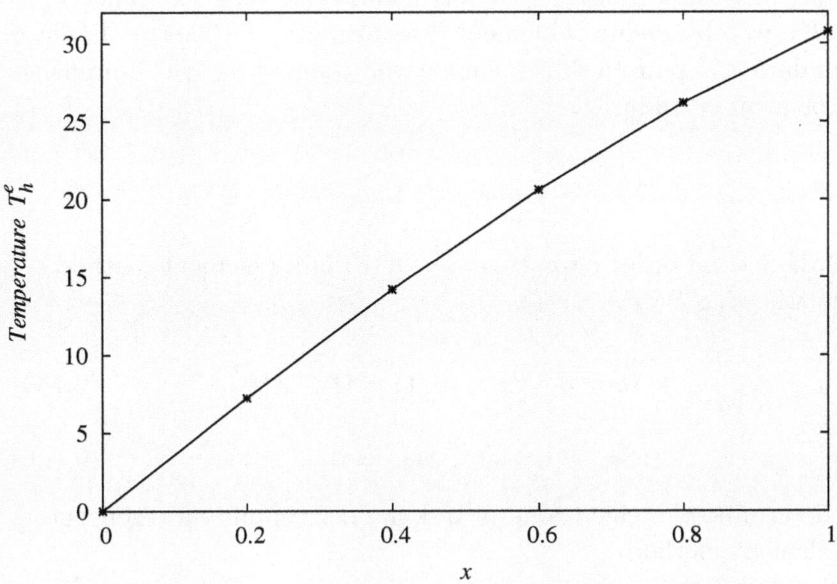

Figure 9.9: T versus x (example 9.2)

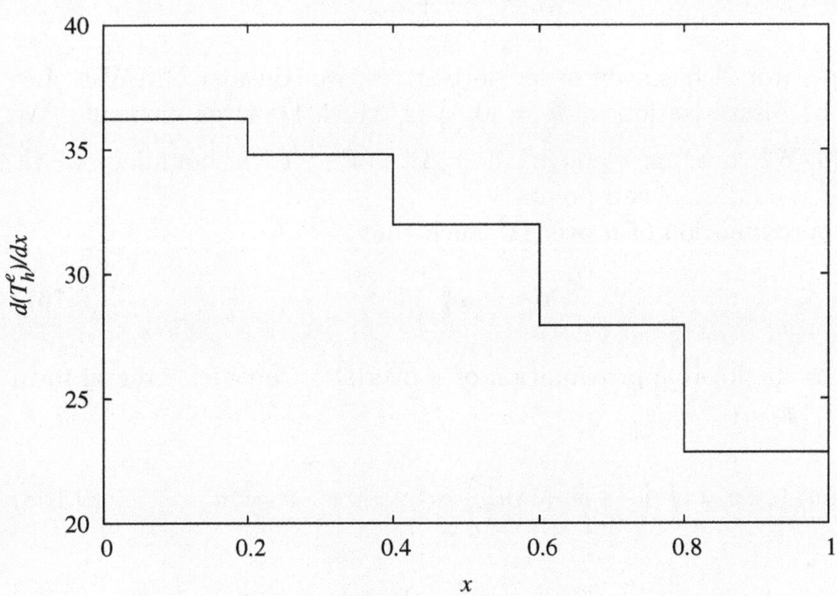

Figure 9.10: ${}^{dT}/_{dx}$ versus x (example 9.2)

Figures 9.9 and 9.10 show plots of T versus x and ${}^{dT}/_{dx}$ versus x. Since T_h^e is of class $C^0(\bar{\Omega}^e)$, we observe inter element discontinuity of ${}^{dT_h^e}/_{dx}$ at the inter element boundaries. Upon mesh refinement the jumps in ${}^{dT_h^e}/_{dx}$ diminishes at the inter element boundaries.

Example 9.3. Second order homogeneous ODE: finite element method
Consider the following BVP.

$$\frac{d^2u}{dx^2} + \lambda u = 0 \quad \forall x \in (0,1) = \Omega \subset \mathbb{R}^1 \tag{9.164}$$

$$\text{BCs:} \;\; u(0) = 0 \;, \;\; u(1) = 0 \tag{9.165}$$

we wish to determine the eigenvalue λ and the corresponding eigenvectors using finite element method.
We can write (9.164) as

$$\begin{aligned} Au &= 0 \quad \forall x \in \Omega \\ A &= \frac{d^2}{dx^2} + \lambda \end{aligned} \tag{9.166}$$

Since the operator A has even order derivatives, we consider GM/WF. Let $\bar{\Omega}^T = \underset{e}{\cup}\bar{\Omega}^e$ be discretization of $\bar{\Omega} = [0,1]$ in which $\bar{\Omega}^e$ is an element. We consider GM/WF over an element $\bar{\Omega}^e = \Omega^e \cup \Gamma^e$; Γ^e is boundary of $\bar{\Omega}^e$ consisting of x_e and x_{e+1} end points.
Let u_h be approximation of u over $\bar{\Omega}^T$ such that

$$u_h = \underset{e}{\cup} u_h^e \tag{9.167}$$

in which u_h^e is the local approximation of u over $\bar{\Omega}^e$. Consider integral form over $\bar{\Omega}^e = [x_e, x_{e+1}]$

$$\int_{\bar{\Omega}^e} (Au_h^e) v \, dx = \int_{\bar{\Omega}^e} \left(\frac{d^2 u_h^e}{dx^2} + \lambda u_h^e \right) v \, dx = 0 \; ; \quad v = \delta u_h^e \tag{9.168}$$

Integrate by parts once in the first term in the integrand

$$\int_{\bar{\Omega}^e} (Au_h^e) v(x) \, dx = \int_{\bar{\Omega}^e} \left(-\frac{dv}{dx}\frac{du_h^e}{dx} + \lambda u_h^e v \right) dx + v \left(\frac{du_h^e}{dx} \right) \bigg|_{x_e}^{x_{e+1}} \tag{9.169}$$

Consider a two node linear element $\bar{\Omega}^e$ with degrees of freedom δ_1^e and δ_2^e at the nodes, then
Let

$$u_h^e = \left(\frac{1-\xi}{2}\right)\delta_1^e + \left(\frac{1+\xi}{2}\right)\delta_2^e = N_1(\xi)\delta_1^e + N_2(\xi)\delta_2^e \tag{9.170}$$

and

$$v = \delta u_h^e = N_j \; ; \quad j = 1,2 \tag{9.171}$$

First, concomitant in (9.169)

$$<Au_h^e, v>_{\Gamma^e} = v(x_{e+1})\left(\frac{du_h^e}{dx}\right)\bigg|_{x_{e+1}} - v(x_e)\left(\frac{du_h^e}{dx}\right)\bigg|_{x_e}$$

Let (9.172)

$$-\left(\frac{du_h^e}{dx}\right)\bigg|_{x_e} = P_1^e \, , \; \left(\frac{du_h^e}{dx}\right)\bigg|_{x_{e+1}} = P_2^e$$

Then,

$$\int_{\bar{\Omega}^e} (Au_h^e)v(x)\,dx = \int_{\bar{\Omega}^e}\left(-\frac{dv}{dx}\frac{du_h^e}{dx} + \lambda u_h^e v\right)dx + v(x_{e+1})P_2^e + v(x_e)P_1^e \tag{9.173}$$

Substituting (9.170) and (9.171) in (9.173) we can write

$$\int_{\bar{\Omega}^e} (Au_h^e)v(x)\,dx = \int_{\bar{\Omega}^e}\left(-\frac{dN_j}{dx}\left(\sum_{i=1}^{2}\frac{dN_i}{dx}\delta_i^e\right) + \lambda\left(\sum_{i=1}^{2}N_i\delta_i^e\right)N_j\right)dx + N_j(x_{e+1})P_2^e + N_j(x_e)P_1^e \tag{9.174}$$

Noting that $x_e \to \xi = -1$, $x_{e+1} \to \xi = +1$ and using properties of N_i; $i = 1,2$, we can write (9.174) in the matrix form

$$\int_{\bar{\Omega}^e} (Au_h^e)v(x)\,dx = [K^e]\{\delta^e\} + \{P^e\} \tag{9.175}$$

$[K^e]$ is calculated using procedure described in example 9.1 and we have

$$K_{ij}^e = -\frac{2}{h_e}\int_{-1}^{+1}\frac{dN_i}{d\xi}\frac{dN_j}{d\xi}\,d\xi + \frac{\lambda h_e}{2}\int_{-1}^{+1}N_iN_j\,d\xi \tag{9.176}$$

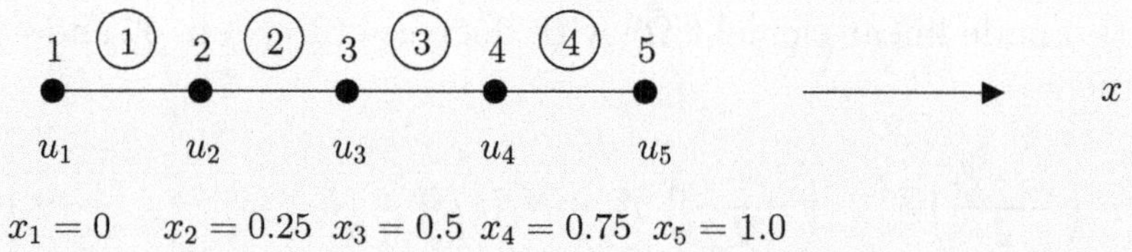

Figure 9.11: A four element uniform discretization using two node elements

Consider a four element uniform discretization shown in Figure 9.11. For an element $\bar{\Omega}^e$ with $h_e = 1/4$. We can obtain the following using (9.176).

$$[K^e] = -\frac{1}{h_e}\begin{bmatrix} 1 & -1 \\ -1 & 1 \end{bmatrix} + \frac{1}{6}h_e\lambda\begin{bmatrix} 2 & 1 \\ 1 & 2 \end{bmatrix} = \begin{bmatrix} -4 & 4 \\ 4 & -4 \end{bmatrix} + \lambda\begin{bmatrix} \frac{1}{12} & \frac{1}{24} \\ \frac{1}{24} & \frac{1}{12} \end{bmatrix} \tag{9.177}$$

Thus, for each element of the discretization of Figure 9.11 we can write

$$\int_{\bar{\Omega}^e} (Au_h^e)v\,dx = [[^1K^e] + \lambda[^2K^e]]\begin{Bmatrix} u_e \\ u_{e+1} \end{Bmatrix} + \begin{Bmatrix} P_1^e \\ P_2^e \end{Bmatrix} \;;\quad e = 1, 2, \dots, 4 \tag{9.178}$$

$$[^1K^e] = \begin{bmatrix} -4 & 4 \\ 4 & -4 \end{bmatrix} \;;\; [^2K^e] = \begin{bmatrix} \frac{1}{12} & \frac{1}{24} \\ \frac{1}{24} & \frac{1}{12} \end{bmatrix} \tag{9.179}$$

Assembly of the element equations can be written as

$$\int_{\bar{\Omega}^T} (Au_h)v\,dx = \sum_{e=1}^{4}\int_{\bar{\Omega}^e} (Au_h^e)v\,dx = [\sum_{e=1}^{4}[^1K^e] + \lambda\sum_{e=1}^{4}[^2K^e]]\{\delta\} + \sum_{e=1}^{4}\{P^e\} = \{0\}$$

$$\{\delta\} = \bigcup_e\{\delta^e\} \tag{9.180}$$

or

$$\begin{aligned} [K]\{\delta\} &= -\{P\} \\ [K] &= \left[\sum_{e=1}^{4}[^1K^e] + \lambda\sum_{e=1}^{4}[^2K^e]\right] \\ &= [[^1K] + \lambda[^2K]] \\ \text{and } \{P\} &= \sum_{e=1}^{4}\{P^e\} \end{aligned} \tag{9.181}$$

Since $u(0) = 0$ and $u(1) = 0$ implies that $u_1 = 0$ and $u_5 = 0$; we order the degrees of freedom in $\{\delta\}$ as follows

$$\{\delta\}^T = \{u_1\; u_5\; u_2\; u_3\; u_4\} \tag{9.182}$$

so that known u_1 and u_5 are together, hence the assembled equations will be in partitioned form. Thus, we label rows and columns of $[K]$ as u_1, u_2, u_3, u_4 and u_5. Assembled $[{}^1K^e]$, $[{}^2K^e]$ and $\{P^e\}$ are shown in the following.

$$[{}^1K^e] = \begin{bmatrix} -4 & 0 & 4 & 0 & 0 \\ 0 & -4 & 0 & 0 & 4 \\ 4 & 0 & (-4-4) & 4 & 0 \\ 0 & 0 & 4 & (-4-4) & 4 \\ 0 & 4 & 0 & 4 & (-4-4) \end{bmatrix} \begin{matrix} u_1 \\ u_5 \\ u_2 \\ u_3 \\ u_4 \end{matrix} \quad \text{(columns: } u_1\ u_5\ u_2\ u_3\ u_4) \tag{9.183}$$

$$[{}^2K^e] = \begin{bmatrix} \frac{1}{12} & 0 & \frac{1}{24} & 0 & 0 \\ 0 & \frac{1}{12} & 0 & 0 & \frac{1}{24} \\ \frac{1}{24} & 0 & (1/12+1/12) & \frac{1}{24} & 0 \\ 0 & 0 & \frac{1}{24} & (1/12+1/12) & \frac{1}{24} \\ 0 & \frac{1}{24} & 0 & \frac{1}{24} & (1/12+1/12) \end{bmatrix} \begin{matrix} u_1 \\ u_5 \\ u_2 \\ u_3 \\ u_4 \end{matrix} \quad \text{(columns: } u_1\ u_5\ u_2\ u_3\ u_4) \tag{9.184}$$

$$\{P\} = \begin{Bmatrix} P_1^1 \\ P_2^4 \\ P_2^1 + P_1^2 \\ P_2^2 + P_1^3 \\ P_2^3 + P_1^4 \end{Bmatrix} = \begin{Bmatrix} P_1 \\ P_5 \\ P_2 \\ P_3 \\ P_4 \end{Bmatrix} \tag{9.185}$$

$$\text{EBCs: } u(0) = u_1 = 0\,,\ u(1) = u_5 = 0 \tag{9.186}$$

$$\begin{aligned} &\text{NBCs: } P_1^1 \text{ and } P_2^4 \text{ are unknown} \\ &P_2^1 + P_1^2 = 0 \\ &P_2^2 + P_1^3 = 0 \\ &P_2^3 + P_1^4 = 0 \end{aligned} \tag{9.187}$$

We partition $[{}^1K^e]$ and $[{}^2K^e]$

$$\left[\begin{bmatrix} [{}^1K_{11}] & [{}^1K_{12}] \\ [{}^1K_{21}] & [{}^1K_{22}] \end{bmatrix} + \lambda \begin{bmatrix} [{}^2K_{11}] & [{}^2K_{12}] \\ [{}^2K_{21}] & [{}^2K_{22}] \end{bmatrix}\right] \begin{Bmatrix} \{\delta\}_1 \\ \{\delta\}_2 \end{Bmatrix} + \begin{Bmatrix} \{P\}_1 \\ \{P\}_2 \end{Bmatrix} = \{0\} \tag{9.188}$$

$$\{\delta\}_1^T = [u_1\ u_5] = [0.0\ 0.0] \tag{9.189}$$

$$\{\delta\}_2^T = [u_2\ u_3\ u_4] \tag{9.190}$$

using (9.185) – (9.187) in (9.188), we obtain the following from the second set of partitioned equations

$$[[^1K_{22}] + \lambda[^2K_{22}]]\{\delta\}_2 = \{0\} \tag{9.191}$$

Equation (9.191) define eigenvalue problem. We note that

$$\begin{aligned}
[^1K_{11}] &= \begin{bmatrix} -4 & 0 \\ 0 & -4 \end{bmatrix} \\
[^1K_{12}] &= \begin{bmatrix} 4 & 0 & 0 \\ 0 & 0 & 4 \end{bmatrix} \\
[^1K_{21}] &= [^1K_{12}]^T \\
[^1K_{22}] &= \begin{bmatrix} -8 & 4 & 0 \\ 4 & -8 & 4 \\ 0 & 4 & -8 \end{bmatrix} \\
[^2K_{11}] &= \begin{bmatrix} \frac{1}{3} & 0 \\ 0 & \frac{1}{3} \end{bmatrix} \\
[^2K_{12}] &= \begin{bmatrix} \frac{1}{6} & 0 & 0 \\ 0 & 0 & \frac{1}{6} \end{bmatrix} \\
[^2K_{21}] &= [^2K_{12}]^T \\
[^2K_{22}] &= \begin{bmatrix} \frac{2}{3} & \frac{1}{6} & 0 \\ \frac{1}{6} & \frac{2}{3} & \frac{1}{6} \\ 0 & \frac{1}{6} & \frac{2}{3} \end{bmatrix}
\end{aligned} \tag{9.192}$$

Using $[^1K_{22}]$ and $[^2K_{22}]$ from (9.192) in (9.191) and changing sign throughout

$$\left[\begin{bmatrix} 8 & -4 & 0 \\ -4 & 8 & -4 \\ 0 & -4 & 8 \end{bmatrix} - \lambda \begin{bmatrix} \frac{2}{3} & \frac{1}{6} & 0 \\ \frac{1}{6} & \frac{2}{3} & \frac{1}{6} \\ 0 & \frac{1}{6} & \frac{2}{3} \end{bmatrix}\right] \begin{Bmatrix} u_2 \\ u_3 \\ u_4 \end{Bmatrix} = \{0\} \tag{9.193}$$

Eigenpairs of (9.193) extracted using inverse iteration with iteration vector deflection technique are given in the following

$$
\begin{aligned}
(\lambda_1, \{\phi\}_1) &= (1.03867, \begin{Bmatrix} 1.0528 \\ 1.4886 \\ 1.0528 \end{Bmatrix}) \\
(\lambda_2, \{\phi\}_2) &= (48.0, \begin{Bmatrix} 1.7321 \\ -0.10165 \times 10^{-3} \\ -1.7320 \end{Bmatrix}) \\
(\lambda_3, \{\phi\}_3) &= (126.75, \begin{Bmatrix} 1.5222 \\ -2.1554 \\ 1.5226 \end{Bmatrix})
\end{aligned}
\tag{9.194}
$$

Theoretical values of λ (given in example 9.6) are $\lambda = 9.8696$, 39.4784 and 88.8264.

9.4 Finite Difference Method for ODEs and PDEs

In Section 9.1 we have shown that the mathematically justifiable approach for solving ODEs and PDEs, approximately or otherwise, is to integrate them. At present there are many other numerical approaches used in attempts to obtain approximate numerical solutions of ODEs and PDEs. The finite difference technique is one such approach. In this method, the derivatives appearing in the statement of the ODEs or PDEs are replaced by their algebraic approximations derived using Taylor series expansions. Thus, all derivative terms in the ODEs and PDEs are replaced by algebraic expressions containing nodal values of the functions (and/or their derivatives) for a discretization containing the desired number of points (nodes). This process yields algebraic equations in the unknown nodal values of the functions and their derivatives. Solution of these algebraic equations yields nodal values of the solution.

The fundamental question in this approach is to examine what mathematical principle is behind this approach that ensures that this process indeed yields approximate solutions of ODEs and PDEs. This is an unresolved issue based on the author's opinion. Nonetheless, since this technique is still commonly used in applications such as CFD, we consider basic details of the method for ODEs and PDEs describing BVPs.

9.4.1 Finite Difference Method for Ordinary Differential Equations

We consider the basic steps in the following, which are then applied to specific model problems.

Let $A\phi - f = 0$ in $\Omega \subset \mathbb{R}^1$ be the boundary value problem, in which A is the differential operator, ϕ is the dependent variable(s) and Ω is the domain of definition of the BVP.

(a) Consider a discretization $\bar{\Omega}^T$ of $\bar{\Omega} = \Omega \cup \Gamma$; Γ being the boundary of Ω, i.e., the end points of Ω in this case. Generally a uniform discretization is simpler. Label grid points or nodes and establish their coordinates.

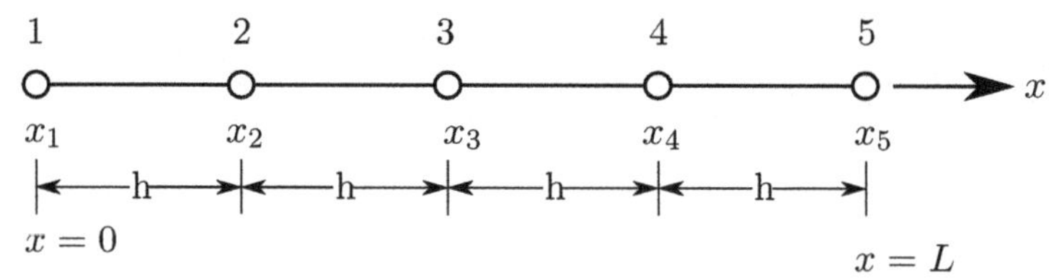

Figure 9.12: Discretization of $\bar{\Omega}^T$ of $\bar{\Omega}$

Figure 9.12 shows a five-node or points discretization $\bar{\Omega}^T$ of $\bar{\Omega} = [0, L]$.

(b) Express the derivatives in the differential equation describing the BVP in terms of their finite difference approximations using Taylor series expansions about the nodes in $[0, L]$ and substitute these in the differential equation.

(c) We also do the same for the derivative boundary conditions if there are any.

(d) As far as possible we use finite difference expressions for the various derivatives that have truncation error of the same order so that the order of the truncation error in the solution is clearly defined. In case of using finite difference expressions for the derivatives that have truncation errors of different orders, it is the lowest order truncation error that controls the order of the truncation error in the numerically computed solution.

(e) In step (b), the differential equation is converted into a system of algebraic equations. Arrange the final equations resulting in (b) in matrix form and solve for the numerical values of the unknown dependent variables at the grid or node points.

We consider some examples in the following.

Example 9.4. Second Order Non-Homogeneous ODE: Finite Difference Method Consider the following ODE.

$$\frac{d^2T}{dx^2} + T = x^3 \quad ; \qquad \forall x \in (0,1) = \Omega \subset \mathbb{R}^1 \tag{9.195}$$

with BCs :

$$T(0) = 0, \qquad T(1) = -0.5 \tag{9.196}$$

Find the numerical solution of the ODE using finite difference method with central differencing and uniform spacing of node points with $h = 0.2$ (Figure 9.13).

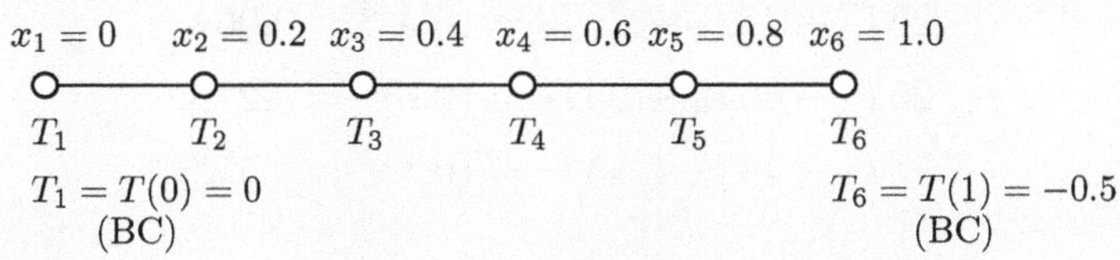

Figure 9.13: Schematic of Example 9.4

Consider 3-node *stencil* (the nodal discretization used to convert the derivatives to algebraic expressions, Figure 9.14).

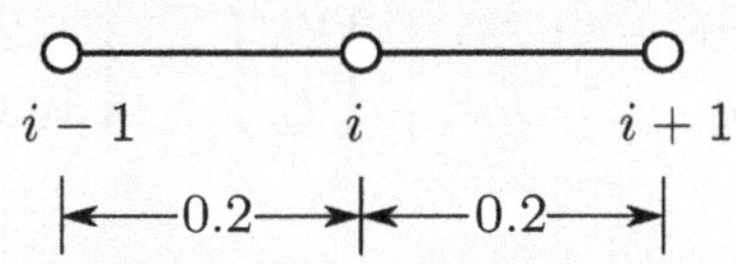

Figure 9.14: A three-node stencil of points in Example 9.4

Using

$$T_i = T(x_i) \qquad ; i = 1, 2, \dots \text{with } h = 0.2 \tag{9.197}$$

we can write:

$$\left.\frac{d^2T}{dx^2}\right|_{x=x_i} = \frac{T_{i+1} - 2T_i + T_{i-1}}{h^2} = \frac{T_{i+1} - 2T_i + T_{i-1}}{(0.2)^2} \tag{9.198}$$

Consider (9.195) at node i.

$$\left.\frac{d^2T}{dx^2}\right|_{x=x_i} + T\big|_{x=i} = x_i^3 \tag{9.199}$$

Substituting for $\left.\frac{d^2T}{dx^2}\right|_{x=x_i}$ from (9.198) into (9.199):

$$\frac{T_{i+1} - 2T_i + T_{i-1}}{(0.2)^2} + T_i = x_i^3 \tag{9.200}$$

or

$$25T_{i-1} - 49T_i + 25T_{i+1} = x_i^3 \tag{9.201}$$

Since $T_1 = T(0) = 0$ and $T_6 = T(1) = -0.5$, the solution T is known at nodes 1 and 6, therefore the finite difference form of (9.195), i.e., (9.200), only needs to be satisfied for $i = 2, 3, 4$, and 5, which gives us:

$$\begin{aligned} 25T_1 - 49T_2 + 25T_3 &= (0.2)^3 = 0.008 \\ 25T_2 - 49T_3 + 25T_4 &= (0.4)^3 = 0.064 \\ 25T_3 - 49T_4 + 25T_5 &= (0.6)^3 = 0.216 \\ 25T_4 - 49T_5 + 25T_6 &= (0.8)^3 = 0.512 \end{aligned} \tag{9.202}$$

Using $T_1 = 0$ and $T_6 = (-0.5)$ and arranging (9.202) in matrix form:

$$\begin{bmatrix} -49 & 25 & 0 & 0 \\ 25 & -49 & 25 & 0 \\ 0 & 25 & -49 & 25 \\ 0 & 0 & 25 & -49 \end{bmatrix} \begin{Bmatrix} T_2 \\ T_3 \\ T_4 \\ T_5 \end{Bmatrix} = \begin{Bmatrix} 0.008 \\ 0.064 \\ 0.216 \\ 13.012 \end{Bmatrix} \tag{9.203}$$

Solution of linear simultaneous equations in (9.203) gives:

$$\begin{aligned} T_2 &= -0.128947, & T_3 &= -0.252416 \\ T_4 &= -0.363228, & T_5 &= -0.450872 \end{aligned} \tag{9.204}$$

The values of T_i in (9.204) is the approximate solution of (9.195) and (9.196) at $x = x_i$; $i = 2, 3, 4$ and 5.

Table 9.5: Temperature values at the grid points (example 9.4)

node	x	T
1	0.0	0.0
2	0.2	-0.128947
3	0.46	-0.25328
4	0.6	-0.36418
5	0.8	-0.45155
6	1.0	-0.5

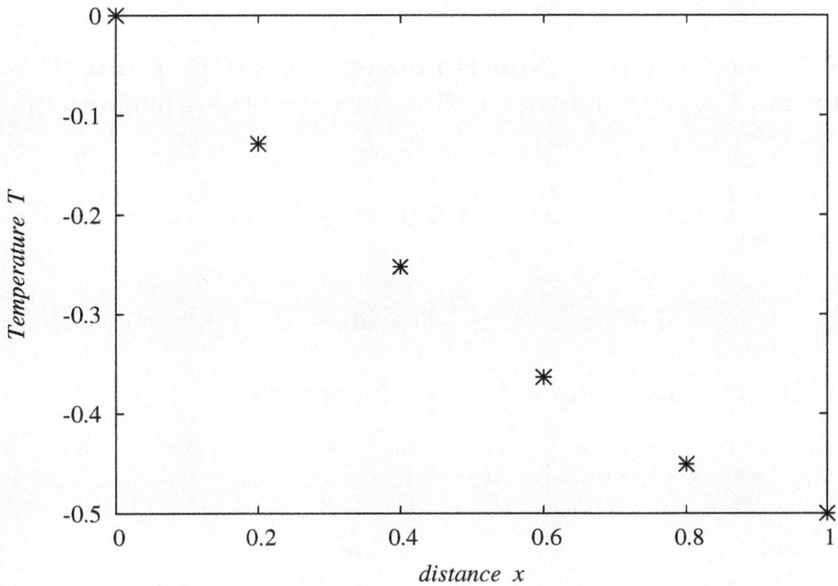

Figure 9.15: T versus x (example 9.4)

Table 9.5 gives values of T at the grid points. Figures 9.15 show plot of (T_i, x_i); $i = 1, 2, \dots, 6$.

Remarks.

(1) We note that the coefficient matrix in (9.203) is in tridiagonal form, hence we can take advantage in storing the coefficients of the matrix as well as in solution methods for calculating T_i, $i = 2, 3, \dots, 5$.

(2) The solution in (9.204) is approximate.

  (a) The finite difference expression for the derivatives are approximate.

  (b) We have only used a finite number of node points (only six) in $\bar{\Omega}^T$.

  (c) If the number of points in $\bar{\Omega}^T$ are increased, accuracy of the computed nodal values of T will improve.

(3) Both boundary conditions in (9.196) are function values, i.e., values of T at the two boundaries $x = 0$ and $x = 1.0$.

(4) We only know the solution at the grid or node points. Between the node points we only know that the solution is continuous and differentiable, but we do not know what it is. This is not the case in finite element method.

Example 9.5. Second Order Non-Homogeneous ODE: Finite Difference Method Consider the same BVP as in Example 9.4 but with different BCs than for Example 9.4.

$$\frac{d^2T}{dx^2} + T = x^3 \quad ; \qquad \forall x \in (0,1) = \Omega \subset \mathbb{R}^1 \tag{9.205}$$

with BCs :

$$T(0) = 0, \qquad \left.\frac{dT}{dx}\right|_{x=1} = 20.0 \tag{9.206}$$

We consider the same discretization $\bar{\Omega}^T$ as in Example 9.4.

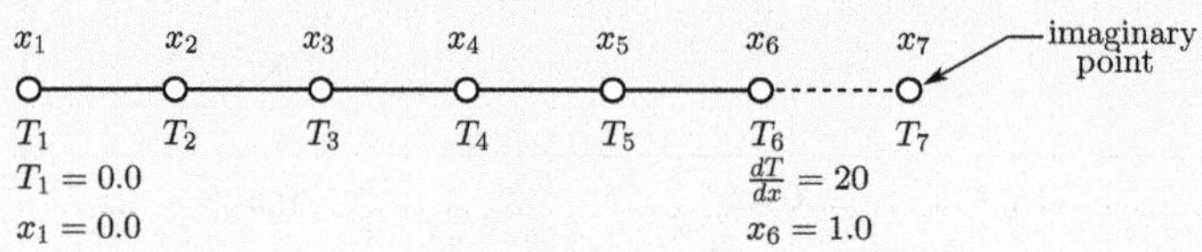

Figure 9.16: Schematic of Example 9.5

Consider central difference to find numerical values of the solution T at the nodes. Consider a 3-node stencil.

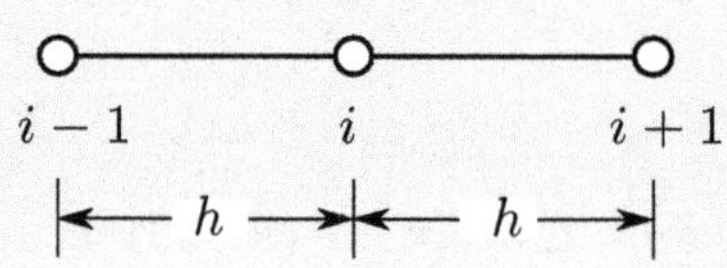

Figure 9.17: A three-node stencil of points in Example 9.5

$$\left.\frac{d^2T}{dx^2}\right|_i \simeq \frac{T_{i+1} - 2T_i + T_{i-1}}{h^2} = \frac{T_{i+1} - 2T_i + T_{i-1}}{(0.2)^2} \tag{9.207}$$

Consider (9.205) for a node i.

$$\left.\frac{d^2T}{dx^2}\right|_{x=x_i} + T|_{x=x_i} = x_i^3 \tag{9.208}$$

Substituting for $\left.\frac{d^2T}{dx^2}\right|_{x=x_i}$ in (9.208) from (9.207):

$$\frac{T_{i+1} - 2T_i + T_{i-1}}{(0.2)^2} + T_i = x_i^3 \tag{9.209}$$

$$\text{or} \quad 25T_{i-1} - 49T_i + 25T_{i+1} = x_i^3 \tag{9.210}$$

Since at $x = 1.0$, $\frac{dT}{dx}$ is given, T at $x = 1.0$ is not known, hence (9.210) must also hold at $x = 1.0$, i.e., at node 6 in addition to $i = 2, 3, 4, 5$. In order to satisfy the BC $\frac{dT}{dx} = 20$ at $x = 1.0$ using a central difference approximation of $\frac{dT}{dx}$, we need an additional node 7 (outside the domain) as shown in Figure 9.16. Using a three-node stencil $i-1$, i and $i+1$, we can write the following (central difference).

$$\left.\frac{dT}{dx}\right|_{x=x_i} = \frac{T_{i+1} - T_{i-1}}{2h} = \frac{T_{i+1} - T_{i-1}}{2(0.2)} = 2.5(T_{i+1} - T_{i-1}) \tag{9.211}$$

Using (9.211) for $i = 6$:

$$\left.\frac{dT}{dx}\right|_{x=x_6} = \left.\frac{dT}{dx}\right|_{x=1} = 20 = 2.5(T_7 - T_5) \tag{9.212}$$

or

$$T_7 = T_5 + 8.0 \tag{9.213}$$

Thus T_7 is known in terms of T_5. Using (9.210) for $i = 2, 3, 4, 5$ and 6:

$$\begin{aligned}
25T_1 - 49T_2 + 25T_3 &= (0.2)^3 = 0.008 \\
25T_2 - 49T_3 + 25T_4 &= (0.4)^3 = 0.064 \\
25T_3 - 49T_4 + 25T_5 &= (0.6)^3 = 0.216 \\
25T_4 - 49T_5 + 25T_6 &= (0.8)^3 = 0.512 \\
25T_5 - 49T_6 + 25T_7 &= (1.0)^3 = 1.000
\end{aligned} \tag{9.214}$$

Substitute $T_1 = 0$ and $T_7 = T_5 + 8.0$ (BCs) in (9.214) and arrange the resulting equations in the matrix form.

$$\begin{bmatrix} -49 & 25 & 0 & 0 & 0 \\ 25 & -49 & 25 & 0 & 0 \\ 0 & 25 & -49 & 25 & 0 \\ 0 & 0 & 25 & -49 & 25 \\ 0 & 0 & 0 & 50 & -49 \end{bmatrix} \begin{Bmatrix} T_2 \\ T_3 \\ T_4 \\ T_5 \\ T_6 \end{Bmatrix} = \begin{Bmatrix} 0.008 \\ 0.064 \\ 0.216 \\ 0.512 \\ -199.0 \end{Bmatrix} \tag{9.215}$$

Solution of linear simultaneous equations in (9.215) gives:

$$\begin{matrix} T_2 = 7.32918 & , & T_3 = 14.36551 & , \\ T_4 = 20.82977 & , & T_5 = 26.46949 & , \end{matrix} \quad T_6 = 31.07091 \tag{9.216}$$

The values of T_i; $i = 1, 2, \ldots, 6$ are the approximate solution of (9.205) and (9.206) at $x = x_i$; $i = 2, 3, \ldots, 6$.

Table 9.6: Temperature values at the grid points (example 9.5)

node	x	T
1	0.0	0.0
2	0.2	7.32918
3	0.46	14.36551
4	0.6	20.82977
5	0.8	26.46949
6	1.0	31.07091

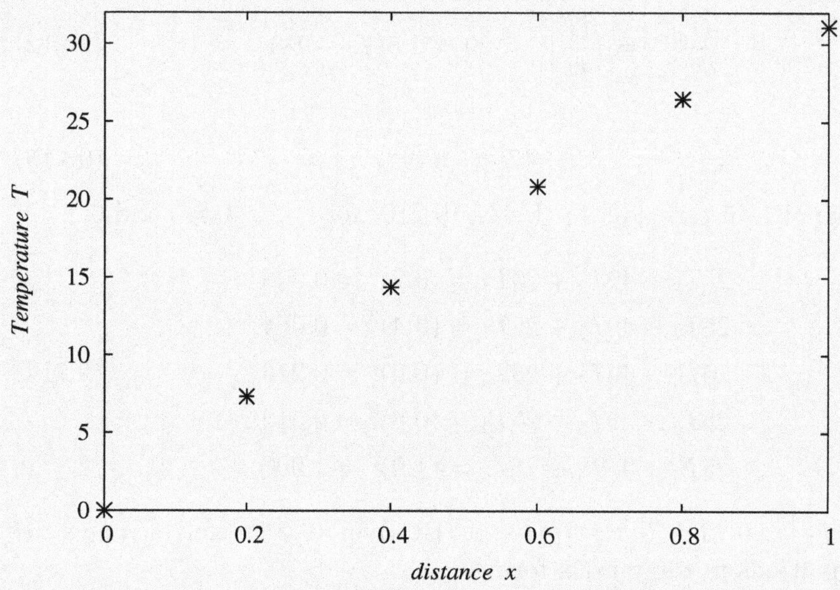

Figure 9.18: T versus x (example 9.5)

Table 9.6 gives values of T at the grid points. Figures 9.18 show plot of (T_i, x_i); $i = 1, 2, \ldots, 6$.

Remarks.

(1) Node points such as point 7 that are outside the domain $\bar{\Omega}$ are called *imaginary points.* These are necessary when the derivative (first or second) boundary conditions are specified at the boundary points.

(2) This example demonstrates how the function value and derivative boundary condition (first derivative in this case) are incorporated in the finite difference solution procedure.

(3) The accuracy of the approximation in general is poorer when the derivative boundary conditions are specified at the boundary points due to the additional approximation of the boundary condition.

(4) Here also, we only know the solution at the grid or node points. Between the node points we only know that the solution is continuous and differentiable, but we do not know what it is. This is not the case in finite element method.

Example 9.6. Second Order Homogeneous ODE: Finite Difference Method Consider the following BVP:

$$\frac{d^2u}{dx^2} + \lambda u = 0 \quad \forall x \in (0,1) = \Omega \subset \mathbb{R}^1 \tag{9.217}$$

with the following BCs :

$$u(0) = 0 \quad , \quad u(1) = 0 \tag{9.218}$$

The quantity λ is a scalar and is unknown. We seek solution of (9.218) with BCs (9.218) using finite difference approximation of the derivatives by central difference method.

Consider a five-node uniform discretization ($h = 0.25$).

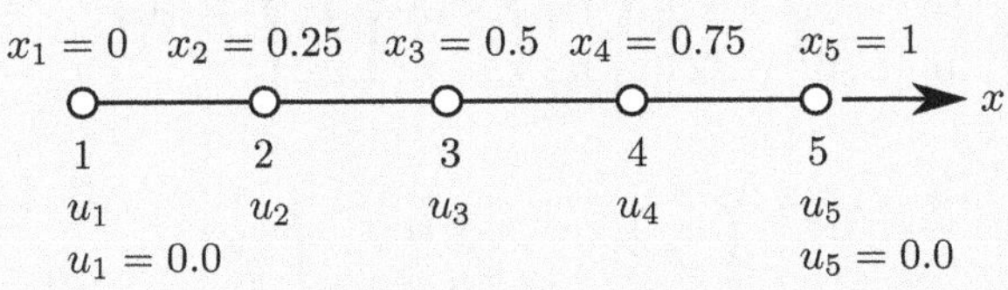

Figure 9.19: Schematic of Example 9.6

In central difference approximation of the derivatives we consider a three node stencil.

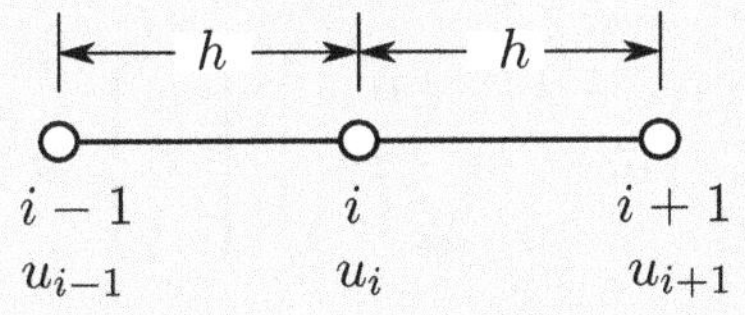

Figure 9.20: A three-node stencil of points in Example 9.6

At node i we have:

$$\left.\frac{d^2u}{dx^2}\right|_{x=x_i} = \frac{u_{i-1} - 2u_i + u_{i+1}}{h^2} = \frac{u_{i-1} - 2u_i + u_{i+1}}{(0.25)^2} \tag{9.219}$$

The differential equation at node i can be written as follows.

$$\left.\frac{d^2u}{dx^2}\right|_{x=x_i} + \lambda u_i = 0 \tag{9.220}$$

Substituting (9.220) in (9.219):

$$\frac{u_{i-1} - 2u_i + u_{i+1}}{(0.25)^2} + \lambda y_i = 0 \tag{9.221}$$

or

$$16u_{i-1} - 32u_i + 16u_{i+1} + \lambda u_i = 0 \tag{9.222}$$

Since nodes 1 and 5 have function u specified, at these locations the solution is known. Thus, (9.222) must only be satisfied at x_i ; $i = 2, 3, 4$.

$$\begin{aligned} 16u_1 - 32u_2 + 16u_3 + \lambda u_2 = 0 \\ 16u_2 - 32u_3 + 16u_4 + \lambda u_3 = 0 \\ 16u_3 - 32u_4 + 16u_5 + \lambda u_4 = 0 \end{aligned} \tag{9.223}$$

Substitute $u_1 = 0$ and $u_5 = 0$ (BCs) in (9.223) and arrange in the matrix and vector form.

$$\begin{bmatrix} -32 & 16 & 0 \\ 16 & -32 & 16 \\ 0 & 16 & -32 \end{bmatrix} \begin{Bmatrix} u_2 \\ u_3 \\ u_4 \end{Bmatrix} + \lambda \begin{Bmatrix} u_2 \\ u_3 \\ u_4 \end{Bmatrix} = \{0\} \tag{9.224}$$

or

$$\begin{bmatrix} 32 & -16 & 0 \\ -16 & 32 & 16 \\ 0 & -16 & 32 \end{bmatrix} \begin{Bmatrix} u_2 \\ u_3 \\ u_4 \end{Bmatrix} - \lambda \begin{bmatrix} 1 & 0 & 0 \\ 0 & 1 & 0 \\ 0 & 0 & 1 \end{bmatrix} \begin{Bmatrix} u_2 \\ u_3 \\ u_4 \end{Bmatrix} = \{0\} \tag{9.225}$$

This is an eigenvalue problem. We can write (9.225) as:

$$\begin{bmatrix} 32-\lambda & -16 & 0 \\ -16 & 32-\lambda & -16 \\ 0 & -16 & 32-\lambda \end{bmatrix} \begin{Bmatrix} u_2 \\ u_3 \\ u_4 \end{Bmatrix} = \{0\} \tag{9.226}$$

The characteristic polynomial of (9.226) is given by:

$$\det \begin{bmatrix} 32-\lambda & -16 & 0 \\ -16 & 32-\lambda & -16 \\ 0 & -16 & 32-\lambda \end{bmatrix} = 0 \tag{9.227}$$

We can find the eigenpairs of (9.225) using any one of the methods discussed in Chapter 4, and we obtain:

$$
\begin{aligned}
(\lambda_1, \{\phi\}_1) &= 9.37, \begin{Bmatrix} 0.5 \\ 0.707 \\ 0.5 \end{Bmatrix} \\
(\lambda_2, \{\phi\}_2) &= 32.0, \begin{Bmatrix} 0.70711 \\ 0.0 \\ -0.70711 \end{Bmatrix} \\
(\lambda_3, \{\phi\}_3) &= 54.62, \begin{Bmatrix} 0.49957 \\ -0.70722 \\ 0.49957 \end{Bmatrix}
\end{aligned}
\tag{9.228}
$$

We note that the eigenvectors are not normalized.

Theoretical Solution and Error in the Numerical Approximation:

The general solution of the BVP (9.217) is given by:

$$u = c_1 \sin \sqrt{\lambda} x + c_2 \cos \sqrt{\lambda} x \tag{9.229}$$

Using BCs:

$$
\begin{aligned}
&\text{at } x = 0, \quad u = u(0) = 0 \quad ; \quad 0 = c_1(0) + c_2(1) \implies c_2 = 0 \\
&\text{at } x = 1, \quad u = u(1) = 0 \quad ; \quad 0 = c_1 \sin \sqrt{\lambda} \implies \sqrt{\lambda} = n\pi \quad ; \quad n = 1, 2, \dots
\end{aligned}
\tag{9.230}
$$

Therefore we have:

$$u = c_n \sin n\pi x \quad ; \quad \sqrt{\lambda} = n\pi \tag{9.231}$$

$$
\begin{aligned}
&\text{for } n = 1 \quad ; \quad \lambda_1 = (\pi)^2 = 9.8696 \\
&\text{for } n = 2 \quad ; \quad \lambda_2 = 4(\pi)^2 = 39.4784 \\
&\text{for } n = 3 \quad ; \quad \lambda_3 = 9(\pi)^2 = 88.8264
\end{aligned}
\tag{9.232}
$$

Comparing the theoretical values of λ_i ; $i = 1, 2, 3$ in (9.232) with the numerically calculated values in (9.228), we find that:

$$
\begin{aligned}
\text{error in } \lambda_1 = e_1 &= 9.8696 - 9.37 = 0.4996 \\
\text{error in } \lambda_2 = e_2 &= 39.4784 - 32.0 = 7.4784 \\
\text{error in } \lambda_3 = e_3 &= 88.82640 - 54.62 = 34.2064
\end{aligned}
\tag{9.233}
$$

Error in λ_i ; $i = 1, 2, 3$ becomes progressively larger for higher eigenvalues.

9.4.2 Finite Difference Method for Partial Differential Equations

Differential mathematical models in $\mathbb{R}^2$ and $\mathbb{R}^3$ result in partial differential equations in which dependent variables exhibit dependence on (x, y) (in $\mathbb{R}^2$) or on (x, y, z) (in $\mathbb{R}^3$). Some of the simplest forms are:

$$\frac{\partial^2 \phi}{\partial x^2} + \frac{\partial^2 \phi}{\partial y^2} = 0 \quad \forall x, y \in \Omega \tag{9.234}$$

$$\frac{\partial^2 \phi}{\partial x^2} + \frac{\partial^2 \phi}{\partial y^2} = f(x,y) \quad \forall x, y \in \Omega \tag{9.235}$$

Equation (9.234) is called Laplace's equation, a homogeneous second-order partial differential equation in dependent variable $\phi = \phi(x, y)$. Equation (9.235) is Poisson's equation, a non-homogeneous Laplace equation. Ω is the xy-domain over which the differential equation holds. We define:

$$\bar{\Omega} = \Omega \cup \Gamma \tag{9.236}$$

$\bar{\Omega}$ is called the *closure* of Ω. Γ is the boundary of Ω. Boundary conditions are defined on all or part of Γ. These may consist of known values of ϕ or $\frac{\partial \phi}{\partial x}, \frac{\partial \phi}{\partial y}, \frac{\partial \phi}{\partial z}$ or $\frac{\partial \phi}{\partial n}$ (derivative of ϕ normal to the boundary Γ) or any combination of these. BCs make the solution of (9.234) or (9.235) (or any BVP) unique.

In the finite difference method of solving BVPs described by partial differential equations, the derivatives in the BVPs must be expressed in terms of algebraic expressions using their finite difference approximations. The outcome of doing so is that the partial differential equation is converted into a system of algebraic equations. After imposing BCs, we solve for the remaining quantities (function values at the nodes). The details are presented in the following for Laplace and Poisson's equations.

9.4.2.1 Laplace's Equation

$$\frac{\partial^2 \phi}{\partial x^2} + \frac{\partial^2 \phi}{\partial y^2} = 0 \quad \forall x, y \in \Omega = (0, L) \times (0, M) \tag{9.237}$$

$$\text{BCs} : \phi(0, y) = \phi(x, 0) = \phi(L, y) = 0 \quad , \quad \phi(x, M) = 100 \tag{9.238}$$

Figure 9.21 shows the domain Ω, its boundary Γ, and the boundary conditions (9.238).

In the finite difference method we approximate $\frac{\partial^2 \phi}{\partial x^2}$ and $\frac{\partial^2 y}{\partial x^2}$ by their finite difference approximation at a countable number of points in $\bar{\Omega}$. Consider the following uniform grid (discretization $\bar{\Omega}^T$ of $\bar{\Omega}$) in which Δx and Δy are spacing in x- and y-directions.

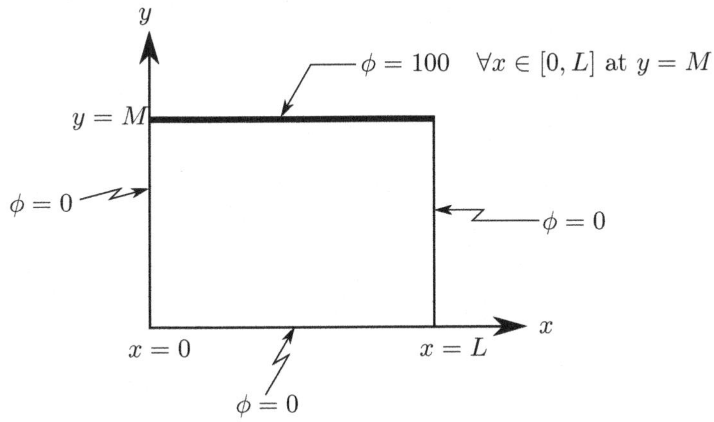

Figure 9.21: Domain $\bar{\Omega}$ of BVP (9.237) and BCs (9.238)

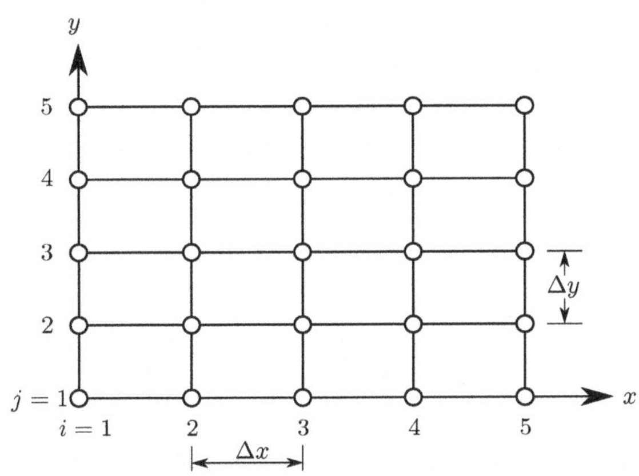

Figure 9.22: Discretization of $\bar{\Omega}^T$ of $\bar{\Omega}$

This discretization establishes 25 grid points or node points. At the grid points located on the boundary Γ, ϕ is known (see (9.238) and Figure 9.21). Our objective is to find function values ϕ at the interior nine points. Along x-axis $i = 1, 2, 3, 4, 5$ and along y-axis $j = 1, 2, 3, 4, 5$ and their intersections uniquely define all grid or node points. We consider (9.237) and approximate $\frac{\partial^2 \phi}{\partial x^2}$ and $\frac{\partial^2 \phi}{\partial y^2}$ at each of the nine interior nodes (or grid points) using the finite difference method. Consider an interior grid point (i, j) and the node stencil shown in Figure 9.23.

At node (i, j), $\frac{\partial^2 \phi}{\partial x^2}$ and $\frac{\partial^2 \phi}{\partial y^2}$ can be approximated using the following (central difference):

$$\left.\frac{\partial^2 \phi}{\partial x^2}\right|_{i,j} = \frac{\phi_{i+1,j} - 2\phi_{i,j} + \phi_{i-1,j}}{(\Delta x)^2} \quad , \quad \left.\frac{\partial^2 \phi}{\partial y^2}\right|_{i,j} = \frac{\phi_{i,j+1} - 2\phi_{i,j} + \phi_{i,j-1}}{(\Delta y)^2} \tag{9.239}$$

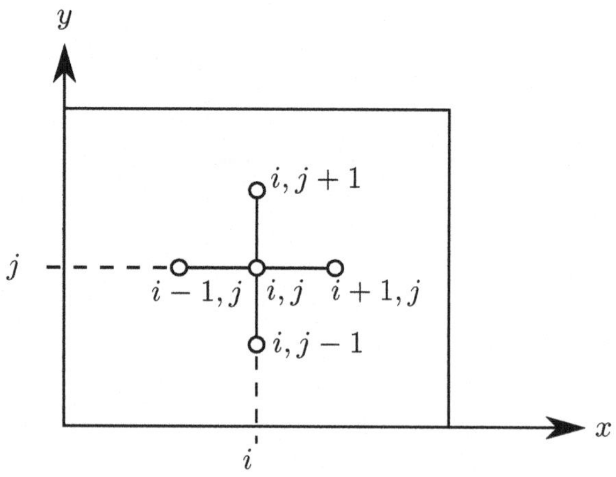

Figure 9.23: Node(i, j) and five-node stencil

Substituting from (9.239) in (9.237), we obtain the following for (9.237) at node i, j.

$$\frac{\phi_{i+1,j} - 2\phi_{i,j} + \phi_{i-1,j}}{(\Delta x)^2} + \frac{\phi_{i,j+1} - 2\phi_{i,j} + \phi_{i,j-1}}{(\Delta y)^2} = 0 \tag{9.240}$$

If we choose $\Delta x = \Delta y$ (hence $L = M$), then (9.240) reduces to:

$$\phi_{i,j+1} + \phi_{i,j-1} + \phi_{i+1,j} + \phi_{i-1,j} - 4\phi_{i,j} = 0 \tag{9.241}$$

Equation (9.241), the finite difference approximation of the BVP (9.237), must hold at all interior grid points (or nodes).

Consider the following (Figure 9.24) for more specific details and numerical calculation of the solution.

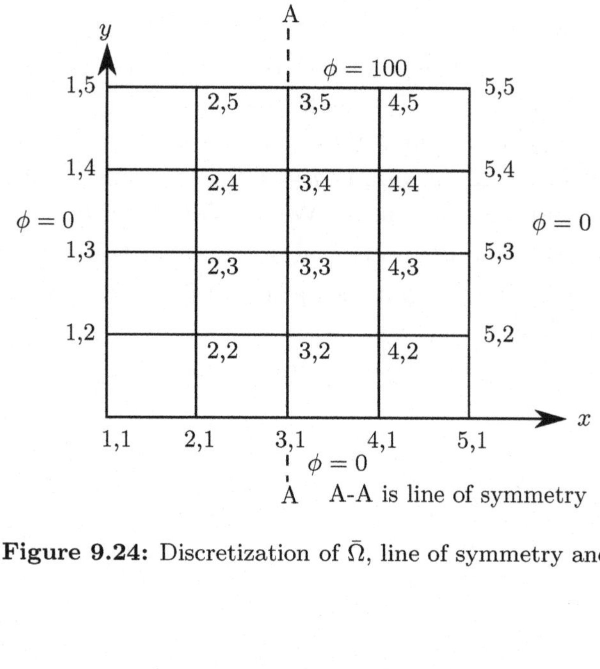

Figure 9.24: Discretization of $\bar{\Omega}$, line of symmetry and BCs

We note that $A-A$ is a line (or plane) of symmetry. $\bar{\Omega}$, BCs, and the solution ϕ are all symmetric about this line, i.e., the left half and right half of line $A-A$ are reflections of each other. Hence, for interior points we have:

$$\begin{aligned}\phi_{4,4} &= \phi_{2,4}\\ \phi_{4,3} &= \phi_{2,3}\\ \phi_{4,2} &= \phi_{2,2}\end{aligned} \tag{9.242}$$

and the following for the boundary nodes.

$$\begin{array}{lllll}\phi_{4,1}=\phi_{2,1} & , & \phi_{4,5}=\phi_{2,5} & , & \phi_{5,1}=\phi_{1,1}\\ \phi_{5,2}=\phi_{1,2} & , & \phi_{5,3}=\phi_{1,3} & , & \phi_{5,4}=\phi_{1,4}\\ \phi_{5,5}=\phi_{1,5} & & & & \end{array} \tag{9.243}$$

These conditions are already satisfied by the boundary conditions of the BVP. From (9.241), we can write:

$$\phi_{i,j}=0.25(\phi_{i,j+1}+\phi_{i,j-1}+\phi_{i+1,j}+\phi_{i-1,j}) \tag{9.244}$$

Using (9.244), we can obtain the following for the nine interior grid points (after substituting the value of ϕ for boundary nodes):

$$\begin{aligned}\phi_{2,2} &= 0.25(\phi_{2,3}+\phi_{2,3})\\ \phi_{2,3} &= 0.25(\phi_{2,2}+\phi_{3,3}+\phi_{2,4})\\ \phi_{2,4} &= 0.25(\phi_{2,3}+\phi_{3,4}+100)\\ \phi_{3,2} &= 0.25(\phi_{4,2}+\phi_{3,3}+\phi_{2,2})\\ \phi_{3,3} &= 0.25(\phi_{3,2}+\phi_{4,3}+\phi_{3,4}+\phi_{2,3})\\ \phi_{3,4} &= 0.25(\phi_{3,3}+\phi_{4,4}+\phi_{2,4}+100)\\ \phi_{4,2} &= 0.25(\phi_{4,3}+\phi_{3,2})\\ \phi_{4,3} &= 0.25(\phi_{4,2}+\phi_{4,4}+\phi_{3,3})\\ \phi_{4,4} &= 0.25(\phi_{4,3}+\phi_{3,4}+100)\end{aligned} \tag{9.245}$$

If we substitute condition (9.242) in the last three equations in (9.245), we obtain:

$$\begin{aligned}\phi_{4,2} &= 0.25(\phi_{2,3}+\phi_{3,2})=\phi_{2,2}\\ \phi_{4,3} &= 0.25(\phi_{2,2}+\phi_{2,4}+\phi_{3,3})=\phi_{2,3}\\ \phi_{4,4} &= 0.25(\phi_{2,3}+\phi_{3,4}+100)=\phi_{2,4}\end{aligned} \tag{9.246}$$

Which establishes existence of symmetry about the line $A-A$. Thus in (9.245), we only need to use the first six equations. Writing these in the matrix form:

$$\begin{bmatrix} 1 & -0.25 & 0 & -0.25 & 0 & 0 \\ -0.25 & 1 & -0.25 & 0 & -0.25 & 0 \\ 0 & -0.25 & 1 & 0 & 0 & -0.25 \\ -0.5 & 0 & 0 & 1 & -0.25 & 0 \\ 0 & -0.5 & 0 & -0.25 & 1 & -0.25 \\ 0 & 0 & -0.5 & 0 & -0.25 & 1 \end{bmatrix} \begin{Bmatrix} \phi_{2,2} \\ \phi_{2,3} \\ \phi_{2,4} \\ \phi_{3,2} \\ \phi_{3,3} \\ \phi_{3,4} \end{Bmatrix} = \begin{Bmatrix} 0 \\ 0 \\ 25 \\ 0 \\ 0 \\ 25 \end{Bmatrix} \tag{9.247}$$

Solution of the linear simultaneous simultaneous equation (9.247) yields:

$$\begin{aligned} \phi_{2,2} &= 7.14286 \quad , \quad \phi_{2,3} = 18.75 \quad , \quad \phi_{2,4} = 42.85714 \\ \phi_{3,2} &= 9.82143 \quad , \quad \phi_{3,3} = 25.0 \quad , \quad \phi_{3,4} = 52.6786 \end{aligned} \tag{9.248}$$

The solution values in (9.248) can be shown schematically (see Figure 9.25). We note that this solution holds for any $L = M$ i.e. for all square domains $\bar{\Omega}$.

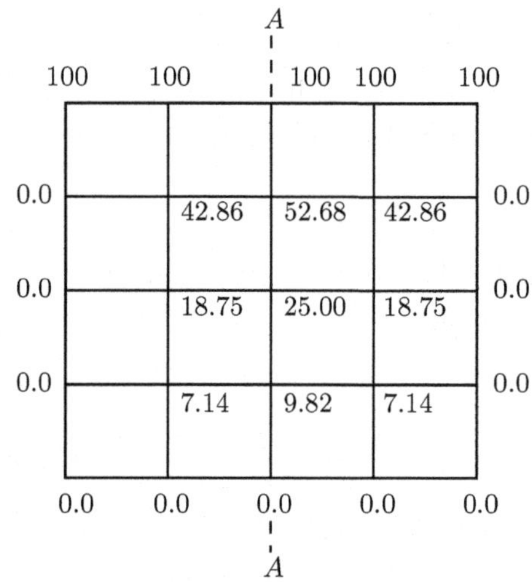

Figure 9.25: Solution values at the grid points

9.4.2.2 Poisson's Equation

Consider:

$$\frac{\partial^2 \phi}{\partial x^2} + \frac{\partial^2 \phi}{\partial y^2} = -2 \quad \forall x, y \in \Omega = (0, L) \times (0, L) \tag{9.249}$$

$$\text{with BCs} : \phi(0, y) = \phi(x, 0) = \phi(L, y) = \phi(x, L) = 0.0 \tag{9.250}$$

In this case Ω is a square domain with side length L. Following the details for Laplace equation, we can write the following for a grid points (i, j) (using $\Delta x = \Delta y = h$).

$$\phi_{i+1,j} + \phi_{i-1,j} + \phi_{i,j+1} + \phi_{i,j-1} - 4\phi_{i,j} = -2h^2 \tag{9.251}$$

Consider the following cases for obtaining numerical solutions.

Case (a): Nine-Node Discretization

Consider the following discretization of $\bar{\Omega}$ with $L = 2$ and $h = 1$:
Using (9.251) for node $(2, 2)$ we obtain (since ϕ is zero on all boundary points):

$$\begin{aligned} -4\phi_{2,2} &= -2(1)^2 \\ \therefore \quad \phi_{2,2} &= 0.5 \end{aligned} \tag{9.252}$$

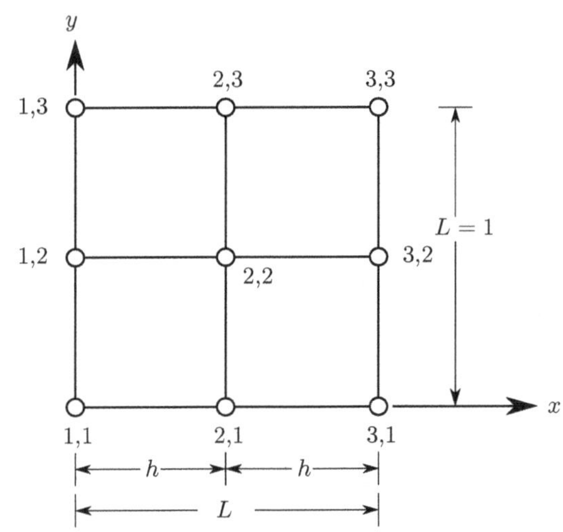

Figure 9.26: A nine grid point discretization of $\bar{\Omega}$

Case (b): 25-Node Discretization

Consider the following discretization of $\bar{\Omega}$. $A - A$, $B - B$ are lines of symmetry and so is $C - C$. Thus we only need to consider the following:
Due to symmetry about $C - C$ we only need to consider nodes $(2, 2)$, $(2, 3)$ and $(3, 3)$. We note that:

$$\phi_{3,2} = \phi_{2,3}$$

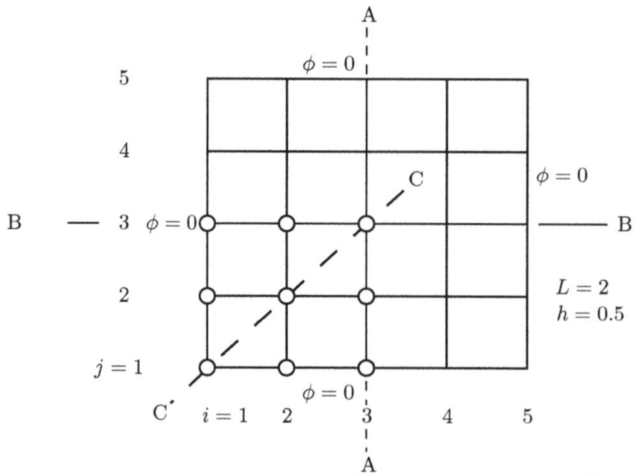

Figure 9.27: A 25-node discretization of $\bar{\Omega}^T$

At node (2,2):

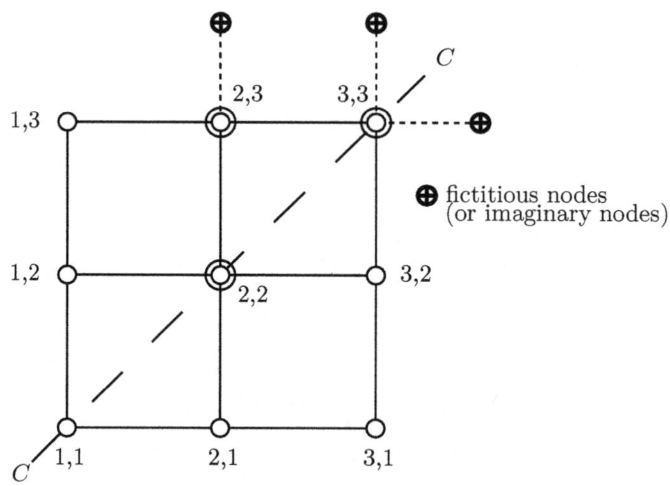

Figure 9.28: Subdomain of $\bar{\Omega}^T$

$$2\phi_{2,3} - 4\phi_{2,2} = -0.5 \tag{9.253}$$

At node (2,3):

$$2\phi_{2,2} + \phi_{3,3} - 4\phi_{2,3} = -0.5 \tag{9.254}$$

At node (3,3):

$$4\phi_{2,3} - 4\phi_{3,3} = -0.5 \tag{9.255}$$

or in matrix form:

$$\begin{bmatrix} -4 & 2 & 0 \\ 2 & -4 & 1 \\ 0 & 4 & -4 \end{bmatrix} \begin{Bmatrix} \phi_{2,2} \\ \phi_{2,3} \\ \phi_{3,3} \end{Bmatrix} = \begin{Bmatrix} -0.5 \\ -0.5 \\ -0.5 \end{Bmatrix} \tag{9.256}$$

Solution of (9.256) gives:

$$\phi_{2,2} = 0.3438 \quad , \quad \phi_{2,3} = 0.4375 \quad , \quad \phi_{3,3} = 0.5625 \tag{9.257}$$

We note that at $x = \frac{L}{2}$, $y = \frac{L}{2}$, ϕ has changed from 0.5 in case (a) with $h = 1$ to $\phi = 0.5625$ for $h = 0.5$. Obviously $\phi = 0.5625$ is a better approximation of ϕ as it is obtained from a more refined discretization. As we continue to refine $\bar{\Omega}^T$ (add more grid points), the solution continues to improve. This is the concept of convergence.

Remarks.

(1) The finite difference method of obtaining approximate numerical solution of BVP is a rather crude method of approximation.

(2) When Γ of Ω in $\bar{\Omega}$ is a curved boundary or a surface, many difficulties are encountered in defining the derivative boundary conditions.

(3) This method, though crude, is simple for obtaining quick solutions that give some idea about the solution behavior.

(4) The finite element method of approximation for BVPs has sound mathematical foundation. This method eliminates or overcomes many of the difficulties encountered in the finite difference method.

9.5 Concluding Remarks

We have shown that solutions of BVPs require that we integrate them over their domain of definition. This is the only mathematically justifiable technique of obtaining their valid solutions numerically or otherwise. Finite element method (FEM) is one such method in which one constructs integral of the BVP over the discretized domain of definition of the BVP. Details of FEM are presented for 1D BVPs in single variable including example problems. In view of the fact the solutions of BVPs require their integration over their domain of definition, finite difference, finite volume and other techniques are not as meritorious as FEM and in some cases may even yield wrong solutions. Nonetheless, since finite difference method is in wide use in Computational Fluid Dynamics (CFD), details of finite difference method are illustrated using the same example problems as considered in the FEM so that some comparisons can be made.

Problems

9.1 Consider the following 1D BVP in $\mathbb{R}^1$

$$\frac{d^2u}{dx^2} = x^2 + 4u \quad \forall x \in (0,1) \tag{1}$$

with boundary conditions:

$$u(0) = 0 \ , \ u(1) = 20 \tag{2}$$

Consider finite difference approximation of the derivatives using central difference method to obtain an algebraic system for (1). Obtain numerical solution of u in (1) with BCs in (2) using a uniform discretization containing five grid points (i.e. $\Delta x = 0.25$). Plots graph of u, du/dx and d^2u/dx^2 versus x. du/dx and d^2u/dx^2 can be calculated using central difference approximation and forward and backward difference approximations at grid points 1 and 5 for du/dx.

9.2 Consider the following 1D BVP in $\mathbb{R}^1$

$$\frac{d^2u}{dx^2} = x^2 + 4u \quad \forall x \in (0,1) \tag{1}$$

with boundary conditions:

$$u'(0) = 0 \ , \ u'(1) = 20 \tag{2}$$

Consider finite difference approximation of the derivatives using central difference method to obtain a system of algebraic equations for (1). Obtain numerical solution of (1) with BCs in (2) using a uniform discretization containing five grid points (i.e. $\Delta x = 0.25$). Plots graph of u, du/dx and d^2u/dx^2 versus x. du/dx and d^2u/dx^2 can be calculated using central difference approximation and forward and backward difference approximations at grid points 1 and 5 for du/dx.

9.3 Consider the following 1D BVP in $\mathbb{R}^1$

$$\frac{d^2u}{dx^2} + 4\frac{du}{dx} + 2u = x^2 \quad \forall x \in (0,2) \tag{1}$$

with boundary conditions:

$$u(0) = 0 \ , \ u'(2) = \left.\frac{du}{dx}\right|_{x=2} = 0.6 \tag{2}$$

Consider finite difference approximation of the derivatives using central difference method to obtain an algebraic system for (1). Obtain numerical solution of (1) with BCs in (2) using a uniform discretization containing five grid points (i.e. $\Delta x = 0.5$). Plots graph of u, du/dx and d^2u/dx^2 versus x. du/dx and d^2u/dx^2 can be calculated using central difference approximation and forward and backward difference approximations at grid points 1 and 5 for du/dx.

9.4 Consider the following 1D BVP in $\mathbb{R}^1$

$$\frac{d^2T}{dx^2} - \left(1 - \frac{x}{5}\right) T = x \quad \forall x \in (1,3) \tag{1}$$

with boundary conditions:

$$T(1) = 10\ ,\ T''(3) = \left.\frac{d^2T}{dx^2}\right|_{x=3} = 6 \tag{2}$$

Consider finite difference approximation of the derivatives using central difference method to obtain an algebraic system for (1). Obtain numerical solution of (1) with BCs (2) using a uniform discretization containing five grid points (i.e. $\Delta x = 0.5$). Plots graph of T, dT/dx and d^2T/dx^2 versus x. dT/dx and d^2T/dx^2 can be calculated using central difference approximation and forward and backward difference approximations at grid points 1 and 5 for dT/dx.

9.5 Consider the same BVP (i.e. (1)) as in problem 9.4, but with the new BCs :

$$\frac{d^2T}{dx^2} - \left(1 - \frac{x}{5}\right) T = x \quad \forall x \in (1,3) \tag{1}$$

with boundary conditions:

$$T''(1) = \left.\frac{d^2T}{dx^2}\right|_{x=1} = 2\ ,\ T'(3) = \left.\frac{dT}{dx}\right|_{x=3} = 1 \tag{2}$$

Consider finite difference approximation of the derivatives using central difference method to obtain an algebraic system for (1). Obtain numerical solution of (1) with BCs (2) using a uniform discretization containing five grid points (i.e. $\Delta x = 0.5$). Plots graph of T, dT/dx and d^2T/dx^2 versus x. dT/dx and d^2T/dx^2 can be calculated using central difference approximation.

9.6 Consider the Laplace equation given below

$$\frac{\partial^2 T}{\partial x^2} + \frac{\partial^2 T}{\partial y^2} = 0 \quad \forall (x,y) \in (\Omega_{xy}) = \Omega_x \times \Omega_y \tag{1}$$

The domain $\bar{\Omega}_{xy}$ and the discretization including boundary conditions are given in the following.

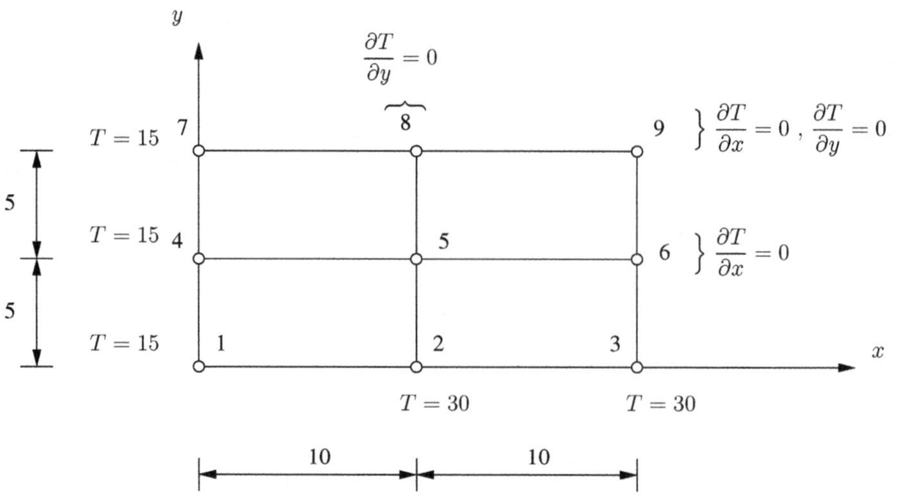

Consider finite difference approximation of the derivatives in the BVP (1) using central difference method. Construct a system of algebraic equations for (1) using this approximation of the derivatives. Grid points are numbered. The value of T at the grid i is T_i. Thus $T_1 = 15$, $T_4 = 15$, $T_7 = 15$, $T_2 = 30$ and $T_3 = 30$ are known due to boundary conditions. Calculate unknown values of T at the grid points 5, 6, 8 and 9, i.e. calculate T_5, T_6, T_8 and T_9.

9.7 Consider the Poisson's equation given below

$$\frac{\partial^2 \phi}{\partial x^2} + \frac{\partial^2 \phi}{\partial y^2} = xy \quad \forall (x, y) \in (\Omega_{xy}) = \Omega_x \times \Omega_y \tag{1}$$

The domain $\bar{\Omega}_{xy}$ and the discretization including boundary conditions are given in the following.

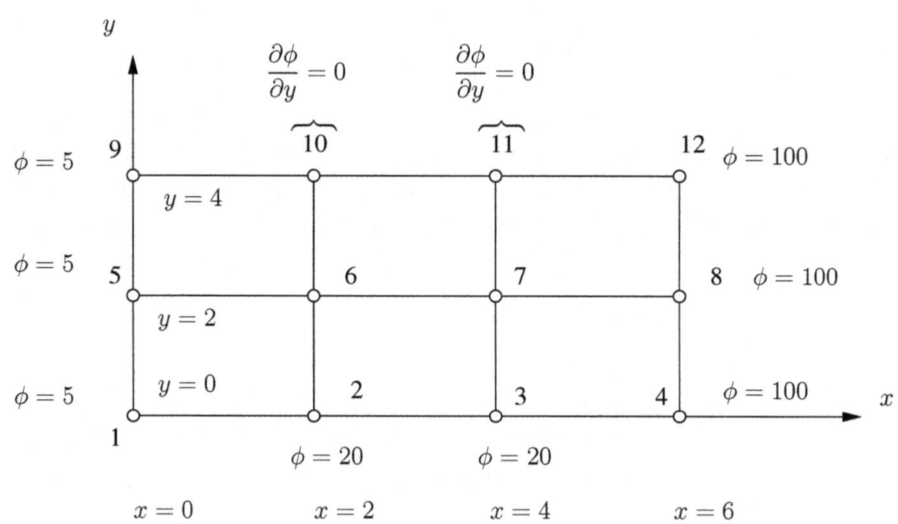

Consider finite difference approximation of the derivatives in the BVP (1) using central difference method. Construct a system of algebraic equations for (1) using this approximation of the derivatives. Grid points are numbered. The value of ϕ at the grid i is ϕ_i. Thus $\phi_1 = 5$, $\phi_5 = 5$, $\phi_9 = 5$, $\phi_2 = 20$, $\phi_3 = 20$, $\phi_4 = 100$, $\phi_8 = 100$ and $\phi_{12} = 100$ are known due to boundary conditions. Calculate unknown values of ϕ at the grid points 6, 7, 10 and 11 i.e. calculate ϕ_6, ϕ_7, ϕ_{10} and ϕ_{11}.

9.8 Consider Laplace equation given in the following

$$\frac{\partial^2 u}{\partial x^2} + \frac{\partial^2 u}{\partial y^2} = 0 \quad \forall (x, y) \in (\Omega_{xy}) = \Omega_x \times \Omega_y \tag{1}$$

The domain $\bar{\Omega}_{xy}$ and the discretization including boundary conditions are shown in Figure (a) in the following.

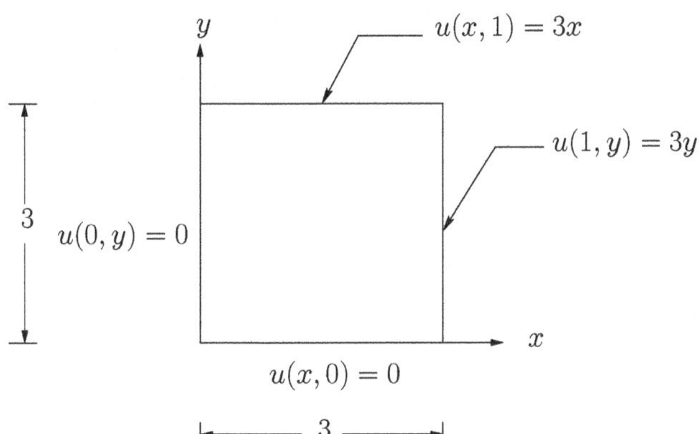

(a) Schematic of $\bar{\Omega}_{xy}$

Consider the following two discretizations.

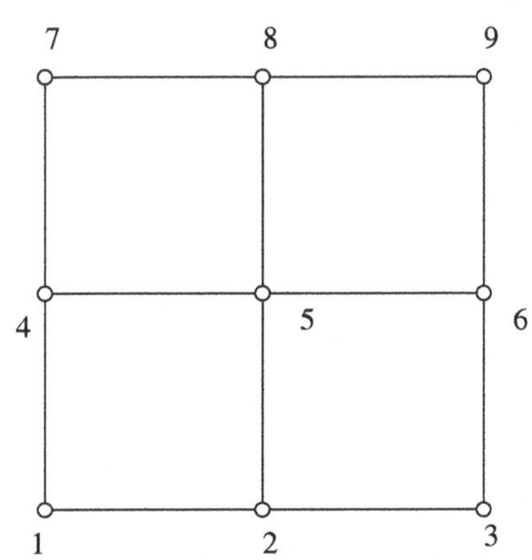

(b) A nine grid point discretization

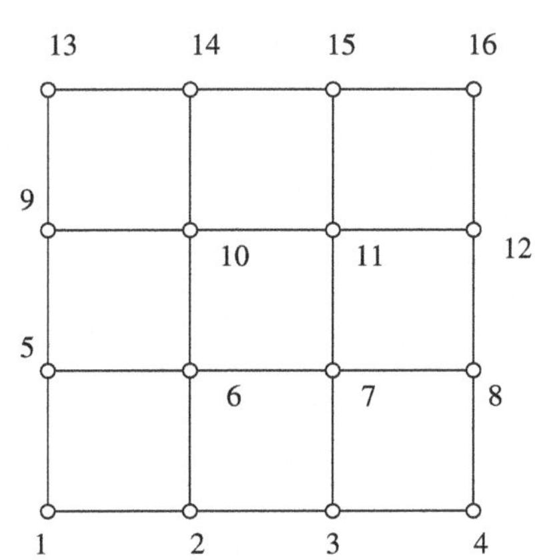

(c) A sixteen grid point discretization

Consider finite difference approximations of the derivative in the BVP (1) using central difference method to construct a system of algebraic equations for (1) using this approximation of the derivatives. Find numerical values of u_5 for discretization in Figure (b) and the numerical values of u_6, u_7, u_{10} and u_{11} for the discretization in Figure (c).

9.9 Consider Laplace equation given in the following

$$\frac{\partial^2 u}{\partial x^2} + \frac{\partial^2 u}{\partial y^2} = 0 \quad \forall (x, y) \in (\Omega_{xy}) = \Omega_x \times \Omega_y \tag{1}$$

The domain $\bar{\Omega}_{xy}$ and the discretization including boundary conditions are shown in Figure (a) in the following.

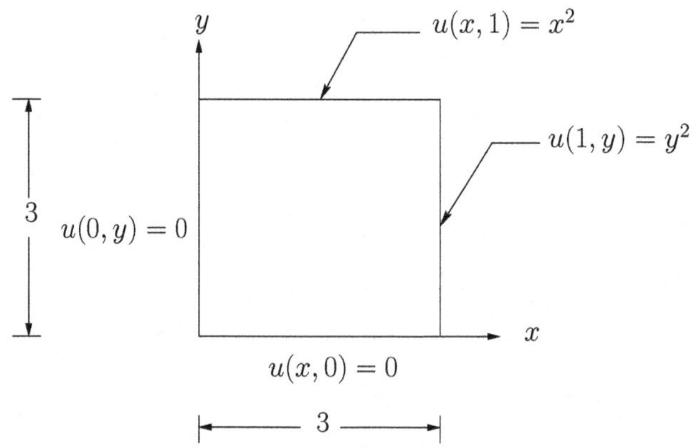

(a) Schematic of $\bar{\Omega}_{xy}$

Consider the following two discretizations. Consider finite difference approx-

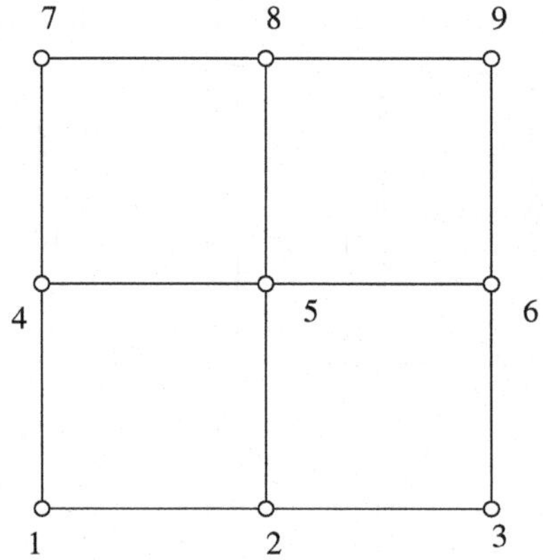

(b) A nine grid point discretization

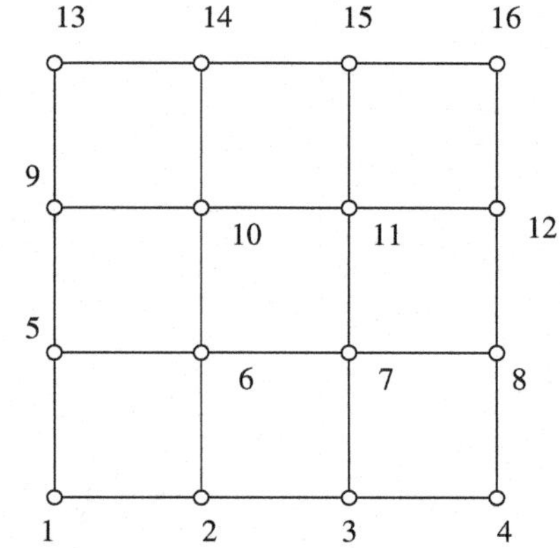

(c) A sixteen grid point discretization

imations of the derivative in the BVP (1) using central difference method to construct a system of algebraic equations for (1) using this approximation of the derivatives. Find numerical values of u_5 for discretization in Figure (b) and the numerical values of u_6, u_7, u_{10} and u_{11} for the discretization in Figure (c).

9.10 Consider 1D convection diffusion equation in $\mathbb{R}^1$.

$$\frac{d\phi}{dx} - \frac{1}{Pe}\frac{d^2\phi}{dx^2} = 0 \quad \forall x \in (\Omega_x) = (0,1) \tag{1}$$

with boundary conditions:

$$\phi(0) = 1\ ,\ \phi(1) = 0 \tag{2}$$

Consider finite difference approximation of the derivatives using central difference method to obtain an algebraic system for (1). Consider a five grid point uniform discretization shown below.

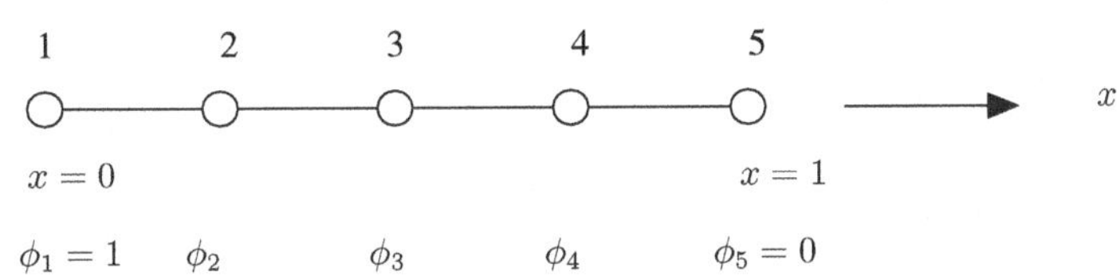

(a) A five grid point uniform discretization

Calculate numerical values of ϕ_2, ϕ_3 and ϕ_4 for Peclet number $Pe = 1$, 5, 10 and 20. For each Pe plot graphs of ϕ, $d\phi/dx$ and $d^2\phi/dx^2$ versus x. Note that $d\phi/dx$ and $d^2\phi/dx^2$ can be calculated using central difference approximation and forward and backward difference approximations at grid points 1 and 5 for $d\phi/dx$.

9.11 Consider 1D Burgers equation in $\mathbb{R}^1$.

$$\phi\frac{d\phi}{dx} - \frac{1}{Re}\frac{d^2\phi}{dx^2} = 0 \quad \forall x \in (\Omega_x) = (0,1) \tag{1}$$

with boundary conditions:

$$\phi(0) = 1\ ,\ \phi(1) = 0 \tag{2}$$

This is a nonlinear BVP, hence the resulting algebraic system will be a system of nonlinear algebraic equations. Consider finite difference approximation of the derivatives using central difference method to obtain an algebraic system for (1). Consider a five grid point uniform discretization shown below.

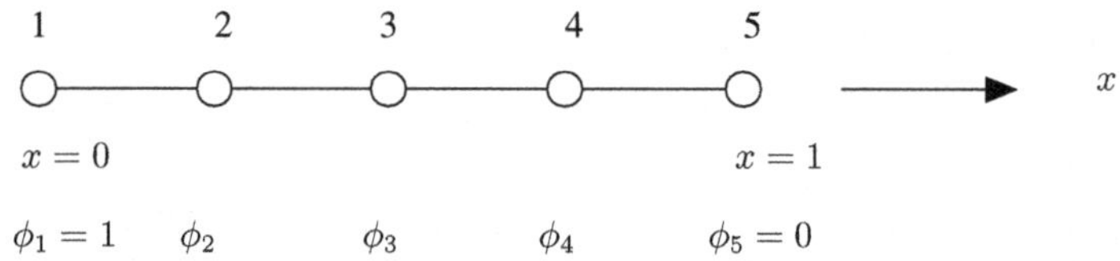

(a) A five grid point uniform discretization

Calculate numerical values of ϕ_2, ϕ_3 and ϕ_4 for $Re = 1, 5, 10$ and 20. For each Reynolds number Re plot graphs of ϕ, $d\phi/dx$ and $d^2\phi/dx^2$ versus x. Note that $d\phi/dx$ and $d^2\phi/dx^2$ can be calculated using central difference approximation and forward and backward difference approximations at grid points 1 and 5 for $d\phi/dx$. Solve nonlinear algebraic equations using any one of the methods described in chapter 3.

9.12 Consider the following BVP

$$\frac{d^2u}{dx^2} = x^2 + 4u \quad \forall x \in (0,1) = \Omega \tag{1}$$

with boundary conditions:

$$u(0) = 0\ ,\ u(1) = 20 \tag{2}$$

Consider a four element uniform discretization $\bar{\Omega}^T = \bigcup_{e=1}^{4} \bar{\Omega}^e$ of $\bar{\Omega}$ in which $\bar{\Omega}^e$ is a two node linear element. Construct a finite element formulation of (1) over an element $\bar{\Omega}^e$ of $\bar{\Omega}^T$ using Galerkin method with weak form (GM/WF). Derive and calculate matrices and vectors of element equations. Assemble element equations and obtain numerical values of u at the grid points where u is unknown using boundary conditions (2). Plot graphs of u versus x and du/dx versus x. Calculate du/dx using element local approximation. Compare this solution with the solution calculated in problem 9.1. Also calculate values of the unknown secondary variables.

9.13 Consider the following BVP

$$\frac{d^2u}{dx^2} = x^2 + 4u \quad \forall x \in (0,1) = \Omega \tag{1}$$

with boundary conditions:

$$u'(0) = 0\ ,\ u'(1) = 20 \tag{2}$$

Consider a four element uniform discretization $\bar{\Omega}^T = \bigcup_{e=1}^{4} \bar{\Omega}^e$ of $\bar{\Omega}$ in which $\bar{\Omega}^e$ is a two node linear element. Construct a finite element formulation of (1) over an element $\bar{\Omega}^e$ of $\bar{\Omega}^T$ using Galerkin method with weak form (GM/WF). Assemble element equations and obtain numerical values of u at the grid points with unknown u using boundary conditions (2). Plot graphs of u versus x and du/dx versus x. Calculate du/dx using element local approximation. Compare this solution with the solution calculated in problem 9.2. Also calculate values of the unknown secondary variables (if any).

9.14 Consider the following BVP

$$\frac{d^2u}{dx^2} - 4\frac{du}{dx} + 2u = x^2 \quad \forall x \in (0,2) = \Omega \tag{1}$$

with boundary conditions:

$$u(0) = 0\ ,\ u'(2) = 0.6 \tag{2}$$

Consider a four element uniform discretization $\bar{\Omega}^T = \bigcup_{e=1}^{4} \bar{\Omega}^e$ of $\bar{\Omega}$ in which $\bar{\Omega}^e$ is a two node linear element. Construct a finite element formulation of (1) over an element $\bar{\Omega}^e$ of $\bar{\Omega}^T$ using Galerkin method with weak form (GM/WF). Assemble element equations and obtain numerical values of u at the grid points with unknown u using boundary conditions (2). Plot graphs of u versus x and du/dx versus x. Calculate du/dx using element local approximation. Compare this solution with the solution calculated in problem 9.3. Also calculate values of the unknown secondary variables.

9.15 Consider the following BVP

$$\frac{d^2T}{dx^2} - \left(1 - \frac{x}{5}\right)T = x \quad \forall x \in (1,3) \tag{1}$$

with boundary conditions:

$$T(1) = 10\ ,\ T'(3) = 5 \tag{2}$$

Consider a four element uniform discretization $\bar{\Omega}^T = \bigcup_{e=1}^{4} \bar{\Omega}^e$ of $\bar{\Omega}$ in which $\bar{\Omega}^e$ is a two node linear element. Construct a finite element formulation of (1) over an element $\bar{\Omega}^e$ of $\bar{\Omega}^T$ using Galerkin method with weak form (GM/WF). Assemble element equations and obtain numerical values of u at the grid points with unknown T using boundary conditions (2). Plot graphs of T versus x and dT/dx versus x. Calculate dT/dx using element local approximation. Also calculate values of the unknown secondary variables.

9.16 Consider 1D convection diffusion equation in $\mathbb{R}^1$.

$$\frac{d\phi}{dx} - \frac{1}{Pe}\frac{d^2\phi}{dx^2} = 0 \quad \forall x \in (\Omega_x) = (0,1) \tag{1}$$

with boundary conditions:

$$\phi(0) = 1 \ , \ \phi(1) = 0 \tag{2}$$

Consider a four element uniform discretization $\bar{\Omega}^T = \overset{4}{\underset{e=1}{\cup}} \bar{\Omega}^e$ of $\bar{\Omega}$ in which $\bar{\Omega}^e$ is a two node linear element. Construct a finite element formulation of (1) over an element $\bar{\Omega}^e$ of $\bar{\Omega}^T$ using Galerkin method with weak form (GM/WF). Assemble element equations and obtain numerical values of ϕ at the grid points where ϕ is unknown using boundary conditions (2) for Pe = 1, 5, 10 and 20. For each Peclet number Pe plot graphs of ϕ versus x and $d\phi/dx$ versus x. Calculate $d\phi/dx$ using element local approximation. Also calculate values of the unknown secondary variables. Compare the solution computed here with the solution calculated in problem 9.10 using finite difference method.

9.17 Consider 1D Burgers equation in $\mathbb{R}^1$.

$$\phi\frac{d\phi}{dx} - \frac{1}{Re}\frac{d^2\phi}{dx^2} = 0 \quad \forall x \in (\Omega_x) = (0,1) \tag{1}$$

with boundary conditions:

$$\phi(0) = 1 \ , \ \phi(1) = 0 \tag{2}$$

Consider a four element uniform discretization $\bar{\Omega}^T = \overset{4}{\underset{e=1}{\cup}} \bar{\Omega}^e$ of $\bar{\Omega}$ in which $\bar{\Omega}^e$ is a two node linear element.
Construct a finite element formulation of (1) over an element $\bar{\Omega}^e$ of $\bar{\Omega}^T$ using Galerkin method with weak form (GM/WF). Assemble element equations and obtain numerical values of ϕ at the grid points where ϕ is unknown using boundary conditions (2) for Re = 1, 5, 10 and 20. For each Reynolds number Re plot graphs of ϕ versus x and $d\phi/dx$ versus x. Calculate $d\phi/dx$ using element local approximation. Also calculate values of the unknown secondary variables. Compare the solution computed here with the solution calculated in problem 9.11 using finite difference method.

9.18 – 9.23 Consider the same boundary value problems as in 9.12 – 9.17 with the corresponding boundary conditions. For each BVP consider a five node uniform discretization $\bar{\Omega}^T = \overset{4}{\underset{e=1}{\cup}} \bar{\Omega}^e$ of $\bar{\Omega}$ in which $\bar{\Omega}^e$ is a three node quadratic element (Lagrange family). Consider a typical element $\bar{\Omega}^e$ to construct finite element formulations using GM/WF. Assemble and compute solutions for each case and plot similar graphs as in problem 9.12 – 9.17. Compare the results computed here with those in 9.12 – 9.17 computed using two node linear elements. For each problem compare and discuss the results.

10
Numerical Solution of Initial Value Problems

10.1 General overview

The physical processes encountered in all branches of sciences and engineering can be classified into two major categories: time-dependent processes and stationary processes. Time-dependent processes describe *evolutions* in which quantities of interest change with time. If the quantities of interest cease to change in an evolution then the evolution is said to have reached a stationary state. Not all evolutions have stationary states. The evolutions without a stationary state are often referred to as *unsteady* processes. Stationary processes are those in which the quantities of interest do not depend upon time. For a stationary process to be valid or viable, it must correspond to the stationary state of an evolution. Every process in nature is an evolution. Nonetheless it is sometimes convenient to consider their stationary state. In this book we only consider non-stationary processes, i.e. evolutions that may have a stationary state or may be unsteady.

A mathematical description of most stationary processes in sciences and engineering often leads to a system of *ordinary* or *partial* differential equations. These mathematical descriptions of the stationary processes are referred to as boundary value problems (BVPs). Since stationary processes are independent of time, the partial differential equations describing their behavior only involve dependent variables and space coordinates as independent variables. On the other hand, mathematical descriptions of evolutions lead to partial differential equations in dependent variables, space coordinates, and time and are referred to as initial value problems (IVPs).

In case of simple physical systems, the mathematical descriptions of IVPs may be simple enough to permit analytical solutions. However, most physical systems of interest may be quite complicated and their mathematical description (IVPs) may be complex enough not to permit analytical solutions. In such cases, two alternatives are possible. In the first case, one could undertake simplifications of the mathematical descriptions to a point that analytical solutions are possible. In this approach, the simplified forms may not be descriptive of the actual behavior and sometimes this simplification may not be possible at all. In the second alternative, we abandon the possibility of theoretical solutions altogether as viable means of solving complex practical problems involving IVPs and instead we resort to *numerical methods* for obtaining *numerical solutions* of IVPs. The finite element method (FEM) is one such method of solving IVPs numerically and constitutes the subject matter of this book. Before we delve deeper into the FEM for IVPs, it is perhaps fitting to discuss a little about the broader classes of available methods for obtaining numerical solutions of IVPs.

The fundamental difference between BVPs and IVPs is that IVPs describe evolutions, i.e. the solution changes at spatial locations as time elapses. This important distinction between BVPs and IVPs necessitates a fundamentally different appraoch(es) for obtaining numerical solutions of IVPs compared to BVPs. Consider an abstract initial value problem

$$A\phi - f = 0 \quad \forall (x,t) \in \Omega_{xt} = (0,L) \times (0,\tau) \tag{10.1}$$

with some boundary and initial conditions. In (10.1), A is a space-time differential operator, $\phi = \phi(x,t)$ is the dependent variable, $f = f(x,t)$ is the non-homogeneous part, and Ω_{xt} is the space-time domain over which (10.1) holds. Time $t = 0$ and $t = \tau$ are initial and final values of time for which we seek $\phi = \phi(x,t)$, the solution of (10.1).

We note that in initial value problems the dependent variables are functions of spatial coordinates and time and their mathematical description contain spatial as well as time derivatives of the dependent variable. Parallel to the material presented in section 9.1 for BVP, we find that solution $\phi(x,t)$ of IVP $A\phi - f = 0 \; \forall x,t \in \Omega_{xt} = \Omega_x \times \Omega_t$ requires that we integrate $A\phi - f = 0$ over $\bar{\Omega}_{xt} = \Omega_{xt} \cup \Gamma$, Γ being the closed boundary of the space-time domain Ω_{xt}. That is we need to consider

$$\int_{\bar{\Omega}_{xt}} (A\phi(x,t) - f(x,t))\, dxdt = 0 \tag{10.2}$$

The integrand in (10.2) is a space-time integral. Thus, space-time coupled methods that consider space-time integrals of $A\phi - f = 0$ are the only mathematically justifiable methods of obtaining solution of $A\phi - f = 0$ numerically or otherwise. Thus, we see that space-time coupled classical and finite element methods (considered subsequently) are meritorious over all other methods. In the following sections we consider these methods as well.

10.2 Space-time coupled methods of approximation for the whole space-time domain $\bar{\Omega}_{xt}$

We note that since $\phi = \phi(x,t)$, the solution exhibits simultaneous dependence on spatial coordinates x and time t. This feature is intrinsic in the mathematical description (10.1) of the physics.

Thus, the most rational approach to undertake for the solution of (10.1) (approximate or otherwise) is to preserve simultaneous dependence of ϕ on x and t. Such methods are known as *space-time coupled methods.* Broadly speaking, in such methods time t is treated as another independent variable in addition to spatial coordinates. Fig. 10.1 shows space-time domain $\bar{\Omega}_{xt} = \Omega_{xt} \cup \Gamma$; $\Gamma = \bigcup\limits_{i=1}^{4} \Gamma_i$ with closed boundary Γ. For the sake of discussion, as an example we could have a boundary condition (BC) at $x = 0$ $\forall t \in [0, \tau]$, boundary Γ_1, as well as at $x = L$ $\forall t \in [0, \tau]$, boundary Γ_2, and an initial condition (IC) at $t = 0$ $\forall x \in [0, L]$, boundary Γ_3. Boundary Γ_4 at final value of time $t = \tau$ is open, i.e. at this boundary only the evolution (the solution of (10.1) subjected to these BCs and IC), will yield the function $\phi(x, \tau)$ and its spatial and time derivatives.

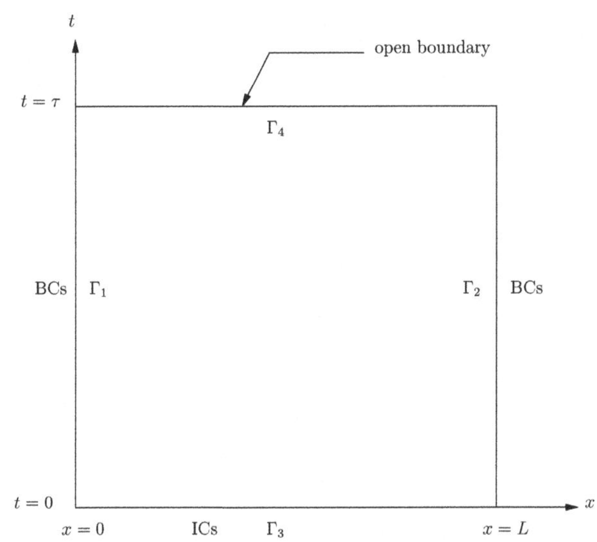

Figure 10.1: Space-time domain $\bar{\Omega}_{xt}$

When the initial value problem contains two spatial coordinates, we have space-time slab $\bar{\Omega}_{xt}$ shown in Fig. 10.2 in which

$$\Omega_{xt} = (0, L_1) \times (0, L_2) \times (0, \tau) \tag{10.3}$$

is a prism. In this case Γ_1, Γ_2, Γ_3, and Γ_4 are faces of the prism (surfaces). For illustration, the possible choices of BCs and ICs could be: BCs on $\Gamma_1 = ADD_1A_1$ and $\Gamma_2 = BCC_1B_1$, IC on $\Gamma_3 = ABCD$, and $\Gamma_4 = A_1B_1C_1D_1$ is the open boundary. This concept of space-time slab can be extended for three spatial dimensions and time. Using space-time domain shown in Fig. 10.1 or 10.2 and treating time as another independent variable, we could consider the following methods of approximation.

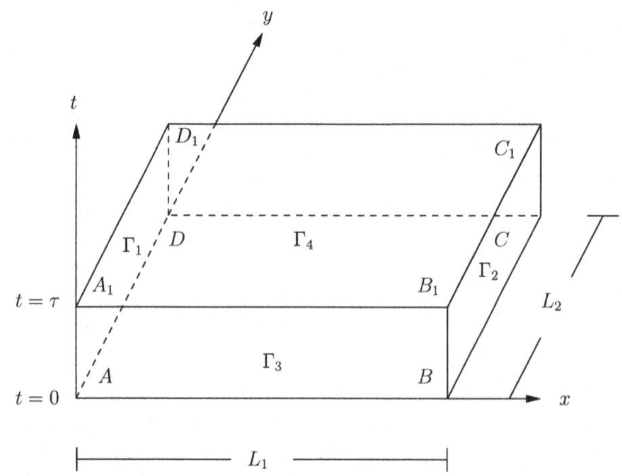

Figure 10.2: Rectangular prism space-time domain

1. Finite difference method
2. Finite volume method
3. Finite element method
4. Boundary element method
5. Spectral element method
6. And possibly others

In all such methods the IVP in dependent variable(s), spatial coordinate(s) x (or x, y or x, y, z), and time t is converted into a system of algebraic equations for the entire space-time domain $\bar{\Omega}_{xt}$ from which numerical solution is computed after imposing BCs and ICs.

In the methods listed here there are two features that are common: (1) partial differential equation or a system of partial differential equations in (10.1) is converted into an algebraic system for the space-time domain $\bar{\Omega}_{xt}$ and (2) in general, the numerical solution over $\bar{\Omega}_{xt}$ obtained from the algebraic system is an approximation of the true solution. The differences in the various methods of approximation lie in the manner in which the PDE or PDEs are converted into the algebraic system.

10.3 Space-time coupled methods using space-time strip or slab with time-marching

In space-time coupled methods for the whole space-time domain $\bar{\Omega}_{xt} = [0, L] \times [0, \tau]$, the computations can be intense and sometimes prohibitive if the final time τ is large. This problem can be easily overcome by using space-time strip or slab for an increment of time Δt and then time-marching to obtain the entire evolution. Consider the space-time domain

$$\bar{\Omega}_{xt} = \Omega_{xt} \cup \Gamma; \quad \Gamma = \bigcup_{i=1}^{4} \Gamma_i$$

shown in Fig. 10.3. For an increment of time Δt, that is for $0 \leq t \leq \Delta t$, consider the first space-time strip $\bar{\Omega}^{(1)}_{xt} = [0, L] \times [0, \Delta t]$. If we are only interested in the evolution up to time $t = \Delta t$ and not beyond $t = \Delta t$, then the evolution in the space-time domain $[0, L] \times [\Delta t, \tau]$ has not taken place yet, hence does not influence the evolution for $\bar{\Omega}^{(1)}_{xt}$, $t \in [0, \Delta t]$. We also note that for $\bar{\Omega}^{(1)}_{xt}$, the boundary at $t = \Delta t$ is open boundary that is similar to the open boundary at $t = \tau$ for the whole space-time domain. We remark that BCs and ICs for $\bar{\Omega}_{xt}$ and $\bar{\Omega}^{(1)}_{xt}$ are identical in the sense of those that are known and those that are not known. For $\bar{\Omega}^{(2)}_{xt}$, the second space-time strip, the BCs are the same as for $\bar{\Omega}^{(1)}_{xt}$ but the ICs at $t = \Delta t$ are obtained from the computed evolution for $\bar{\Omega}^{(1)}_{xt}$ at $t = \Delta t$. Now, with the known ICs at $t = \Delta t$, the second space-time strip $\bar{\Omega}^{(2)}_{xt}$ is exactly similar to the first space-time strip $\bar{\Omega}^{(1)}_{xt}$ in terms of BCs, ICs, and open boundary. For $\bar{\Omega}^{(1)}_{xt}$, $t = \Delta t$ is open boundary whereas for $\bar{\Omega}^{(2)}_{xt}$, $t = 2\Delta t$ is open boundary. Both open boundaries are at final values of time for the corresponding space-time strips.

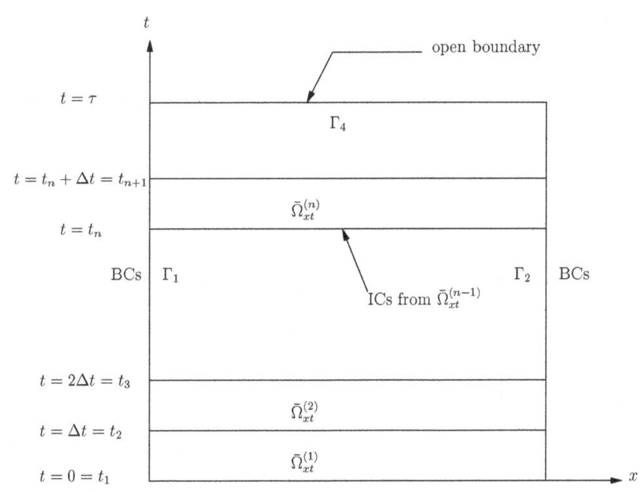

Figure 10.3: Space-time domain with 1st, 2nd, and n^{th} space-time strips

In this process the evolution is computed for the first space-time strip $\bar{\Omega}^{(1)}_{xt} = [0, L] \times [0, \Delta t]$ and refinements are carried out (in discretization and p-levels in the sense of finite element processes) until the evolution for $\bar{\Omega}^{(1)}_{xt}$ is a converged solution. Using this converged solution for $\bar{\Omega}_{xt}^{(1)}$, ICs are extracted at $t = \Delta t$ for $\bar{\Omega}_{xt}^{(2)}$ and a converged evolution is computed for the second space-time strip $\bar{\Omega}_{xt}^{(2)}$. This process is continued until $t = \tau$ is reached.

Remarks.

(1) In this process, evolution is computed for an increment of time Δt and the time-marched to obtain the entire evolution. This allows the computation of entire evolution through solutions of relatively small problems associated with each space-time strip (or slab) corresponding to an increment of time Δt, resulting in significant efficiency in the computations compared to computing the evolution for entire space-time domain simultaneously.

(2) Since the initial conditions for the nth space-time strip ($\bar{\Omega}_{xt}^{(n)}$) are extracted from the $(n-1)^{th}$ space-time strip ($\bar{\Omega}_{xt}^{(n-1)}$), it is necessary to have accurate evolution for the space-time strip $\bar{\Omega}_{xt}^{(n-1)}$ otherwise the initial condition for the space-time strip $\bar{\Omega}_{xt}^{(n)}$ will be in error.

(3) Accuracy of the evolution is ensured for each space-time strip (or slab) before time-marching, hence ensuring accuracy of the entire evolution for the entire space-time domain $\bar{\Omega}_{xt}$. This feature of space-time strip with time-marching is absent in the first approach in which the solution is obtained simultaneously for the entire space-time domain $\bar{\Omega}_{xt}$. It is only after we have the entire evolution that we can determine its accuracy in $\bar{\Omega}_{xt}$, either element by element or for the whole space-time domain.

(4) When using space-time strip with time-marching there are no assumptions or approximations, only added advantages of assurance of accuracy and significant increase in computational efficiency. However, care must be exercised to ensure sufficiently converged solution for the current space-time strip before moving on to the next space-time strip as the initial conditions for the next space-time strip are extracted from the computed evolution for the current space-time strip.

(5) In constructing the algebraic system for a space-time strip or slab, the methods listed for the first approach in Section 10.2 are applicable here as well.

10.4 Space-time decoupled or quasi methods

In space-time decoupled or quasi methods the solution $\phi = \phi(x, t)$ is assumed not to have simultaneous dependence on space coordinate x and time t. Referring to the IVP (10.1) in spatial coordinate x (i.e. $\mathbb{R}^1$) and time t, the solution $\phi(x, t)$ is expressed as the product of two functions $g(x)$ and

$h(t)$:

$$\phi(x,t) = g(x)h(t) \tag{10.4}$$

where $g(x)$ is a known function that satisfies differentiability, continuity, and the completeness requirements (and others) as dictated by (10.1). We substitute (10.4) in (10.1) and obtain

$$A\left(g(x)h(t)\right) - f(x,t) = 0 \quad \forall x,t \in \Omega_{xt} \tag{10.5}$$

Integrating (10.5) over $\bar{\Omega}_x = [0,L]$ while assuming $h(t)$ and its time derivatives to be constant for an instant of time, we can write

$$\int_{\bar{\Omega}_x} \left(A\left(g(x)h(t)\right) - f(x,t)\right) dx = 0 \tag{10.6}$$

Since $g(x)$ is known, the definite integral in (10.6) can be evaluated, thereby eliminating $g(x)$, its spatial derivatives (due to operator A), and more specifically spatial coordinate x altogether. Hence, (10.6) reduces to

$$\underset{\sim}{A}h(t) - \underset{\sim}{f}(t) = 0 \quad \forall t \in (0,\tau) \tag{10.7}$$

in which $\underset{\sim}{A}$ is a time differential operator and $\underset{\sim}{f}$ is only a function of time. In other words, (10.7) is an ordinary differential equation in time which can now be integrated using explicit or implicit time integration methods or finite element method in time to obtain $h(t)$ $\forall t \in [0,\tau]$. Using this calculated $h(t)$ in (10.4), we now have the solution $\phi(x,t)$:

$$\phi(x,t) = g(x)h(t) \quad \forall x,t \in \bar{\Omega}_{xt} = [0,L] \times [0,\tau] \tag{10.8}$$

Remarks.

(1) In this approach decoupling of space and time occurs in (10.4).

(2) A partial differential equation in ϕ, x, and t as in (10.1) is reduced to an ordinary differential equation in time as in (10.7).

(3) $\phi(x,t)$ in (10.4) must satisfy all BCs and ICs of the initial value problem (10.1). When seeking theoretical solution $\phi(x,t)$ using (10.4) it may be difficult or may not even possible to find $g(x)$ and $h(t)$ such that $\phi(x,t)$ in (10.4) satisfies all BCs and ICs of the IVP.

(4) However, when using methods of approximation in conjunction with (10.4) this difficulty does not arise as BCs and ICs are imposed during time integration of the ordinary differential equation in time (10.7). Specifically, in context with space-time decoupled finite element processes, (10.4) is used over an element $\bar{\Omega}_x^e$ of the spatial discretization $\bar{\Omega}_x^T$ of $\bar{\Omega}_x = [0,L]$, hence $g(x)$ are merely local approximation functions over

$\bar{\Omega}_x^e$ that are obtained using interpolation theory irrespective of BCs and ICs.

(5) In principle, (10.4) holds for all of the methods of approximation listed in Section 10.2. In all these methods spatial coordinate is eliminated using (10.4) for discretization in space that may be followed by integration of $A(g(x)h(t)) - f(x,t)$ over $\bar{\Omega}_x^T$ depending upon the method chosen. In doing so the IVP (10.1) reduces to a system of ordinary differential equations in time which are then integrated simultaneously using explicit or implicit time integration methods or finite element method in time after imposing BCs and the ICs of the IVP.

In the following we present two example model problems of decoupling space and time using a time-dependent convection diffusion equation, a linear initial value problem, and using a time-dependent Burgers equation, a nonlinear initial value problem.

Example 10.1 (1D convection diffusion equation). Consider 1D convection diffusion equation

$$\frac{\partial \phi}{\partial t} + \frac{\partial \phi}{\partial x} - \frac{1}{Pe}\frac{\partial^2 \phi}{\partial x^2} \quad \forall (x,t) \in \Omega_{xt} = (0,1) \times (0,\tau) = \Omega_x \times \Omega_t \tag{10.9}$$

with some BCs and ICs. Equation (10.9) is a linear partial differential equation in dependent variable ϕ, space coordinate x, and time t. Pe is the Péclet number. Let

$$\phi(x,t) = g(x)h(t) \tag{10.10}$$

in which $g(x) \in V \subset H^3(\bar{\Omega}_x)$. Substituting (10.10) in (10.9)

$$g(x)\frac{dh(t)}{dt} + h(t)\frac{dg(x)}{dx} - \frac{1}{Pe}h(t)\frac{d^2g(x)}{dx^2} = 0 \tag{10.11}$$

Integrating (10.11) with respect to x over $\bar{\Omega}_x = [0,1]$ while treating $h(t)$ and its time derivatives as constant

$$\frac{dh(t)}{dt}\int_{\bar{\Omega}_x} g(x)\,dx + h(t)\int_{\bar{\Omega}_x} \frac{dg(x)}{dx}dx - \frac{h(t)}{Pe}\int_{\bar{\Omega}_x} \frac{d^2g(x)}{dx^2}dx = 0 \tag{10.12}$$

Let

$$C_1 = \int_{\bar{\Omega}_x} g(x)\,dx \; ; \quad C_2 = \int_{\bar{\Omega}_x} \frac{dg(x)}{dx}dx \; ; \quad C_3 = \int_{\bar{\Omega}_x} \frac{d^2g(x)}{dx^2}dx \tag{10.13}$$

Using (10.13) in (10.12)

$$C_1\frac{dh(t)}{dt} + \left(C_2 - \frac{C_3}{Pe}\right) h(t) = 0 \quad \forall t \in (0, \tau) \tag{10.14}$$

We note that (10.14) is a linear ordinary differential equation in dependent variable $h(t)$ and time t. Thus, decoupling of space and time due to (10.10) has reduced a linear partial differential equation (10.9) in $\phi(x,t)$, space x, and time t to a linear ordinary differential equation in $h(t)$ and time t.

Example 10.2 (1D Burgers equation). Consider 1D Burgers equation

$$\frac{\partial \phi}{\partial t} + \phi\frac{\partial \phi}{\partial x} - \frac{1}{Re}\frac{\partial^2 \phi}{\partial x^2} \quad \forall (x,t) \in \Omega_{xt} = (0,1) \times (0,\tau) = \Omega_x \times \Omega_t \tag{10.15}$$

with some BCs and ICs. Equation (10.15) is a non-linear partial differential equation in dependent variable ϕ, space coordinate x, and time t. Re is Reynolds number. Let

$$\phi(x,t) = g(x)h(t) \tag{10.16}$$

in which $g(x) \in V \subset H^3(\bar{\Omega}_x)$. Substituting (10.16) into (10.15),

$$g(x)\frac{dh(t)}{dt} + g(x)h(t)\left(h(t)\frac{dg(x)}{dx}\right) - \frac{1}{Re}h(t)\frac{d^2g(x)}{dx^2} = 0 \tag{10.17}$$

Integrating (10.17) with respect to x over $\bar{\Omega}_x = [0,1]$ while treating $h(t)$ and its derivatives constant

$$\frac{dh(t)}{dt}\int\limits_{\bar{\Omega}_x} g(x)\,dx + (h(t))^2\int\limits_{\bar{\Omega}_x} g(x)\frac{dg(x)}{dx}\,dx - \frac{1}{Re}h(t)\int\limits_{\bar{\Omega}_x}\frac{d^2g(x)}{dx^2}\,dx = 0 \tag{10.18}$$

Let

$$C_1 = \int\limits_{\bar{\Omega}_x} g(x)\,dx \; ; \quad C_2 = \int\limits_{\bar{\Omega}_x} g(x)\frac{dg(x)}{dx}dx \; ; \quad C_3 = \int\limits_{\bar{\Omega}_x}\frac{d^2g(x)}{dx^2}dx \tag{10.19}$$

Using (10.19) in (10.18),

$$C_1\frac{dh(t)}{dt} + C_2(h(t))^2 - C_3h(t) = 0 \quad \forall t \in (0,\tau) \tag{10.20}$$

Equation (10.20) is a non-linear ordinary differential equation in $h(t)$ and time t. Thus, the decoupling of space and time due to (10.16) has reduced a non-linear partial differential equation (10.15) in $\phi(x,t)$, space x, and time t into a non-linear ordinary differential equation in $h(t)$ and time t.

10.5 General remarks

From the material presented in Sections 10.2 – 10.4 it is clear that one could entertain any of the methods of approximation listed in Section 10.2, space-time coupled, or space-time decoupled approaches for obtaining numerical solutions of the IVPs.

In this book we only consider finite element method in conjunction with space-time coupled and space-time decoupled approaches for obtaining numerical solutions of the IVPs. The finite element method for both approaches has rigorous mathematical foundation, hence in this approach it is always possible to ascertain feasibility, stability, and accuracy of the resulting computational processes. Error estimation, error computation, convergence, and convergence rates are additional meritorious features of the finite element processes for IVPs compared to all other methods listed in Section 10.2.

In the following sections we present a brief description of space-time coupled and space-time decoupled finite element processes, their merits and shortcomings, time integration techniques for ODEs in time resulting from decoupling space and time, stability of computational processes, error estimation, error computation, and convergence.

Some additional topics related to linear structural and linear solid mechanics such as mode superposition techniques of obtaining time evolution are also discussed.

10.6 Space-time coupled finite element method

In the initial value problem (10.1), the operator A is a space-time differential operator. Thus, in order to address STFEM for totality of all IVPs in a problem- and application-independent fashion we must mathematically classify space-time differential operators appearing in all IVPs into groups. For these groups of space-time operators we can consider space-time methods of approximation such as space-time Galerkin method (STGM), space-time Petrov-Galerkin method (STPGM), space-time weighted residual method (STWRM), space-time Galerkin method with weak form (STGM/WF), space-time least squares method or process (STLSM or STLSP), etc., thereby addressing totality of all IVPs. The space-time integral forms resulting from these space-time methods of approximation are necessary conditions. By making a correspondence of these integral forms to the space-time calculus of variations we can determine which integral forms lead to unconditionally stable computational processes. The space-time integral forms that satisfy all elements of the space-time calculus of variations are termed *space-time variationally consistent* (STVC) integral forms. These integral forms result in unconditionally stable computational processes during the entire evolution. The integral forms in which one or more aspects of the space-time calculus of variations is not satisfied are termed *space-time variationally inconsistent* (STVIC) integral forms. In STVIC integral forms, unconditional stability of the computations is not always ensured.

Using the space-time operator classifications and the space-time methods of approximation, space-time finite element processes can be considered for the entire space-time domain $\bar{\Omega}_{xt} = [0, L] \times [0, \tau]$ or for a space-time strip (or slab) $\bar{\Omega}_{xt}^{(n)} = [0, L] \times [t_n, t_{n+1}]$ for an increment of time Δt with time-marching. In both approaches, simultaneous dependence of ϕ on x and t is maintained (in conformity with the physics) and the finite elements are space-time finite elements. Determination of STVC or STVIC of the space-time integral forms for the classes of operators decisively establishes which methods of approximation are worthy of consideration for which classes of space-time differential operators for unconditional stability of computations.

This space-time finite element methodology with either $\bar{\Omega}_{xt}$ or $\bar{\Omega}_{xt}^{(n)}$ with time-marching is most meritorious as it preserves the physics in the description of the IVP in the computational process and permits consistent and rigorous mathematical treatment of the process including establishing correspondence with space-time calculus of variations. In the next section, we consider space-time decoupled approach, where a two-stage approximation is used to obtain the solution to the original IVP.

10.7 Space-time decoupled finite element method

In this methodology space and time are decoupled, i.e. $\phi(x,t)$ does not have simultaneous dependence on x and t. Consider the IVP (10.1) in which A is a linear operator in space and time (for simplicity). The spatial domain $\bar{\Omega}_x = [0, L]$ is discretized (in this case in $\mathbb{R}^1$), that is, we consider discretization $\bar{\Omega}_x^T = \bigcup_e \bar{\Omega}_x^e$ of $\bar{\Omega}_x$ in which $\bar{\Omega}_x^e$ is the eth finite element in the spatial domain $\bar{\Omega}_x = [0, L]$. We consider local approximation $\phi_h^e(x,t)$ of $\phi(x,t)$ over $\bar{\Omega}_x^e$ using

$$\phi_h^e(x,t) = \sum_{i=1}^{n} N_i(x)\delta_i^e(t) \tag{10.21}$$

in which $N_i(x)$ are local approximation functions and $\delta_i^e(t)$ are nodal degrees of freedom for an element e with spatial domain $\bar{\Omega}_x^e$. Using (10.1) we construct integral form over $\bar{\Omega}_x^T$ using any of the standard methods of approximation. Let us consider Galerkin method with weak form:

$$(A\phi_h - f, v)_{\bar{\Omega}_x^T} = \int_{\bar{\Omega}_x^T} (A\phi_h - f)v(x)\,dx = 0; \quad v = \delta\phi_h \tag{10.22}$$

in which $\phi_h = \underset{e}{\cup}\phi_h^e$ is the approximation of ϕ over $\bar{\Omega}_x^T$. The integral in (10.22) can be written as

$$(A\phi_h - f, v)_{\bar{\Omega}_x^T} = \sum_e (A\phi_h^e - f, v)_{\bar{\Omega}_x^e}; \quad v = \delta\phi_h^e \tag{10.23}$$

Consider $(A\phi_h^e - f, v)_{\bar{\Omega}_x^e}$ for an element $\bar{\Omega}_x^e$ in x. We transfer half of the differentiation from ϕ_h^e to v only for those terms that contain even order derivatives of ϕ_h^e with respect to x. Using definition of secondary variables, etc., we arrive at

$$(A\phi_h^e - f, v)_{\bar{\Omega}_x^e} = B^e(\phi_h^e, v) - l^e(v) \tag{10.24}$$

$l^e(v)$ is concomitant that contains secondary variables in addition to other terms related to nonhomogeneous part f. We substitute local approximation (10.21) into (10.24) and note that

$$\begin{aligned} \frac{d\phi_h^e}{dx} &= \sum_{i=1}^{n} \frac{dN_i}{dx}\delta_i^e(t); \quad & \frac{d^2\phi_h^e}{dx^2} &= \sum_{i=1}^{n} \frac{d^2N_i}{dx^2}\delta_i^e(t) \\ \frac{d\phi_h^e}{dt} &= \sum_{i=1}^{n} N_i(x)\dot{\delta}_i^e(t); \quad & \frac{d^2\phi_h^e}{dt^2} &= \sum_{i=1}^{n} N_i(x)\,\ddot{\delta}_i^e(t) \end{aligned} \tag{10.25}$$

to obtain (noting that $v = N_j(x)$; $j = 1, 2, \dots, n$),

$$(A\phi_h^e - f, v)_{\bar{\Omega}_x^e} = B^e\Big(\sum_{i=1}^{n} N_i(x)\delta_i^e(t), N_j\Big) - l^e(N_j)\ ; \quad j = 1, 2, \dots, n \tag{10.26}$$

After performing integration with respect to x in (10.26), (10.26) reduces to a system of ordinary differential equations in time in $\{\delta^e\}$, $\{\dot{\delta}^e\}$, etc., load vector $\{f^e\}$, and the secondary variables $\{P^e\}$:

$$(A\phi_h^e - f, v)_{\bar{\Omega}_x^e} = [C_1^e]\{\delta^e\} + [C_2^e]\{\dot{\delta}^e\} + \dots - \{f^e\} - \{P^e\} \tag{10.27}$$

If

$$\{\delta\} = \underset{e}{\cup}\{\delta^e\}\ ; \quad \{\dot{\delta}\} = \underset{e}{\cup}\{\dot{\delta}^e\}\ \dots\dots \tag{10.28}$$

then, the assembly of the element equations over $\bar{\Omega}_x^T$ yields

$$\sum_e (A\phi_h^e - f, v)_{\bar{\Omega}_x^e} = (A\phi_h - f, v)_{\bar{\Omega}_x^T} = [C_1]\{\delta\} + [C_2]\{\dot{\delta}\} + \dots - \{f\} - \{P\} = 0 \tag{10.29}$$

The order of the time derivatives of $\{\delta^e\}$ and $\{\delta\}$ in (10.27) and (10.29) depend on the orders of the time derivatives in (10.1). Equations (10.29) are a system of ordinary differential equations in time. We note that the choice of $N_i(x)$ is straightforward (interpolation theory) as opposed to the choice of $g(x)$ in $\phi(x,t) = g(x)h(t)$. This is a significant benefit of space-time decoupling using finite element discretization in space. Equations (10.29) are then integrated using explicit or implicit time integration methods or finite element method in time after imposing BCs and ICs of the IVP.

10.8 Time integration of ODEs resulting from space-time decoupled FEM

Using BCs and ICs of the IVP, $\{\delta(t)\}, \{\dot{\delta}(t)\}, \dots$ for spatial locations in $\bar{\Omega}_x^T$ are calculated by integrating (10.29) in time for an increment of time and then by time-marching the integration process for subsequent values of time. For this purpose explicit or implicit time integration methods or finite element method in time can be employed. Once $\{\delta\}$, $\{\dot{\delta}\}$, etc. are known for an increment of time, the solution $\{\delta^e\}$, $\{\dot{\delta}^e\}$, etc. are known for each $\bar{\Omega}_x^e$ in space.

We note that since ODEs in time only result in space-time decoupled methods, the time integration schemes are neither needed nor used in space-time coupled methods.

Remarks.

(1) A detailed study of various methods of approximation briefly discussed here is beyond the scope of study in this book.

(2) In this chapter we only consider methods of approximation primarily based on finite difference approach using Taylor series, for obtaining solution of ODEs in time resulting from the PDEs describing IVPs after decoupling of space and time.

(3) The finite element method for ODEs in time are similar to those for BVPs presented in chapter 9. See reference [50], textbook for finite element method for IVPs.

10.9 Some time integration methods for ODEs in time

Mathematical description of time dependent processes result in PDEs in dependent variables, space coordinates and time. If we use space-time decoupled methods of approximation and consider discretization in spatial domain, then the result is a system of ODEs in time. On the other hand if we consider lumping in space coordinates, then this also results in a single or a system of ODEs in time. The subject of study here is to find numerical methods of solution of ODE(s) in time.

Consider a simple ODE in time.

$$\frac{d\phi}{dt} = f(\phi, t) \quad \forall t \in (t_1, t_2) = (0, \tau) = \Omega_t \tag{10.30}$$

We refer to (10.30) as IVP in which as time t elapses, ϕ changes, as $\phi = \phi(t)$. Equation (10.30) (IVP) must have initial condition(s) (just like BVPs have

BCs). In case of (10.30), a first order ODE in time, we need one initial condition at the commencement of the evolution (at $t = t_1$ or simply $t_1 = 0$).

$$\phi(0) = \phi\big|_{t=0} = \phi_0 \tag{10.31}$$

In (10.30), ϕ is only a function of time t and (10.31) represents the state of the solution at $t = 0$, at commencement of evolution.

In this chapter we consider methods to find numerical solutions of IVP (10.30) subject to initial conditions (10.31). Let $\Delta t = h$ represent an increment of time. Consider evolution described by (10.30) between times $t = t_i$ and $t = t_{i+1}$, where $t_{i+1} - t_i = \Delta t = h$.

Integrate (10.30) for time interval $[t_i, t_{i+1}]$.

$$\int_{t_i}^{t_{i+1}} \frac{d\phi}{dt}\, dt = \int_{t_i}^{t_{i+1}} f(\phi, t)\, dt \tag{10.32}$$

$$\text{or} \quad \int_{t_i}^{t_{i+1}} d\phi = \int_{t_i}^{t_{i+1}} f(\phi, t)\, dt \tag{10.33}$$

$$\text{or} \quad \phi\big|_{t_{i+1}} - \phi\big|_{t_i} = \int_{t_i}^{t_{i+1}} f(\phi, t)\, dt \tag{10.34}$$

$$\text{or} \quad \phi_{i+1} = \phi_i + \int_{t_i}^{t_{i+1}} f(\phi, t)\, dt \tag{10.35}$$

$$\text{where} \quad \phi\big|_{t_{i+1}} = \phi_{i+1} \quad \text{and} \quad \phi\big|_{t_i} = \phi_i \tag{10.36}$$

Equation (10.35) defines evolution at time t_{i+1} in terms of the solution ϕ at time t and the integral of $f(\phi, t)$ between the limits t_i and t_{i+1}, which is area under the curve $f(\phi, t)$ versus t between $t = t_i$ and $t = t_{i+1}$ (see figure 10.4).

Knowing ϕ_{i+1} from (10.35), we can reuse (10.35) to find ϕ_{i+2} and continue doing so until the desired time is reached. In various methods of approximation for finding $\phi(t)$ numerically for (10.30), we use (10.35) in which the integral of $f(\phi, t)$ over the interval $[t_i, t_{i+1}]$ is approximated.

10.9.1 Euler's Method

In this method we approximate the integral in (10.35) by the area of a rectangle of width h and height ϕ_i, the function value at the left endpoint

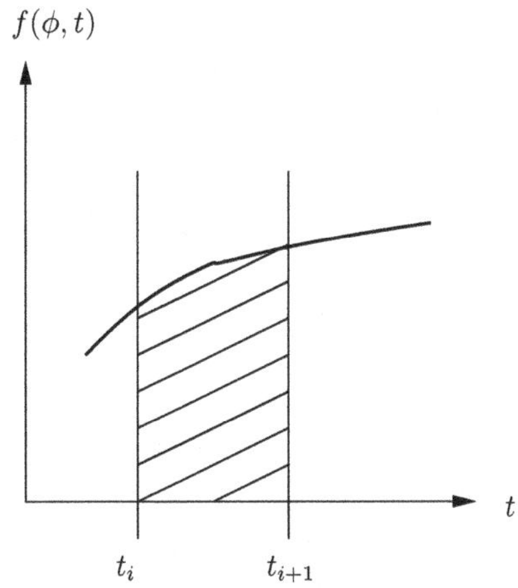

Figure 10.4: $f(\phi, t)$ versus t

of the interval (see figure 10.5).

$$\int_{t_i}^{t_{i+1}} f(\phi, t)\, dt \simeq hf(\phi_i, t_i) = (t_{i+1} - t_i) f(\phi_i, t_i) \tag{10.37}$$

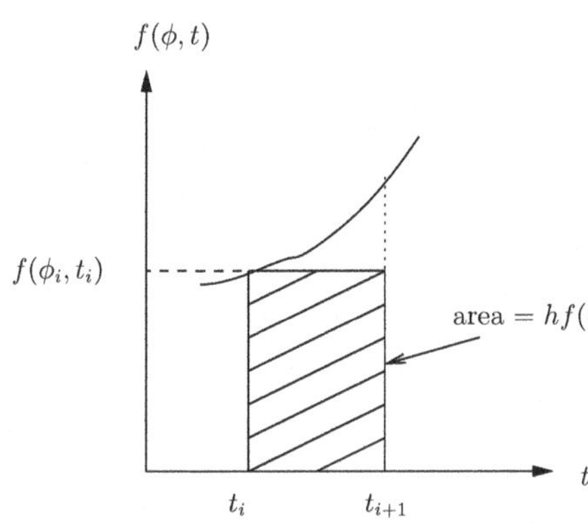

Figure 10.5: Euler's method

Hence, the evolution for ϕ is computed using

$$\phi_{i+1} = \phi_i + hf(\phi_i, t_i) \tag{10.38}$$

The error made in doing so is illustrated by the empty area bound by the dotted line which is neglected in the approximation (10.37) and hence (10.38). In Euler's method we begin with $i = 0$ and ϕ_0 defined using initial condition at time $t = 0$ and march the solution in time using (10.38). It is obvious that smaller values of $h = \Delta t$ will yield more accurate results. Euler's method is one of the simplest and most crude approximation techniques for ODEs in time. This method is called explicit method as in this method the solution at new value of time is explicitly expressed in terms of the solution at current value of time. Thus, computations of solution at new value of time is simply a matter of substitution in (10.38). By using more accurate approximation of $\int_{t_i}^{t_{i+1}} f(\phi,t)\,dt$ we can devise methods that will yield more accurate numerical solution of $\phi(t)$ of (10.30).

Example 10.3 (Euler's Method). Consider the IVP

$$\frac{d\phi}{dt} - t - \phi = 0 \qquad \text{for } t > 0 \tag{10.39}$$

$$\text{IC:} \qquad \phi(0) = 1 \tag{10.40}$$

We consider numerical solution of (10.39) with IC (10.40) using $\Delta t = 0.2$, 0.1 and 0.05 for $0 \leq t \leq 1$.

We rewrite (10.39) in the standard form (10.30).

$$\frac{d\phi}{dt} = t + \phi = f(\phi,t) \tag{10.41}$$

Thus using (10.38) for (10.41), we have

$$\begin{aligned} \phi_{i+1} &= \phi_i + h(t_i - \phi_i) \qquad ; \qquad i = 0,1,... \\ \text{with} \quad \phi_0 &= 1 = \phi\big|_{t=0} \qquad ; \qquad t_0 = 0 \end{aligned} \tag{10.42}$$

We calculate numerical values of ϕ using (10.42) with $h = 0.2\,,0.1$ corresponding to 5 and 10 time steps for $0 \leq t \leq 1$. Calculated values of ϕ for various values of time using $h = 0.2\,,0.1$ are given in tables 10.1 and 10.2 (using (10.42), for $i = 0,1,...$).

Table 10.1: Results of Euler's method for (10.42), $h = 0.2$

h = 0.2, $0 \leq t \leq 1$

step number	time, t	function value, ϕ	time derivative $\left.\frac{d\phi}{dt}\right\|_i = f(\phi_i, t_i)$
0	0.000000E+00	0.100000E+01	0.100000E+01
1	0.200000E+00	0.120000E+01	0.140000E+01
2	0.400000E+00	0.148000E+01	0.188000E+01
3	0.600000E+00	0.185600E+01	0.245600E+01
4	0.800000E+00	0.234720E+01	0.314720E+01
5	0.100000E+01	0.297664E+01	0.397664E+01

Table 10.2: Results of Euler's method for (10.42), $h = 0.1$

h = 0.2
0 ≤ t ≤ 1

step number	time, t	function value, ϕ	time derivative $\left.\frac{d\phi}{dt}\right\|_i = f(\phi_i, t_i)$
0	0.000000E+00	0.100000E+01	0.100000E+01
1	0.100000E+00	0.110000E+01	0.120000E+01
2	0.200000E+00	0.122000E+01	0.142000E+01
3	0.300000E+00	0.136200E+01	0.166000E+01
4	0.400000E+00	0.152820E+01	0.192820E+01
5	0.500000E+00	0.172102E+01	0.222102E+01
6	0.600000E+00	0.194312E+01	0.254312E+01
7	0.700000E+00	0.219743E+01	0.289743E+01
8	0.800000E+00	0.248718E+01	0.328718E+01
9	0.900000E+00	0.281590E+01	0.371590E+01
10	0.100000E+01	0.318748E+01	0.418748E+01

From tables 10.1 and 10.2, we note that even for $h = 0.2$ and 0.1, rather large time increments, the values of ϕ and $\frac{d\phi}{dt}$ are quite reasonable. Plots of ϕ and $\frac{d\phi}{dx}$ versus t are shown in figures 10.6 and 10.7 illustrate this.

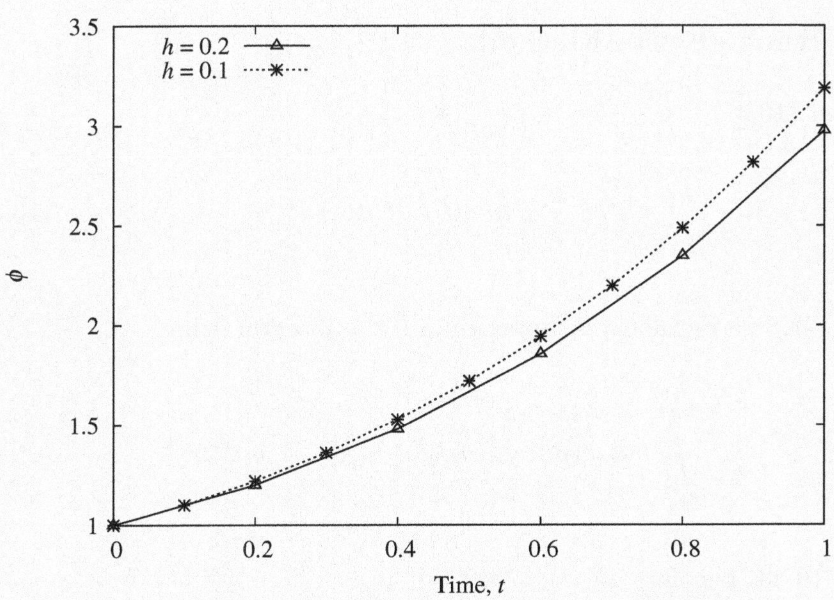

Figure 10.6: Solution ϕ versus time t

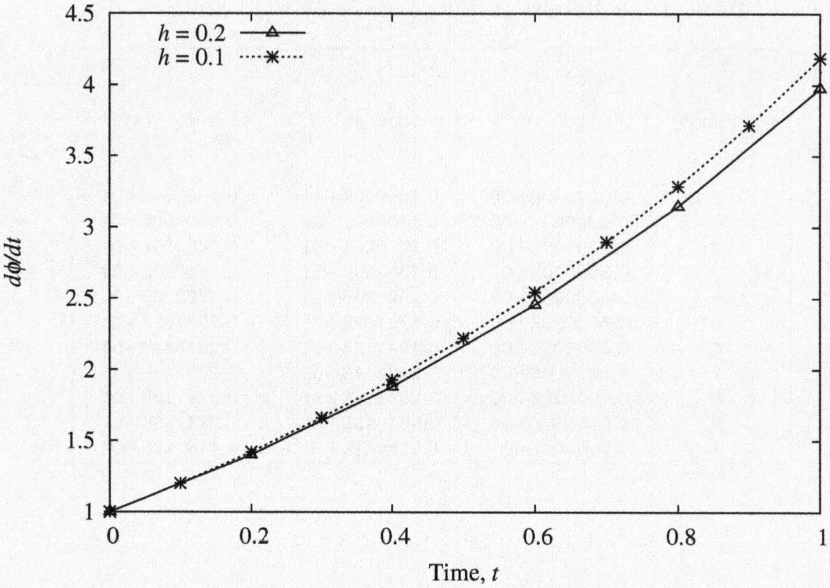

Figure 10.7: $\frac{d\phi}{dt}$ versus time t

10.9.2 Runge-Kutta Methods

Recall (10.35)

$$\phi_{i+1} = \phi_i + \int_{t_i}^{t_{i+1}} f(\phi, t)\, dt \tag{10.43}$$

In Runge-Kutta methods we approximate the integral using

$$\int_{t_i}^{t_{i+1}} f(\phi, t)\, dt = h(a_1k_1 + a_2k_2 + ...a_nk_n) \tag{10.44}$$

Hence, (10.43) becomes

$$\phi_{i+1} = \phi_i + h(a_1k_1 + a_2k_2 + ... + a_nk_n) \tag{10.45}$$

This is known as n^{th} order Runge-Kutta method in which k_1, k_2, ... k_n are given by

$$\begin{aligned}
k_1 &= f(\phi_i, t_i) \\
k_2 &= f(\phi + q_{11}k_1h, t_i + p_1h) \\
k_3 &= f(\phi + q_{21}k_1h + q_{22}k_2h, t_i + p_2h) \\
k_4 &= f(\phi + q_{31}k_1h + q_{32}k_2h + q_{33}k_3, t_i + p_3h) \\
&\vdots \\
k_n &= f(\phi_i + q_{n-1,1}k_1h + q_{n-1,2}k_2h + ...q_{n-1,n-1}k_{n-1}h)
\end{aligned} \tag{10.46}$$

In (10.46) p and q are constants. Note the recurrence relationship in k, i.e. k_2 contains k_1, k_3 contains k_1 and k_2, and so on. p and q are determined by using Taylor series expansions. We consider details in the following.

10.9.2.1 Second Order Runge-Kutta Methods

Consider a Runge-Kutta method with $n = 2$ (second order).

$$\phi_{i+1} = \phi_i + h(a_1k_1 + a_2k_2) \tag{10.47}$$

$$\text{where} \qquad k_1 = f(\phi_i, t_i) \tag{10.48}$$

$$k_2 = f(\phi_i + q_{11}k_1h, t_i + p_1h) \tag{10.49}$$

Our objective is determine a_1, a_2, q_{11} and p_1. We do this using Taylor series expansions. Consider Taylor series expansion of ϕ_{i+1} in terms ϕ_i and $f(\phi_i, t_i)$ and retain only up to quadratic terms in h.

$$\phi_{i+1} = \phi_i + f(\phi_i, t_i)h + f'(\phi_i, t_i)\frac{h^2}{2!} \tag{10.50}$$

$$\text{where} \quad f'(\phi, t) = \frac{\partial f(\phi, t)}{\partial t} + \frac{\partial f(\phi, t)}{\partial \phi}\frac{\partial \phi}{\partial t} \tag{10.51}$$

$$\therefore \quad \phi_{i+1} = \phi_i + f(\phi_i, t_i)h + \left(\frac{\partial f(\phi, t)}{\partial t} + \frac{\partial f(\phi, t)}{\partial t}\frac{d\phi}{dt}\right)\bigg|_{t_i}\frac{h^2}{2!} \tag{10.52}$$

But $\frac{d\phi}{dt} = f(\phi, t)$, hence (10.52) becomes

$$\phi_{i+1} = \phi_i + f(\phi_i, t_i)h + \left(\frac{\partial f(\phi_i, t_i)}{\partial t} + f(\phi_i, t_i)\frac{\partial f(\phi_i, t_i)}{\partial \phi}\right)\frac{h^2}{2!} \tag{10.53}$$

Consider Taylor series expansion of $f(\cdot)$ in (10.49), using it as a function of two variables as in

$$g(x + u, y + v) = g(x, y) + \frac{\partial g}{\partial x}u + \frac{\partial g}{\partial y}v + ... \tag{10.54}$$

$$\therefore \quad k_2 = f(\phi_i + q_{11}k_1h, t + p_1h) = f(\phi_i, t_i) + p_1h\frac{\partial f}{\partial t} + q_{11}k_1h\frac{\partial f}{\partial \phi} + O(h^2) \tag{10.55}$$

Substituting from (10.55) and (10.48) into (10.47)

$$\phi_{i+1} = \phi_i + ha_1f(\phi_i, t_i) + a_2hf(\phi_i, t_i) + a_2p_1h^2\frac{\partial f}{\partial t} + a_2q_{11}k_1h^2\frac{\partial f}{\partial \phi} + O(h^2) \tag{10.56}$$

Rearranging terms in (10.56)

$$\phi_{i+1} = \phi_i + (a_1h + h_2h)f(\phi_i, t_i) + a_2p_1h^2\frac{\partial f}{\partial t} + a_2q_{11}k_1h^2\frac{\partial f}{\partial t} + O(h^2) \tag{10.57}$$

For ϕ_{i+1} in (10.53) and (10.57) to be equivalent, the following must hold.

$$\begin{aligned} a_1 + a_2 &= 1 \\ a_2p_1 &= \frac{1}{2} \\ a_2q_{11} &= \frac{1}{2} \end{aligned} \tag{10.58}$$

Equation (10.58) are three equations in four unknowns (a_1, a_2, p_1 and q_{11}), hence they do not have a unique solution. There are infinitely many solutions, so an arbitrary choice must be made at this point.

10.9.2.2 Heun Method

If we choose $a_2 = \frac{1}{2}$, then (10.58) give $a_1 = \frac{1}{2}$, $p_1 = q_{11} = \frac{1}{2}$ and we have the following for the 2nd order Runge-Kutta method.

$$\begin{aligned} \phi_{i+1} &= \phi_i + h(\frac{1}{2}k_1 + \frac{1}{2}k_2) \\ \text{where:} \quad k_1 &= f(\phi_i, t_i) \\ k_2 &= f(t_i + h, \phi_i + k_1h) \end{aligned} \tag{10.59}$$

This set of constants a, p, and q is known as Heun's method.

10.9.2.3 Midpoint Method

If we choose $a_2 = 1$, then (10.58) gives $a_1 = 0$, $p_1 = q_{11} = \frac{1}{2}$ and we have the following for the 2nd order Runge-Kutta method.

$$\begin{aligned} \phi_{i+1} &= \phi_i + hk_2 \\ &\text{where:} \\ k_1 &= f(\phi_i, t_i) \\ k_2 &= f(t_i + \frac{1}{2}h, \phi_i + \frac{1}{2}k_1h) \end{aligned} \tag{10.60}$$

This is known as the midpoint method.

Using the derivation similar to second order Runge-Kutta method, we can also derive higher order Runge-Kutta methods.

10.9.2.4 Third Order Runge-Kutta Method

Consider a Runge-Kutta method with $n = 3$ (third order).

$$\frac{d\phi}{dt} = f(\phi, t) \quad \forall t \in (t_1, t_2) = (0, \tau) = \Omega_t \tag{10.61}$$

Then

$$\phi_{i+1} = \phi_i + h(a_1 k_2 + a_2 k_2 + a_3 k_3) \tag{10.62}$$

where

$$a_1 = \frac{1}{6}, \quad a_2 = \frac{4}{6}, \quad a_3 = \frac{1}{6} \tag{10.63}$$

and

$$\begin{aligned} k_1 &= f(\phi_i, t_i) \\ k_2 &= f\left(\phi_i, \frac{k_1}{2} h, t_i + \frac{h}{2}\right) \\ k_3 &= f(\phi_i + 2hk_2 - hk_1, t_i + h) \end{aligned} \tag{10.64}$$

10.9.2.5 Fourth Order Runge-Kutta Method

Consider a Runge-Kutta method with $n = 4$ (fourth order).

$$\frac{d\phi}{dt} = f(\phi, t) \quad \forall t \in (t_1, t_2) = (0, \tau) = \Omega_t \tag{10.65}$$

Then

$$\phi_{i+1} = \phi_i + h(a_1 k_1 + a_2 k_2 + a_3 k_3 + a_4 k_4) \tag{10.66}$$

where

$$a_1 = a_4 = \frac{1}{6} \quad , \quad a_2 = a_3 = \frac{1}{3} \tag{10.67}$$

$$\text{and} \quad \begin{aligned} k_1 &= f(t_i, \phi_i) \\ k_2 &= f(t_i + \frac{h}{2}, \phi_i + h\frac{k_1}{2}) \\ k_3 &= f(t_i + \frac{h}{2}, \phi_i + h\frac{k_2}{2}) \\ k_4 &= f(t_i + h, \phi_i + hk_3) \end{aligned} \tag{10.68}$$

10.9.2.6 Runge-Kutta Method for a System of ODEs in Time

The concepts described for a single ODE in time can be extended to a system of ODEs in time. Consider the following ODEs in time. We consider 4th order Runge-Kutta method.

$$\frac{du}{dt} = f_1(u, v, t) \qquad \text{and} \qquad \frac{dv}{dt} = f_2(u, v, t) \quad \forall t \in (t_1, t_2) = (0, \tau) = \Omega_t \tag{10.69}$$

$$u_{i+1} = u_i + \frac{h}{6}(k_1 + 2k_2 + 2k_3 + k_4) \qquad v_{i+1} = v_i + \frac{h}{6}(l_1 + 2l_2 + 2l_3 + l_4) \tag{10.70}$$

$$k_1 = f_1(u_i, v_i, t_i) \qquad l_1 = f_1(u_i, v_i, t_i)$$

$$k_2 = f_1(u_i + \frac{hk_1}{2}, v_i + \frac{hl_1}{2}, t_i + \frac{h}{2}) \qquad l_2 = f_1(u_i + \frac{hk_1}{2}, v_i + \frac{hl_1}{2}, t_i + \frac{h}{2})$$

$$k_3 = f_1(u_i + \frac{hk_2}{2}, v_i + \frac{hl_2}{2}, t_i + \frac{h}{2}) \qquad l_3 = f_1(u_i + \frac{hk_2}{2}, v_i + \frac{hl_2}{2}, t_i + \frac{h}{2})$$

$$k_4 = f_1(u_i + hk_3, v_i + hl_3 + t_i + h) \qquad l_4 = f_1(u_i + hk_3, v_i + hl_3 + t_i + h)$$

Remarks.

(1) For each we introduce k_i, l_i, ..., $i = 1, 2, 3, 4$ when we have more than one ODE in time.

(2) Similar to fourth order Runge-Kutta method described above, we can also use 2nd and 3rd Runge-Kutta methods for a system of ODEs.

10.9.2.7 Runge-Kutta Method for Higher Order ODEs in Time

When ODEs in time contain time derivatives of order higher than one, the Runge-Kutta methods can also be used. In such cases we recast the higher order ODE as a system of first order ODEs using *auxiliary variables* and *auxiliary equations* and then use the method described in section 10.9.2.6. Consider

$$\frac{d^2v}{dt^2} = f_1(v, \frac{dv}{dt}, t) \quad \forall t \in (t_1, t_2) = (0, \tau) = \Omega_t \tag{10.71}$$

a second order ODE in time. Let $\frac{dv}{dt} = u$ be an auxiliary equation in which u is the auxiliary variable. Substituting this in (10.71), we have the following.

$$\begin{aligned} &\frac{du}{dt} = f_1(v, u, t) \\ \text{and} \\ &\frac{dv}{dt} = u = f_2(u) \end{aligned} \tag{10.72}$$

Equations (10.72) are a pair of first order ODEs in u and v, hence we can use the method described in section 10.9.2.6. In this case also we can use

2nd and 3rd order Runge-Kutta methods as we see fit. For (10.72), we give details for 4th order Runge-Kutta method.

$$\frac{du}{dt} = f_1(u,v,t) \qquad \frac{dv}{dt} = f_2(u) = u \tag{10.73}$$

$$\begin{aligned}
u_{i+1} &= u_i + \frac{h}{6}(k_1 + 2k_2 + 2k_3 + k_4) & v_{i+1} &= v_i + \frac{h}{6}(l_1 + 2l_2 + 2l_3 + l_4)\\
k_1 &= f_1(u_i, v_i, t_i) & l_1 &= f_2(u_i) = u_i\\
k_2 &= f_1(u_i + \frac{hk_1}{2}, v_i + \frac{hl_1}{2}, t_i + \frac{h}{2}) & l_2 &= f_2(u_i + \frac{hk_1}{2}) = u_i + \frac{hk_1}{2}\\
k_3 &= f_1(u_i + \frac{hk_2}{2}, v_i + \frac{hl_2}{2}, t_i + \frac{h}{2}) & l_3 &= f_2(u_i + \frac{hk_2}{2}) = u_i + \frac{hk_2}{2}\\
k_4 &= f_1(u_i + hk_3, v_i + hl_3, t_i + h) & l_4 &= f_2(u_i + hk_3) = u_i + hk_3
\end{aligned}$$

10.9.3 Numerical Examples

In this section we consider a few numerical examples to illustrate the computational details.

Example 10.4. Consider a first order ODE

$$\frac{du}{dt} = t + 4 \qquad \text{with} \qquad u(0) = 1 \quad \forall t \in (t_1, t_2) = (0, \tau) = \Omega_t,\ \tau > 0$$

We calculate u at $t = 0.2$ and $t = 0.4$ using $\Delta t = 0.2$ using 2nd, 3rd, and 4th order Runge-Kutta methods.

(a) Second Order Runge-Kutta Method (Huen Method)

$$\begin{aligned}
u_{i+1} &= u_i + h(\frac{1}{2}k_1 + \frac{1}{2}k_2)\\
k_1 &= f(u_i, t_i)\\
k_2 &= f(u_i + hk_1, t_i + h)
\end{aligned}$$

In this case $f = (t + u)$.

For $i = 1$:
$t = t_1 = 0, \ \ u_1 = u(0) = 1$

For $i = 2$:

$t = t_2 = \Delta t = 0.2$ using $\Delta t = h = 0.2$

$$k_1 = f(t_1 + u_1) = 0 + 1 = 1$$
$$k_2 = f(t_1 + h, u_1 + k_1 h) = 0.2 + (1 + 1(0.2)) = 1.4$$
$$u_2 = 1 + 0.2(\frac{1}{2} + \frac{1.4}{2}) = 1 + 0.24 = 1.24$$

For $i = 3$:

$t = t_3 = 2\Delta t = 0.4$ using $\Delta t = h = 0.2$

$$k_1 = f(t_2, u_2) = 0.2 + 1.24 = 1.44$$
$$k_2 = f(t_2 + h, u_2 + k_1 h) = (0.2 + 0.2) + 1.24 + (0.2(1.44)) = 1.928$$
$$u_3 = u_2 + h(\frac{1}{2}k_1 + \frac{1}{2}k_2) = 1.24 + 0.2(1.44 + 1.928) = 1.5768$$

Thus we have

t	u	$\frac{du}{dt} = f(u,t)$
0	1	1
0.2	1.24	1.44
0.4	1.5768	1.9768

(b) Third Order Runge-Kutta Method

For $i = 1$:

$t = t_1 = 0, \quad u_1 = 1, \quad f(u,t) = u + t, \quad h = 0.2$

For $i = 2$:

$t = t_2 = \Delta t = 0.2$ using $\Delta t = h = 0.2$

$$k_1 = f(u_1, t) = 1 + 0 = 1$$
$$k_2 = f(u_1 + \frac{k_1 h}{2}, t_1 + \frac{h}{2}) = (1 + \frac{1(0.2)}{2}) + (0 + \frac{0.2}{2}) = 1.2$$
$$k_3 = f(u_1 + 2hk_2 - hk_1, t_1 + h) = (1 + 2(0.2)(1.2) - 0.2(1)) + 0 + 0.2 = 1.48$$
$$u_2 = u_1 + \frac{h}{6}(k_1 + uk_2 + k_3)$$
$$u_2 = 1 + \frac{0.2}{6}(1 + 4(1.2) + 1.48) = 1.24267$$

For $i = 3$:

$$t = t_3 = 2\Delta t = 0.4 \text{ using } \Delta t = h = 0.2$$
$$k_1 = f(u_2, t_2) = u_2 + t_2 = 1.24267 + 0.2 = 1.44267$$
$$k_2 = f(u_2 + \frac{k_1 h}{2}, t_2 + \frac{h}{2}) = (u_2 + k_1\frac{h}{2}) + (t_2 + \frac{h}{2})$$
$$= (1.24267 + 1.44267(\frac{0.2}{2})) + (0.2 + \frac{0.2}{2}) = 1.686937$$
$$k_3 = f(u_2 + 2hk_2 - hk_1, t_2 + h) = (u_2 + 2hk_2 - hk_1, t_2 + h)$$
$$= (1.24267 + 2(0.2)(1.686937) - (0.2)(1.44267)) + (0.2 + 0.2)$$
$$= 2.0289$$
$$u_3 = u_2 + \frac{h}{6}(k_1 + 4k_2 + k_3) = 1.24267 + \frac{0.2}{6}(1.44267 + 4(1.686937) + 2.0289)$$
$$= 1.583315$$

Thus we have

t	u	$\frac{du}{dt} = f(u,t)$
0	1	1
0.2	1.24267	1.44267
0.4	1.583315	1.983315

(c) Fourth Order Runge-Kutta Method

For $i = 1$:

$$t = t_1 = 0, \quad u_1 = 1, \quad f(u,t) = u + t, \quad h = 0.2$$

For $i = 2$:

$$t = t_2 = \Delta t = 0.2 \text{ using } \Delta t = h = 0.2$$

$$k_1 = f(u_1, t_1) = 1 + 0 = 1$$
$$k_2 = f(u_1 + \frac{hk_1}{2}, t_1 + \frac{h}{2}) = (u_1 + \frac{hk_1}{2}) + (t_1 + \frac{h}{2})$$
$$= (1 + \frac{(0.2)(1)}{2}) + (0 + \frac{0.2}{2}) = 1.22$$
$$k_3 = f(u_1 + \frac{hk_2}{2}, t_1 + \frac{h}{2}) = (u_1 + \frac{hk_2}{2}) + (t_1 + \frac{h}{2})$$
$$= (1 + \frac{(0.2)(1.2)}{2}) + (0 + \frac{0.2}{2}) = 1.22$$
$$k_4 = f(u_1 + hk_3, t_1 + h) = (u_1 + hk_3) + (t_1 + h)$$
$$= 1 + (0.2)(1.22) + (0 + 0.2) = 1.444$$
$$u_2 = u_1 + \frac{h}{6}(k_1 + 2k_2 + 2k_3 + k_4)$$
$$= 1 + \frac{1}{6}(0.2)(1 + 2(1.2) + 2(1.22) + 1.444) = 1.2428$$

For $i = 3$:
$t = t_3 = 2\Delta t = 0.4$ using $\Delta t = h = 0.2$

$$\begin{aligned}
k_1 &= 1.3328 \\
k_2 &= 1.68708 \\
k_3 &= 1.71151 \\
k_4 &= 1.9851 \\
u_3 &= u_2 + \frac{h}{6}(k_1 + 2k_2 + 2k_3 + k_4) \\
&= 1.2428 + \frac{0.2}{6}(1.3328 + 2(1.68708) + 2(1.71151) + 1.9851) \\
&= 1.58364
\end{aligned}$$

Similarly, we find that for $t = t_4 = 3\Delta t = 0.6$ we have

$$u_4 = 2.044218$$

Thus, we have

t	u	$\frac{du}{dt} = f(u,t)$
0	1	1
0.2	1.2428	1.4428
0.4	1.58364	1.98314
0.6	2.044218	2.644218

Remarks.

(1) Obviously Euler's method has the poorest accuracy as it is an extremely crude approximation of the area under $f(\phi, t)$ versus t between $[t_i, t_{i+1}]$.

(2) The accuracy of Runge-Kutta methods improve as the order increases.

(3) 4th order Runge-Kutta method has very good accuracy. This method is used widely in practical applications.

Example 10.5 (4th order Runge-Kutta Method for a System of First Order ODEs in Time). Consider the following system of ODEs in time.

$$\frac{dx}{dt} = xy+t = f(x,y,t) \quad ; \quad \frac{dy}{dt} = x+yt = g(x,y,t) \quad \forall t \in (t_1, t_2) = (0, \tau) = \Omega_t$$

in which $x = x(t)$, $y = y(t)$
with

$$x(0) = x\Big|_{t=0} = 1 = x_0 \qquad ; \qquad t = 0 = t_0$$
$$y(0) = y\Big|_{t=0} = -1 = y_0$$

Calculate the solutions x and y at $t = 0.2$ using $\Delta t = h = 0.2$. Let k_j ; $j = 1, 2, ..., 4$ and l_j ; $j = 1, 2, ..., 4$ be the area constants for the two ODEs. For $i = 0, 1, 2, ...$ we have

$$x_{i+1} = x_i + \frac{h}{6}(k_1 + 2k_2 + 2k_3 + k_4) \qquad ; \qquad y_{i+1} = y_i + \frac{h}{6}(l_1 + 2l_2 + 2l_3 + l_4)$$

For $i = 0$:

$$\begin{aligned} k_1 &= f(x_0, y_0, t_0) \\ &= (1(-1) + 0) = -1 \\ l_1 &= g(x_0, y_0, t_0) \\ &= (0(-1) + 1) = 1 \end{aligned}$$

$$\begin{aligned} k_2 &= f(x_0 + \frac{k_1 h}{2}, y_0 + \frac{l_1 h}{2}, t_0 + \frac{h}{2}) \\ &= (1 + \frac{(-1)(0.2)}{2})(-1 + \frac{(1)(0.2)}{2}) + \frac{0.2}{2} \\ &= 0.71 \\ l_2 &= g(x_0 + \frac{k_1 h}{2}, y_0 + \frac{l_1 h}{2}, t_0 + \frac{h}{2}) \\ &= (1 + \frac{(-1)(0.2)}{2}) + (-1 + \frac{1(0.2)}{2})(\frac{0.2}{2}) \\ &= 0.81 \end{aligned}$$

$$\begin{aligned} k_3 &= f(x_0 + \frac{k_2 h}{2}, y_0 + \frac{l_2 h}{2}, t_0 + \frac{h}{2}) \\ &= (1 + \frac{(0.71)(0.2)}{2})(-1 + \frac{(0.81)(0.2)}{2} + \frac{0.2}{2} \\ &= -0.754 \\ l_3 &= g(x_0 + \frac{k_2 h}{2}, y_0 + \frac{l_2 h}{2}, t_0 + \frac{h}{2}) \\ &= (1 + \frac{(-0.71)(0.2)}{2}) + (-1 + \frac{(0.81)(0.2)}{2}) + (\frac{0.2}{2}) \\ &= 0.837 \end{aligned}$$

$$\begin{aligned}
k_4 &= f(x_0 + k_3h, y_0 + l_3h, t_0h) \\
&= (1 + (-0.754)(0.2)(-1 + 0.837(0.2)) + 0.2 \\
&= -0.507 \\
l_4 &= g(x_0 + k_3h, y_0 + l_3h, t_0 + h) \\
&= (1 + (-0.754)(0.2)) + (-1 + (0.837)(0.2)) + (\frac{0.2}{2}) \\
&= 0.68
\end{aligned}$$

$$\begin{aligned}
x_1 &= x_0 + \frac{h}{6}(k_1 + 2k_2 + 2k_3 + k_4) \\
&= 1 + \frac{0.2}{6}(-1 + 2(-0.71) + 2(-0.754) - 0.507) \\
x_1 &= 0.8522
\end{aligned}$$

$$\begin{aligned}
y_1 &= y_0 + \frac{h}{6}(l_1 + 2l_2 + 2l_3 + l_4) \\
&= -1 + \frac{0.2}{6}(1 + 2(0.81) + 2(0.837) + 0.68) \\
y_1 &= -0.834
\end{aligned}$$

Hence solution at $t = 0.2$ is $(x_1, y_1) = (0.8522, -0.8341)$.

Example 10.6 (4th Order Runge-Kutta Method for a Second Order ODE in Time). Consider the following second order ODE in time.

$$\begin{aligned}
&\frac{d^2\theta}{dt^2} + \frac{32.2}{r}\sin\theta = 0 \quad \forall t \in (t_1, t_2) = (0, \tau) = \Omega_t \\
&\theta = \theta(t)
\end{aligned}$$

with

$$\begin{aligned}
\theta(0) = \theta\big|_{t=0} &= \theta_0 = 0.8 \quad \text{in radians} \\
\frac{d\theta}{dt}\Big|_{t=0} &= 0\,; \quad r = 2
\end{aligned}$$

Use fourth order Runge-Kutta method to calculate θ, $\frac{d\theta}{dt}$, and $\frac{d^2\theta}{dt^2}$ for $t = 0.1$ using $\Delta t = h = 0.1$. Convert the 2nd order ODE to a system of first order ODEs in time.

Let

$$\frac{d\theta}{dt} = u = f(u) \qquad \frac{du}{dt} = -16.1\sin\theta = g(\theta)\,; \quad (r = 2)$$

Hence

$$u\big|_{t=0} = u_0 = 0$$

Let k_j ; $j = 1, 2, ..., 4$ and l_j ; $j = 1, 2, ..., 4$ be the area constants for the two ODEs in time. $\theta_{i+1} = \frac{h}{6}(k_1 + 2k_2 + 2k_3 + k_4)$ and $u_{i+1} = u_i + \frac{h}{6}(l_1 + 2l_2 + 2l_3 + l_4)$.

For $i = 0$:

$$\begin{aligned}
k_1 &= f(u_0) & l_1 &= g(\theta_0) \\
&= 0 & &= -16.1\sin(0.8) = -11.55 \\
k_2 &= f(u_0 + \frac{l_1 h}{2}) & l_2 &= g(\theta_0 + \frac{k_1 h}{2}) \\
&= 0 + (\frac{-11.55(0.1)}{2}) & &= -16.1\sin(0.8 + \frac{9(0.1)}{2}) \\
&= -0.578 & &= -11.55 \\
k_3 &= f(u_0 + \frac{l_2 h}{2}) & l_3 &= g(\theta_0 + \frac{k_2 h}{2}) \\
&= 0 + (-11.55)(\frac{0.1}{2}) & &= -16.1\sin(0.8 + \frac{(-0.578)(0.1)}{2}) \\
&= -0.578 & &= -11.22 \\
k_4 &= f(u_0 + l_3 h) & l_4 &= g(\theta_0 + k_3 h) \\
&= 0 + (-11.22)(0.1) & &= -16.1\sin(0.8 + (-0.578)(0.1)) \\
&= -1.122 & &= -10.882
\end{aligned}$$

$$\begin{aligned}
\theta_1 &= \theta_0 + \frac{h}{6}(k_1 + 2k_2 + 2k_3 + k_4) \\
&= 0.8 + \frac{0.1}{6}(0 + 2(-0.578) + 2(-0.578) - 1.122) \\
\theta_1 &= 0.7429
\end{aligned}$$

$$\begin{aligned}
u_1 &= u_0 + \frac{h}{6}(l_1 + 2l_2 + 2l_3 + l_4) \\
&= 0 + \frac{0.1}{6}(-11.55 + 2(-11.55) + 2(-11.55) - 10.882) \\
u_1 &= -1.133 = \frac{d\theta}{dt}
\end{aligned}$$

$$\frac{d^2\theta}{dt^2}\Big|_{t=0.1} = \frac{-32.2}{2}\sin(\theta_1) = -16.1\sin(0.7429) = -10.89$$

Therefore the solution at $t = 0.1$ is given by

$$\theta\big|_{t=0.1} = 0.7429$$
$$\frac{d\theta}{dt}\Big|_{t=0.1} = -1.133$$
$$\frac{d^2\theta}{dt^2}\Big|_{t=0.1} = -10.89$$

10.9.4 Further Remarks on Runge-Kutta Methods

(1) In case of single first order ODE in time the computations of area constants k_i ; $i = 1, 2, \dots$ must be in the order $k_1, k_2, k_3, k_4, \dots$ up to the order of the Runge-Kutta method.

(2) In case of two first order simultaneous ODEs the area constants k_i ; $i = 1, 2, ..., 4$ and l_i ; $i = 1, 2, ..., 4$ associated with the two ODEs must be computed in the following order.

$$(k_1, l_1) \quad , \quad (k_2, l_2) \quad , \quad (k_3, l_3) \quad , \quad (k_4, l_4)$$

This is due to the fact that (k_2, l_2) contain (k_1, l_1) and (k_3, l_3) contain (k_2, l_2) and so on.

(3) When there are more than two first order ODEs, we also follow the rule (2).

$$(k_1, l_1, m_1, ...) \quad \text{first followed by} \quad (k_2, l_2, m_2, ...)$$

10.10 Concluding Remarks

In this chapter we have presented a general overview of various methods of approximations that can be used to obtain approximate solutions of PDEs describing initial value problem. Out of all of the methods mentioned here the *space-time coupled* leading to unconditionally stable computations are by far the most meritorious. The space-time finite element method is one such method of approximation. This method can be applied to any IVP regardless of complexity and the nature of the space-time differential operator. Unfortunately the limited scope of study here only permit consideration of the time integration methods for ODEs in time resulting from decoupling of space and time in the IVPs.

Problems

10.1 Consider the following ordinary differential equation in time

$$\frac{du}{dt} = tu^2 \quad \forall t \in (1,2) = (t_1, t_2) = \Omega_t \tag{1}$$

$$\text{with IC} : u(1) = 1 \tag{2}$$

(a) Use Euler's method to calculate the solution $u(t)$ and $u'(t) \quad \forall t \in (1,2]$ using integration step of 0.1. tabulate your results and plot graphs of u versus t and u' versus t.

(b) Repeat the calculations for step size of 0.05. Tabulate and plot graphs and computed solution and compare the solution computed here with the results obtained in (a). Write short discussion of the results calculated in (a) and (b).

10.2 Consider a system of ordinary differential equations

$$\frac{du}{dt} = u + v \quad ; \quad \frac{dv}{dt} = -u + v \quad \forall t \in (t_1, t_2) = (0, \tau)\,;\ \tau > 0 \tag{1}$$

$$\text{with ICs} : u(0) = 0\,,\ v(0) = 1 \tag{2}$$

(a) Calculate u, u', v and v' $\forall t \in (0, 1.0]$ with time step of 0.1 using second order and fourth order Runge Kutta methods. Plot graphs of u, u', v and v' versus t. Tabulate your computed solution. Compare two solutions from the second and the fourth order methods.

(b) Repeat the calculations in (a) using time step of 0.05. Tabulate your computed solution and plot similar graphs in (a). Compare the computed solution here with that in (a). Write a short discussion. Also compare the two solutions obtained here from second and fourth order Runge-Kutta method.

10.3 Consider a system of ordinary differential equations

$$\frac{d^2\phi}{dt^2} + \frac{1}{t}\frac{d\phi}{dt} + \left(1 - \frac{1}{4t^2}\right)\phi = 0 \quad \forall t \in (t_1, t_2) = \Omega_t \tag{1}$$

$$\text{with ICs} : \phi(\pi/2) = 0\,,\ \phi'(\pi/2) = -1 \tag{2}$$

(a) Calculate $\phi(t)$, $\phi'(t)$ $\forall t \in (\pi/2, \pi/2 + 3]$ using second order and fourth order Runge Kutta methods with time step of 0.1. Tabulate your calculations and plot graphs of $\phi(t)$, $\phi'(t)$ versus t. Compare the two sets of solutions from second and fourth order Runge-Kutta methods. Provide a short discussion.

(b) Repeat the calculations and details in (a) using time step of 0.05. Compare the computed the solution calculated here with that in (a). Provide a short discussion of your findings.

10.4 Consider the following ordinary differential equations in time

$$\frac{d^2u}{dt^2} + 0.1\left(\frac{du}{dt}\right)^2 + 0.6u = 0 \quad \forall t \in (t_1, t_2) = (0, \tau) = \Omega_t \tag{1}$$

$$\text{with ICs} : u(0) = 1\ ,\ u'(0) = 0 \tag{2}$$

(a) Calculate $u(t)$, $u'(t)$ $\forall t \in (0, 2.0]$ with time step of 0.1 and using second and fourth order Runge Kutta methods. Plot graphs of $u(t)$ and $u'(t)$ versus t using calculated solutions. Compare the solutions obtained using second and fourth order Runge Kutta methods.

(b) Repeat the calculations and all the details in (a) using time step of 0.05. Compare the solution obtained here with that obtained in (a). Provide a short discussion.

10.5 Consider the following ordinary differential equations in time

$$\frac{d^2u}{dt^2} + 4\frac{du}{dt} + 2u - t^2 = 0 \quad \forall t \in (t_1, t_2) = (0, \tau) = \Omega_t \tag{1}$$

$$\text{with ICs} : u(0) = 1\ ,\ u'(0) = 4 \tag{2}$$

(a) Calculate $u(t)$, $u'(t)$ $\forall t \in (0, 2.0]$ using Runge Kutta methods of second and fourth orders using an integrating time step of 0.2. Tabulate your results and plot graphs of $u(t)$ and $u'(t)$ versus t using calculated solutions. Compare the two sets of solutions.

(b) Repeat the calculations and all other details in (a) using time step of 0.01. Compare these results with those in (a). Write a short discussion.

10.6 Consider a system of ordinary differential equations

$$\frac{d^2\phi}{dt^2} - \left(1 - \frac{t}{5}\right)\phi = t \quad \forall t \in (t_1, t_2) = \Omega_t \tag{1}$$

$$\text{with ICs} : \phi(1) = 10\ ,\ \phi'(1) = 0.1 \tag{2}$$

(a) Calculate $\phi(t)$, $\phi'(t)$ $\forall t \in (1, 3]$ using first order and second order Runge Kutta methods with time step of 0.2. Tabulate your calculated solution and plot graphs of $\phi(t)$, $\phi'(t)$ versus t. Compare the two sets of solutions.

(b) Repeat the calculations and details in (a) using integration time step of 0.1. Compare this computed solution with the one calculated in (a). Write a short discussion.

11
Fourier Series

11.1 Introduction

In many applications such as initial value problems often the periodic forcing functions i.e. periodic non-homogeneous part may not be analytical. Such forcing functions are not continuous and differential every where in the domain of definition. Rectangular or square waves, triangular waves, saw tooth waves, etc. are a few examples. In such cases solutions of the initial value problems may be difficult to obtain. Fourier series provides means of approximate representation of such functions that are continuous and differentiable everywhere in the domain of definitions, hence are meritorious in the solutions of the IVPs.

11.2 Fourier series representation of arbitrary periodic function

In the Fourier series representation of arbitrary periodic function $f(t)$ with a time period T, we represent $f(t)$ as an infinite series of sinusoids of harmonically related frequencies. The fundamental frequency ω corresponding to the time period T is $\omega = {}^{2\pi}/_{T}$. $2\omega, 3\omega, \dots$ etc. are called harmonics. We represent $f(t)$ using a constant term a_0 and as linear combinations of $\sin(k\omega t)$; $k = 1, 2, \dots, \infty$ and $\cos(k\omega t)$; $k = 1, 2, \dots, \infty$. Thus we can write $f(t)$ as

$$f(t) = a_0 + \sum_{k=1}^{\infty} \big(a_k \sin(k\omega t) + b_k \cos(k\omega t)\big) \tag{11.1}$$

in which $f(t)$ is the given periodic function, a_0, a_k, b_k; $k = 1, 2, \dots, \infty$ are to be determined. We proceed as follows

(i) Determination of a_0
Integrate (11.1) with respect to time with limits $[0, T]$. Since

$$\int_0^T \sin(k\omega t)\, dt = 0\ , \quad \int_0^T \cos(k\omega t)\, dt = 0\ ; \quad k = 1, 2, \dots, \infty \tag{11.2}$$

we obtain from (11.1)

$$\int_0^T f(t)\, dt = \int_0^T a_0\, dt = a_0 T \tag{11.3}$$

Hence,

$$a_0 = \frac{1}{T}\int_0^T f(t)\,dt \tag{11.4}$$

(ii) Determination of a_k; $k = 1, 2, \dots, j, \dots, \infty$
To determine a_j, we multiply (11.1) by $\sin(j\omega t)$ and integrate with respect to t with limits $[0, T]$.

$$\int_0^T f(t)\sin(j\omega t)\,dt = \int_0^T a_0 \sin(j\omega t)\,dt + \sum_{k=1}^{\infty}\int_0^T \sin(j\omega t)\big(a_k \sin(k\omega t) + b_k \cos(k\omega t)\big)\,dt \tag{11.5}$$

we note that

$$\begin{aligned}
&\int_0^T \sin(j\omega t)\,dt = 0\\
&\int_0^T \sin(j\omega t)\sin(k\omega t)\,dt = 0\ ; \quad k = 1, 2, \dots, \infty \ \ k \neq j\\
&\int_0^T \sin(j\omega t)\cos(k\omega t)\,dt = 0\ ; \quad k = 1, 2, \dots, \infty \ \ k \neq j\\
&\int_0^T \sin(j\omega t)\sin(j\omega t)\,dt = \int_0^T \sin^2(j\omega t)\,dt = \frac{T}{2}
\end{aligned} \tag{11.6}$$

Using (11.6) in (11.5) we obtain

$$\int_0^T f(t)\sin(j\omega t)\,dt = a_j \frac{T}{2} \tag{11.7}$$

Hence,

$$a_j = \frac{2}{T}\int_0^T f(t)\sin(j\omega t)\,dt\ ; \quad j = 1, 2, \dots, \infty \tag{11.8}$$

(iii) Determination of b_k; $k = 1, 2, \dots, j, \dots, \infty$
To determine b_j, we multiply (11.1) by $\cos(j\omega t)$ and integrate with respect to time with limits $[0, T]$.

$$\int_0^T f(t)\cos(j\omega t)\,dt = \int_0^T a_0 \cos(j\omega t)\,dt + \sum_{k=1}^{\infty}\int_0^T \cos(j\omega t)\big(a_k \sin(k\omega t) + b_k \cos(k\omega t)\big)\,dt \tag{11.9}$$

we note that

$$\begin{aligned}
&\int_0^T \cos(j\omega t)\,dt = 0 \\
&\int_0^T \cos(j\omega t)\sin(k\omega t)\,dt = 0\ ; \quad k = 1, 2, \dots, \infty\ \ k \neq j \\
&\int_0^T \cos(j\omega t)\cos(k\omega t)\,dt = 0\ ; \quad k = 1, 2, \dots, \infty\ \ k \neq j \\
&\int_0^T \cos(j\omega t)\sin(j\omega t)\,dt = \int_0^T \cos^2(j\omega t)\,dt = \frac{T}{2}
\end{aligned} \tag{11.10}$$

Using (11.10) in (11.9) we obtain

$$\int_0^T f(t)\cos(j\omega t)\,dt = b_j \frac{T}{2} \tag{11.11}$$

Hence,

$$b_j = \frac{2}{T}\int_0^T f(t)\cos(j\omega t)\,dt\ ; \quad j = 1, 2, \dots, \infty \tag{11.12}$$

Equations (11.4), (11.8) and (11.12) completely define a_0, a_k; $k = 1, 2, \dots, \infty$ and b_k; $k = 1, 2, \dots, \infty$.

Remarks.

(1) Regardless whether $f(t)$ is analytic or not its Fourier series approximation (11.1) is always analytic.

(2) Fourier series approximation of $f(t)$ is an infinite series in sinusoids containing fundamental frequency and its harmonics. As the number of terms are increased in the Fourier approximation, the proximity of the Fourier approximation to actual $f(t)$ improves.

(3) It is possible to design L_2-norm of error between actual $f(t)$ and its Fourier approximation to quantitatively judge the accuracy of the approximation.

In the following we present numerical example.

Example 11.1. A rectangular wave.

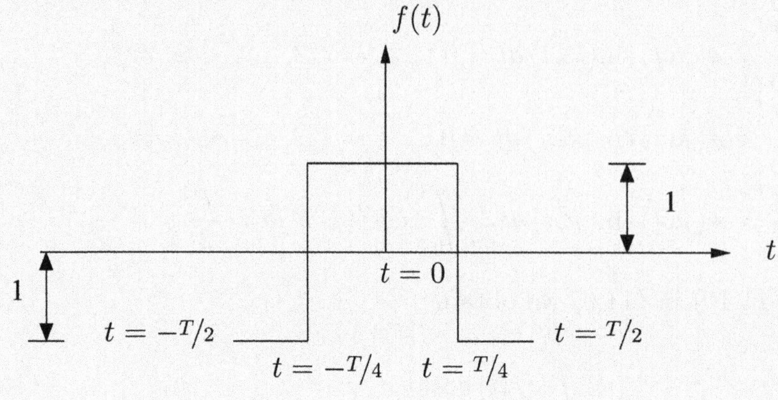

Figure 11.1: One time period of a rectangular wave

One time period of a rectangular wave is shown in Figure 11.1. Referring to Figure 11.1, we have the following.

$$f(t) = \begin{cases} -1; & t \in [-T/2, -T/4] \\ 1; & t \in [-T/4, T/4] \\ -1; & t \in [T/4, T/2] \end{cases}$$

$$\begin{aligned} a_0 = \frac{1}{T}\int_0^T f(t)\,dt &= \frac{1}{T}\left(\int_{-T/2}^{-T/4}(-1)\,dt + \int_{-T/4}^{T/4}(1)\,dt + \int_{T/4}^{T/2}(-1)\,dt\right) \\ &= \frac{1}{T}((-T/2 + T/4) + (T/4 - T/4) + (T/2 - T/4)) \\ &= \frac{1}{T}(0) = 0 \end{aligned}$$

$$a_j = \frac{2}{T}\int_{-T/2}^{T/2} f(t)\sin(j\omega t)\,dt$$

or

$$\begin{aligned} a_j &= \frac{2}{T}\left(\int_{-T/2}^{-T/4}(-1)\sin(j\omega t)\,dt + \int_{-T/4}^{T/4}(1)\sin(j\omega t)\,dt + \int_{T/4}^{T/2}(-1)\sin(j\omega t)\,dt\right) \\ &= \frac{2}{T}\left(\frac{1}{j\omega}\left(\cos(j\omega t)\Big|_{-T/2}^{-T/4} - \cos(j\omega t)\Big|_{-T/4}^{T/4} + \cos(j\omega t)\Big|_{T/4}^{T/2}\right)\right) \end{aligned}$$

or

$$a_j = \frac{2}{T}\left(\frac{1}{j\omega}\Big(\cos(j\omega T/4) - \cos(j\omega T/2) - \cos(j\omega T/4) + \cos(j\omega T/4) + \cos(j\omega T/2) - \cos(j\omega T/4)\Big)\right) = \frac{2}{T}(0) = 0$$

and

$$b_j = \frac{2}{T}\int_{-T/2}^{T/2} f(t)\cos(j\omega t)\,dt$$

or

$$b_j = \frac{1}{T}\left(\int_{-T/2}^{-T/4}(-1)\cos(j\omega t)\,dt + \int_{-T/4}^{T/4}(1)\cos(j\omega t)\,dt + \int_{T/4}^{T/2}(-1)\cos(j\omega t)\,dt\right)$$

or

$$b_j = \frac{2}{T}\left(\frac{1}{j\omega}\left(-\sin(j\omega t)\Big|_{-T/2}^{-T/4} + \sin(j\omega t)\Big|_{-T/4}^{T/4} - \sin(j\omega t)\Big|_{T/4}^{T/2}\right)\right)$$

From which we obtain

$$b_j = \begin{cases} 4/j\pi; & j = 1, 5, 9, \ldots \\ -4/j\pi; & j = 3, 7, 11, \ldots \\ 0; & j = 2, 4, 6, \ldots \end{cases}$$

Thus, the Fourier series approximation can be written as

$$f(t) = \frac{4}{\pi}\cos(\omega t) - \frac{4}{3\pi}\cos(3\omega t) + \frac{4}{5\pi}\cos(5\omega t) - \frac{4}{7\pi}\cos(7\omega t) + \ldots$$

11.3 Concluding Remarks

In this chapter Fourier series representation of periodic function has been presented. When the periodic functions are not analytic every where (i.e. not continuous and differentiable) the Fourier series representation is helpful. Though Fourier series representation of the actual function is often approximate, the representation is analytic i.e. continuous and differentiable every where. We have seen that Fourier series is an infinite series in sinusoids of fundamental frequency and its harmonious, thus in this approximation the accuracy of approximation improves as the number of terms are increased.

Problems

11.1 Figure (a) below shows a rectangular wave.

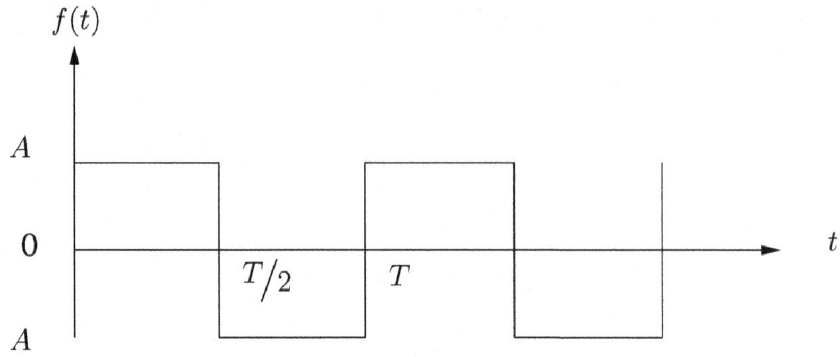

(a) A rectangular wave of period T

$f(t)$ of Figure (a) can be described by:

$$f(t) = \begin{cases} A; & t \in [0, T/2] \\ -A; & t \in [T/2, T] \end{cases}$$

Derive Fourier series approximation of $f(t)$ in the form (11.1).

11.2 Consider the wave shown in Figure (a) below.

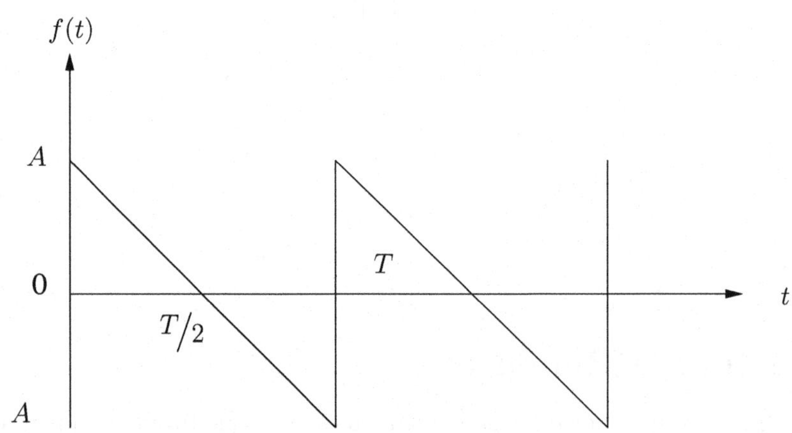

(a) A sawtooth wave of period T

$f(t)$ of Figure (a) can be described by:

$$f(t) = (-2A/T)\,t + A \quad \forall\, t \in [0, T]$$

Derive Fourier series approximation of $f(t)$ in the form (11.1).

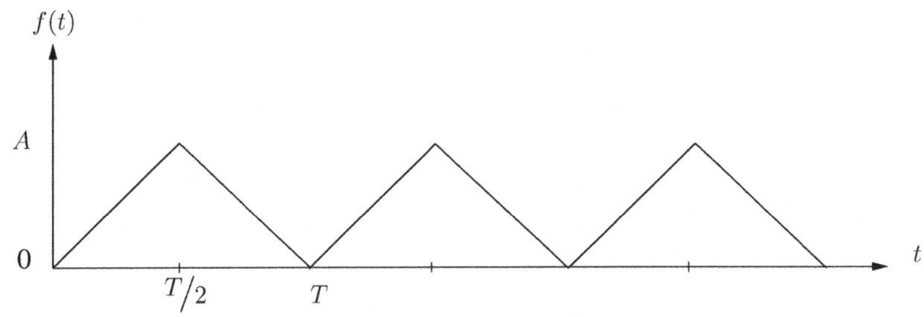

(a) A triangular wave of time period T

11.3 Consider triangular wave shown in figure below.

$f(t)$ of Figure (a) can be described by:

$$f(t) = \begin{cases} (-2A/T)\,t & ; \quad t \in [0, T/2] \\ (-2A/T)\,t + 2A; & t \in [T/2, T] \end{cases}$$

Derive Fourier series approximation of $f(t)$ in the form (11.1).

BIBLIOGRAPHY

[1] Allaire, F.E.: Basics of the Finite Element Method. William C. Brown, Dubuque, IA (1985)

[2] Ames, W.F.: Numerical Methods for Partial Differential Equations. Academic Press, New York (1977)

[3] Atkinson, K.E.: An Introduction to Numerical Analysis. Wiley, New York (1978)

[4] Baker, A.J.: Finite Element Computational Fluid Mechanics. McGraw-Hill, New York (1983)

[5] Bathe, K.J., Wilson, E.L.: Numerical Methods in Finite Element Analaysis. Prentice-Hall, Englewood Cliffs, NJ (1976)

[6] Belytschko, T., Hughes, T.J.R.: Computational Methods for Transient Analysis, Volume 1. North-Holland (1992)

[7] Burden, R.L., Faires, J.D.: Numerical Analysis, 5th edn. PWS Publishing, Boston (1993)

[8] Carnahan, B., Luther, H.A., Wilkes, J.O.: Applied Numerical Methods. Wiley, New York (1969)

[9] Chapra, S.C., Canale, R.P.: Introduction to Computing for Engineers, 2nd edn. McGraw-Hill, New York (1994)

[10] Cheney, W., Kincaid, D.: Numerical Mathematics and Computing, 2nd edn. Brooks/Cole, Monterey, CA (1994)

[11] Collatz, L.: The Numerical Treatment of Differential Equations. Springer-Verlag (1966)

[12] Crandall, S.H.: Engineering Analysis. McGraw-Hill (1956)

[13] Crandall, S.H., Karnoff, D.C., Kurtz, E.F.: Dynamics of Mechanical and Electromechanical Systems. McGraw-Hill (1967)

[14] Davis, P.J., Rabinowitz, P.: Methods of Numerical Integration. Academic Press, New York (1975)

[15] Fadeev, D.K., Fadeeva, V.N.: Computational Methods of Linear Algebra. Freeman, San Francisco (1963)

[16] Ferziger, J.H.: Numerical Methods for Engineering Application. Wiley, New York (1981)

[17] Forsythe, G.E., Malcolm, M.A., Moler, C.B.: Computer Methods for Mathematical Computation. Prentice-Hall, Englewood Cliffs, NJ (1977)

[18] Froberg, C.E.: Introduction to Numerical Analysis. Addison-Wesley Publishing Company (1969)

[19] Gear, C.W.: Numerical Initial-Value Problems in Ordinary Differential Equations. Prentice-Hall, Englewood Cliffs, NJ (1971)

[20] Gear, C.W.: Applied Numerical Analysis, 3rd edn. Addison-Wesley, Reading, MA (1989)

[21] Hamming, R.W.: Numerical Methods for Scientists and Engineers. Wiley, New York (1973)

[22] Hartley, H.O.: The modified gauss-newton method for fitting non-linear regression functions by least squares. Technometrics **3**, 269–280 (1961)

[23] Henrici, P.H.: Elements of Numerical Analysis. Wiley, New York (1964)

[24] Henrick, P.: Error Propagation for Finite Difference Methods. John Wiley & Sons (1963)

[25] Hildebrand, F.B.: Introduction to Numerical Analysis, 2nd edn. McGraw-Hill, New York (1974)

[26] Hoffman, J.: The Theory of Matrices in Numerical Analysis. Blaisdell, New York (1964)

[27] Hoffman, J.: Numerical Methods for Engineers and Scientists. McGraw-Hill, New York (1992)

[28] Householder, A.S.: Principles of Numerical Analysis. McGraw-Hill, New York (1953)

[29] Hurty, W.C., Rubinstein, M.F.: Dynamics of Structures. Prentice-Hall (1964)

[30] Isaacson, E., Keller, H.B.: Analysis of Numerical Methods. Wiley, New York (1966)

[31] Lapidus, L., Pinder, G.F.: Numerical Solution of Partial Differential Equations in Science and Engineering. Wiley, New York (1981)

[32] Lapidus, L., Seinfield, J.H.: Numerical Solution of Ordinary Differential Equations. Academic Press, New York (1971)

[33] Maron, M.J.: Numerical Analysis, A Practical Approach. Macmillan, New York (1982)

[34] Moursund, D.G., Duris, C.S.: Elementary Theory and Applications of Numerical Analysis. McGraw-hill (1967)

[35] Na, T.Y.: Computational Methods in Engineering Boundary Value Problems. Academic Press, New York (1979)

[36] Ortega, J., Rheinboldt, W.: Iterative Solution of Nonlinear Equations in Several Variables. Academic Press, New York (1970)

[37] Paz, M.: Structural Dynamics: Theory and Computations. Van Nostrand Reinhold Company (1984)

[38] Ralston, A., Rabinowitz, P.: A First Course in Numerical Analysis, 2nd edn. New York (1978)

[39] Reddy, J.N.: An Introduction to the Finite Element Method, 3rd edn. McGraw-Hill (2006)

[40] Rice, J.R.: Numerical Methods, Software and Analysis. McGraw-Hill, New York (1983)

[41] Rubinstein, M.F.: Structural Systems – Statics, Dynamics, and Stability. Prentice-Hall (1970)

[42] Shampine, L.F., Jr., R.C.A.: Numerical Computing: An Introduction. Saunders, Philadelphia (1973)

[43] Stark, P.A.: Introduction to Numerical Methods. Macmillan, New York (1970)

[44] Stasa, F.L.: Applied Finite Element Analysis for Engineers. Hold, Rinehard and Winston, New York (1985)

[45] Stewart, G.W.: Introduction to Matrix Computations. Academic Press, New York (1973)

[46] Surana, K.S., Ahmadi, A.R., Reddy, J.N.: The k-version of finite element method for self-adjoint operators in BVP. International Journal of Computational Engineering Science **3**(2), 155–218 (2002)

[47] Surana, K.S., Ahmadi, A.R., Reddy, J.N.: The k-version of finite element method for non-self-adjoint operators in BVP. International Journal of Computational Engineering Science **4**(4), 737–812 (2003)

[48] Surana, K.S., Ahmadi, A.R., Reddy, J.N.: The k-version of finite element method for non-linear operators in BVP. International Journal of Computational Engineering Science **5**(1), 133–207 (2004)

[49] Surana, K.S., Reddy, J.N.: The Finite Element Method for Boundary Value Problems: Mathematics and Computations. CRC Press/Taylor & Francis Group (2016)

[50] Surana, K.S., Reddy, J.N.: The Finite Element Method for Initial Value Problems: Mathematics and Computations. CRC Press/Taylor & Francis Group (2017)

[51] Surana, K.S., Reddy, J.N., Allu, S.: The k-version of finite element method for initial value problems: Mathematical and computational framework. International Journal for Computational Methods in Engineering Science and Mechanics **8**(3), 123–136 (2007)

[52] Wilkinson, J.H.: The Algebraic Eigenvalue Problem. Oxford University Press, Fair Lawn, NJ (1965)

[53] Wilkinson, J.H., Reinsch, C.: Linear Algebra: Handbook for Automatic Computation, vol. 11. Springer-Verlag, Berlin (1971)

[54] Young, D.M.: Iterative Solution of Large Linear Systems. Academic Press, New York (1971)

[55] Zienkiewicz, O.C.: The Finite Element Method in Engineering Science. McGraw-Hill, London (1971)

INDEX

A

B

C

F

G

H

I

J

K

L

M

N

O

T

V

W

Z